Technology Engineering and Management in Aviation:

Advancements and Discoveries

Evon Abu-Taieh
Arab International University, Syria

Asim El Sheikh
The Arab Academy for Banking & Financial Sciences, Jordan

Mostafa Jafari
Islamic Republic of Iran Meteorological Organization (IRIMO), Iran

Senior Editorial Director:	Kristin Klinger
Director of Book Publications:	Julia Mosemann
Editorial Director:	Lindsay Johnston
Acquisitions Editor:	Erika Carter
Development Editor:	Joel Gamon
Production Editor:	Sean Woznicki
Typesetters:	Natalie Pronio, Milan Vracarich, Jr.
Print Coordinator:	Jamie Snavely
Cover Design:	Nick Newcomer

Published in the United States of America by
    Information Science Reference (an imprint of IGI Global)
    701 E. Chocolate Avenue
    Hershey PA 17033
    Tel: 717-533-8845
    Fax: 717-533-8661
    E-mail: cust@igi-global.com
    Web site: http://www.igi-global.com

Library of Congress Cataloging-in-Publication Data

Technology engineering and management in aviation: advancements and discoveries / Evon Abu-Taieh, Asim El-Sheikh and Mostafa Jafari, editors.
    p. cm.
  Includes bibliographical references and index.
  Summary: "This book details the essential new developments in technology and management in the aviation industry, specifically important advances in navigation, air traffic control, and environmental impact"--Provided by publisher.
  ISBN 978-1-60960-887-3 (hbk.) -- ISBN 978-1-60960-888-0 (ebook) 1. Aeronautics. I. Abu-Taieh, Evon M. O. II. El-Sheikh, Asim A, 1956- III. Jafari, Mostafa, 1956-
  TL552.T43 2012
  629.13--dc22
              2011005920

British Cataloguing in Publication Data
A Cataloguing in Publication record for this book is available from the British Library.

Editorial Advisory Board

Table of Contents

Detailed Table of Contents

Chapter 1

This chapter was drafted based on the Special Report that was prepared by the Intergovernmental Panel on Climate Change (IPCC) following a request from the International Civil Aviation Organization (ICAO) and the Parties to the Montreal Protocol on Substances that Deplete the Ozone Layer (IPCC, 1999). In this context, the state of understanding of the relevant science of the atmosphere, aviation technology, and socio-economic issues associated with mitigation options is assessed and reported for both subsonic and supersonic fleets. The potential effects that aviation has had in the past and may have in the future on both stratospheric ozone depletion and global climate change are covered; environmental impacts of aviation at the local scale, however, are not addressed.

Chapter 2

Aviation-related emissions and their impacts are comprehensively discussed in this chapter. Previous studies are described in section two, and their relevance is discussed throughout the chapter. In section three, five common emissions species with contrail formation are described along with qualitative and quantitative results from important investigations. Relationships between aviation emissions and fuel usage are also illustrated. In the mitigation strategies section, emission abatement methods are investigated, focusing on three main areas: technology, flight procedures, and alternative fuels. Both theoretical and practical methods with the potential to decrease emissions are discussed. In the legislation section, the status of emissions regulations is discussed, and emissions charges applied by airports are identified. Examples are provided throughout the chapter to illustrate the points addressed. To complement the main body of the chapter, much detailed information related to aircraft emissions compiled from various sources are provided in the appendix.

Effective Human Factors Engineering (HFE) has provided the aerospace industry with design considerations that promote aviation safety in the development of complex aircraft systems, as well as the operators and maintainers that utilize those systems. HFE is an integral aspect within the systems engineering process. Measuring the effectiveness of Human Systems Integration (HSI) in the research & development stage is critical for the design of new and modified systems. This chapter focuses on how providing effective HFE design solutions enhances product design and system safety. Providing the customer with safe and reliable products augments mission capabilities throughout the product lifecycle.

Chapter 4

The purpose of this chapter is to describe important aspects of how to manage human factors in the development of fighter aircraft, by presenting prominent features of the domain, corresponding design concepts to cope with what is special, and guidance in the approach to managing human factors in the development of fighter aircraft. The chapter includes a description of the domain of fighter aircraft development, followed by a section on how to design for what is special about a fighter pilot, a fighter aircraft, and the flight environment. Examples of design concepts from the domain, such as HOTAS and Basic-T are discussed in light of domain specific demands. Later, human factor considerations, such as evaluation of human factors and usable systems, are discussed together with how to manage human factors in development.

Chapter 5

This chapter discusses the historical approaches to monitoring aviation fuel quality, and the current industry movement towards continuous monitoring electronic measurement systems. The application of mature technology from other industries is reviewed and found to be inadequate. A new type of sensing system designed specifically for the needs of aviation fuel quality is introduced, showing advanced user features and proven to be far more accurate than any other method current available. This article also discusses a typical problem in real world applications where a combination of contaminant is encountered, and how only the new type of sensing system can properly measure the contamination to aviation fuel quality specifications.

Chapter 6

Salih N. Akour, Sultan Qaboos University, Oman

Mohammad Al-Husban, Civil Aviation Regulatory Commission, Jordan

Musa O. Abdalla, The University of Jordan, Jordan

Reducing stress and weight of structures are most important to structural designers. Most engineering structures are an assembly of different parts. On most of these structures, parts are assembled by bolts, rivets, et cetera. This riveted and bolted structure is highly used in aerospace industry. These holes that are drilled for bolts and rivets work as stress raisers. Defense hole theory deals with introducing auxiliary holes beside the main hole to reduce the stress concentration by smoothing the stress trajectories in the vicinity of the main hole. These holes are introduced in the areas of low stresses that appear near the main circular hole. Defense hole system under uniaxial loading is investigated. The optimum defense hole system parameters for a circular hole in an infinite laminated composite plate are unveiled. This study is conducted using finite element method by utilizing commercial software package. The optimum design of the defense hole system is reached by utilizing the redesign optimization technique and univariate search optimization technique. The finite element model is verified experimentally using RGB-Photoelasticity and by reproducing some well known cases that are available in the literature. Digital Image Processing is utilized to analyze the photoelastic images. Stress concentrations associated with circular holes in pure Uniaxial-loaded laminates can be reduced by up to 24.64%. This significant reduction is made possible by introducing elliptical auxiliary holes along the principal stress direction. The best reduction is achieved when four elliptical defense holes are introduced in the vicinity of the main hole. The effect of the fiber orientation, as well as the stiffness of both the fiber and the matrix are investigated.

Chapter 7

Salih N. Akour, Sultan Qaboos University, Oman

Mohammad Al-Husban, Civil Aviation Regulatory Commission, Jordan

Jamal F. Nayfeh, Prince Mohammad Bin Fahd University, Kingdom of Saudi Arabia

Stress reduction and increasing the load to weight ratio are two of the main goals of designers. Fiber reinforced composite materials are preferred by aerospace industry because of their high load to weight carrying capacity. Many parts of aircraft structures are assembled by bolts and rivets. These holes that are drilled for bolts and rivets work as stress raisers. Reducing the stress in the vicinity of these holes is a need. Defense hole theory deals with introducing auxiliary holes beside the main hole to reduce the stress concentration by smoothing the stress trajectories in the vicinity of the main hole. These holes are introduced in the areas of low stresses that appear near the main circular hole. Defense hole system under hybrid (shear and tension) loading is investigated. The optimum defense hole system parameters for a circular hole in an infinite laminated composite plate are unveiled. The study has been conducted using finite element method by utilizing commercial software package. The finite element model is verified experimentally using RGB-Photoelasticity. Digital Image Processing is utilized to analyze the photoelastic images.

Stress concentration associated with circular holes in hybrid loading (i.e., tension-compression ratios of 0.25, 0.50, and 0.75) achieved maximum reduction of 31.7%. This reduction is obtained by introducing elliptical defense holes along the principal stress direction. Finite element analysis is used to optimize the size and location for defense hole system. The effect of the stacking sequence, the fiber orientation, and the stiffness of both the fiber and the matrix are investigated.

Chapter 9

Reducing stress and weight of structures are most important to structural designers. Most engineering structures are an assembly of different parts. On most of these structures, parts are assembled by bolts, rivets, et cetera; this riveted and bolted structure is highly used in aerospace industry. Defense hole theory deals with introducing auxiliary holes beside the main hole to reduce the stress concentration by smoothing the stress trajectories in the vicinity of the main hole. These holes are introduced in the areas of low stresses that appear near the main circular hole. Defense hole system under shear loading is investigated. The optimum defense hole system parameters for a circular hole in an infinite laminated composite plate are unveiled. This study is conducted using finite element method by utilizing commercial software package. The finite element model is verified experimentally using RGB-Photoelasticity. Digital Image Processing is utilized to analyze the photoelastic images. Stress concentrations associated with circular holes in pure biaxial shear-loaded laminates can be reduced by up to 20.56%. This significant reduction is made possible by introducing elliptical auxiliary holes along the principal stress direction. The effect of the stacking sequence, the fiber orientation, and the stiffness of both the fiber and the matrix are investigated.

Chapter 10

Even during the economic crisis, air traffic demand has continued to increase in certain areas of the world, such as the Middle East. Other regions are on their way to recover to pre-crisis traffic demands and will shortly be back to previous growth rates. Airport operators and air traffic control service providers face the challenge to handle this traffic in an expeditious, environmentally friendly, and safe way without generating delays. Conventional ATC concepts in many parts of the world need to be augmented with next generation surveillance technology, in order to keep pace with the required level of safety in those regions. Conventional technologies, such as primary radar and secondary radar, are not able to deliver the required cost-performance ratios for these increasing demands and need to be replaced by multilateration and ADS-B surveillance techniques. This chapter outlines the recent achievements in worldwide operational deployments in the fields of ADS-B and multilateration for airport and air traffic control applications and discusses the integration into larger aviation system applications.

Today, wireless technology forms the communications backbone of many industries—including aviation, transportation, government, and defense. Security breaches since 9/11 have confirmed the need for a discreet wireless communications device onboard commercial aircraft. Real time, discreet communication devices are needed to improve communication between the pilots, flight attendants, and air marshals- a concept that is essential in today's age of terrorism. Flight attendants and Federal Air Marshalls (FAMs) need to be able to alert the flight deck discreetly of such dangers, and thereby pre-warn the pilots of possible attempts to enter the flightdeck or security breaches in the cabin. This chapter will study the effectiveness of discreet, secure, hands-free, wireless communications methods for enhancing coordination during security incidents among cabin crewmembers, between the cabin and flight compartment, ground support personnel, and report these findings. It will identify breakthrough technologies to mitigate the likelihood of individual radical and/or violent behavior, resulting in catastrophic airline casualties, and it will also improve communications and overall safety in-flight.

This chapter examines how airport and airline managers could review their incident and command plans to enhance security counter-measures for terrorist attacks through the use of a well constructed plan-do-check-act (PDCA) tool, in the context of a Safety Management System (SMS), and incorporating a structured field survey into their emergency incident plan and command plan reviews. Thus, through the examination of actual emergency incident plan and command plan survey, airport managers are given the opportunity to work issues through the trials and tribulations of refining their incident and command plans on a recurring basis. It is suggested that a PDCA tool be implemented as a SMS model for the enhancement of these plans in the airline environment.

EPlanAirport is a Web-based tool that allows running complex studies based on airport systems. The primary goal of the tool is helping to the airport stakeholders and policy makers in the decision-support processes. Nowadays, there is a lack of tools and systems that may help the targeted users in such a process. Otherwise, this tool could guide them in the current global scenario. So, ePlanAirport fills this gap detected allowing a non-expert user in complex tools to fulfill successfully his mission. For that purpose, a relevant set of data got in a simple and fast way via Web will be available, and they will help in the planning of airport infrastructures and operations. For instance, the planner will know how a change in an operational procedure or a change in the fleet characteristics or a change in the current infrastructure of the airport will impact on key performance indicators such as capacity, delay, or environment.

Chapter 14

This chapter introduces the different aviation and airport Information Technology systems. Also, this chapter provides architecture based on the Service Oriented Architecture (SOA) that improves the information accessibility and sharing across the different airport departments, integrating the existing legacy systems with other applications, and improving and maximizing the system's reliability, adaptability, robustness, and availability using the self-healing agent and virtual Web service connector to guarantee the quality of service (QoS).

Chapter 15

This chapter will focus on the role of pilot/flightcrew training and performance evaluation in the identification and management of risk, especially while aloft and in changing conditions. The chapter will integrate different- but we posit interrelated, topic areas: First, a decision-making paradigm for flight crew's use in the operational environment. Second, training and performance evaluation in flight simulators (FS), as well as the design and development of FS scenarios to test decision performance. Third, Relevant Federal Aviation regulations (FAR's) and approved programs in current pilot/flightcrew training. Fourth, accident investigations; the role and use-value of accident investigation data in flying safety. Finally, the authors will present recommendations for the next steps in the development and use of new and emerging technologies for maximum pilot/flight crew decision performance and safety. This will be done via a collaborative ground-air, automated system and is what we propose to achieve our goal, increasing safety of flight.

Chapter 16

The global positioning satellite system (GPS) has been utilized for commercial use after the year 2000. Since then, GPS receivers have been integrated for accurate positing of ground as well as space vehicles. Almost all aircrafts nowadays rely on GPS based system for their take off, landing, and en-route navigation. Relying on GPS alone does note provide the meter level accuracy needed to guarantee safe operation of aircrafts. Thus several augmentation systems have been deployed worldwide to enhance the accuracy of the GPS system. Several augmentation systems that serve local as well as wide coverage areas are discussed in this chapter, specifically the LAAS system, the WAAS system as well as the EGNOS system. The architecture as well the performance metrics for each of these augmentation systems are presented and discussed.

This chapter presents the certification standards applied with the simulation study steps, in addition to the Confidence Grid which is used to assets the quality (Reliability and Accuracy) of the data and the process of the simulation study step which will be the base for the validation and verification.

Congestion in networks is considered a serious problem; in order to manage and control this phenomena in early stages before it occurs, a derivation of a new discrete-time queuing network analytical model based on dynamic random early drop (DRED) algorithm is derived to present analytical expressions to calculate three performance measures: average queue length ($Q_{avg,j}$), packet-loss rate ($P_{loss,j}$), and packet dropping probability ($p_d(j)$). Many scenarios can be implemented to analyze the effectiveness and flexibility of the model. We compare between the three queue nodes of the proposed model using the derived performance measures to identify which queue node provides better performance. Results show that queue node one provides highest $Q_{avg,j}$, $P_{loss,j}$, and ($p_d(j)$) than queue nodes two and three, since it has the highest priority than other nodes. All the above results of performance measure are obtained only based on the queuing network setting parameters.

A number of disciplines have approached the concept of knowledge. None of the existing definitions of knowledge can be generalized to other disciplines, and most importantly, none of such attempts fit the requirements of information systems (IS). This chapter suggests to perceive knowledge from the point of view of IS, as an attempt to answer IS requirements better. The proposed vision of knowledge is based on Information Systems' layers.

Virtual Reality (VR) is a technology which has various application fields (from video games to psychiatry). It is indispensable in critical simulation, for instance in military training, in surgical operation simulation, in creation of environments which could set off phobias (in psychiatry), or in realization of virtual prototypes, for instance in industrial design. The aim of this chapter is to present how the VR also finds excellent application fields in architecture and in engineering, for instance, in the teaching of the basic concepts, in techniques of graphic rebuilding for the building restoration, in realization of virtual visits inside buildings, and in urban generative processes simulated by computer. Another use of the virtual reality is in the introduction of a new kind of architecture: Virtual Architecture, strongly connected to the Information and Communication Technology (ITC), to the Internet, and in the virtual prototyping in engineering.

Many analytical models have been developed to evaluate the performance of the transport control protocol (TCP) in wireless networks. This chapter presents a description, derivation, implementation, and comparison of two well-known analytical models, namely, the PFTK and PLLDC models. The first one is a relatively simple model for predicting the performance of the TCP protocol, while the second model is a comprehensive and realistic analytical model. The two models are based on the TCP Reno flavor, as it is one of the more popular implementations on the Internet. These two models were implemented in a user-friendly TCP performance evaluation package (TCP-PEP). The TCP-PEP was used to investigate the effect of packet-loss and long delay cycles on the TCP performance measured in terms of sending rate, throughput, and utilization factor. The results obtained from the PFTK and PLLDC models were compared with those obtained from equivalent simulations carried-out on the widely used NS-2 network simulator. The PLLDC model provides more accurate results (closer to the NS-2 results) than the PFTK model.

Preface

As computer and Information Systems technology advances, industries such as aviation stand to benefit from the overwhelming new advances in hardware, software, and best practices. *Technology Engineering and Management in Aviation: Advancements and Discoveries* details the essential new developments in technology and management in the aviation industry. Specific and important advances in navigation, air traffic control, and environmental impact all make their way into this volume, which also focuses on management policies keeping up with new technology. This volume is a vital reference for practitioners, managers, students, and all those interested in the field of aviation.

Technology Engineering and Management in Aviation: Advancements and Discoveries is composed of 21 chapters written by highly qualified scholars discussing a wide range of topics spanning from aviation environmental crises to technological solution in the aviation industry. The first two chapters discuss aviation and environment; the first chapter is written by Mostafa Jafari, a lead author of IPCC, Nobel Peace Prize Winner for 2007; the second chapter is written by Enis T. Turgut & Marc A. Rosen.

The third and fourth chapters came from scholars affiliated with Boeing Company and Saab Aeronautics bother major airline manufacturers to discuss the topic of *Human Factors*. The third chapter covers *Enhancing Product Safety through Effective Human Factors Engineering Design Solutions* and the fourth discusses *Managing Human Factors in the Development of Fighter Aircraft*.

Chapters five through nine discuss airplane hardware design. The fifth chapter is written by scholars affiliated with *Velcon Filters*, discussing *State-Of-The-Art Real-Time Jet Fuel Quality Measurement*. Chapter six discusses *Design and Optimization of Defense Hole System for Uniaxially Loaded Laminates*. Chapter seven suggests *Design and Optimization of Defense Hole System for Hybrid Loaded Laminates*. Next, chapter eight propose *Design and Optimization of Defense Hole System for Shear Loaded Laminates.*" Chapter nine discusses *Effect of Core Thickness on Load Carrying Capacity of Sandwich Panel Behavior Beyond Yield Limit*.

Chapters ten to twelve cover the safety and security issues in aviation. Chapter ten, *Next Generation Surveillance Technology for Airport and ATC Operations*, outlines the recent achievements in worldwide operational deployments in the fields of ADS-B and multilateration for airport and air traffic control applications and discusses the integration into larger aviation system applications. Furthermore, chapter eleven relates the study of *The Evaluation of Wireless Communication Devices: To Improve In-Flight Security Onboard Commercial Aircraft*. The chapter studies the effectiveness of discreet, secure, hands-free, wireless communications methods for enhancing coordination during security incidents among cabin crewmembers, between the cabin and flight compartment, ground support personnel, and reporting these findings. Chapter twelve, *Terrorist Attacks: A Safety Management System Training Tool for Airport and Airline Managers*, examines how airport and airline managers could review their incident and command plans to enhance security counter-measures for terrorist attacks through the use of a well constructed plan-do-check-act (PDCA) tool, in the context of a Safety Management System (SMS), and incorporating a structured field survey into their emergency incident plan and command plan reviews.

Chapters thirteen to sixteen suggest computerized solution to existing aviation problems.

Chapter thirteen is titled *EPlanAirport: A Web-based tool to user-friendly decision-support systems for airport stakeholders and policy-makers.* EPlanAirport is a Web-based tool that allows running complex studies based on airport systems.

Chapter fourteen, *Airport Enterprise Service Bus with Self-Healing Architecture (AESB-SH),* introduces the different aviation and airport Information Technology systems. Also, it provides architecture based on the Service Oriented Architecture (SOA) that improves the information accessibility and sharing across the different airport departments, integrates the existing legacy systems with other applications, and improves and maximizes the system's reliability, adaptability, robustness, and availability using the Self-Healing Agent and Virtual Web Service Connector to guarantee the Quality of Service (QoS).

Chapter fifteen, *Integrating Decision-Making Methodology, Flight Simulation and Computerized Systems to Advance Civil Aviation Safety,* focuses on the role of pilot/flightcrew training and performance evaluation in the identification and management of risk, especially while aloft and in changing conditions.

Chapter sixteen is titled: *Augmentation Systems: Use of Global Positioning System (GPS) in Aviation.* Several augmentation systems that serve local as well as wide coverage areas are discussed in this chapter, specifically the LAAS system, the WAAS system, as well as the EGNOS system. The architecture, as well the performance metrics for each of these augmentation systems, are presented and discussed.

Chapter seventeen through twenty one offer solutions to embedded problems in the IT world. The chapters discuss: simulation, networks congestion, Knowledge management, Virtual reality, and network analytical models.

Chapter seventeen, *Applying the Certification's Standards to the Simulation Study Steps,* presents the certification standards applied with the simulation study steps, In addition to the Confidence Grid which is used to assets the quality (reliability and accuracy) of the data and the process of the simulation study step which will be the base for the validation and verification.

Chapter eighteen is titled: *Derivation A Discrete-time Analytical Model Based on Dynamic Random Early Drop Algorithm.* Congestion in networks considered a serious problem, and in order to manage and control this phenomena in early stages before it occurs, a derivation of a new discrete-time queuing network analytical model based on dynamic random early drop (DRED) algorithm is derived to present analytical expressions to calculate three performance measures.

Chapter nineteen covers *Knowledge: an Information Systems Practical Perspective.* This chapter suggests to perceive knowledge from the point of view of IS, as an attempt to answer IS requirements better.

Chapter twenty is titled: *Virtual Reality in Architecture, Engineering and Beyond.* The aim of this chapter is to present how the VR also finds excellent application fields in architecture and in engineering. Examples include: in the teaching of the basic concepts, in techniques of graphic rebuilding for the building restoration, in realization of virtual visits inside buildings, and in urban generative processes simulated by computer.

Chapter twenty one is called: *Effects of Packet-Loss and Long Delay Cycles on the Performance of the TCP Protocol in Wireless Networks.* This chapter presents a description, derivation, implementation, and comparison of two well-known analytical models, namely, the PFTK and PLLDC models. The two models are based on the TCP Reno flavor as it is one of the more popular implementation on the Internet. These two models were implemented in a user-friendly TCP performance evaluation package (TCP-

PEP). The TCP-PEP was used to investigate the effect of packet-loss and long delay cycles on the TCP performance measured in terms of sending rate, throughput, and utilization factor. The results obtained from the PFTK and PLLDC models were compared with those obtained from equivalent simulations carried-out on the widely used NS-2 network simulator. The PLLDC model provides more accurate results (closer to the NS-2 results) than the PFTK model.

Mohamed Said Safadi
Dean of Faculty of Informatics Engineering, Arab International University

Mohamed Said Safadi *has been a professor in Electronic engineering since 1975. Prof Safadi earned his PhD, DEA (Diploma of Profound Studies), from Paul-Sabatier University – France. Currently, he is the Dean of the faculty of Informatics & Communication Engineering at Arab-International University. Prior to that he was Vice president of AIU, Visiting professor- Electronic & Computer Engineering Department, Faculty of technology- University of Portsmouth- UK, director of scientific research department. at Alkalamoon Private University-Syria, Chief-editor of the journal of Engineering Sciences at Damascus University, and a Member of the Arabic Academy (Majmaa-Allogha Alarabiah). He has supervised many Master and PhD students' research. He is also the author of four books in communications systems, and has translated one book. He participated in many international IEEE conferences as committee member and chaired many sessions, as well as authored more than 14 original research papers.*

Chapter 1
Aviation Industry and Environment Crisis:
A Perspective of Impacts on the Human, Urban and Natural Environments

Mostafa Jafari
Islamic Republic of Iran Meteorological Organization (IRIMO), Iran

ABSTRACT

This chapter was prepared based upon an invitation made by the Conference organizers, to be presented by the author as opening keynote speaker to highlight "Aviation Industry and Environment Crisis" and focus on the impacts on the human, urban and natural environments. The importance and various dimensions of the issue have been reported by the IPCC following a request from the ICAO and the Parties to the Montreal Protocol on Substances that Deplete the Ozone Layer in 1999.

INTRODUCTION AND BACKGROUND

This paper was drafted based on the Special Report that was prepared by the Intergovernmental Panel on Climate Change (IPCC) following a request from the International Civil Aviation Organization (ICAO) and the Parties to the Montreal Protocol on Substances that Deplete the Ozone Layer (IPCC, 1999). In this context, the state of understanding of the relevant science of the atmosphere, aviation technology, and socio-economic issues associated with mitigation options is assessed and reported for both subsonic and supersonic fleets. The potential effects that aviation has had in the past and may have in the future on both stratospheric ozone depletion and global climate change are covered; environmental impacts of aviation at the local scale, however, are not addressed.

DOI: 10.4018/978-1-60960-887-3.ch001

Furthermore, based on the aforementioned report, the paper takes into consideration all the gases and particles emitted by aircraft into the upper atmosphere and the role that they play in modifying the chemical properties of the atmosphere and initiating the formation of condensation trails (contrails) and cirrus clouds. Subsequently, the paper considers (a) how the radiative properties of the atmosphere can be modified as a result, possibly leading to climate change, and (b) how the ozone layer could be modified, leading to changes in ultraviolet radiation reaching the Earth's surface. The paper also considers how potential changes in aircraft technology, air transport operations, and the institutional, regulatory, and economic framework might affect emissions in the future. The paper does not deal with the effects of engine emissions on local air quality near the surface (IPCC, 1999).

In view that airports constitute considerable part of the communities within which they operate, as such, reducing their impact on the environment is a major focus for many around the world. While much of the current attention is on climate change and reduction of greenhouse gas emissions, it is just one of a number of areas that airports and the rest of the aviation industry are active in the environment (ACI, 2009).

Although the environmental stresses to which man is subjected on the ground are less than those commonly encountered in aviation or underwater, they may still exceed an individual's powers of adaptation (Sloan, 1975). Accordingly, several meetings and summits related to the "Aviation & Environment" were held over the past few years around the world, in order to discuss this important issue.

AVIATION: DEVELOPMENT AND IMPROVEMENT

The oldest testimonies about man's efforts to learn how to fly dates from the time of ancient civilizations, accordingly, aviation development leads to engine burning, and when aircraft engines burn fuel, they produce emissions that are similar to other emissions resulting from fossil fuel combustion. However, aircraft emissions are unusual in that a significant proportion is emitted at altitude. These emissions give rise to important environmental concerns regarding their global impact and their effect on local air quality.

Development

The results show that due to the high growth rates of international transport expected under the chosen scenario, by 2050 the share of unabated emissions from international aviation and shipping in total greenhouse gas emissions may increase significantly from 0.8% to 2.1% for international aviation (excluding non-CO_2 impacts on global warming) and from 1.0% to 1.5% for international shipping. Although these shares may still seem rather modest, compared to total global allowable emissions in 2050 in a 450 ppm stabilization scenario, unabated emissions from international aviation may have a 6% share (for CO_2 only) and unabated international shipping emissions have a 5% share. Thus, total unregulated bunker emissions account for about 11% of the total global allowable emissions of a 450 ppm scenario (European Commission, 16 May 2007).

Furthermore, the incorporation of the non-CO_2 impacts of aviation on climate change into the UNFCCC accounting scheme for GHG emissions could be considered, since aviation is a special case in this respect where the non-CO_2 impacts make a significant contribution. The inclusion of the global warming impact of non-CO_2 emissions, of which a significant fraction originates from NOx emissions (through ozone formation), would increase the share of international aviation emissions in 2050 from 6% to 17% (European Commission, 16 May 2007).

Improvement

ICAO provides a framework to ensure interoperability between NextGen and other international air traffic modernization efforts, such as Europe's SESAR initiative. The environmental benefit of NextGen and other international modernization initiatives will be reduced fuel burn and carbon dioxide emissions through the elimination of airport congestion and en route delay through an evolving system that is safe, secure, and efficient (AIA, 2008).

ENVIRONMENTAL CRISIS

In light of the aforementioned, in the next two sections the definition of "environment crisis" will be discussed, in addition to the "climate change" as environmental crisis.

What is Environmental Crisis?

It is argued that the current environmental crisis from the perspective of pragmatist philosophy is at least in part a result of an ancient split in western thinking between the physical and human worlds. If progress is to be made toward realistic solutions to this crisis, the irrational aspects of human experience must be made part of the calculus. While scientific understandings of the environment certainly help us to identify environmental problems it must be remembered that solutions to these problems will be forged not only from the facts but also from the scientifically incommensurable yet important facets of human experience – emotion, patriotism, faith, etc. (Jerry Williams, Austin State University).

Little doubt exists as to the immediate threat posed by global environmental problems. Resource depletion, global warming, and unprecedented levels of species extinction are evidence that human societies are pushing the limits of the natural world. Two questions, however, seem apparent: how did this happen and what might be done about it? (Jerry Williams, Austin State University). Our plant is facing with different global, regional and local problems which will lead to some kind of environmental disruptions and disturbances (Figure 1).

Figure 1. Plant problems in global, regional and local levels

Downgraded planets

Climate Change as Main Environmental Crisis

The Earth is rapidly getting warmer. This change in the climate threatens serious and even catastrophic disruption to our societies and to the natural environment on which we depend for food and other vital resources. It is being caused mainly by a build-up of 'greenhouse gases' that are released by human activities, in particular the burning of fossil fuels (coal, oil and gas), deforestation and certain types of agriculture. These gases trap the sun's heat in the atmosphere in the same way as a greenhouse does. Over the course of the 20th century the average surface air temperature increased by around 0.6 °C globally, by almost 1 °C in Europe and by no less than 5 °C in the Arctic. This man-made warming is already having many discernible impacts around the globe. Hence, Climate change can be defined as a change in the "average weather" of a given region experiences, including such factors as storm frequency, temperature, wind patterns and precipitation. The rate and magnitude of global climate changes over the long term have many implications for natural ecosystems. As society becomes increasingly reliant on the energy consumption, in work and at home and for mobility, which manifested in the heat-trapping nature of the atmosphere to increase.

Climate change will affect all countries but developing countries are particularly vulnerable while being least able to afford the cost of adapting to it (European Commission, August 2005). As our scientific understanding of this situation increases, so does public concern and the requirement for a policy response. Aviation contributes a small but growing proportion to this problem (less than 4% of man-made atmospheric emissions). A key factor however, is that some of aviation's emissions are emitted in the upper atmosphere and may have a more direct effect. The science of climate change is still relatively new and the future is uncertain. However, there is a broad consensus that policy needs to be enacted now if climate change related problems and costs are to be avoided (EUROCONTROL).

In the same context, historically, aviation's biggest environmental issues have been associated with airports. These remain a major impediment to achieving maximum airport throughput, and without their successful resolution it will be impossible to deliver sufficient capacity. However, when dealing with an average of 25,000 flights per day in European airspace, a large proportion of which could generate contrails, this is no longer a simple problem to solve. It is unlikely, therefore, that fiscal or operational measures will be introduced before 2010 to combat aviation's climate change impact (Mr. Andrew Watt, EUROCONTROL Environment Domain Manager).

Ozone layer sensitivity to GHGs is one of the important targets in climate change negotiations (Figure 2).

Causative Source of Pollution: Air Pollution (GHGs, Aerosol, Smoke and Particulate, Dust)

Emission

There are some main gases emissions need to be considered: Carbon Dioxide, Tropospheric ozone, stratospheric ozone, Sulphur and nitrogen compounds, Smoke and particulates.

Air Quality

Aviation air quality concerns are principally related to the areas on and around airports. Further, for most airports the most significant air quality related emissions presently come from ground transport (cars, buses, trains etc). However, because of factors such as growth in demand, more public transport access to airports, and the long service life of aircraft, it is widely expected that aircraft will eventually become the dominant air quality related pollution source for many airports. The significance of aviation's impact on air quality

Figure 2. Ozone layer sensitivity to GHGs

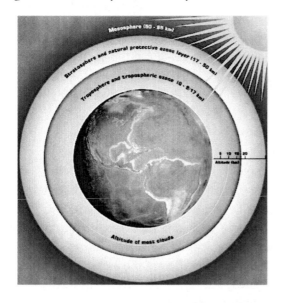

will vary depending on many other factors such as, background pollution levels, other sources of pollution, weather and proximity of residential areas. Around many airports some large emission sources already exist (power stations, factories) that are not related to the airport at all. Also local roads and motorways, even roads associated with an airport, may be heavily used by non-airport traffic.

The chief local air quality relevant emissions attributed to aircraft operations at airports are as follows: Oxides of Nitrogen (NOx), Carbon Monoxide (CO), Unburnt hydrocarbons (CH_4 and VOCs), Sulphur Dioxide (SO_2), Fine Particulate Matter (PM10 and PM2.5), and Odour.

These are produced by aircraft engines, auxiliary power units, apron vehicles, de-icing, and apron spillages of fuel and chemicals. Local factor influence the significance of individual emissions species for each airport, but often NOx is by far the most abundant and is often considered the most significant pollutant from an air quality standpoint.

Water Pollution

1. **Sea level:** One of the key factors to evaluate for many impact studies in low lying coastal regions is the current level of the sea relative to the land. Globally, Eustatic sea level (the volume of water in the oceans) appears to have been rising during the past century. However, there are large regional deviations in relative sea level from this global trend due to local land movements. Subsidence, due to tectonic movements, sedimentation, or human extraction of groundwater or oil, enhances relative sea-level rise. Uplift, due to post glacial isostatic rebound or tectonic processes, reduces or reverses sea level rise. As a reference, most studies of vulnerability to sea-level rise use the mean sea-level at a single date. For instance, studies employing the IPCC Common Methodology use the level in 1990. However, to assess coastal vulnerability to sea-level effects, baseline tide gauge and wave height observations are required. These reflect tidal variations in combination with the effects of weather such as severe storms and atmospheric pressure variations.

2. **Inland water levels:** The levels of lakes, rivers and groundwater also vary with time, usually for reasons related to the natural balance between water inflow (due to precipitation and runoff) and losses (due to evaporation and seepage). Human intervention can also affect water levels, through flow regulation

5

and impoundment, land use changes, water abstraction and effluent return and large scale river diversions. Sometimes these fluctuations in levels can be very large (often much larger than mean changes anticipated in the future). Thus, where time series are available, it is important to be able to identify the likely causes of fluctuations (i.e. natural or anthropogenic), as this information could influence the selection of an appropriate baseline period.

3. **Other impacted sectors include:** Land cover and land use, Soil, Agricultural practices, and Biodiversity.

Hazardous Materials

Hazardous Materials in the Vancouver International Airport (As an Example)

The Airport Authority, airlines, fuellers, car rental companies, couriers, maintenance shops, construction companies and a number of other tenants located on Sea island use hazardous chemical products in their operations. Hazardous materials are also produced as waste products of some airport-related operations.

Chemical products and wastes considered hazardous materials may include:

- Flammable liquids (aviation fuel, jet fuel, solvent, paint)
- Compressed gases (propane, natural gas, nitrogen, oxygen)
- Corrosives (batteries, battery acid, sodium hypochlorite)
- Poisonous or infectious chemicals (medical samples, syringes)
- Others (PCBs, waste oil, and asbestos)

The majority of hazardous wastes generated by the Airport Authority include waste oil, waste paint, antifreeze, waste fuel, batteries and oil filters. These materials are generated during spill clean-ups, vehicle preventative maintenance and line painting, among other things.

The Airport Authority has designated areas where hazardous materials can be stored. All wastes are inventoried and labeled prior to being shipped offsite for disposal or recycling.

Noise

A major concern for communities surrounding many airports is the noise that aircraft make, particularly during take-off and landing. This is a focus for ACI and their member airports and, even though noise from new aircraft has been substantially reduced in the past 10 years (and is expected to be further reduced in the next decade), it remains an important issue.

Aeronautical Noise

Noise associated with an airport can be attributed to a number of sources or activities, such as:

- Aircraft take-offs and landings
- Aircraft over-lights of residential neighborhoods
- Engine run-ups, which are tests performed on aircraft engines and systems after maintenance to ensure they are functioning safely
- Reverse thrust, which is used to slow an aircraft when landing on the runway
- General noise from ground service equipment

LINK BETWEEN AVIATION IMPACTS AND ENVIRONMENTAL CRISIS

Recognizing the relationship between aviation and the environment, Association of Asia Pacific Airlines (AAPA) strives to continually consider solutions to mitigate the environmental impacts. Environmental impacts are seen as systemic beyond the control of the operators. Inefficient management of airspace, restrictive operational procedures and inadequate infrastructure can inadvertently offset the investments by airlines to mitigate its effects on the environment (AAPA, 2006).

The Aviation Challenge

The Asia Pacific Region is predicted to be the largest and fastest growing aviation market in the world, outstripping the United States and Europe. Notwithstanding this the aviation industry is facing enormous challenges. Volatile oil prices, a slowing world economy, falling revenue, rising fuels costs and increasing pressures due to environmental considerations such as global warming and climate change (Figure 3), all point to the need for a major review of the way we plan for, not just aviation needs, but for our transportation systems as a whole (WSROC LIMITED, 2009).

Aviation's fuel consumption and emission production had an increasing rate in the past decades (Figures 4, 5, 6, 7 and 8) and it has been forecasted which it will be increasing in the future (Figures 9 and 10).

Different Perspectives

Human Dimension

People living near airports have long suffered from aircraft noise, traffic congestion and air pollution. Indeed communities around airports have been concerned about these issues for years. However new evidence shows that air travel is contributing towards a far greater threat as Climate Change (Friends of the Earth).

Global warming could lead to the displacement of millions of people. Rising sea levels, floods and drought could make former land inhabitable. Changing weather patterns could effect food crops and accelerate water shortages. According to a Red Cross report in 1999 for the first time environmental refugees out numbered those displaced by war (Friends of the Earth).

Aircraft emissions can also have a significant effect at ground level. Air and ground traffic at major airports can lead to pollution levels as high as city centers. A recent study of Gatwick airport predicts that NOx emissions from cars could decrease by 75% by 2000 due largely to cleaner vehicles, but aircraft emissions of NOx are expected to double by 2008. As a result the National Air Quality standards for nitrogen dioxide (NO_2) may be exceeded in nearby towns (Friends of the Earth).

A report undertaken for the Health Council of the Netherlands reveals airports have a negative impact on public health. The Health Council has

Figure 3. Aviation impact on wildlife

Figure 4. Selected greenhouse gases and other emissions from aircraft at cruising altitude

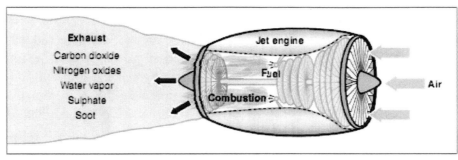

Source: GAO

Figure 5. Source of EU GHGs emissions (European Commission, August 2005, Environment fact sheet: Climate Change)

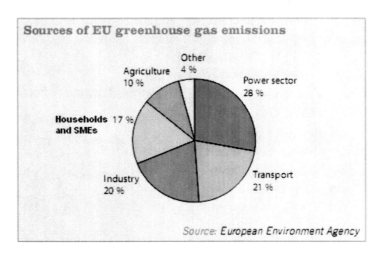

Figure 6. Global transportation's and global aviation's contributions to carbon dioxide emissions, 2004

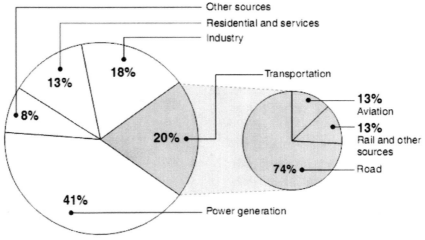

Figure 7. Air emissions from mobile sources at YVR (Vancouver International Airport Authority, 2004, an example)

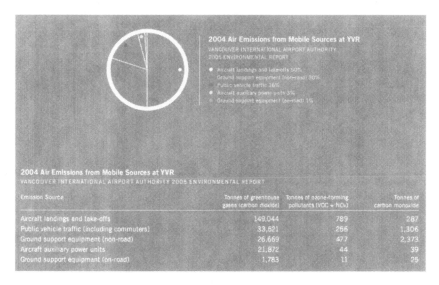

Figure 8. Estimated relative contribution of aviation emissions to positive radiative forcing

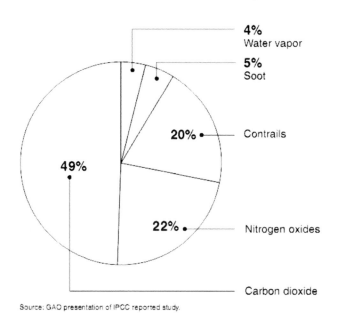

called for public health impact assessments of airports that would assess the cumulative way people are exposed to hazards including air pollution, noise and safety from airport operations (Friends of the Earth).

Urban Environment (Local, Regional, Global)

Concerns about the environmental consequences of aviation have increasingly focused on emissions from airport operations - including emissions

Figure 9. Total fuel consumption and fuel efficiency of U.S. airlines (an example)

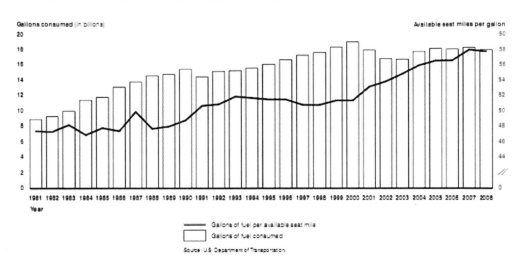

Figure 10. Forecasted fuel consumption by U.S. airlines (an example)

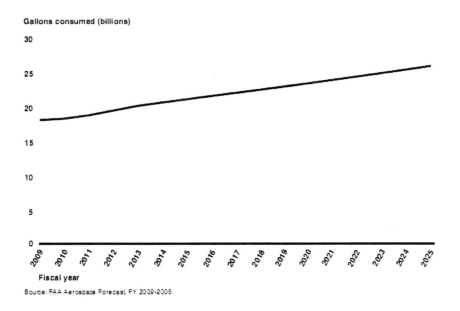

from aircraft; the ground equipment that services aircraft; and the vehicles that transport passengers to, from, and within airport grounds. According to the Environmental Protection Agency (EPA), aviation activities result in the emission of pollutants that account for less than 1 percent [note: This estimate pertains to aircraft emissions, and it does not include emissions from other sources at airports, such as vehicles and equipment that service aircraft. According to EPA, in areas that do not meet federal Clean Air Act requirements for ozone (which is formed from nitrogen oxides and volatile organic compounds), aircraft emissions are estimated to contribute as much as 3 percent of this pollutant.] of the total local air pollution in the United States, but the contribution of these pollutants in areas surrounding airports can be much larger.

Likewise, aviation-related pollutants such as Nitrogen Oxide, which contributes to ozone formation, are expected to increase based on forecasted growth in the aviation sector. Better scientific understanding of the potential health effects of certain aviation emissions and the contribution of aviation emissions, such as Carbon Dioxide, to climate change have also intensified the concerns about the overall impact of aviation emissions. As communities have gained more awareness of the health and environmental effects of aviation emissions, opposition to airport expansion projects, which has thus far focused primarily on aviation noise, has broadened to include emissions.

In addition, airport expansion projects, which can result in increased emissions, must comply with federal Clean Air Act (CAA) requirements. Expanding airport capacity will be necessary to accommodate both the predicted increases in air traffic that are envisioned for the coming decades and the development of the Next Generation Air Transportation System, which is intended to handle those increases. Addressing the effects of airport ground emissions and other types of aviation emissions is expected to be a major challenge to aviation growth in the coming decades (GAO-09-37, Nov. 2008).

Aircraft engines produce emissions that are similar to other emissions resulting from any oil based fuel combustion. These, like any exhaust emissions, can affect local air quality at ground level. It is emissions from aircraft below 1,000 ft above the ground (typically around 3 kilometers from departure or, for arrivals, around 6 kilometers from touchdown) that are chiefly involved in influencing local air quality. These emissions disperse with the wind and blend with emissions from other sources such as domestic heating emissions, factory emissions and transport pollution.

Natural Environments (Terrestrial, Aquatic, Atmospheric)

Pollutants and climate change with affect on all types of environments namely terrestrial, aquatic and atmospheric. These effects include human, animals, plants and all nonliving materials.

Results of studies in the California have emphasized the strong linkage between levels of air pollution-related atmospheric nitrogen (N) inputs into mountain watersheds and levels of nitrate in surface and subsurface drainage waters (Fenn et.al., 2005).

Due to the interaction of N deposition with land management activities, it is possible that past, present, and future land management practices (including fire suppression, introduction of invasive species, and forestry practices) could minimize or exacerbate the adverse effects of N deposition on terrestrial and aquatic ecosystems. Hydrologic flow paths in a watershed also influence the impact of atmospheric N deposition on aquatic ecosystems.

In summary, chronic N deposition results in excess N in terrestrial, riparian, and aquatic habitats. This dramatic change in the chemical environment of these habitats has high potential to upset the normal communities of vegetation, microbes, and micro- and macro-flora and fauna either via direct effects on sensitive organisms or via cascading effects on the food chain (Fenn et.al, 2005).

Birds Killed by Intervention

In 2005, approximately 1.6 million birds were moved away from aircraft operating areas using a variety of harassment techniques, including pyrotechnics, sirens, lights, propane cannons and specially trained Border Collies. This represents a 7% increase over 2004.

While habitat management and harassment techniques are the primary tools used, killing occurs when the officer perceives wildlife behavior to be a safety risk. This may consist of an immediate risk to an approaching aircraft, or a potential or chronic risk that has increased to unacceptable levels. In 2005, 1.060 birds were killed by control officers.

In 2005, 222 birds were killed in 155 bird-strikes with aircraft, a 34% increase over 2004. However, compared with 2004, a larger portion of the bird-strikes in 2005 involved barn swallows, which, because of their small size, pose less of a safety risk than larger bird species.

Factors that contribute to bird-strikes include aircraft operations, environmental conditions and variability in bird population. In 2005, ducks, dunlin, starlings and swallows accounted for more than 86% of birds killed by aircraft and control officers at YVR.

CONCLUDING REMARKS

Aviation is a global enterprise that requires uniform international product acceptance and operating procedures. However, recent European actions threaten the ability of the International Civil Aviation Organization (ICAO) to establish global standards and practices that foster continued growth while reducing the impact of aviation on the environment (AIA, 2008).

The United States provides 25 percent of ICAO's budget, which enables U.S. specialists to fill a large number of ICAO technical leadership and staff positions. U.S. leadership in ICAO,

combined with the technical expertise of the Committee on Aviation Environmental Protection (CAEP), provides a framework to ensure that U.S. aviation environmental issues are well represented in the global aviation community (AIA, 2008).

The International Energy Agency (IEA) estimates that world energy demand will increase by over 50% between now and 2030 if policies remain unchanged, with more than 60% of the increase coming from developing and emerging countries. This would mean an increase of 52% in emissions of carbon dioxide (CO_2), the main greenhouse gas (European Commission, March 2006).

Aviation releases gases and particulates which alter the atmospheric composition, thus contributing to climate change. Although aviation's contribution is still small compared to other sources of human emissions, the rapid growth of air traffic is increasing the impact of aviation on climate. Even though there has been significant improvement in aircraft technology and operational efficiency, this has not been enough to neutralize the effect of increased traffic, and the growth in emissions is likely to continue in the next decades. If the present trend continues, it is expected that emissions from international flights from EU airports will increase by 150% by 2012 in comparison to 1990 levels. One of the effects of aircrafts is the emission of water vapor, which at high altitude often triggers the formation of condensation trails, i.e. line-shaped ice clouds that are also called "contrails", which tend to warm the Earth's surface by trapping outgoing heat emitted by the Earth and the atmosphere. Furthermore, such contrails may develop into cirrus clouds, which are suspected of having a significant warming effect, but this remains uncertain. It became necessary to improve the understanding of the resulting impact of contrails on climate (European Commission, 13 July 2006).

Acting responsibly in concert with ICAO, international aviation has demonstrated a history of reducing aviation's environmental impact. For example, over the past 40 years, carbon dioxide emissions have been reduced by 70 percent. An international approach remains critical; and, because of ICAO's leadership role, national, regional, and local solutions have not been successful (AIA, 2008).

Demand for air transport is continually growing and, if this demand is to be met with all the attendant benefits, society must also accept the costs (noise, pollution, climate change, risk, resource use etc). Thus, if aviation is to continue to play its role in our present concept of sustainability, where possible it must achieve a balance of social, economic and environmental imperatives. It is also clear therefore, that all practical opportunities to minimise these adverse costs should be achieved, otherwise aviation will not achieve the required balanced. And if the balance cannot be achieved, society will then face difficult decisions regarding the global economy and global mobility (EUROCONTROL).

Benefits

Aviation brings several sustainability issues related to the benefits including: Freedom of mobility, Leisure, Improvement to health through poverty reduction, Cultural enrichment and diversity, Employment, Technology transfer, Major direct, secondary and indirect economic improvement, Global business links, Military security, and Positive globalization effects.

Impacts

Aviation also provides costs including: Finite resource depletion, Noise, Atmospheric emissions (air quality, ozone depletion, acid rain and climate change), Water and land pollution, Waste products, Negative globalization effects, Associated adverse health impacts, and Accidents (EUROCONTROL).

In conclusion, it is the goal of the airport to keep these aspects in a balance that secures future operations (Frantz Buch Knudsen, 2004).

Considering the costs and benefits of different types of transportations may provide a suitable ground to be able to reduce emissions and increase efficiency (Figures 11 and 12).

RECOMMENDATIONS

The cap on emission allowances for the sectors covered by the system - power generation, energy-intensive manufacturing industry and, from 2012, aviation - will be cut in a linear fashion every year from 2013, with the result that the number of emission allowances available in 2020 will be 21% below 2005 levels. The international aviation is large and rapidly growing source of GHG emissions yet it is not covered by the Kyoto Protocol. The post-2012 agreement must include emission reduction targets for this industry. In addition, countries should work together through the International Civil Aviation Organization (ICAO) to agree global measures by 2010, which should be approved by 2011. Market-based instruments, including emissions trading, can ensure that emission reductions from this sector are achieved cost-effectively (European Commission, 2009).

The international aviation and shipping sectors are projected to contribute significantly to global emissions of greenhouse gases (GHGs), in particular carbon dioxide (CO_2). These so-called bunker emissions are, however, not yet regulated

Figure 11. Airport operations: benefits and impacts

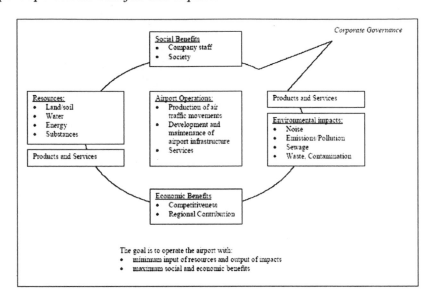

Figure 12. CO_2 emissions per passenger per return flight

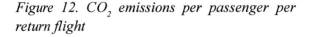

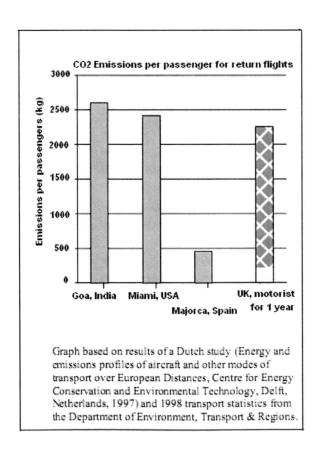

Graph based on results of a Dutch study (Energy and emissions profiles of aircraft and other modes of transport over European Distances, Centre for Energy Conservation and Environmental Technology, Delft, Netherlands, 1997) and 1998 transport statistics from the Department of Environment, Transport & Regions.

by international policies formulated by the United Nations Framework Convention on Climate Change (UNFCCC) or the Kyoto Protocol. One of the reasons why international bunker emissions are not yet regulated is due to the unclear situation regarding who is responsible for these emissions. In this regard, the European Union (EU) indicated in its Environmental Council decision in 2004 that international bunker emissions should be included in climate policy arrangements for the post-2012 period (European Commission, 16 May 2007).

Aviation emissions from developed countries should be capped at 2005 levels. While "Emissions trading and offsetting offer useful short to medium term flexibility for meeting aviation targets", in the long run the industry will need to make deep cuts in its own emissions, according to the Committee on Climate Change (CCC) (Climate Committee, Sep 9, 2009). In December last year the Committee reported that to achieve an overall cut in UK emissions of 80% while allowing aviation emissions to grow in the short term, other sectors would need to make even greater reductions. In January the government announced a special target for aviation - that by 2050 emissions should be brought down to 2005 levels (Note: The Department for Transport's UK Air Passenger Demand and CO_2 Forecasts, published in January 2009 gave aviation emissions figures of 16.9 $MtCO_2$ in 1990, and 37.5 $MtCO_2$ in 2005). This in fact allows aviation emissions to increase by 120% compared with 1990 levels, while other sectors are required to make 90% reductions.

To have a chance of bringing the UK's aviation emissions back down to 2005 levels by 2050 the government will need to look again at the expansion plans they set out for UK airports in 2003. Aviation emissions have more than doubled since 1990 and are set to carry on rising under government growth projections.

Emissions Trading

Emissions Trading Aviation plays a significant role in the economic and social development of the European Union. The industry also acknowledges its impact on the environment and is committed to delivering an ongoing program of environmental improvement. It aims to ensure the reduction of its environmental impact through research and development, technological innovation and revised operational procedures.

Environmental Performance

ADS members are involved in a range of initiatives and programs aimed at reducing the impact of their operations and products on the environment. Much of the information they gather in relation to environmental performance is published through their annual environmental reports.

A|D|S wants to demonstrate that its members are making ongoing environmental improvements, through the sustainable manufacture and consumption of their products, as well as improvements at their sites.

Technological Developments

UK aerospace is working towards the 2020 targets set by ACARE which challenge the European aerospace industry to reduce fuel consumption and CO_2 emissions by 50%, NOx emissions by 80% and perceived external noise by 50%. The ACARE targets represent a doubling of the historical rate of improvement. Delivery against the ACARE targets will require a series of step changes in the industry's ability to design, manufacture and operate aircraft.

The sector continues to make good progress, having improved fuel efficiency by 50% and reduced noise by 75% in the last 30 years. Current products reflect this ongoing commitment to further reductions in noise and emissions. For example, the Airbus A380 has NOx emissions 31% lower than those currently set by ICAO. The A380 has enabled further aerodynamic improvements to be realized with a noise footprint of half that of the Boeing 747-200 (ACARE).

In the United States, the Next Generation Air Transportation System, or NextGen, developed by the Joint Planning and Development Office (JPDO) will pull together operational and technological advancements to reduce the environmental effects of aviation. Successful deployment of NextGen is the key to U.S. leadership in the global aviation community and ICAO (AIA, 2008).

ICAO provides a framework to ensure interoperability between NextGen and other international air traffic modernization efforts, such as Europe's SESAR initiative. The environmental benefit of NextGen and other international modernization initiatives will be reduced fuel burn and carbon dioxide emissions through the elimination of airport congestion and en route delay through an evolving system that is safe, secure, and efficient (AIA, 2008).

In the coming century, the impact of air travel on the environment will become an increasingly powerful influence on aircraft design. Unless the impact per passenger kilometre can be reduced substantially relative to today's levels, environmental factors will increasingly limit the expansion of air travel and the social benefits that it brings. The three main impacts are noise, air pollution around airports and changes to atmospheric composition and climate as a result of aircraft emissions at altitude. The Air Travel - Greener by Design programme to assess the technological, design and operational possibilities for reducing these impacts. If these opportunities are pursued, the aircraft in production in 2050 could be very different from those of 2005 (Green, 2006).

Aviation is a growing contributor to climate change, with unique impacts due to the altitude of emissions. If existing traffic growth rates continue, radical engineering solutions will be required to prevent aviation becoming one of the dominant contributors to climate change.

The engineering options for mitigating the climate impacts of aviation using aircraft and airspace technologies can be reviewed. These options include not only improvements in fuel efficiency, which would reduce CO_2 emissions, but also measures to reduce non-CO_2 impacts including the formation of persistent contrails. Integrated solutions to optimize environmental performance will require changes to airframes, engines, avionics, air traffic control systems and airspace design (Williams, 2007).

A recent British study has analyzed the most important factors influencing the warming effect on climate from condensation trails formed from the water vapor emitted by aircrafts at high altitude. The results of the study suggest that shifting air traffic from night-time to daytime may help to minimize the climate effect of aircraft condensation trails, thus reducing the climate impact of aviation (European Commission, 13 July 2006).

Recommendations by UK Friends of the Earth (Friends of the Earth):

- Choose to fly less frequently whether for business or pleasure

- Consider taking a train as an alternative to domestic or short hop flights
- Investigate teleconferencing as an alternative to business flights
- Support the domestic tourist industry and plan more holidays in the UK

Recommendations by AIA urge the candidates to:

- Continue U.S. commitment to ICAO as the pre-eminent global body responsible for all aviation environmental matters.
- Ensure strong public-private partnership engagement in the definition and execution of U.S. international aviation programs within the ICAO framework.

REFERENCES

AAPA. (2006). *Environment*. Association of Asia Pacific Airlines (AAPA). Retrieved from http://www.aapairlines.org/ Environment.aspx

ACARE. (n.d.). *Home page*. Retrieved from http://www.acare4europe.org

ACI. (2009). *ACI World Director General, Angela Gittens gives a speech at the Aviation & Environment Summit*. 4th Aviation & Environment Summit. Geneva, 31 March, 2009, Opening Session. Retrieved from http://www.aci.aero/ cda/ aci_common/ display/ main/ aci_content07_c.jsp ?zn=aci& cp=1-7-3475^ 28415_666_2

A|D|S. (n.d.). *Home page*. Retrieved from http://www.adsgroup.org.uk/

AIA. (2008). *Aerospace Industries Association, the environment and civil aviation-Ensure environmental standards and policies that are global in development and application. Election 2008 Issues*. Keeping America Strong.

Climate Committee. (2009). *The Committee on Climate Change* (CCC). Retrieved from http:// www.theccc.org.uk/ topics/ international-action-on- climate-change/ international-aviation

EEA. (n.d.). *European Environment Agency website*. Retrieved from http://www.eea.europa.eu/

EUROCONTROL. (n.d.). *Homepage*. Retrieved from http://www.eurocontrol.int/

European Commission. (August 2005). *Environment fact sheet: Climate change.*

European Commission. (13 July 2006). *Science for environment policy, DG environment news alert service: Minimizing the climate impact of aviation.*

European Commission. (March 2006). *Environment fact sheet: Energy for sustainable development.*

European Commission. (16 May 2007). *Science for environment policy, DG environment news alert service: Regulating international aviation and shipping emissions.*

European Commission. (2009). *EU action against climate change, leading global action to 2020 and beyond.*

Fenn, M., Poth, M., & Meixner, T. (2005). *Atmospheric nitrogen deposition and habitat alteration in terrestrial and aquatic ecosystems in Southern California: Implications for threatened and endangered species*, (pp. 269-271). (USDA Forest Service Gen. Tech. Rep. PSW-GTR-195).

Friends of the Earth. (n.d.). *Aviation and global climate change*. Retrieved from www.foe.co.uk

GAO. (n.d.). *United States Government Accountability Office website*. Retrieved from http://www.gao.gov/

GAO-09-37. (November 2008). *Aviation and the environment.*

Green, J. E. (2006). Civil aviation and the environment - The next frontier for the aerodynamicist. *Aeronautical Journal, 110*(1110), 469–486.

IPCC. (1999). *Intergovernmental Panel on Climate Change- IPCC special report, aviation and the global atmosphere, summary for policymakers* (Eds. J. E. Penner, D. H. Lister, D. J. Griggs, D. J. Dokken, & M. McFarland, p. 13). Retrieved from http://www.ipcc.ch/ ipccreports/ sres/ aviation/ index.php?idp=0

Knudsen, F. B. (2004). *Defining sustainability in the aviation sector. EEC/SEE/2004/003*. EUROCONTROL Experimental Centre.

Sloan, A. W. (1975). Adaptation and failure of adaptation to extreme natural environments. *Forensic Science, 5*(1), 81–89. doi:10.1016/0300-9432(75)90091-6

Watt, A. (n.d.). *EUROCONTROL Environment Domain Manager.*

Williams, J. (2007). *Pragmatist philosophy and the global environmental crisis.*

Williams, V. (2007). The engineering options for mitigating the climate impacts of aviation. *Philosophical Transactions of the Royal Society A- Mathematical, Physical, and Engineering Sciences, 365*(1861), 3047-3059.

WSROC Ltd. (2009). *Response to the national aviation policy green paper*. February 2009, Western Sydney Regional Organisation of Councils Ltd. ISBN 186271 0193

YVR. (2004). *Vancouver International Airport Authority website*. Retrieved from http://www.yvr.ca/

GLOSSARY, ACRONYMS AND ABBREVIATIONS

Term	Description
AAPA	Association of Asia Pacific Airlines
A|D|S	A|D|S is the trade body advancing UK AeroSpace, Defence and Security industries with Farnborough International Limited as a wholly-owned subsidiary. A|D|S also encompasses the British Aviation Group (BAG). It is formed from the merger of the Association of Police and Public Security Suppliers (APPSS), the Defence Manufacturers Association (DMA) and the Society of British Aerospace Companies (SBAC).
ACARE	the Advisory Council on Aeronautics Research in Europe
ACI	Airports Council International
AIA	Aerospace Industries Association
APPSS	the Association of Police and Public Security Suppliers
BAG	the British Aviation Group
CAA	Clean Air Act
CAEP	the Committee on Aviation Environmental Protection (CAEP) of ICAO
CCC	the Committee on Climate Change. The Committee on Climate Change (CCC) is an independent body established under the Climate Change Act to advise the UK Government on setting carbon budgets, and to report to Parliament on the progress made in reducing greenhouse gas emissions. http://www.theccc.org.uk/home
CO	Carbon Monoxide
CO_2	Carbon dioxide
Contrails	Condensation trails
DMA	the Defense Manufacturers Association
EEA	European Environment Agency
EPA	United States Environmental Protection Agency
EU	the European Union
EUROCONTROL	The European Organization for the safety of Air Navigation
EC	European Commission
FoE	Friends of the Earth,
GAO	United States Government Accountability Office
GHGs	The Greenhouse Gases
ICAO	the International Civil Aviation Organization
IEA	The International Energy Agency (IEA) of EU
IPCC	Intergovernmental Panel on Climate Change
JPDO	the Joint Planning and Development Office
NextGen,	the United States, the Next Generation Air Transportation System
NOx	Oxides of Nitrogen
PM10	particulate matter or fine particles, thoracic fraction, <=10 μm
PM2.5	particulate matter or fine particles, respirable fraction, <=2.5 μm
PM1	particulate matter (PM) or fine particles, PM1, <=1 μm PM10-PM2.5 (coarse fraction), 2.5 μm - 10 μm Ultrafine (UFP or UP), <=0.1 μm
SBAC	the Society of British Aerospace Companies
SESAR	Single European Skies (SESAR), Europe's SESAR initiative

continued on following page

Glossary, Acronyms and Abbreviations. Continued

Term	Description
SMEs	Small- and Medium Size Enterprises
SO_2	Sulphur Dioxide
UNFCCC	the United Nations Framework Convention on Climate Change
VOCs	Volatile Organic Compounds (VOCs) are organic chemical compounds that have high enough vapor pressures under normal conditions to significantly vaporize and enter the atmosphere
WSROC LIMITED	Western Sydney Regional Organization of Councils Ltd,
YVR	Vancouver International Airport

Chapter 2
Emission Assessment of Aviation

Enis T. Turgut
Anadolu University, Turkey

Marc A. Rosen
University of Ontario Institute of Technology, Canada

ABSTRACT

Aviation-related emissions and their impacts are comprehensively discussed in this chapter. Previous studies are described in "Previous Studies," and their relevance is discussed throughout the chapter. In "Jet Engine Emissions," five common emissions species with contrail formation are described along with qualitative and quantitative results from important investigations. Relationships between aviation emissions and fuel usage are also illustrated. In the mitigation strategies section, emission abatement methods are investigated, focusing on three main areas: technology, flight procedures, and alternative fuels. Both theoretical and practical methods with the potential to decrease emissions are discussed. In the legislation section, the status of emissions regulations is discussed, and emissions charges applied by airports are identified. Examples are provided throughout the chapter to illustrate the points addressed. To complement the main body of the chapter, much detailed information related to aircraft emissions compiled from various sources are provided in the appendix.

INTRODUCTION

In July 2010, the first manned solar airplane, SolarImpulse, having four 10 hp electrical engines and weighing 1600 kg, performed a nonstop 26 hour flight using only solar power to provide power both day and night. This significant development is believed to have opened a new era in aviation in terms of fuel and environmental concerns.

During the last decade in air transport (1999-2008), it can be seen that, except 2001, air travel indicators showed incremental trends. Of these indicators, the annual increase in passenger-kilometers has averaged 5.8%. Moreover, the

DOI: 10.4018/978-1-60960-887-3.ch002

passenger indicator increase is an annual average of 7.5% (ICAO, 2008). According to Boeing, passenger traffic will continue to increase over the next twenty years, by about 4.1% annually, which implies the introduction of a large number of single and twin aisle new aircraft (Boeing, 2010a).

Regarding year to date (YTD) aircraft movements (as of March 2010), the maximum movement is recorded in Atlanta airport in the U.S., with a total count of 227,388 landings and takeoffs. Considering aircraft movements at the top 30 airports for the same YTD period, it can be seen that there are 20 airports from the U.S., totaling 2,394,422 movements, while there are six airports from Europe totaling 619,401 movements. The remaining four airports are from China, Canada and Mexico and accounted for 389,291 aircraft movements (ACI, 2010).

Fuel cost is one of the key direct operating costs for air transport. A variation in fuel price has a direct effect on flight cost. Fuel prices have increased annually by an average of 28.2% in the U.S. between 2003 and 2008. In 2009, however, fuel prices declined by 38.1%, ending at 0.5 $ liter^{-1} by 2010 June (BTS, 2010). Despite the fact that efficient engines have been developed in recent decades, increasing the capacity and performance of these aircraft requires higher amounts of fuel consumption compared to those built in the previous decades. The increase in fuel consumption leads to a corresponding increase in emissions and relevant atmospheric impacts.

The first jet airline, the de Havilland Comet, made its first flight in 1949. It had four engines, each capable of producing a thrust of 22 kN (Anonymous, 2010a). The specific fuel consumption (SFC) of each of the four Olympus 593 engines of the Concorde, first flown in 1969, was 33.8 g (kN s)$^{-1}$ (1.195 lb (lb$_f$-h)$^{-1}$) at cruise. The J79 is a popular military turbojet engine, which is still being used on F4, F5 and IAI Kfir and was used on the F104, B58 and some versions of Convair airline in the past, has a thrust of 53 kN (79 kN with afterburner) and a specific fuel consumption

as 24.0 g (kN s)$^{-1}$ (0.85 lb (lb$_f$-h)$^{-1}$) at military thrust and 55.5 g (kN s)$^{-1}$ (1.96 lb (lb$_f$-h)$^{-1}$) with afterburner (Anonymous, 2010b).

With the advent of the turbofan, the efficiency of the jet engines increased dramatically. The first turbofan engine, the Conway, was certified in 1963. It had a by-pass ratio of 0.25-0.3 and produced 73 kN of thrust. Since that time, turbofan engines evolved continuously to provide the needs of larger aircraft carrying heavier loads over longer ranges. For instance, one of the large jet engines which is going to power the B787, the Rolls-Royce Trent 1000, has a by-pass ratio of almost 11 and a thrust of 328 kN. Current aircraft are about 70% more fuel efficient per passenger-km than those of 40 years ago (Ribeiro et al., 2007).

The radiative force (RF) is defined as the change in net radiative flux due to a change in either atmospheric composition or solar irradiance. According to the fourth assessment report of Intergovernmental Panel on Climate Change (IPCC) (Forster et al., 2007), the radiative force is a parameter that is used to assess and compare anthropogenic and natural drivers of climate change. A positive value of RF leads to global warming (Marquart et al., 2001), while a negative value leads to global cooling. The contribution of common emissions on positive radiative forcing is shown in Figure 1. While the largest contributor is CO_2, it can be seen that NO_x emissions and contrails also play significant roles in global warming.

Investigations indicate that the radiative forcing resulting from aircraft is 0.05 W m^{-2} or about 3.5% of the total radiative forcing from anthropogenic activities in 1992 (Schumann, 2000). Another estimation (Lee et al., 2009), which is obtained using the same approach (Schumann, 2000) but excluding aviation-induced contrails, indicates that aircraft are responsible for 4.9% of all radiative forcing including aviation-induced contrails. To provide a broader and historical perspective, the global annual mean radiative forcing from the year 1750 to around 2000 is

Figure 1. Estimated relative contribution of aircraft emissions to positive radiative forcing. Modified from GOA (2009).

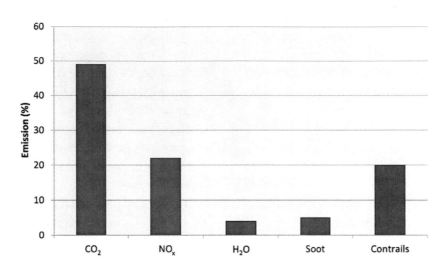

shown in Figure 2, while the historical changes in atmospheric concentrations of commonly emitted substances are shown in Figure 3.

The composition of the atmosphere is determined primarily by a balance between human-related emissions, solar radiation in ultraviolet, visible and near-infrared wavelengths and thermal radiation flows from the earth at infrared wavelengths (Isaksen et al., 2009). The impact mechanism of atmospheric composition on climate is explained in the same source as a kind of radiation budget. Incoming solar radiation can be reflected by clouds (the albedo effect) and outgoing thermal radiation can be absorbed by clouds and reradiate to the surface (greenhouse effect).

In evaluating this budget, it is noted that the thermal structure of the atmosphere is influenced by the presence of small amounts of H_2O, CO_2, CH_4, O_3 and aerosols. Compounds like CH_4, O_3, and different types of secondary particles (sulfates, organic particles and nitrates) are active chemical compounds in the troposphere; they also have important radiative effects on climate (Isaksen et al., 2009). Not all of the produced particles result from anthropogenic activities, as natural events such as plant growth and lightning also contribute.

The concentration of CO_2 in the atmosphere is subject to seasonal variations, with investigations having shown for northern latitudes that the lowest CO_2 concentration occur in the months of April and May, while the peak months are August and September (Florides & Christodoulides, 2009).

Other particles such as CO, VOC, NO_x (which is comprised of NO and NO_2) and SO_2 have an important indirect effect on climate, since they are capable of changing the abundance of radiatively active gases such as O_3 and CH_4 (Isaksen et al., 2009).

PREVIOUS STUDIES

Much global research into climate impacts of air transport have been conducted at past (Baughcum et al., 1996; Gardner et al., 1997; Sutkus et al., 2001; Eyers et al., 2004; Kim et al., 2005). Currently much of the research focus has shifted to the vicinity of airports. Such research includes investigations on NO_x emission measurements at John F. Kennedy Airport in New York (Herndon et al., 2004), emissions at Boston's Logan Airport (Herndon et al., 2006), odor emissions

Figure 2. Global mean radiative forcing of the climate system for the year 2000, relative to 1750. H, M, L, and VL denote high, medium, low and very low levels, respectively. Reported by Folland et al. (2001). (Reproduced by permission from Climate Change 2001: The Scientific Basis. Contribution of Working Group I to the Third Assessment Report of the Intergovernmental Panel on Climate Change, Figure 6.6. Cambridge University Press.)

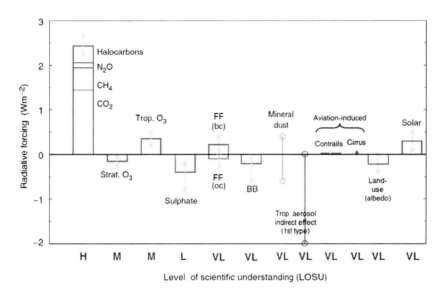

Figure 3. Historical change of atmospheric concentrations of commonly emitted substances. Reported by Forster et al. (2007). (Reproduced by permission from Climate Change 2007: The Physical Science Basis. Working Group I Contribution to the Fourth Assessment Report of the Intergovernmental Panel on Climate Change, FAQ 2.1, Figure 1. Cambridge University Press.)

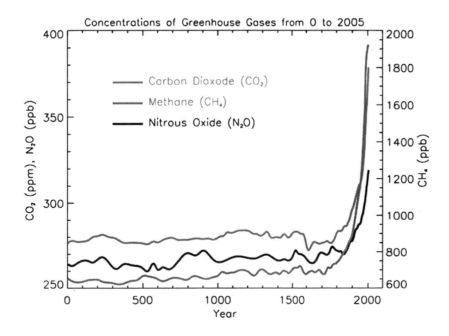

measurements at Copenhagen airport (Winther et al., 2006), airport emission impacts on air quality at the airports of Atlanta (Unal et al., 2005) and Zurich (Schurmann et al., 2007), ultrafine particles at Los Angeles airport (Westerdahl et al., 2008), emissions measurements for idle and taxi conditions at Heathrow, Frankfurt and Vienna airports (Schafer et al., 2003), and others. In the following paragraphs, several of these investigations are discussed.

In Mazaheri et al. (2009), engine emissions such as NO_x, CO_2 and particulate matter less than 2.5 micrometers in aerodynamic diameter ($PM_{2.5}$) were measured for over 280 aircraft, including the B737, B767, B777, A320 and A330 during landing and takeoff (LTO) cycles. It is reported that the emissions and NO_x emission factors are dependent on the engine power setting. The results also highlight that International Civil Aviation Organization (ICAO) emission standards can be modified since the power setting during taxi conditions is considerably higher than during idle. In general, it is noted that emission factors from this study compare well with ICAO results, except that the emission factors for takeoff are overestimated by ICAO (which accepts as 100% power setting) since the takeoff power settings of aircraft can be less than 100% depending on aircraft takeoff weight.

In another study, NO_x emissions were predicted in the vicinity of Manchester airport for two commercial aircraft, B737 and B747 (Graham & Raper, 2006). In that work, a novel model was developed to assess the impact of vortex transport on NO_x concentration. According to the study, exhaust emissions are subject to vortex-mediated transport to ground level when an aircraft is at an altitude between 0.5 and 1.5-2.8 wingspans (which implies an altitude as great as 200 m height for a B747).

Emission measurements generally exclude aircraft having turboprop engines. However, (Sutkus et al., 2001) stated that, in 1999, the global departure ratio of turboprop aircraft was around 30% of all aircraft. Hence, when aircraft

emissions are investigated globally, one should take into account turboprop aircraft, as well as the auxiliary power unit (APU). This circumstance is addressed in the following two studies.

Schafer et al. (2003) investigated idle emissions released from over 140 aircraft at Frankfurt/Main, London and Vienna airports using remote sensing methods. An important aspect of this study is its consideration of turboprop engines and APUs, as well as turbofan engines. In this research NO_x and CO emission indices measured by remote sensing systems are compared with those given in ICAO emission data banks. Three remote sensing systems are utilized: (1) Fourier transform infrared (FTIR) absorption spectrometry for CO_2 and CO emissions, (2) FTIR emission spectrometry for CO_2, CO and NO, and (3) differential optical absorption spectrometry (DOAS) for NO and NO_2. NO_x emission indices are found to be lower by about 50% and CO emission indices to be slightly higher than those provided by the ICAO. The authors note that part of the deviation might result from variations at idle $N1^1$ RPM due to various takeoff weights, while others may be caused by age and level of engine maintenance. Another interesting aspect of this study is the investigation of the emissions from auxiliary power units. Based on measurements of emissions for eight APUs used for different aircraft, Schafer et al. (2003) point out that APU emissions are not negligible in comparison to those for the main engine. As reported by the authors, the measured CO emissions were generally found to be slightly higher than those given by the ICAO, while the minimum and maximum differences of the measured values obtained are 82% lower and 302% higher than those given by ICAO for the A340-211 and B747-236 aircraft, respectively. Measured NO_x emission indices were found to be significantly lower than those reported by ICAO (see Table 2 in the Appendix). The emission indices of the APU were also measured in the same study. Some other results of this study are given in the Appendix (Figure 18).

In Winther et al. (2006), hydrocarbon (HC) emissions, which are sometimes called unburned hydrocarbon (UHC) emissions and which are linked with odor, and NO_x emissions from the main engines and APUs are measured for the LTO activities in Copenhagen airport (see Table 3 and Figure 19 in the Appendix). It was observed that the taxi and take off power settings were found to be lower than those given by ICAO. Since less fuel is correspondingly consumed, the HC emissions obtained exceeded those given by ICAO.

Morrel (2009) investigated the relationship between emissions/fuel efficiency and aircraft size or capacity, and determined that for every 1% increase in seat capacity or 1% increase in payload, a reduction in fuel efficiency of 0.83% or 0.65%, respectively, may be obtained. It was concluded that there is a strong relationship between fuel efficiency and aircraft size, with a higher coefficient of determination (R2) for single-aisle than twin-aisle aircraft, although double decker aircraft (A380, B747) do not fit the pattern.

Kohler et al. (1997) studied the relative contribution of aircraft NO_x emissions for various latitudes and heights. January and July mean distributions were examined of different NO_x sources such as fossil fuel combustion, biomass burning, soil microbial activity, lightning, and degradation of NO_2 in the stratosphere, with aircraft emissions. During winter, represented by January, 60% of NO_x emissions at latitudes between 30° and 60° and at vertical regions between 175-325 hPa[2] were found to arise from aircraft, while in July the contribution was calculated as only 20% for the same region. The maximum NO_x contribution of aircraft for the North America flight corridor was found to be 85%.

Schurmann et al. (2007) investigated NO, NO_2, CO and CO_2 emissions in the vicinity of Zurich airport. The study particularly focused on speciation for volatile organic compounds (VOCs) and the correspondence of real in-use emissions of NO_x and CO with those obtained from emission inventories. Two methods are used in the study:

Fourier transform infrared (FTIR) spectroscopy in the infrared region and DOAS in the UV/visible spectral range. The authors note that, during idle, predictions are underestimated of CO emissions and overestimated for NO, relative to those in the ICAO emission database for certain engine types. This phenomenon can likely be attributed to lower power use in real idling. It is reported that up to 98% of NO emissions in the vicinity of the airport are caused by aircraft, but that ground support activities cannot be neglected.

Pham et al. (in press) investigated the total aircraft emissions from Australian air transport over a period of six months. Real time data are obtained, unlike the data in other inventories such as the System for Assessing Aviation's Global Emissions (SAGE) and the Global Aviation Emissions Inventories for 2002 and 2025 (AERO2K). In that study, a base of aircraft data (BADA) is used to obtain fuel consumption data, while the Boeing Emission Method 2 and ICAO emission database are utilized for obtaining emission indices. The emission indices accompanied by fuel consumption data allow the determination of total aircraft emissions. The quantities of various emission species are given in the Appendix (see Figure 20).

Turgut & Rosen (2010) discuss the effect of factors such as LTO cycle, aircraft type and engine type on the cumulative emissions for eight busy airports around the world for a single day. One of their findings indicates that the LTO cycle is not sufficient for comparing airports in terms of total emissions. In addition to LTO cycle, emissions exhibit notable dependences on aircraft and engine type, number of engines and age. The lowest NO_x emissions are observed at Los Angeles airport and the highest at London Heathrow airport, among given eight airports considered. This result is mainly attributed to aircraft type, since the major aircraft types operating in Los Angeles are single aisle aircraft, such as the B737 (25%) and the B725 (25%). In addition, the share of B747 aircraft is only 4% and the share of other heavy aircraft is less than 1%. At London's airport, however, although the main types of aircraft operating are the A319 (18%), A320 (18%) and A321 (13%), there are significant numbers of heavy aircraft such as the A340 (5%), B747 (9%) and B777 (9%).

Among the two global emission inventories, AERO2K includes global aircraft emissions for the year 2002 and forecasts for 2025. In addition to civil aviation, military aviation is also considered. The AERO2K inventory was developed by the European Commission as part of the FP5 Project. In the inventory, significant indicators are listed, including fuel consumed, emissions and distance flown for each month in 2002 and 2025. In addition, diurnal variations in global emissions are shown at six-hour intervals (Eyers et al., 2004). Some of the important results are compiled in the Appendix (Table 4 and Figures 21 and 22).

The second inventory, SAGE version 1.5, has been developed by the Federal Aviation Administration Office of Environment and Energy in the U.S. in 2005 (Lee et al., 2007). The objective of that effort is to list global inventories of fuel burn and emissions for the years 2000 through 2004. Some of the model capabilities are summarized in Kim et al. (2005) as follows:

"The three basic inventories generated by SAGE are: (1) four-dimensional (4D) variable world grids currently generated in a standardized 1° latitude by 1° longitude by 1 km altitude format; (2) modal results of each individual flight worldwide; and (3) individual chorded (flight segment) results for each flight worldwide."

Although the entire results are not available to the public, some findings are shared openly available (see Table 5 and Figure 23 in the Appendix).

JET ENGINE EMISSIONS

The composition of aviation fuels has generally been determined by requirements such as energy content, fluidity, density, corrosion protection, stability, low temperature resistance and cost (Maurice et al., 2001). In commercial aviation the main fuel is kerosene, an oil distillate which is more volatile and less viscous than residual oils (Gaffney & Marley, 2009). The amount of sulfur in kerosene is generally less than 0.3%, while the nitrogen content is negligible. Despite the fact that there is no single chemical formula for kerosene, a widely accepted typical one is $C_{12}H_{23}$ (Lee et al., in press; Nojoumi et al., 2009; Soares, 2008), which has a lower heating value of 43.1 MJ kg^{-1}. Further reading regarding to kerosene and other types of aviation fuel can be obtained elsewhere (Maurice et al., 2001).

Although the same general power cycle, the Brayton cycle, has been the basis of aircraft gas turbine engines since the first gas turbine engines were patented in 1939, the fuel efficiency of aircraft has been improved continuously over the decades, particularly with the advent of turbofan engines. These improvements are mainly associated with improvements in aircraft and engine technologies (airframe, material, and aircraft systems), cost effective air traffic management and well-organized seating schemes with cooperation between airlines. As a result, despite increases in oil costs, the fuel cost of air transportation has improved and associated environmental impacts reduced. In Figure 4, average unit fuel consumptions are shown for various common commercial aircraft. As seen from Figure 4, the B787 has become the most advantageous option in terms of fuel economy, exhibiting a fuel consumption per 100 passenger-kilometer (pax-km) of 2.80 liters and the maximum available tonne kilometers (ATK) per US gallon. Even though it is currently in the test stage, these indicators seem to be reasonable considering its light airframe structure, advanced aircraft systems and efficient engine.

A different perspective is provided by the fuel consumption per seat for different phases of flight. Such data are given for several aircraft types in Figure 5, along with emissions. The minimum total seat fuel cost is observed from that figure

Figure 4. Operating data for several commercial aircraft: (a) specific fuel consumption per 100 pax-km, (b) available tonnes kilometers per liter fuel. Modified from Morrel (2009).

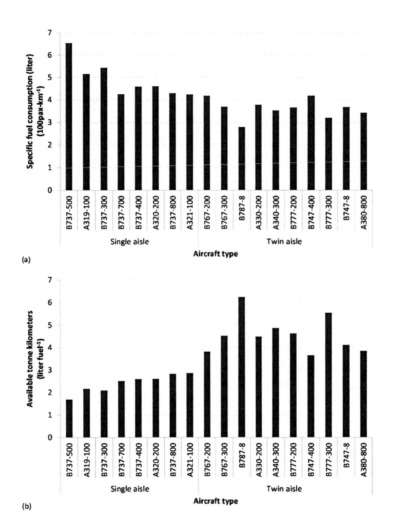

(a)

(b)

for the A320-200, the aircraft with the higher seat capacity. However, it is worth noting that flight range plays an important role on fuel cost per seat.

Aircraft with gas turbine engines generally operate in upper troposphere and lower stratosphere, except during takeoff and landing. These engines influence climate as a consequence of the changes they impart on the composition and structure of the atmosphere. The impacts of subsonic and supersonic flights on climate, as a function of altitude and ambient conditions, are detailed elsewhere (Brasseur et al., 1998; Lee et al., in press). The causes of climatic impacts for subsonic flights are given as follows:

1. Emissions of CO_2 (+ RF warming)
2. Emissions of NO_x
    a. resulting in the formation of O_3 (ozone) in the troposphere (+ RF warming)

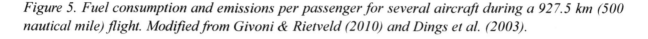

Figure 5. Fuel consumption and emissions per passenger for several aircraft during a 927.5 km (500 nautical mile) flight. Modified from Givoni & Rietveld (2010) and Dings et al. (2003).

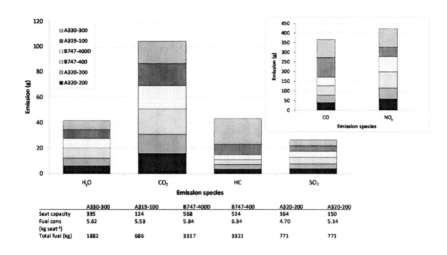

	A330-300	A319-100	B747-400D	B747-400	A320-200	A320-200
Seat capacity	335	124	568	524	164	150
Fuel cons (kg seat⁻¹)	5.62	5.53	5.84	6.34	4.70	5.14
Total fuel (kg)	1882	686	3317	3322	771	771

b. resulting in the destruction of CH_4 (methane) in the troposphere (- RF cooling) accompanied by loss of tropospheric O_3.

3. Emissions of sulphate particles resulting from the sulphur in fuels (- RF cooling)
4. Emissions of soot particles (+ RF warming)
5. Emissions of H_2O, which results in the formation of persistent contrails (+ RF warming) and the formation of polar stratospheric clouds (ozone depletion)
6. Emissions of hydrocarbons, which lead to ozone formation

The causes of impacts on climate for supersonic flights follow:

1. Emissions of CO_2 (+ RF warming)
2. Emissions of NO_x
   a. resulting in the destruction of O_3 at altitudes above around 20 km (- RF cooling)
   b. resulting in the destruction of CH_4 in the stratosphere (- RF cooling, negligible)
3. Emissions of sulphate particles resulting from the sulphur in fuels (- RF cooling),

in addition indirectly contributing to O_3 formation
4. Emissions of soot particles (+ RF warming)
5. Emissions of H_2O, which results in the formation of persistent contrails (+ RF warming) and the formation of polar stratospheric clouds (ozone depletion)

The variation with altitude of fuel combustion and various emission species is shown in Figure 6, on a fractional basis for a single month. The highest share (17.2%) of global fuel usage via scheduled aircraft in 1999 is attributable to the B747-400 aircraft, followed by earlier versions of the B737(300/400/500) and B747(100/200/300), which contributed 8.9% and 8.8%, respectively. The shares contributed by other aircraft are presented elsewhere (Sutkus et al., 2001).

Aircraft emissions can be expressed on a normalized basis by a general emission index EI_x, which denotes the mass of pollutant x per mass of fuel. The emission indices for the most abundant outputs of gas turbine engines, H_2O and CO_2, can be determined from combustion balances and are widely accepted to be around as 3.15 kg of CO_2 per kg of fuel and 1.26 kg of H_2O per kg of fuel

Figure 6. Variation with altitude of fuel combustion and various emission species, on a fractional basis, for Jan. 1999. Modified from Sutkus et al. (2001).

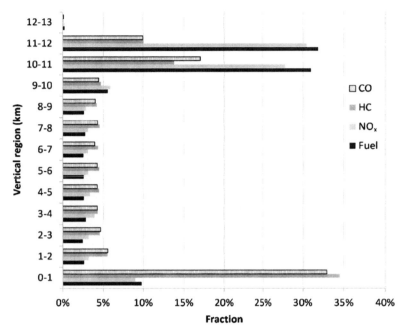

(Baughcum et al., 1996; Sutkus et al., 2001; Penner et al., 2001).

NO$_x$

NO$_x$ emissions are continuing to rise, at an approximately linear rate of 0.26% yr^{-1}, and reached an atmospheric concentration of 319 ppb in 2005 (Forster et al., 2007). Some NO$_x$ emissions are from aircraft in atmosphere. The contribution of NO$_x$ emissions from aircraft for altitudes of 9-13 km has been studied by Kohler et al. (1997) considering two emission inventories and is given in Figure 7. Approximately 31% of the fuel burned and 35% of NO$_x$ emissions occur below 9 km. The remainder of fuel consumption and NO$_x$ emissions occur at altitudes of 10-12 km (Sutkus et al., 2001). According to Sutkus et al. (2001), fuel use and NO$_x$ emissions increased between 1992 and 1999 by 33% and 35%, respectively. Further details are presented in the Appendix for ANCAT (Table 6 and Figure 24), for NASA 1996 (Figure 25) and for NASA 1999 (Tables 7 and 8 and Figures 26 and 27).

NO$_x$ emissions from aircraft mainly affect the atmosphere via ozone. Various studies have found that there is almost a linear relationship between the global annual aircraft NO$_x$ emission rate and the increase of O$_3$ in the troposphere of the northern hemisphere (Marquart et al., 2001). Another impact of aircraft NO$_x$ emissions is relates to CH$_4$ abundance in the atmosphere. Increasing O$_3$ leads to an increase in OH which easily reacts with CH$_4$, resulting in a decrease in CH$_4$ (Marquart et al., 2001). Therefore at cruise altitude, NO$_x$ emissions basically increase ozone and reduce methane concentrations in the atmosphere (Williams & Noland, 2006).

Jacobson and coworkers point out that NO$_x$ includes both NO$_2$ and NO, the latter of which is the less toxic (Wood et al., 2008). NO$_2$ absorbs solar radiation readily, thereby contributing to RF. Due to its short lifetime, the distribution of NO$_2$ is strongly heterogeneous, with the highest

Figure 7. Absolute and relative contributions of NO_x sources to the total atmospheric burden of NO_x, as a mean over the region 30-60° north latitude and 175-325 hPa (9-13 km). ANCAT/EC emission inventory suggests the NO_x burden of aircraft emissions as 13.7x10⁶ and 10.7x10⁶ kg for January and July, respectively. Modified from Kohler et al. (1997).

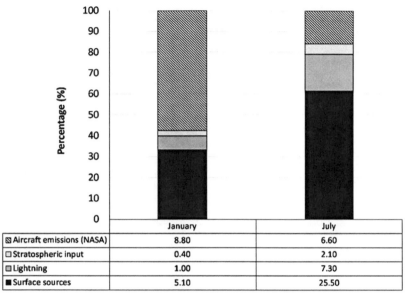

	January	July
▨ Aircraft emissions (NASA)	8.80	6.60
▢ Stratospheric input	0.40	2.10
▨ Lightning	1.00	7.30
▪ Surface sources	5.10	25.50

NO_x emission (10⁶ kg)

concentrations in industrialized regions (Isaksen et al., 2009). This observation is more relevant for assessing the climate impact of industrialized megacities and the high air traffic in some flight corridors, such as some of those over North America and Europe.

After combustion in a gas turbine, NO_x is generally emitted as NO with small amounts of NO_2 depending on engine power setting (Gaffney & Marley, 2009). The NO_2 fraction of such NO_x decreases with power setting, from over 98% at a power setting of 4% to under 10% at power settings of 65-100% (Wood et al., 2008). A similar result is reported by Lee (2004), who indicates that NO_x is emitted primarily in the form of NO_2 at idle and NO at high thrust.

The engine thermal efficiency is related to the combustion temperature, which is the turbine inlet temperature. As this temperature is increased the engine thermal efficiency increases, leads to bet-

ter fuel economy. NO_x formation is also related to combustion temperature. Gupta (1997) reports that the NO_x emissions from gas turbine engines increases exponentially with temperature, according to the following form: $\propto e^{0.009T}$. Nevertheless, NO_x formation from combustion is small below a temperature of 1200°C, while it increases notably above 1400-1527°C (Gupta, 1997; Hunecke, 2003; Daggett, 2004). Hence, an almost linear relationship exists between engine power setting and NO_x emission. For NO_2 emissions from a CFM56-3B1 aircraft, for instance, Wood et al. (2008) notes that 50% of the NO_x is emitted below an altitude of 152 m (500 ft) while 25% of the remaining NO_x emission is released when aircraft reach 914 m (3000 ft). Engine manufacturers need to take into account this inherent tradeoff between fuel efficiency and NO_x emissions.

The factors that lead to the formation of NO_x are as follows (Giampaolo, 2003; Sutkus et al., 2001):

- temperature of reaction
- residence time of exhaust gases in the high-temperature combustion region
- fuel/air ratio at the inlet and outlet of the combustion chamber
- fuel composition
- combustion chamber design
- combustion pressure

These factors highlight that the key driver in the formation of NO_x is combustion. Much research has been reported on developing combustion systems that achieve NO_x emissions well below emission regulations, including studies on low and ultra-low NO_x combustion (Gupta, 1997; Aida et al., 2005; McDonell, 2008), catalytic combustion (Vatcha, 1997; Kuper et al., 1999; Forzatti, 2003), flameless oxidation (Wunning & Wunning, 1997; Levy et al., 2004; Flamme, 2004; Wang et al., 2006), as well as lean premixed prevaporized combustion, and variable geometry combustion, staged combustion and rich-burned, quick quench, lean burn (RQL) combustion. Besides combustion-related causes of NO_x formation, Wulff & Hourmouziadis (1997) reports that the ambient temperature at the cruise altitude affects NO_x emission rate, suggesting that this effect could lead an increase in NO_x emissions of 10-20% for a temperature rise of only 10 K.

One of the challenges associated with the application of the low-NO_x combustion concepts to gas turbine combustors is related to the trade-off between CO and NO_x emissions (Hayashi et al., 2000; Gupta, 1997). Reducing the fuel air ratio, which implies a making the combustion mixture leaner, is seen to decrease NO_x emissions while increasing the CO and HC emissions.

NO_x formation is derived via three main mechanisms: thermal, prompt and fuel NO_x. Thermal NO_x is produced in post flame conditions by the oxidation of nitrogen in air due to the high-temperature conditions. The oxidation process is described by the Zeldovich mechanism, which involves three reactions (Wulff & Hourmouziadis, 1997):

$$N_2 + O \leftrightarrow NO + N$$
$$N + O_2 \leftrightarrow NO + O$$
$$N + OH \leftrightarrow NO + H$$

Increasing temperature and residence time shift the reactions toward to the right side, and are both key parameters affecting NO_x formation. Also, more than 90-95% NO_x emissions are in the form of NO (Gohlke et al., 2010; Boyce, 2002; Giampaolo, 2003). The presence of O and OH enhances NO_x formation. The OH is produced from the reaction of oxygen and water vapor in the presence of ultraviolet (UV) radiation and can initiate a large number of chemical reactions, in addition to NO_x forming ones, which affect the abundance of ozone, methane and secondary particles (Isaksen et al., 2009).

Unlike thermal NO_x formation, prompt NO_x is produced at low temperature in fuel-rich flames (Wulff & Hourmouziadis, 1997). The reaction involves atmospheric nitrogen and low molecular weight products of the thermal decomposition of hydrocarbons.

The last type of NO_x formation is associated with the N_2 molecules existing in the fuel. However, considering current fuel compositions, the effect of fuel NO_x on total NO_x formation is negligible.

Carslaw et al. (2008) investigated the NO_x emissions resulting from several commercial aircraft using various types of engines in the vicinity of London's Heathrow airport. They reported that good agreement between their results and those obtained from ICAO, but noted for some engine types that the emission outputs are overestimated in the ICAO engine emission database. They also reported that larger engines emit lower NO_x concentrations compared to those from smaller engines. One explanation is attributed to the higher total thermal energy of the exhaust gases from the larger engines, which leads to the dispersal of the plume over wider areas since they rise to higher levels due to bouyancy. NO_x emissions sampled for several engine types and aircraft are given in the Appendix (Table 9). With this data, Carslaw et al.

(2008) highlight the fact that NO_x emission outputs can vary significantly even when sampled for the same type of engine. For instance, the RR Trent 892 engine, which is used on both the B777-200 and B777-300 aircraft, emits different amounts of NO_x emissions in these applications. Similarly, NO_x emissions resulting from the IAE V2533-A5 engine are sampled for two types of aircraft, A320 and A321, and the amount of NO_x emissions is observed to be different. Consequently, Carslaw et al. (2008) conclude that the airframe is an important factor related to NO_x emissions.

Another interesting result of the study by Carslaw et al. (2008) is the observed differences in NO_x emissions when both engine and airframe types are same. In this case, the difference is attributed to the weight of the aircraft; longer routes require more fuel which results in higher aircraft weight and takeoff power. This effect can be observed for several aircraft types and different airlines in the Appendix (Table 9).

NO_x emissions in the troposphere have the potential to increase the quantity of ozone molecules, while those in the stratosphere lead to a decrease in the quantity of ozone molecules, affecting the energy balance associated with incoming UV radiation. Since the radiative efficiency of ozone is strongest near the tropopause, ozone production via NO_x emissions amplifies greenhouse effects (Brasseur et al., 1998; Groob et al., 1998). Additionally, ozone can cause harmful effects on human health (e.g., respiratory diseases including asthma), particularly for children and elderly people, and on agricultural crops (Slanina, 2008; Wilson, 2009). The amount of ozone in the troposphere has more than doubled since the early of 1900s and this is increase is almost certainly a result of increasing emissions of such gases as NO_x and CO (Staehelin et al., 2001).

The greenhouse effect caused by H_2O and NO_x (which leads to ozone formation or destruction depending on altitude (Groob et al., 1998) released from aircraft at cruise altitudes is larger than that when they are released at the earth's surface, because of such factors as residence time, temperature, concentration and radiative efficiency (Schumann, 1997).

The effect of cruise altitude on the NO_x emission index (EI) is investigated by Schumann (1997). Based on test results for the RB-211 and PW-305 aircraft engines, it is concluded that there is an inverse proportionality between the EI for NOx and cruise altitude, with the NO_x EI greater for releases at low cruise altitudes and higher for those at higher altitudes. Schumann (1997) also identified a relationship between the EI for NO_x and combustor inlet temperature; an exponential relation is observed for the RB-211 engine, while an almost a linear relationship can be seen for the PW305 gas turbine engine. It is also reported in this study that almost all the NO_x is emitted as NO, a result in agreement with that of (Wood et al., 2008).

Another factor that affects NO_x emissions, besides flame temperature, is combustion chamber inlet temperature. When the inlet temperature is high, the required fuel to generate sufficient heat could lead to high outlet temperatures from combustion chamber, which can in turn increase NO_x emissions. Water injection in the compressor inlet can be used to reduce NO_x emissions since it allows cooler air to enter the combustion chamber. However, increasing the pressure ratio in gas turbine engines has the potential to increase NO_x emissions. The effect can be explained noting the relationship between pressure ratio and temperature rise in the compressor, which is illustrated in Figure 8 for a 100% compressor efficiency. As the efficiency of the compressor decreases, the combustion chamber inlet temperature increases further.

CO_2

CO_2 and H_2O are the main chemical compounds formed from hydrocarbon combustion. Moreover, the amount of CO_2 observed in the exhaust from the combustion of a hydrocarbon fuel indicates

Figure 8. Variation of combustion inlet temperature with compressor pressure ratio for several ambient temperatures (aircraft types are identified corresponding to the overall pressure ratios for their engines).

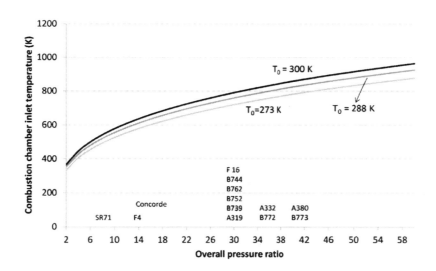

how efficiently combustion is occurring. All C atoms react to CO_2 during efficient combustion.

Forster et al. (2007) have examined atmospheric carbon dioxide, and note that the concentration of carbon dioxide in the atmosphere was 379 ppm in 2005. The RF calculated by the concentration of CO_2 is +1.66 [±0.17] W m^{-2}, and three-quarters of this amount results from utilization of fossil fuels and cement production. The largest change in atmospheric CO_2 concentration observed for any decade in the last 200 years was for 1995 to 2005, which exhibited a growth rate of 1.9 ppm yr^{-1}.

Sources of global CO_2 emissions, highlighting the contributions of transportation in general and aviation in particular are shown in Figure 9. The CO_2 emission share of aviation is about 13% of all transportation emissions, which corresponds to 2.6% of global CO_2 emissions. This value has tended to increase over time as air transport increased. CO_2 is the main greenhouse gas and the primary contributor to climate change. Carbon dioxide in the atmosphere does not allow the outgoing thermal radiation to pass to space, and supports the greenhouse effect against the albedo

effect, leading to warming. Surface temperature and atmospheric CO_2 concentration are related. For instance, Florides & Christodoulides (2009) notes that doubling the atmospheric CO_2 concentration from 0.035% to 0.070% increases the temperatures of the surface and the atmosphere at a 10 km altitude by 0.01°C and 0.03°C, respectively. This relationship is linked to the CO_2 solubility of sea water, which increases as the surface temperature decreases.

According to Brasseur et al. (1998), the worldwide aircraft fleet produced 2-3% of all fossil fuel-based CO_2 in 1998. This percentage declined to 2% in the following decade (Ponater et al., 2006; Boeing, 2010b; GOA, 2009), likely as a result more fuel efficient engines being used (despite the fact that utilization air transport increased). According to Boeing, for instance, an increase in fuel efficiency of 16-20% is anticipated with its new B747-8 and B787 aircraft. The decrease is also reported by the International Air Transport Association (IATA), which noted that aviation CO_2 emissions decreased from 671 million tonnes in 2007 to 666 million tonnes in 2008 (IATA, 2009).

Figure 9. Sources of global CO_2 emissions, highlighting the contribution of aviation. Modified from GOA (2009).

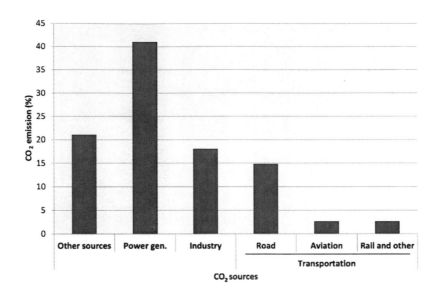

In Pejovic et al. (2009), the effects of the closure of an airport due to severe air transport system disruptions are assessed in terms of fuel consumption, CO_2 emissions and cost related to both fuel and environmental aspects. Two scenarios are investigated for a one hour closure of London Heathrow airport. In the first scenario, 43 aircraft planning to land at Heathrow airport are diverted to alternative airports. In the second scenario, some of these 43 aircraft are diverted to alternate airports (as in the first scenario), but others are forced to hold for 20 min. The fuel consumption and CO_2 emissions, respectively, are found to be 33 tonnes and 103 tonnes for first scenario, and 22 tonnes and 70 tonnes for second scenario. This result indicates that instead of diverting all aircraft to alternative airports, diversion combined with holding aircraft can lead to lower amounts of fuel consumption and CO_2 emissions.

This study (Pejovic et al., 2009) suggests by extension that running similar scenarios for all airports in the world could lead to substantial fuel savings and emissions reductions in such circumstances. Considering the significant of sav-

ings and reductions for only one airport, further research appears to be merited related to this topic, perhaps with a view to optimize the response to airport closures.

Flight range has an effect on CO_2 emissions. It is noted by Williams & Noland (2006) that long-haul flights exhibit lower CO_2 emissions per km than short-haul flights. A likely explanation is that, since the duration of the cruise phase of long-haul aircraft is larger compared to short-hauled aircraft, the duration of flying time at peak fuel efficiency is greater for long-haul aircraft, leading to a decrease in emissions. In addition to the effects on fuel efficiency, long-haul aircraft have higher seating capacities, which also tend to lower CO_2 emissions per passenger-km. Miyoshi & Mason (2009) assess the effect on air transport CO_2 emissions of such factors as aircraft type, load factor and seat configuration. In this investigation, CO_2 emissions are calculated using the approach of DEFRA (Department for Environment, Food and Rural Affairs, UK). Accordingly, around half of air transport fuel consumption is attributable to flights over 2200 km (1200 nautical miles), or greater than 2½ to 3 hours flight time (Ruffles's study (as cited in Morrell, 2009)), which implies potentially less carbon dioxide emissions for shorter distance flights.

H_2O

Water vapor is a greenhouse gas has a warming effect on climate, but its residence time in the atmosphere is relatively low compared to other emission species. Several behavioral characteristics of H_2O in the atmosphere are worth noting:

- Water vapor trends are significant in the troposphere, as they represent a feedback to the climate system. Forster et al. (2007) suggests an increase in water vapor concentration in the troposphere of 1.2% per decade for the 1988–2004 period, and an increase on the order of 5% over the 20th century (Isaksen et al., 2009).
- H_2O emissions increase the occurrence frequency of polar stratospheric clouds which leads to ozone destruction (Brasseur et al., 1998).
- Water vapor emissions in the troposphere are typically removed by precipitation within 1 to 2 weeks, while water vapor concentrations in the lower stratosphere can accumulate to levels (Schumann et al., 2000).
- Changes in hydroxyl radical (OH) concentrations directly affect the rate of methane oxidation. Higher levels of water vapor lead to higher levels of OH. The reaction of methane with OH is strongly temperature dependent, and proceeds more rapidly at higher temperatures (Isaksen et al., 2009).

The amount of water vapor released from aircraft is much smaller than the amount of water evaporating from surface of the earth (Schumann, 2000). However, it has a direct effect on radiative forcing, particularly considering its role on contrail formation at high altitudes. The oxidation of CH_4 is a source of mid-stratospheric H_2O and has led to a slight increase in H_2O concentrations in the stratosphere since 1750, i.e., the concentrations ranged from 3 to 5 ppm in 1750 to from 3 to 6 ppm in 1998 (Ehhalt et al., 2001). Since the existence of H_2O in the stratosphere is known to cause a significant greenhouse effect, activities leading to increased stratospheric levels of H_2O, through the use of alternative fuels such as hydrogen and supersonic aircraft fleets, should be carefully designed and monitored.

CO and HC

Carbon monoxide (CO) and hydrocarbon (HC) emissions associated with aviation generally arise from incomplete combustion and/or lower power engine operation, such as that used during starting and idle phases (Mazaheri et al., 2009; Sutkus et al., 2001; Schurmann et al., 2007; Anderson et al., 2006). Combustion efficiency thus has an important role on CO and HC production. As seen from Figure 10, aircraft engines exhibit relatively low combustion efficiencies at lower power settings, and this trend is particularly evident for old technology. In these engines, the efficiency is stabilized at somewhat higher power settings than at idle. The effect of new technology can be observed, in that newer engines have higher combustion efficiencies, even at lower power settings. However, the current technology does not completely eliminate the production of CO and HC.

Fuel air ratios that are rich are likely to induce CO production, while local flame extinctions in the combustion chamber tend to be responsible for unburned HC (Brasseur et al., 1998). The design of the primary zone in a combustion chamber also affects CO formation (Giampaolo, 2003). For example, the conversion of carbon atoms to CO increases as residence time in the combustion chamber is increased (Hunecke, 2003). Residence time has also impacts NO_x emissions from com-

Figure 10. Variation of combustion efficiency with engine thrust for various types of engines. Modified from Eyers et al. (2004).

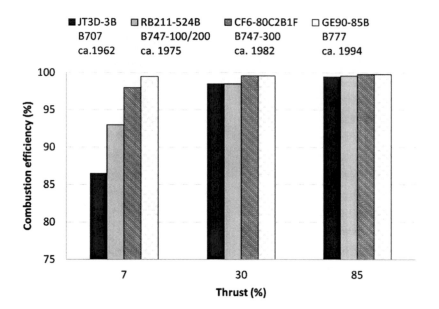

bustion. In order to reduce such emissions, research is ongoing to enhance combustion chamber designs and several investigations are discussed in "Technology."

The chemical composition of the gaseous exhaust plumes varies significantly during the LTO cycle since the engine power changes for various flight phases (Wood et al., 2008). Since carbon monoxide and hydrocarbons are both produced at lower engine power settings, the main flight phases in which these emissions are likely to be observed are the LTO cycles. According to Sutkus et al. (2001), approximately 70% of CO and HC releases are emitted below 9 km. A large part of this percentage leads to considerable degradation in air quality in the vicinities of airports.

Contrails

One of the main products of combustion in a jet engine is H_2O. Since it is discharged at relatively high temperatures compared to ambient air, H_2O in exhaust gases mixes with ambient air and in-

creases its relative humidity. Much of the water vapor condenses to small water droplets. The relative humidity is a function of atmospheric water vapor concentration and can be defined as the ratio of the partial pressure of local water vapor to the pressure of saturated water vapor, at a given temperature. Schumann points out that the threshold temperature for contrail formation decreases with cruise altitude and increases with ambient humidity, while the radiative properties of contrails depend on their coverage area and optical depth (Schumann, 2000).

The relative humidity (RH) can also be defined as the ratio of H_2O condensation rate to evaporation rate. When the RH is below 100%, the rate of evaporation is observed to be greater than the rate of condensation, which leads to no water droplets formation (Williams et al., 2002).

When the ambient air temperature is sufficiently low, the water droplets turn into ice crystals, which form the visible trails behind aircraft. A contrail (condensation trail) forms when saturation with respect to liquid water is reached or surpassed in the exhaust plume (Rogers et al., 2002). Contrails are one of the most significant anthropogenic effects which are visible (Marquart et al., 2001).

Contrails are short-lived when the ambient air is dry (Schumann, 2005). At typical cruise altitudes, contrails can be expected to persist for only a few second to about one minute in the range of one kilometer and tens of kilometers behind the aircraft. The relative humidity of the atmosphere is a key factor in contrail formation. The stratosphere is generally composed of relatively dry air which does not lead to contrail formation (Schumann, 2005). The RH is around 40-90% in the higher troposphere and 0-60% in the stratosphere (Shellard, 1949). Similarly, it is reported by (Rogers et al., 2002) that the troposphere is rich in water vapor and poor in ozone, while the stratosphere is rich in ozone and poor in water vapor. According to Brasseur et al. (1998), contrails covered about

0.4% of the sky over Central Europe and about 1% of the sky over the north Atlantic Ocean in 1998.

In addition to the relative humidity of the ambient air, contrail formation also depends on cruise altitude and the propulsive efficiency of the aircraft engine. When the ambient air is dry, contrails evaporate quickly. However, if the relative humidity is sufficiently high, contrails persist and lead to the formation of artificial cirrus clouds, mainly in the upper troposphere (Penner et al., 2001). This phenomenon is explained in Figure 11 by Flippone (2010).

In Figure 11, the structure of contrail formation is shown. An important factor on contrail formation, relative humidity, is denoted by two curves which provide 0% and 100% limits. Above the 100% RH curve (solid curve) contrail formation is not observed, while below the 0% RH curve contrail formation is unavoidable (Flippone, 2010; Schumann, 2005). By contrast, despite the fact that the air below the 0% curve is completely dry,

Figure 11. Relation between contrail formation and such factors as temperature, pressure and relative humidity. Adapted from Flippone (2010). (Reproduced by permission; the source article was published in Aerospace Science and Technology, 14, Flippone, A., Cruise altitude flexibility of jet transport aircraft, 283–294, Copyright Elsevier 2010).

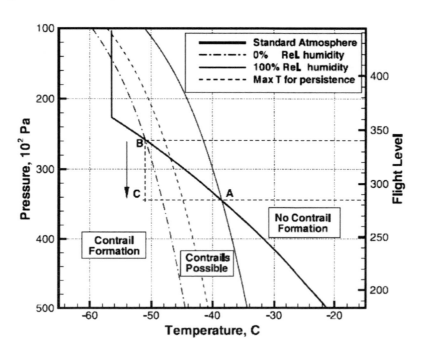

water vapor from engine exhaust adds large amounts of moisture which rapidly condense to clouds because of the very low temperatures at flight altitudes (Flippone, 2010). It can be seen from the graph the avoidance of contrail formation can be attained by descending to lower altitudes where the temperature is higher. Even at lower altitudes which correspond to between 0% and 100% relative humidity, contrail avoidance is not absolute. A reduction in altitude of least 1524 m (5000 ft) is needed to avoid contrail formation during flight, according to ISA conditions. In an analysis of this effect by Flippone (2010), a reduction in altitude of 610 m (2000 ft) with a slight reduction in speed yielded no additional fuel consumption. This case did not completely avoid contrails during the flight, but it did reduce contrail formation compared to a flight at higher altitudes. For a completely contrail free flight, the aircraft needs to reduce its altitude by 1219 m

(4000 ft) which leads to slightly longer flight with 1.5% additional fuel consumption.

Propulsion efficiency also has an impact on contrail formation. Engines with higher propulsion efficiencies use more of the thermal energy of combustion to propel the aircraft. Hence, a lower fraction of this heat is discharged with the exhaust, leading to an increase in relative humidity at higher ambient temperatures (lower altitudes). Therefore, high propulsion efficiency results in contrail formation at lower altitudes. In Figure 12, this case is observed in the side by side flights of two commercial aircraft, B707 and A340, at the same altitude. Details for the engines of the two aircraft are given in (Schumann, 2000). As seen in the figure, despite the fact that the cruise conditions for the two aircraft are almost the same, contrails are observed for the A340 and not for the B707.

Since current propulsion efficiencies are between 0.3-0.4, Schumann et al. (2000) investigated the impacts of likely increases in propulsion

Figure 12. Effect of propulsion efficiency on contrail formation for two aircraft. Adapted from Schumann et al. (2000). (Reproduced by permission; the source article was published in Aerospace Science and Technology, 4, Schumann, U., Influence of propulsion efficiency on contrail formation, 391–401, Copyright Elsevier 2000).

efficiency on contrail formation from a global perspective. They note that for an increase of propulsion efficiency from 0.3 to 0.5, the threshold temperature for contrail formation increases by 4.2-4.9 K, which corresponds to a decrease in the lower limit of contrail formation threshold altitude of 650-670 m. This correspondingly leads to an increase in the vertical region where contrail formation occurs.

The climate problems resulting from contrail formation are linked to the greenhouse effect of contrail-based cirrus clouds (Schumann, 2005). To mitigate the greenhouse effect caused by contrail formation, one focus point that has emerged is the cruise altitude. In ISA conditions two vertical altitude regions are reported where contrails can form due to levels of relative humidity. These include altitudes between 8-12 km for 100% RH (liquid saturation) and 10.2-14 km for 0% RH (Schumann, 2005). This situation along with the temperature and other factors such as propulsion efficiency and humidity, provides the opportunity for contrail formation abatement by changing cruise altitude. The flexibility in cruise altitude is described in several studies, which are discussed below.

In Flippone (2010), a contrail avoidance procedure is developed for commercial aircraft. One of the key findings of this study is the suggestion of decreasing cruise altitude by 1219 m (4000 ft) to avoid contrail formation, with a 1.5% increase in fuel consumption.

In another study, the alleviation of contrail formation is performed by temporarily changing the propulsion efficiency utilizing variable guide vanes for the fan (Haglind, 2008). Since there is a significant inverse relationship between propulsion efficiency and contrail formation, the study suggests a controlled propulsion efficiency reduction when the aircraft flies through regions where persistent contrails are likely to occur. The results suggest that there is a potential for decreasing the contrail threshold temperature by 1.5 K in the troposphere and 1.4 K in the stratosphere.

The reduction in threshold temperature implies an increase in the vertical region where contrail free cruise is likely to occur. This decrease in threshold temperature leads to an additional cruise allowance of 300 m in the troposphere and 1000 m in the stratosphere. However, to obtain these achievements an increase in specific fuel consumption is required (42-48%), since the propulsion efficiency is decreased by about 10-11%.

Another aspect of contrail formation is the effect of range of flight, for which several factors are of relevance. First, the proportion of the high engine power usage in long-haul flights is less than in short-haul flights. This factor is also beneficial for NO_x emissions. Second, cruise altitudes for short-haul flights are relatively lower than those for long-haul flights, suggesting that the contrail formation is less likely during short-haul flights. Williams & Noland (2006) report that, in addition to aircraft type and number of operations, the balance between long- and short-haul flights has a significant climatic impact. A key result from this study is that, for short-hauled flights, the shortest routes produce the lowest contrails per kilometer. It is also highlighted for long-hauled flights that lower contrails per kilometer are observed for routes to North America in January and to Asia in July.

Other results from Williams & Noland (2006) are as follows:

- larger aircraft general produce less contrail per passenger kilometer than smaller aircraft,
- the highest contrail per km is observed for flights in January, considering flights at Heathrow Airport, and
- the contrail per passenger km increases with route length for short- and medium-haul flights.

In this study, the timing of flights is also discussed. In daytime, contrail-based cirrus clouds provide a radiation shield that decreases the solar radiation reaching lower atmosphere levels, while the same shield also hinders longwave terrestrial radiation from leaving the earth. The first of these observations leads to cooling effect, while the second leads to a warming effect. Also, both effects occur in day, but only the second occurs at night.

MITIGATION STRATEGIES

Much effort has been expended by manufacturers, operators, air traffic authorities and the engineering and scientific communities to reduce aircraft emissions, substantially through enhanced fuel efficiency. Studies have also been performed on the following other measures:

- eliminating unnecessary aircraft weight,
- implementing smart air traffic management systems and providing most advantageous performance aircraft conditions during all flight phases,
- scheduling fleets properly and optimizing routes to obtain the maximum load factor,
- reducing the duration of ground operation of engines and auxiliary power units (APUs),
- implementing single engine taxiing,
- reducing thrust reverse usage while considering the tradeoff of that measure with taxi time,
- using blends of alternative fuels, and
- using ground support equipment rather than APUs when sensible.

Some of these measures can be easily applied by operators, while others require substantial mechanical and design modifications. As an example of the former, investigations of the relation between fuel efficiency and seating capacity have shown that for every 1% increase in seating capacity a 0.83% reduction in fuel consumption is possible. Replacing a B767-200 engine with a larger B777-200 would lead to a 66% increase in

seats and a 26% increase in fuel efficiency (Morrell, 2009). Studies on other measures to reduce aircraft emissions are described in the following subsections, in three categories: technology, flight procedures and alternative fuels.

Technology

There is a strong relation between emissions from aircraft engine (for most emission species) and fuel consumption. Hence, technology development, which often increases fuel efficiency, also reduces such emissions. The advent of turbofan engines is considered one of the major developments in both of these areas. Compared to previous decades, more efficient and environmentally benign engines have resulted from such advances as high by-pass ratios, high turbine inlet temperatures through improved cooling systems, and advanced heat-resistant and light weight materials with full electronic control systems. However, aircraft technology is still far from zero emission engines.

Emission abatement methods related to technology include winglets, laminar flow technology (reduced airframe drag through control of the boundary layer), blended wing-body systems (unconventional airframes), advanced engine components, lighter materials, advanced engines (e.g., geared turbofan engines, open rotor engines, counter-rotating fans), and energy savings in lighting and fuel cell systems for onboard energy production. With the recent advent of these technologies, IATA airlines determined some targets in 2009 such as a cap on aviation CO_2 emissions starting in 2020, an average fuel efficiency increase of 1.5% per year from 2009 to 2020, and a reduction in CO_2 emission in 2050 relative to 2005 levels. Although there appear to be significant difficulties related to targets, targets at this level indicate that environmental concerns are being considered by commercial organizations (IATA, 2009; Ribierio, 2007). Recent technological improvements are discussed in the following paragraphs, along with some conventional mitigation approaches.

Figure 13. Variations of emissions with equivalence ratio and power setting. Adapted from Wulff & Hourmouziadis (1997). (Reproduced by permission; the source article was published in Aerospace Science and Technology, 8, Wulff, A., & Hourmouziadis, J., Technology review of aeroengine pollutant emissions, 557–572, Copyright Elsevier 1997).

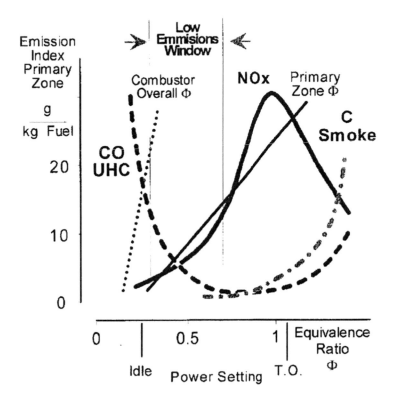

An important technological advance related to emissions abatement has been new combustion chamber designs (briefly mentioned in "CO_2"). One of the main challenges in combustion chamber design is deciding on the proper tradeoff between such parameters as flame temperature, residence time, equivalence ratio, fuel-air ratio and cooling. The relation between emissions, equivalence ratio and power setting is shown in Figure 13. As seen in that figure, to obtain lower emissions there is a relatively narrow window which does not involve the major factors like power settings during idle, cruise, takeoff and climb. CO and HC emissions increase during idle since that condition involves a relatively low equivalence ratio, while NO_x emissions increase during takeoff due to the higher equivalence ratio. Also, increasing pressure ratio and in turn increasing the combustion chamber inlet temperature also increases NO_x emissions.

Numerous combustion chamber designs have been developed to achieve efficient and stable combustion while suppressing emissions, several of which are discussed in depth by Wulff & Hourmouziadis (1997). The variable geometry combustor was developed to eliminate the difficulty of fixed air for primary and dilution zones, which does not allow any change in equivalence ratio. The staged combustor was designed to overcome the tradeoff between NO_x and CO/HC. In this system, there are two combustors with their own set of fuel nozzles, and the air flow is divided into these two combustors, leaving each with half of the total air mass flow. At lower power settings, only one combustor is operated and all fuel is burned in it. Since half the air flow is used, there is a higher equivalence ratio leading to suppression of the CO and HC emissions normally observed at lower power settings. When higher power is required and after the temperature and the equivalence ratio reach a critical level, the second combustor is switched on. This combustor designs are successfully used in several engines, e.g., CFM56 (A320), GE90 (B777), V2500 (A320) and BR700 (B717).

In lean premixed combustion, the lean mixture (which has a lower equivalence ratio) is suggested for high power settings. The objective is to provide a homogeneous prevaporized and premixed combination of fuel and air in the primary zone at an equivalence ratio within the lower emission band of equivalence ratios. By avoiding a localized rich mixture, NO_x emissions are suppressed at high power settings. Another combustor design, the rich-burn, quick-quench, lean-burn (RQL), is based on a lower flame temperature at higher equivalence ratios as well as lower equivalence ratios. Accordingly, a relatively small amount of air is burned in the primary zone, yielding a higher equivalence ratio. While discharging from the primary zone, the gas is rapidly mixed with more air in a quench zone, decreasing the equivalence ratio for the subsequent burning process. In the lean burn zone, combustion occurs in the low emission equivalence ratio band, due to the lower equivalence ratio.

Another strategy to reduce NO_x emissions is catalytic combustion. In this combustion system, a catalyst is used to promote fuel oxidation rather than flame combustion. Oxidation takes place at a very low equivalence ratio compared to conventional combustion. To extract the chemical energy from the fuel molecules, additional energy is required which can be obtained from the recirculation of exhaust gases (McDonell, 2008). Conventional and catalytic combustion systems are shown in Figure 14. As seen in that figure, the turbine inlet temperature is the same for conventional and catalytic combustion. Therefore, there is no loss in thermal efficiency for the catalytic combustion system. Furthermore, elimination of the requirement for turbine cooling leads to an additional increase in power. Since, the combustion temperature is relatively low than for conventional combustion, NO_x emissions are also decreased. The NO_x emissions for the unit in Figure 14 it is reported to be lower than 2.5 ppm.

In recent years, Pratt & Whitney has focused on new turbofan engine technology, the Geared Turbofan engine (see Figure 15). The main concept of this design is that the fan speed is reduced by a gear so that the fan and the low pressure compressor and turbine each run at their optimum speed. The manufacturer notes that this new technology has three main advantages: reduced fuel consumption, emissions and noise. Specifically, this engine is expected to provide an increase in fuel efficiency of 16% compared to the current best engines, a reduction in NO_x emission to 50% below CAEP/6 standards, and a noise reduction of 20 decibels below today's most stringent standard (Pratt & Whitney, 2010a).

A widely known emission abatement (and thrust augmentation) technology is water injection, which involves the injection of a small amount of water into the compressor or combustion chamber. This concept has been used for aviation for a long time, e.g., a paper published in 1945, entitled 'Water Injection for Aircraft Engines' states

Water injection was the subject of experiment in the automobile industry as long as twenty years ago... (Rowe & Ladd, 1945).

Presently, water injection has lost popularity due to factors such as the advent of efficient engines, the potential impacts on engine materials and the desire to avoid additional weight. Nevertheless, water injection (with other versions as steam injection or fogging) is still efficiently used in terrestrial industrial gas turbine engines.

Figure 14. Comparison of conventional and catalytic combustion systems. Fuel molecules are broken down by the catalyst at relatively high temperatures in the XONON module where the flameless reaction between oxygen and fuel molecules takes place. Modified from Cusson (2000).

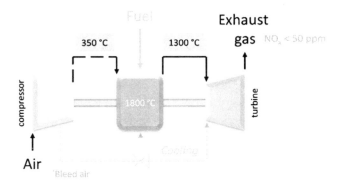

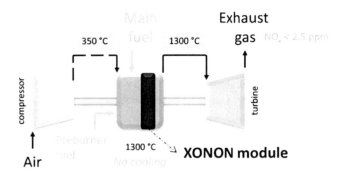

Figure 15. Geared turbofan engine fan drive gear system. Reproduced by permission from Pratt & Whitney (2010b).

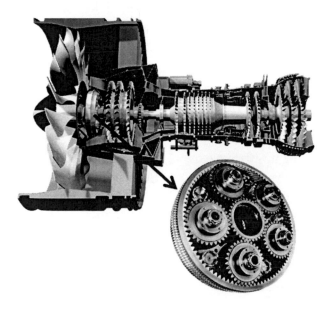

In aviation, several examples were produced in the late 1950s and early 1960s.

The most well known example of a water injection engine is Pratt & Whitney JT3C-6 used with the B707 aircraft (Daggett, 2004). When the ambient temperature is 32°C, water injection provides 35% more thrust. The level of thrust increase depends mainly on the ambient temperature; thrust does not increase for ambient temperatures below -5°C. A modified military version of the B707 for refueling, the KC-135, also has a water injection system and utilizes a 2500 liter water tank. This system provided an increase in thrust of 30%. Other examples of commercial aircraft using water injection were the B747-100 and 200. Water injection was used with the JT9D-3A, 7, 7A and 7F engine types, yielding an increase in thrust of 3.3-4%. In these engines, water is injected into the high pressure compressor discharge air. The water injection system was then adapted to a JT9D-20 engine, which was used in the DC-10 aircraft. A thrust increase was obtained of approximately 7%. Other examples include the B-52A (which use PW-J57 engines with a 1363 liter water tank, but only three were built), the BAC-111 (for which the water injection system caused a fatal accident because the water injection system inadvertently filled with fuel), the Harrier AV/8B (which attained a 7% thrust increase), and the Metro III (turboprop TPE-331 engine which achieved a 10% increase in thrust).

Water injection is first emerged for thrust augmentation, which is achieved by two mechanisms. First, in hot ambient temperatures it cools the air inside the compressor, increasing the air density of and mass flow. Second, the compressor work is reduced, increasing thrust. Since the air before combustion is cooler, the potential amount of fuel for combustion without reaching the limiting turbine inlet temperature is increased. Therefore, more chemical energy can be converted to heat. Positive environmental effects are also observed, including a significant reduction in NO_x. In aeroderivative gas turbine engines, 85%

of NO_x emission can be eliminated by adding water injection systems to combustion chambers (Boyce, 2002).

A variation of water injection is steam injection. Heat is not required for water evaporation is this system, since the steam is already a vapor. Furthermore, since the ratio of the specific heat of steam to the specific heat of air is approximately 1.9, it is clear that more heat is absorbed by steam than air, leading to more power output for a given temperature. It is noted by Giampaolo (2003) that for every 1% of steam injection (compared to main air flow in turbine), power output could increase by 4%.

The injection location in water injection systems has been investigated by Daggett (2004). Low pressure compressor, high pressure compressor and combustion chamber water injection systems are considered in that research, and the main results are shown in Figure 16.

The best result, according to locations and indicators in Figure 16, is observed for the low pressure compressor water injection system. This system achieves reductions in combustion chamber outlet temperature (T_4) and specific fuel consumption (SFC) of 15.6% and 3.4%, respectively, relative to the baseline. Furthermore, a 46.5% decrease in NO_x emission index is observed. Water injection to the high pressure compressor provides somewhat lower gains, with decreases in T4, SFC and NO_x emission index of 11.9%, 1.5% and 44.3%, respectively, relative to the baseline. The effect of water injection on combustion chamber inlet temperature (T_3) is noteworthy, since significant decreases are obtained (19.8% for the low pressure compressor and 18.2% for the high pressure compressor). Note that the decrease in SFC results not only from less fuel consumption but also from decreased cooling air requirements. This effect is attributable to the temperature reduction obtained in T_4, which also leads to longer turbine materials service lifetimes.

The findings of Daggett (2004) suggest that the difference between low and high pressure com-

Figure 16. Comparison of various locations for water injection systems in aircraft gas turbine engine: (a) EI NO$_x$, SFC and pressure at the combustion chamber inlet (P$_3$); (b) temperatures at the combustion chamber inlet (T$_3$) and outlet (T$_4$). LPC and HPC denote low pressure and high pressure compressors, respectively. The altitude is sea level, the ambient temperature is 20.7°C, the Mach number is 0.25 and the thrust is 331 kN. Modified from Daggett (2004).

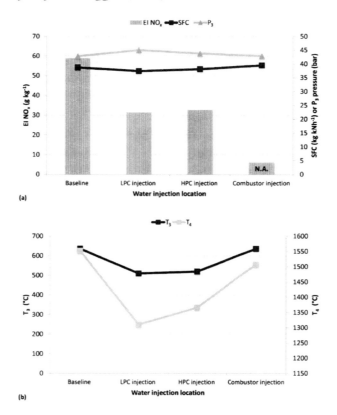

pressor water injection systems affects strongly NO$_x$ emissions and the temperature at combustion chamber outlet, and water injection for both compressor systems provides significant decreases in these parameters, while water injection for the low pressure compressor system provide better results than those for the high pressure compressor.

In Benini et al. (2009), numerical and experimental analyses aimed at reducing emissions of NO and CO are performed for low power turbojet engines. In this study, the effects of water and steam injection on emissions are investigated. It was determined that a 16% reduction in NO is possible using steam at a flow rate double that of the fuel. However, this measure compromised

the amount of CO and unburned hydrocarbons, which both were observed to increase due to the temperature drop and incomplete oxidation. It is also noted that the effect of incomplete oxidation could be decreased by reducing the steam flow. When the steam flow rate equals the fuel flow rate, although less NO reduction is observed, a slight reduction in CO is possible. With water injection, there appears to be less NO reduction and more CO formation compared to steam injection. A detailed explanation of the tradeoff between NO$_x$ and CO in low NO$_x$ combustion is discussed in (Hayashi et al., 2000).

Flight Procedures

Another mitigation strategy is to optimize flight procedures, which involves efficient routes, weight reductions and state-of-art air traffic management systems such as single European sky (SESAR), reduced vertical separation minima (RVSM), continuous descent approaches (CDA), performance based navigation (PBN), etc. According to the 2007 (fourth) IPCC assessment report (Riberio et al., 2007), decreasing the vertical separation from 610 m to 305 m (2000 ft to 1000 ft) can decrease the total fuel burned by 1.6-2.3% yr^{-1} for airlines operating in the European RVSM area. Also, according to IATA (2009), using CDA rather than the conventional stepped approach could save up to 630 kg of CO_2 per landing. IATA and other partners estimate that implementing CDA in 100 airports by the end of the 2013 could avoid emissions of up to 500,000 tonnes of CO_2.

Currently, environmental concerns regarding aircraft emissions are considered mainly in the vicinity of airports, due to factors such as uncertainty regarding the impact of aircraft emissions on climate, measurement challenges at cruise altitudes and experiences by people living around airports with the impacts of emissions on local air quality. The most explicit example is that the ICAO, the authority implementing engine certification standards, has been undertaken emission measurements for only LTO cycles.

In Mannstein et al. (2005), another sound approach (flexible free flight) is considered to avoid contrail formation. Although it was earlier determined that a cruise altitude reduction of 1829 m (6000 ft) eliminates contrail formation by 50% with an increase in fuel consumption of 6% (Rogers et al., 2002). Mannstein et al. (2005) developed a more efficient approach, referred to as selective flight level, when relative humidity detectors (hygrometers) indicate that there is high relative humidity at the cruise altitude. In such cases, a reduction in flight altitude of 610 m (2000 ft) can reduce contrail formation by 50%. With sufficiently precise hygrometers at cold ambient temperatures, this reduction can be observed with smaller altitude reductions, as small as 305 m (1000 ft). Besides hygrometers, Mannstein et al. (2005) also suggest simple camera systems for judging the situation and tracking the contrails produced by aircraft flying earlier in routes with high air traffic. However, neither the camera system nor contrail production tracking of earlier aircraft appear feasible since they increase the burden on the pilot. Therefore, these two suggestions might be used for standby control.

Significant research is also being devoted to decreasing fuel consumption during descent. Despite the fact that less fuel is consumed during descent than other flight phases, substantial fuel savings are possible with proper descent procedures. The main factor which negatively affects fuel consumption during descent is the stepped vertical profile due to terminal air space requirements or preferences of airlines or pilots. A novel alternative descent procedure has been developed, the continuous descent approach (CDA). With this new procedure, a descent with a constant flight path angle occurs rather than a stepped descent. The main advantage of this procedure is the elimination of low level flight and fluctuations in power settings during descent. Since CDA allows to aircraft descend using its own potential energy at minimum power settings (idle power or slightly higher than idle power), this method provides abatement of not only fuel and emissions but also noise. The main drawback of the procedure is that it requires complex computer system supports to avoid increasing the work load on pilots and air traffic controllers. With current technology CDA is likely available only for the airports with low traffic or at nights, but numerous efforts are ongoing to overcome this weakness, as described below.

Wilson & Hafner (2005) conducted three CDA scenario simulations for landings at Atlanta's airport and measured the impacts of these scenarios on descent time, fuel cost and flown distance. The greatest savings achieved are 45 h, $80,000 and 16,600 km per day. These findings correspond to reductions in flight time and flown distance of 1 min and 5.5 km per aircraft, respectively, and annual fuel savings of $29 million.

Clarke et al. (2004) state that it is possible to achieve significant noise reductions via CDA procedures. They predict noise reductions of 3.9 and 6.5 dBA using simulation and demonstrated flights of B757 and B767 aircraft, respectively. They also report fuel savings of 180-225 kg per flight by switching from conventional approach procedures to CDA.

Turgut et al. (2010) investigated CDA for a B757 commercial aircraft for Istanbul terminal airspace, using real flight data for descent and BADA performance tables for low level flight fuel consumption. The emission indices are acquired from SAGE and ICAO emission databases. It is concluded that up to 44.3 kg of fuel savings per landing are possible by switching from conventional to CDA procedures and that the descent time can be decrease by 2.8 minutes since the aircraft can fly faster at higher altitudes.

Alternative Fuels

Some alternative fuels can assist efforts to reduce emissions. A widely known study involves using liquid hydrogen as an aircraft fuel, an idea which has been considered for decades. A B57 twin-engine bomber flew in 1956 with one engine modified to use hydrogen fuel. Afterwards, motivated mainly by the oil crises of the 1970, interest grew in hydrogen fuel (particularly in liquid form) usage in air transport, and many research results were attained by the end of 1980s. A small reciprocating aircraft Grumman-American Cheetah was modified in 1988 to use solely liquid hydrogen fuel, making it the first aircraft to use only liquid hydrogen fuel. In the same year, a Russian commercial aircraft, the TU-154, was modified to use liquid hydrogen energy in one of three engines. In 2000, the Cryoplane project,

funded by European Community, began investigating the "conceptual basis for applicability, safety, and full environmental compatibility, and to investigate medium/long term scenarios for a smooth transition from kerosene to hydrogen in aviation" (Airbus Deutschland GmbH, 2003). A key feature of the Cryoplane project is that it took into account environmental concerns.

Other related studies have also been reported. In Ponater et al. (2006), the reduction potential of climate impact is investigated from switching from kerosene fuel to liquid hydrogen. That author states that the impacts of NO_x, water vapor and particle emissions with contrail formation lead to more important climate changes than CO_2. In a cyroplane aircraft, NO_x emissions can be below those for conventional kerosene fuelled engines due to the extended air/fuel ratio ranges providing lean combustion. Furthermore, excluding the likely CO_2 emissions during the production and liquefaction of hydrogen fuel, it is possible to obtain zero CO_2 emissions from burning hydrogen fuel. However, water vapor emission from hydrogen combustion is higher than for conventional kerosene fuel[3]. According to the results of Ponater et al. (2006), reductions of up to 30% for radiative forcing and 10% for surface temperature increase are possible by transitioning from kerosene fuel to liquid hydrogen fuel by 2050. The key advantages of this switch are pointed out to be complete elimination of CO_2 emissions and significant reductions in NO_x emissions from air transport.

Another study compared conventional kerosene fuelled aircraft with liquid hydrogen fuelled aircraft (Svensson et al., 2004). They determined that, from an environmental point of view, a reduction in altitude as 2-3 km is possible with an increase in fuel consumption of 10% and in takeoff weight of a few percent. However, this reduction of altitude from the optimum altitude for a conventional aircraft flight causes an increase in NO_x emissions due to the increased drag, which increases engine power in order to maintain speed. Nevertheless, the increase in NO_x emissions related to conventional kerosene fuelled aircraft is higher than for the liquid hydrogen fuelled aircraft since the former has an advantage provided by the wider flammability range of hydrogen fuel and its lower lean limit compared to kerosene fuel.

Cryogenic fuel aircraft have been assessed in various studies. In Janic (2008), the potential of cryogenic aircraft in terms of emissions are modeled, considering emissions of three species (CO_2, H_2O and NO_x) for the years between 2006 and 2065. Two scenarios are considered involving annual increases in cryogenic aircraft of 1% and 2%, respectively. According to the developed models, CO_2 emissions related to conventional kerosene fuelled air transport activities will increase 3.5 times compared to 2006, but should cryoplanes been introduced to the global fleet at a rate of 1% and 2% of all aircraft, aviation related CO_2 emissions may increase compared to 2006 by only 2.8 and 1.8 times, respectively. A similar result is observed for NO_x emissions, which will increase by about 3.5 times during 2006-2065 assuming continued use of conventional kerosene fuelled air transport. Introducing cryoplanes rates of 1% and 2% will result in NO_x emissions increases compared to 2006 of about 2.8 and 2 times, respectively. However, it is noted that there is no decrease in H_2O emissions associated with the transition from conventional kerosene fuelled air transport to 1% and 2% of cryoplane usage. On the contrary, an increase of about 4.2% in H_2O emissions compared to conventional kerosene fuelled air transport (3.5%) is predicted for the period 2006-2065 for cryoplane introductions of 1% or 2%.

Nojoumi et al. (2009) compared emissions (in addition to safety and storage factors) for hydrogen and kerosene fuelled flights between several cities and calculated the emissions of NO_x, CO and HC.

A life cycle assessment of kerosene and hydrogen fuels is reported by Koroneos et al. (2005), considering various production methods for hydrogen (e.g., natural gas reforming and production from renewable energy forms). It is found that,

despite the fact that hydrogen fuel is regarded as a clean fuel compared to kerosene, the production method of hydrogen plays an important role in the level of environmental impacts. Excluding kerosene, of the hydrogen production methods, wind and hydropower systems appear to have the lowest environmental impacts followed by biomass and solar thermal production methods.

Regarding alternatives to kerosene fuel, significant efforts have been reported in recent years to introduce fuel cells into aviation. Even though there are major challenges, the technology is promising. It seems unlikely that fuel cells can power the main engines of large aircraft based on current technology, but researchers have focused on relatively lower power applications. These include two light motor gliders and a fuel cell APU.

In 2009, the Antares DLR-H2 motor glider flew, equipped with a 25 kW fuel cell and an electrical engine. This flight demonstrated was the world's first manned and completely fuel cell-driven flight. Another motor glider, the Diamond HK-36, was developed by Boeing. Its hybrid engine is rated at 20 kW and it first flew in 2006. The fuel cell APU has been studied by Boeing. With a 440 kW fuel cell unit, it is believed that fuel savings of 40-75% can be achieved at air and ground levels compared to conventional APUs driving electricity generation and cabin air conditioning (Glover, 2005). The fuel cell APU also reduces the quantities of fuel and water to be carried, and the electricity conversion efficiency is reported to be higher than for a conventional APU (Brady, 2010).

According to IATA, utilizing sustainable biofuels could reduce CO_2 emissions by 80%, considering the full carbon life cycle. Blending biofuels with kerosene has attracted increasing attention recently, with a key advantage being that significant modifications in engine components or systems are not needed in such instances. Consequently, IATA set a target of 10% of kerosene to be obtained from biofuels by 2017.

There have been significant experimental flights performed by large airlines. These test flights have shown substantial fuel efficiency increases and emissions reductions. Several of the test flights are as follows:

- L-29 military aircraft used 100% biofuel in a demonstration flight in 2007.
- An Airbus 380, the world's largest commercial aircraft utilized a fuel mixture of 40% biofuel and 60% kerosene in a demonstration flight in 2008 and became the first aircraft to power its engines with biofuel.
- Virgin Atlantic flew a B747 (with a GE CF6 engine) using a blend of biofuel and kerosene in 2008.
- An Air New Zeeland B747-400 utilized a 50-50 blend of biofuel and kerosene for one Rolls-Royce RB211 engine in a demonstration flight in 2008.
- A Continental Airlines B738-800 (with a CFM56-7B engine) flew using a 50-50 blend of biofuel and kerosene for one engine in a demonstration flight in 2009 and an increase in fuel efficiency of 1.1% was attained.
- A Japan Airlines B747-300 flew with 50-50 blend of biofuel and kerosene in 2009.
- A KLM B747 (with a GE CF6-80C2 engine) flew with a 50-50 biofuel and kerosene blend in 2009 and became the first biofuel flight with passengers.

LEGISLATION

With increasing concerns related to the environment in recent years, gaseous emissions of aircraft have attracted attention. Noise has always been of concern. The later introduction of emissions regulations, compared to noise regulations, is likely attributable to the perceived level of perturbation, which is often perceived as more relevant for noise.

Even though the impacts of air transport on climate are not fully understood, many of the relevant mechanisms have been identified. However, corresponding legislation and enforcement procedures are mainly implemented for a limited segment of the air transport system: the landing and takeoff cycle. Unfortunately, few if any regulations exist devoted to mitigating the climatic impacts of aircraft emissions at cruise altitudes.

Prompted by the Environment Protection Agency (EPA) in 1973 (ATA, 2010), environmental concerns about air transport have received increased focus, particularly over the last two decades. The Committee on Aviation Environmental Protection (CAEP), founded in 1983, is dedicated to environmental protection and conducts the environmental activities of ICAO. The first CAEP regulation, CAEP/1, took effect in 1986 (Lee, 2004) and was followed by CAEP/4 (January 2004) and CAEP/6 (from January 2008). CAEP/8 has been effective since February 2010 (ICAO, 2010a) and proposes up to 15% more stringent NO_x emissions than the current levels and is applicable to aircraft certified after 31 December 2013. The allowance level related to NO_x emissions is available from ICAO (ICAO, 2010b). An extensive history of emissions regulation in aviation can be seen in (ATA, 2010).

Compared to past years, considerable focus is being placed on environmental concerns at present. It is reported in the Emission Trading Scheme, which has included aviation since 2009 through legislation introduced by the European Union, that the total emissions allowance for aircraft operators for the period between Jan. 1, 2012 and Dec. 31, 2013 will be equivalent to 97% of historical aviation emissions (Anonymous, 2009) followed by a lower allowance for future periods (95% for 2013-2020). This legislation will affect operators flying in the area of the emission trading scheme (ETS), and will provide competitive advantages to operators which have recent aircraft and engines and/or efficient schedule and fleet procedures.

Vespermann & Wald (in press) investigated the financial burden of ETS on the aviation industry and determined it to be €3 billion per year over the next ten years. They also reported that the share of carbon permit costs in the early years of ETS implementation could be as high as 1.25% of total aviation industry costs. Despite this high cost, the authors do not expect a significant decline in emissions from air transport.

Another investigation regarding the impacts of EU-ETS on costs was performed by Scheelhaase & Grimme (2007) for various kinds of airline companies, including low cost, full service, holiday and regional. The economic impact is calculated to be as high as €100 million annually for full service airlines, with Lufthansa cited as a specific example. The maximum economic impact for a low cost carrier is predicted to be €90 million annually. Both assessments are for 2012. The authors also reported that the impact of the EU-ETS allowance on ticket price is expected to be as high as €2.97 for 2012.

Another regulation aimed at emissions abatement is the local emission charge, which is determined mainly by aircraft category (Scheelhaase, 2010). This charge is based on an equivalent NO_x emission value, which is identified by the Emission Related Landing Charges Investigation Group (ERLIG). First introduced in the Zurich, Switzerland airport in 1997, there currently are more than twenty airports in Europe which implement emission charges (see Table 1). It is clear from that table that there could be a significant emissions burden for an airline having a high rate of movement. However, this burden is mitigated by using new aircraft since the emission indices for new technology are substantially lower compared to those for older technology engines. The load factor could further mitigate the emissions burden for low cost carriers which have higher landing rates.

As mentioned above, emission charges are usually determined by ERLIG using a formula with the same name. This formula is applied to turbojet and turbofan engines having power ratings higher than 26.7 kN. This formula is intended to provide an emission value for each type

Table 1. Local emission charges at some European airports for July 2010

Airport	Year of effect	Country	Charge per emission value (€)
Zurich	1997	Switzerland	1.86
Geneva	1998	Switzerland	1.04
Berne	2000	Switzerland	2.46
Basel-Mullhouse	2003	Switzerland	0.93-1.50[a]
Lugano	2007	Switzerland	2.53
Arlanda	1998	Sweden[b]	5.31
Heathrow	2004	England	3.26
Gatwick	2005	England	5.39
Luton	2009	England	0.006[c]
Frankfurt	2008	Germany	3.00
Munich	2008	Germany	3.00
Cologne Bonn	2008	Germany	3.00
Hamburg	2010	Germany	3.00
Copenhagen	2010	Denmark	2.21

[a] The limits are given for Class 5 and 1 aircraft, respectively. [b] The emission charges of other airports in Sweden (Bromma, Jonkoping, Kalmar, Karlstad, Kiruna, Landvetter, Malmo, Sundsvall-Harnosand, Umea and Visby) are the same (€5.31). [c] € 0.006 per gram of engine NO_x emission above 400 gram per passenger.

of aircraft based on NO_x and HC emissions as a function of thrust, and has the following form:

Emission value = a × $NO_{x,aircraft}$

where a is a dimensionless coefficient calculated using thrust and HC emissions data for an engine. The formula incorporates the term Dp/F_{oo}, defined in Civil Aviation Authority (2010), as the ratio of the mass (in grams) of a pollutant released from an aircraft during certain LTO phases to the rated output of the engine (in kN). If the average value of Dp/F_{oo} for HCs is less than or equal to the current ICAO standard of 19.6 g kN^{-1} then the coefficient equals 1, but if Dp/F_{oo} for HCs is larger than the ICAO standard then a is found by $(HC\ Dp/F_{oo})/19.6$ with a maximum value of 4. The quantity of NO_x emissions (kg) is then determined as follows:

$$NO_{x,aircraft} = [\text{engines}] \times [\sum_{\text{LTO modes}} (60) \\ \times(\text{time in mode}) \times (\text{fuel flow}) \times EI_{NOx} \div (1000)]$$

where

• engines denotes number of engines,

• LTO modes (ICAO, 2007) is given in Figure 17,

• time in mode is given in Figure 17,

• fuel flow denotes the fuel flow in the operating mode (in kg s^{-1}), and

• EI_{NOX} (in g kg^{-1} of fuel) is the emissions index for NO_x and is obtained from the ICAO emissions database.

Also, (1000) is a conversion factor from g kg^{-1} of EI_{NOX} to kg kg^{-1} of EI_{NOX}, while (60) is a conversion factor for minutes to seconds.

In addition to local emission charges, global environmental cost concerns have been investigated. In Schipper (2004), environmental cost estimations are discussed considering aircraft noise and emissions, and it is found that 77% of the total cost is attributable to noise and 18% to emissions (10% for LTO and 8% for cruise). It is also shown that the environment cost represents a small fraction (2.5%) of the ticket price, and that Chapter 2[4] aircraft are more environmentally costly than Chapter 3 aircraft by a factor of four. The latter result emphasizes the differences between old and new engine technology. In another study, Givoni and Rietveld (2010) investigate environmental

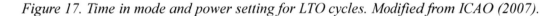

Figure 17. Time in mode and power setting for LTO cycles. Modified from ICAO (2007).

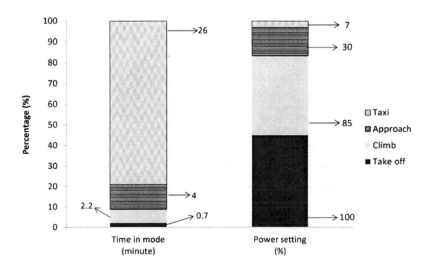

costs, accounting for local air pollution (LAP), climate change and noise impacts, as a function of aircraft size and frequency selection. The results indicate that the total cost of three common emissions (CO_2, NO_x and H_2O) is €3-5 and that the minimum total cost is obtained for a B747-400D aircraft. One of the important findings from the climate change analysis, which runs contrary to the LAP analysis, is that increasing aircraft size is beneficial from a climate change perspective.

CLOSING REMARKS

According to Boeing, there will be a 4.1% annual increase in passenger traffic over the next two decades and, as a consequence, the global commercial aircraft fleet will be almost doubled to 35,600 at the end of the next twenty years, through the addition of 29,000 new aircraft. The breakdown by size of aircraft in 2028 is expected to be 68% single aisle aircraft, 22% twin aisle aircraft, 3% large aircraft and the remainder regional jets. On a regional basis, Asia Pacific is expected to be the top contributor to the new aircraft fleet which implies a gradual increase in air traffic in locations other than the North Atlantic air corridor. A similar forecast is made by Airbus, which suggests a 4.7% annual increase in passenger traffic over the next twenty years, leading to almost 25,000 new aircraft in the global air fleet. Airbus also predicts that low-cost carriers will grow around the world, leading to an increase in frequency of LTO cycles as well as maximum utilization of existing routes and/or introduction of new routes. It is also highlighted that the average size of aircraft is expected to increase by 26% over the next twenty years. Although this change may improve emissions per passenger, larger engines will likely have greater detrimental effects in the vicinity of aircraft.

These predictions suggest that the impacts of aircraft emissions on atmosphere and climate might increase, even with the continued imple-mentation of stringent legislation, unless significant advances in aircraft technology and systems are made. There is a close relationship between emissions and fuel efficiency, but a good design for emissions is often not advantageous for fuel efficiency and vice versa. Nonetheless, growing demand in air transport as well as increases in oil prices are likely to force manufacturers to build larger and more efficient engines, both to maintain the industry and to deal with costs. Additionally, efficient air traffic management can provide substantial fuel savings and corresponding emissions reductions, at least until a critical level of fleet is reached. Biodiesel blends and related technology may provide temporary benefits while other clean and inexpensive fuel energy options are developed.

REFERENCES

Aida, N., Nishijima, T., Hayashi, S., Yamada, H., & Kawakami, T. (2005). Combustion of lean prevaporized fuel–air mixtures mixed with hot burned gas for low-NO_x emissions over an extended range of fuel–air ratios. *Proceedings of the Combustion Institute*, *30*, 2885–2892. doi:10.1016/j.proci.2004.08.040

Airbus Deutschland GmbH. (2003). *Cryoplane: Liquid hydrogen fuelled aircraft – System analysis.* Final Technical Report.

Airports Council International (ACI). (2010). *Year to date aircraft movements, April 2010*. Retrieved June 20, 2010, from http://www.aci.aero/ cda/ aci_common/ display/main/ aci_content07_c.js p?zn=aci&cp=1-5-212-231_666_2__

Anderson, B. E., Chen, G., & Blake, D. R. (2006). Hydrocarbon emissions from a modern commercial airliner. *Atmospheric Environment*, *40*, 3601–3612. doi:10.1016/j.atmosenv.2005.09.072

Anonymous. (2009). Directive 2008/101/EC of the European Parliament and of the Council. *Official Journal of the European Union. L&C, 52*(L8), 3–21.

Anonymous. (2010a). *de Havilland ghost*. Wikipedia. Retrieved June 23, 2010, from http:// en.wikipedia.org/ wiki/ De_Havilland_Ghost

Anonymous. (2010b). *General Electric J79*. Wikipedia. Retrieved June 23, 2010, from http://en.wikipedia.org/ wiki/ J79#Specifications_.28J79- GE-17.29

ATA. (2010). *Emission limits*. Air Transport Association of America, Washington, DC. Retrieved July 25, 2010, from http://www.airlines.org/ Environment/ LocalAirQuality/ Pages/ EmissionsLimits.aspx

Baughcum, S. L., Tritz, T. G., Henderson, S. C., & Pickett, D. C. (1996). *Scheduled aircraft emission inventories for 1992: Database development and analysis. NASA contract report no 4700*. Hanover, MD, U.S.: NASA Center for Aerospace Information.

Benini, E., Pandolfo, S., & Zoppellari, S. (2009). Reduction of NO emissions in a turbojet combustor by direct water/steam injection: Numerical and experimental assessment. *Applied Thermal Engineering, 29*, 3506–3510. doi:10.1016/j.applthermaleng.2009.06.004

Boeing. (2010a). *Current market outlook 2009-2028*. Market analysis, Boeing Commercial Airplanes, Seattle, WA. Retrieved July 25, 2010, from http://www.boeing.com/ commercial/ cmo/ pdf/ Boeing_Current_Market_ Outlook_2009_to _2028.pdf

Boeing. (2010b). *2010 environment report*. Retrieved July 25, 2010, from http://www.boeing. com/ aboutus/ environment/ environment_report_10/ boeing-2010-environment- report.pdf

Boyce, M. P. (2002). *Gas turbine engineering handbook* (2nd ed.). Houston, TX: Gulf Professional Publishing.

Brady, C. (2010). Auxiliary power unit. In *The Boeing 737 technical guide* (57th ed., pp. 62–68). Tech Pilot Services Ltd.

Brasseur, G. P., Cox, R. A., Hauglustaine, D., Isaksen, I., Lelieveld, J., & Lister, D. H. (1998). European scientific assessment of the atmospheric effects of aircraft emissions. *Atmospheric Environment, 32*(13), 2329–2418. doi:10.1016/S1352-2310(97)00486-X

Bureau of Transportation Statistics (BTS). (2010). *Airline fuel cost and consumption (U.S. carriers-Scheduled): January 2000 - May 2010*. Research and Innovative Technology Administration, U.S. Department of Transportation, Washington, DC. Retrieved July 4, 2010, from http://www.transtats. bts.gov/ fuel.asp? pn=0& display=data4

Carslaw, D. C., Ropkins, K., Laxen, D., Moorcroft, S., Marner, B., & Williams, M. L. (2008). Near-field commercial aircraft contribution to nitrogen oxides by engine, aircraft type, and airline by individual plume sampling. *Environmental Science & Technology, 42*, 1871–1876. doi:10.1021/ es071926a

Civil Aviation Authority. (2010). *ICAO aircraft engine emissions databank definitions*. Retrieved July 25, 2010, from http://www.caa.co.uk/ docs/ 02/ 070716% 20Introduction.pdf

Cusson, J. (2000). *The Xonon combustion system*. Presented at Industrial Center - TMAC Meeting, San Francisco. Retrieved July 25, 2010, from http://www.energysolutionscenter.org/ distgen/ AppGuide/ DataFiles/ Xonon.pdf

Daggett, D. L. (2004). *Water misting and injection of commercial aircraft engines to reduce airport NO_x. NASA/CR—2004-212957*. Hanover, MD: NASA Center for Aerospace Information.

Dings, J. M. W., Wit, R. C. N., Leurs, B. A., Davidson, M. D., & Fransen, W. (2003). *External costs of aviation. (Research Report 299 96 106: UBA-FB 000411), Federal Environmental Agency.* Berlin: Umweltbundesamt.

Ehhalt, D., Dentener, F., Derwent, R., Dlugokencky, E., Holland, E., Isaksen, I., et al. Wang, M. (2001). *Atmospheric chemistry and greenhouse gases.* Climate Change 2001: The Scientific Basis. Cambridge, United Kingdom: Cambridge University Press.

Eyers, C. J., Addleton, D., Atkinson, K., & Broomhead, M. J. Christou, R., Elliff, T., ... Stanciou, N. (2004). *AERO2k global aviation emissions inventories for 2002 and 2025.* European Commission, Contract No. G4RD-CT-2000-00382, QinetiQ ltd.

Flamme, M. (2004). New combustion systems for gas turbines (NGT). *Applied Thermal Engineering, 24,* 1551–1559. doi:10.1016/j.applthermaleng.2003.10.024

Flippone, A. (2010). Cruise altitude flexibility of jet transport aircraft. *Aerospace Science and Technology, 14,* 283–294. doi:10.1016/j.ast.2010.01.003

Florides, G. A., & Christodoulides, P. (2009). Global warming and carbon dioxide through sciences. *Environment International, 35,* 390–401. doi:10.1016/j.envint.2008.07.007

Folland, C. K., Karl, T. R., Christy, J. R., Clarke, R. A., Gruza, G. V., & Jouzel, J. ... Wang, S. W. (2001). *Observed climate variability and change.* Climate Change 2001: The Scientific Basis. Cambridge, United Kingdom: Cambridge University Press.

Forster, P., Ramaswamy, V., Artaxo, P., Berntsen, T., & Betts, R. Fahey, ... Dorland, R. (2007). *Changes in atmospheric constituents and in radiative forcing.* Climate Change 2007: The Physical Science Basis. Cambridge, United Kingdom: Cambridge University Press.

Forzatti, P. (2003). Status and perspectives of catalytic combustion for gas turbines. *Catalysis Today, 83,* 3–18. doi:10.1016/S0920-5861(03)00211-6

Gaffney, J. S., & Marley, N. A. (2009). The impacts of combustion emissions on air quality and climate – From coal to biofuels and beyond. *Atmospheric Environment, 43,* 23–36. doi:10.1016/j.atmosenv.2008.09.016

Gardner, R. M., Adams, K., Cook, T., Deidewig, F., Ernedal, S., & Falk, R. (1997). The ANCAT/EC global inventory of NO_x emissions from aircraft. *Atmospheric Environment, 31*(12), 1751–1766. doi:10.1016/S1352-2310(96)00328-7

Giampaolo, T. (2003). *The gas turbine handbook: Principles and practices* (2nd ed.). Lilburn, GA: Fairmont Press.

Givoni, M., & Rietveld, P. (2010). The environmental implications of airlines' choice of aircraft size. *Journal of Air Transport Management, 16,* 159–167. doi:10.1016/j.jairtraman.2009.07.010

Glover, B. (2005). *Fuel cell opportunity.* Presented at AIAA/AAAF Aircraft Noise and Emissions Reduction Symposium, Monterey, CA. Retrieved July 25, 2010, from http://www.aiaa.org/ events/ aners/ Presentations/ ANERS-Glover.pdf

GOA. (2009). *Aviation and climate change.* Report to congressional committees, Number GAO-09-554, United States Government Accountability Office, June.

Gohlke, O., Weber, T., Seguin, P., & Laborel, Y. (2010). A new process for NO_x reduction in combustion systems for the generation of energy from waste. *Waste Management (New York, N.Y.), 30,* 1348–1354. doi:10.1016/j.wasman.2010.02.024

Graham, A., & Raper, D. W. (2006). Transport to ground of emissions in aircraft wakes. Part II: Effect on NO_x concentrations in airport approaches. *Atmospheric Environment, 40,* 5824–5836. doi:10.1016/j.atmosenv.2006.05.014

Groob, J., Bruhl, C., & Peter, T. (1998). Impact of aircraft emissions on tropospheric and stratospheric ozone. Part 1: Chemistry and 2D model results. *Atmospheric Environment, 32*(18), 3173–3184. doi:10.1016/S1352-2310(98)00016-8

Gupta, A. K. (1997). Gas turbine combustion: Prospects and challenges. *Energy Conversion and Management, 38*(10-13), 1311–1318. doi:10.1016/S0196-8904(96)00160-4

Haglind, F. (2008). Potential of lowering the contrail formation of aircraft exhausts by engine re-design. *Aerospace Science and Technology, 12*(6), 490–497. doi:10.1016/j.ast.2007.12.001

Hayashi, S., Yamada, H., & Makida, M. (2000). Short-flame/quick quench: A unique ultralow emissions combustion concept for gas turbine combustors. *Proceedings of the Combustion Institute, 28,* 1273–1280. doi:10.1016/S0082-0784(00)80340-9

Herndon, S. C., Rogers, T., Dunlea, E. J., Jayne, J. T., Miake-Lye, R., & Knighton, B. (2006). Hydrocarbon emissions from in-use commercial aircraft during airport operations. *Environmental Science & Technology, 40,* 4406–4413. doi:10.1021/es0512091

Herndon, S. C., Shorter, J. H., Zahniser, M. S., Nelson, D. D., Jayne, J., & Brown, R. C. (2004). NO and NO_2 emission ratios measured from in-use commercial aircraft during taxi and takeoff. *Environmental Science & Technology, 38,* 6078–6084. doi:10.1021/es049701c

Hunecke, K. (2003). *Jet engines, fundamentals of theriory, design and operation* (6th ed.). Osceola, FL: Motorbooks International Publishers & Wholesalers.

IATA. (2009). *A global approach to reducing aviation emissions. First stop: carbon-neutral growth from 2020.* Report, International Air Transport Association.

ICAO. (2007). *ICAO airport air quality guidance manual.* Doc. 9889, Preliminary Edition, International Civil Aviation Organization, Montreal. Retrieved July 25, 2010, from http://www.icao.int/ icaonet/ dcs/ 9889/ 9889_en.pdf

ICAO. (2008). *ICAO annual report of the council.* Doc. 9916, International Civil Aviation Organization, Montreal. Retrieved June 20, 2010, from http://www.icao.int/ icaonet/ dcs/ 9916/ 9916_en.pdf

ICAO. (2010a). *CAEP.* International Civil Aviation Organization, Air Transport Bureau, Environment Branch. Retrieved July 5, 2010, from http:// www.icao.int/ icao/ en/ nv/ caep.htm

ICAO. (2010b). *Aircraft engine emissions.* International Civil Aviation Organization, Air Transport Bureau, Environment Branch. Retrieved June 25, 2010, from http://www.icao.int/ icao/ en/ env/ aee.htm

ICAO. (2010c). *Aircraft noise.* International Civil Aviation Organization, Air Transport Bureau, Environment Branch. Retrieved June 25, 2010, from http://www.icao.int/ icao/ en/ env/ noise.htm

Isaksen, I. S. A., Granier, C., Myhre, G., Berntsen, T. K., Dalsoren, S. B., & Gauss, M. (2009). Atmospheric composition change: Climate-chemistry interactions. *Atmospheric Environment, 43*(33), 5138–5192. doi:10.1016/j.atmosenv.2009.08.003

Janic, M. (2008). The potential of liquid hydrogen for the future "carbon-neutral" air transport system. *Transportation Research Part D, Transport and Environment, 13*, 428–435. doi:10.1016/j.trd.2008.07.005

Kim, B., Fleming, G., Balasubramanian, S., Malwitz, A., Fleming, G., & Lee, J. ... Gillette, W. (2005). *System for assessing aviation's global emissions (SAGE), version 1.5, global aviation emissions inventories for 2000 through 2004.* Report FAA-EE-2005-02, Federal Aviation Administration, Office of Environment and Energy, Washington, DC.

Kohler, I., Sausen, R., & Reinberger, R. (1997). Contributions of aircraft emissions to the atmospheric NO_x content. *Atmospheric Environment, 31*(12), 1801–1818. doi:10.1016/S1352-2310(96)00331-7

Koroneos, C., Dompros, A., Roumbas, G., & Moussiopoulos, N. (2005). Advantages of the use of hydrogen fuel as compared to kerosene. *Resources, Conservation and Recycling, 44*, 99–113. doi:10.1016/j.resconrec.2004.09.004

Kuper, W. J., Blaauw, M., Berg, F., & Graaf, G. H. (1999). Catalytic combustion concept for gas turbines. *Catalysis Today, 47*, 377–389. doi:10.1016/S0920-5861(98)00320-4

Lebedev, A. B., Secundov, A. N., Starik, A. M., Titova, N. S., & Schepin, A. M. (2009). Modeling study of gas-turbine combustor emission. *Proceedings of the Combustion Institute, 32*, 2941–2947. doi:10.1016/j.proci.2008.05.015

Lee, D. S. (2004). The impact of aviation on climate. *Issues in Environmental Science and Technology* [Royal Society of Chemistry.]. *Transport and the Environment, 20*, 1–23. doi:10.1039/9781847552211-00001

Lee, D. S., Fahey, D. W., Forster, P. M., Newton, P. J., Wite, R. C. N., & Lim, L. L. (2009). Aviation and global climate change in the 21st century. *Atmospheric Environment, 43*, 3520–3537. doi:10.1016/j.atmosenv.2009.04.024

Lee, D. S., Pitari, G., Grewe, V., Gierens, K., Penner, J. E., & Petzold, A.... Sausen, R. (in press). Transport impacts on atmosphere and climate: Aviation. *Atmospheric Environment.* doi:. doi:10.1016/j.atmosenv.2009.06.005

Levy, Y., Sherbaum, V., & Arfi, P. (2004). Basic thermodynamics of FLOXCOM, the low-NO_x gas turbines adiabatic combustor. *Applied Thermal Engineering, 24*, 1593–1605. doi:10.1016/j.applthermaleng.2003.11.022

Mannstein, H., Spichtinger, P., & Gierens, K. (2005). A note on how to avoid contrail cirrus. *Transportation Research Part D, Transport and Environment, 10*(5), 421–426. doi:10.1016/j.trd.2005.04.012

Marquart, S., Sausen, R., Ponater, M., & Grewe, V. (2001). Estimate of the climate impact of cryoplanes. *Aerospace Science and Technology, 5*, 73–84. doi:10.1016/S1270-9638(00)01084-1

Maurice, L. Q., Lander, H., Edwards, T., & Harrison, W. E. (2001). Advanced aviation fuels: A look ahead via a historical perspective. *Fuel, 80*, 747–756. doi:10.1016/S0016-2361(00)00142-3

Mazaheri, M., Johnson, G. R., & Morawska, L. (2009). Particle and gaseous emissions from commercial aircraft at each stage of the landing and takeoff cycle. *Environmental Science & Technology, 43*, 441–446. doi:10.1021/es8013985

McDonell, V. (2008). Lean combustion technology and control. In Rankin, D. D. (Ed.), *Lean combustion in gas turbines* (pp. 121–160). Elsevier.

Miyoshi, C., & Mason, K. J. (2009). The carbon emissions of selected airlines and aircraft types in three geographic markets. *Journal of Air Transport Management, 15*, 138–147. doi:10.1016/j.jairtraman.2008.11.009

Morrell, P. (2009). The potential for European aviation CO_2 emissions reduction through the use of larger jet aircraft. *Journal of Air Transport Management, 15*, 151–157. doi:10.1016/j.jairtraman.2008.09.021

Nojoumi, H., Dincer, I., & Naterer, G. F. (2009). Greenhouse gas emissions assessment of hydrogen and kerosene-fueled aircraft propulsion. *International Journal of Hydrogen Energy, 34*, 1363–1369. doi:10.1016/j.ijhydene.2008.11.017

Pejovic, T., Noland, R. B., Williams, V., & Toumi, R. (2009). A tentative analysis of the impacts of an airport closure. *Journal of Air Transport Management, 15*, 241–248. doi:10.1016/j.jairtraman.2009.02.004

Penner, J. E., Lister, D. H., Griggs, D. J., Dokken, D. J., & McFarland, M. (2001). *Aviation and the global atmosphere*. Special Report of Working Groups I and III of the Intergovernmental Panel on Climate Change.

Pham, V. V., Tang, J., Alam, S., Lokan, C., & Abbass, H. A. (in press). Aviation emission inventory development and analysis. *Environmental Modelling & Software*. doi:.doi:10.1016/j.envsoft.2010.04.004

Ponater, M., Pechtl, S., Sausen, R., Schumann, U., & Huttig, G. (2006). Potential of the cryoplane technology to reduce aircraft climate impact: A state-of-the-art assessment. *Atmospheric Environment, 40*, 6928–6944. doi:10.1016/j.atmosenv.2006.06.036

Pratt & Whitney. (2010a). *PurePower PW1000G engine*. Retrieved July 5, 2010, from http://www.purepowerengine.com/ noise.html

Pratt & Whitney. (2010b). *PurePower PW1000G engine: Photos. Engine cross section and fan drive gear system*. Retrieved July 5, 2010, from http://www.purepowerengine.com/ photos.html

Ribeiro, K. S., Kobayashi, S., Beuthe, M., Gasca, J., Greene, D., & Lee, D. S. … Zhou, P. J. (2007). *Transport and its infrastructure*. Climate Change 2007: Mitigation. Cambridge, United Kingdom: Cambridge University Press.

Rogers, H. L., Lee, D. S., Raper, D. W., Forster, P. M. F., Wilson, C. W., & Newton, P. (2002). *The impact of aviation on the atmosphere*. Report QINETIQ/FST/CAT/TR021654, Centre for Aerospace Technology, Cody Technology Park, Farnborough, United Kingdom.

Rowe, M. R., & Ladd, G. T. (1945, May 23). Water injection for aircraft engines. *Flight*, 517-518. Retrieved July 25, 2010, from http://www.flightglobal.com/ pdfarchive/ view/ 1946/ 1946% 20-% 201007.html

Ruffles, P. C. (2003). Aero engines of the future. *Aeronautical Journal, 107*(1072), 307–321.

Schafer, K., Jahna, C., Sturmb, P., Lechnerb, B., & Bacher, M. (2003). Aircraft emission measurements by remote sensing methodologies at airports. *Atmospheric Environment, 37*, 5261–5271. doi:10.1016/j.atmosenv.2003.09.002

Scheelhaase, J. D., & Grimme, W. G. (2007). Emissions trading for international aviation—An estimation of the economic impact on selected European airlines. *Journal of Air Transport Management, 13*, 253–263. doi:10.1016/j.jairtraman.2007.04.010

Schipper, Y. (2004). Environmental costs in European aviation. *Transport Policy, 11*, 141–154. doi:10.1016/j.tranpol.2003.10.001

Schumann, U. (1997). The impact of nitrogen oxides emissions from aircraft upon the atmosphere at flight altitudes: Result from the AERONOX project. *Atmospheric Environment, 31*(12), 1723–1733. doi:10.1016/S1352-2310(96)00326-3

Schumann, U. (2000). Influence of propulsion efficiency on contrail formation. *Aerospace Science and Technology, 4*, 391–401. doi:10.1016/S1270-9638(00)01062-2

Schumann, U. (2005). Formation, properties and climatic effects of contrails. *C.R. Physique, 6*, 549–565. doi:10.1016/j.crhy.2005.05.002

Schumann, U., Busen, R., & Plohr, M. (2000). Experimental test of the influence of propulsion efficiency on contrail formation. *Journal of Aircraft, 37*(6), 1083–1087. doi:10.2514/2.2715

Schurmann, G., Schafer, K., Jahn, C., Hoffmann, H., Bauerfeind, M., Fleuti, E., & Rappengluck, B. (2007). The impact of NO_x, CO and VOC emissions on the air quality of Zurich airport. *Atmospheric Environment, 41*, 103–118. doi:10.1016/j.atmosenv.2006.07.030

Shellard, H. C. (1949). Humidity of the lower stratosphere. *Meteorological Magazine, 78*(390), 341–349.

Slanina, S. (2008). Impact of ozone on health and vegetation. In C. J. Cleveland (Ed.), *Encyclopedia of Earth*. Environmental Information Coalition, National Council for Science and the Environment, Washington, DC. Retrieved June 21, 2010, from http://www.eoearth.org/ article/ Impact_of_ozone_ on_health_and _vegetation

Soares, C. (2008). *A handbook of air, land, and sea applications*. Elsevier.

Staehelin, J., Harris, N. R. P., Appenzeller, C., & Eberhard, J. (2001). Ozone trends: A review. *Reviews of Geophysics, 39*, 231–290. doi:10.1029/1999RG000059

Sutkus, D. J., Baughcum, S. L., & DuBois, D. P. (2001). *Scheduled civil aircraft emission inventories for 1999: Database development and analysis*. Report NASA/CR-2001-211216, NASA Center for Aerospace Information, Hanover, MD.

Svensson, F., Hasselrot, A., & Moldanova, J. (2004). Reduced environmental impact by lowered cruise altitude for liquid hydrogen-fuelled aircraft. *Aerospace Science and Technology, 8*, 307–320. doi:10.1016/j.ast.2004.02.004

Unal, A., Hu, Y., Chang, M. E., Odman, M. T., & Russell, A. G. (2005). Airport related emissions and impacts on air quality: Application to the Atlanta International Airport. *Atmospheric Environment, 39*, 5787–5798. doi:10.1016/j.atmosenv.2005.05.051

Vatcha, S. R. (1997). Low-emission gas turbines using catalytic combustion. *Energy Conversion and Management, 38*(10-13), 1327–1334. doi:10.1016/S0196-8904(96)00162-8

Vespermann, J., & Wald, A. (in press). Much ado about nothing? – An analysis of economic impacts and ecologic effects of the EU-emission trading scheme in the aviation industry. *Transportation Research Part A*. doi:.doi:10.1016/j.tra.2010.03.005

Wang, Y. D., Huang, Y., Wright, D. M., McMullan, J., Hewitt, N., Eames, P., & Rezvani, S. (2006). A techno-economic analysis of the application of continuous staged-combustion and flameless oxidation to the combustor design in gas turbines. *Fuel Processing Technology, 87*(8), 727–736. doi:10.1016/j.fuproc.2006.02.003

Westerdahl, D., Fruin, S. A., Fine, P. L., & Sioutas, C. (2008). The Los Angeles International Airport as a source of ultrafine particles and other pollutants to nearby communities. *Atmospheric Environment, 42*, 3143–3155. doi:10.1016/j.atmosenv.2007.09.006

Williams, V., & Noland, B. R. (2006). Comparing the CO_2 emission and contrail formation from short and long haul air traffic routes from London Heathrow. *Environmental Science & Policy, 9,* 487–495. doi:10.1016/j.envsci.2005.10.004

Williams, V., Noland, R. B., & Toumi, R. (2002). Reducing the climate change impacts of aviation by restricting cruise altitudes. *Transportation Research Part D, Transport and Environment, 7,* 451–464. doi:10.1016/S1361-9209(02)00013-5

Wilson, E. K. (2009). Ozone's health impact. *Chemical and Engineering News, 87*(11), 9. doi:10.1021/cen-v087n011.p009a

Winther, M., Kousgaard, U., & Oxbel, A. (2006). Calculation of odour emissions from aircraft engines at Copenhagen Airport. *The Science of the Total Environment, 366*(1), 218–232. doi:10.1016/j.scitotenv.2005.08.015

Wood, E. C., Herndon, S. C., Timko, M. T., Yelvington, P. E., & Miake-Lye, R. C. (2008). Speciation and chemical evolution of nitrogen oxides in aircraft exhaust near airports. *Environmental Science & Technology, 42,* 1884–1891. doi:10.1021/es072050a

Wulff, A., & Hourmouziadis, J. (1997). Technology review of aeroengine pollutant emissions. *Aerospace Science and Technology, 8,* 557–572. doi:10.1016/S1270-9638(97)90004-3

Wunning, J. A., & Wunning, J. G. (1997). Flameless oxidation to reduce thermal NO-formation. *Progress in Energy and Combustion Science, 23,* 81–94. doi:10.1016/S0360-1285(97)00006-3

ENDNOTES

[1]   The shafts are called N1 and N2 in two shafted turbofan engines. Low pressure turbines drive the low pressure compressor and fan via N1, while high pressure turbines drive the high pressure compressors via N2. Thrust is measured mainly by the value of N1 in revolution per minute (RPM).

[2]   According to ISA conditions, the atmospheric pressure 175-325 hectapascals (hPa) denotes a vertical altitude region between 9-13 km.

[3]   The amount of H_2O emission varies when the jet fuel is switched from kerosene to liquid hydrogen (LH_2) since the fuels have different heating values based on weight. The ratio of lower heating values for kerosene-liquid hydrogen can be calculated as 2.78 (120.1 MJ kg^{-1}/43.1 MJ kg^{-1}). Therefore, in order to obtain the same heat, 0.36 kg of hydrogen can be used in place of 1 kg of kerosene. Assuming $C_{12}H_{23}$ as the chemical formula of kerosene (Lee et al., in press.; Lebedev et al., 2009; Nojoumi et al., 2009), the H_2O emissions resulting from the complete combustion of 1 kg kerosene is found to be 1.23 kg as follows:

$$C_{12}H_{23} + \left(\frac{35.5}{2}\right)O_2 \rightarrow (12)CO_2 + \left(\frac{23}{2}\right)H_2O.$$

1 kg fuel + 6.80 kg O_2 → 3.16 kg CO_2 + 1.23 kg H_2O. For the same amount of heat, the H_2O emissions from the complete combustion of hydrogen can be calculated as 3.23 kg as follows: $H_2 + \frac{1}{2}O_2 \rightarrow H_2O$. 0.36 kg fuel + 5.74 kg O_2 → 3.23 kg H_2O. A difference of 2.6 times (3.23/1.23) in H_2O emissions is thus observed due to switching from kerosene to hydrogen fuel.

[4]   According to ICAO Annex 16 – Environmental Protection Volume 1 – Aircraft Noise (ICAO, 2010c), aircraft certified before 1977 were included in Chapter 2 of Annex 16 (B727, DC-9, so far). Aircraft certified after 1977 were included in Chapter 3 (B737/300/400, B767, A319, so far). Aircraft certified after 2006 were included in Chapter 4.

APPENDIX

In this appendix, selected important results from various investigations (Schafer et al., 2003; Winther et al., 2006; Pham et al., in press; Carslaw et al., 2008; Eyers et al., 2004; Kim et al., 2005; Baughcum et al., 1996; Sutkus et al., 2001; Gardner et al., 1997) are compiled for comparative purposes. Among these are data on fuel consumption, emission indices and total emissions, broken down by aircraft and engine model, aircraft category, measurement method, engine power setting, number of flights/departures, distance flown, altitude interval, season, month, year, region, and purpose (civil or military).

Table 2. Measured mean CO (FTIR emission spectrometry and FTIR absorption spectrometry) and NO$_x$ (FTIR emission spectrometry and DOAS) emission indices for main engines of various aircraft at idle thrust (16-21% N1[i])

Aircraft	Engine type	No.	EI CO (g kg^{-1}) FTIR em. spectr.	No.	EI CO (g kg^{-1}) FTIR abs. spectr.	ICAO (g kg^{-1})	No.	EI NO$_x$ (g kg^{-1}) FTIR em. spect.	No.	EI NO$_x$ (g kg^{-1}) DOAS	ICAO (g kg^{-1})
A320-211	CFM56-5A1	1	15.50	1	28.82 ± 4.86	17.60	1	bdl	1	1.68	4.0
A320	CFM56-5A3	—	—	1	33.3 ± 3.1	16.20	—	—	—	—	—
A320-214	CFM56-5B4/2	2	48.8 (30.5 – 62.3)	1	53.01 ± 16.45	37.10	—	—	—	—	—
A320-214	CFM56-5B4/2P	7	50.5 (21.3 – 72.6)	24	45.85 ± 3.8	40.10	7	0.9 (bdl – 1.2)	22	2.10 ± 0.24	3.9
A320-231	V2500-A1	1	7.2 (2.5 – 13.0)	1	10.09 ± 5.2	12.43	1	1.3 (bdl–3.8)	3	1.56 ± 0.41	5.91
A321-111	CFM56-5B1	3	49.9 (23.0 – 71.9)	10	42.09 ± 4.35	28.40	3	0.9 (0.7–1.1)	10	2.24 ± 0.32	4.6
A321-211	CFM56-5B3/P	1	55.7 (50.7 – 63.9)	9	43.1 ± 5.03	19.20	1	0.7 (bdl–1.0)	6	1.97 ± 0.36	4.7
A340-211	CFM56-5C2	1	6.00	2	19.03 ± 3.82	34.00	1	bdl	2	1.67 ± 0.47	4.2
A340	CFM56-5C4	—	—	1	15.33 ± 1.51	30.93	—	—	—	—	—
B737-300	CFM56-3C1	1	29.8 (19.9 – 37.1)	—	—	26.80	1	2.1 (1.9–2.3)	1	2.17 ± 0.49	4.3
B737-306	CFM56-3B1	1	37.60	3	38.39 ± 10.5	34.40	—	—	—	—	—
B737-382	CFM56-3B2	1	27.40	2	33.85 ± 3.05	30.10	1	0.50	1	1.39 ± 0.11	4.1
B737-406	CFM56-3B2	1	33.70	—	—	30.10	—	—	—	—	—
B737-600	CFM56-7B22/2	1	59.6 (45.9 – 73.4)	1	48.7 ± 15.05	45.35	1	1.4 (1.0–1.8)	1	2.05 ± 0.11	3.94
B737	CFM56-7B24	—	—	1	45.0 ± 16.36	22.00	—	—	—	—	—
B737-800	CFM56-7B26	1	17.60	—	—	18.80	—	—	—	—	—
B737-800	CFM56-7B27	1	25.7 (17.3 – 33.8)	5	30.54 ± 3.71	17.90	1	1.0 (0.7–1.4)	5	2.18 ± 0.34	4.8
B737-8K2	CFM56-7B27	3	20.3 (8.9 – 34.5)	—	—	17.90	—	—	—	—	—
B747-200	GE CF 6-50E2	1	32.60	—	—	24.04	—	—	—	—	—
B747-236	RB211-524D4	1	26.0 (21.0 – 31.2)	3	37.41 ± 15.25	9.30	1	bdl	2	2.01 ± 1.00	4.11
B747-436	RB211-524H2	2	12.2 (10.4 – 18.1)	6	19.48 ± 6.51	11.75	2	0.3 (bdl–3.8)	5	1.99 ± 0.64	4.78
B757-236	RB211-535C	2	11.8 (11.7 – 12.0)	11	15.41 ± 2.00	18.79	2	0.70	10	1.57 ± 0.23	3.44
B757-236	RB211-535C-37	5	7.5 (3.0 – 11.0)	—	—	18.79	—	—	—	—	—
B757-236	RB211-535E4	1	6.7 (6.6 – 13.4)	4	15.67 ± 5.35	15.44	1	0.6 (0.5–0.7)	3	1.35 ± 0.23	4.3
B757-236	RB211-535E4-37	2	9.2 (5.5 – 13.4)	—	—	15.44	—	—	—	—	—
B767-336	RB211-524H-36	1	7.30	—	—	11.75	—	—	—	—	—
B777-236	GE90-85B	7	39.1 (2.7 – 64.1)	6	43.35 ± 6.21	13.67	7	0.4 (bdl–2.0)	5	1.80 ± 0.75	6.01
MD-87	JT8D-217C	1	10.3 (9.4 – 11.2)	4	24.46 ± 12.45	17.89	1	bdl	4	1.68 ± 0.49	4.05

continued on following page

Table 2. Continued

Aircraft	Engine type	No.	EI CO (g kg⁻¹) FTIR em. spectr.	No.	EI CO (g kg⁻¹) FTIR abs. spectr.	ICAO (g kg⁻¹)	No.	EI NO$_x$ (g kg⁻¹) FTIR em. spect.	No.	EI NO$_x$ (g kg⁻¹) DOAS	ICAO (g kg⁻¹)
MD-87	JT8D-219	—	—	—	—		—	—	1	2.66	4.06
MD-87	GE CF6-80C2D1F	—	—	1	43.8 ± 14.4	18.02	—	—		—	—
Fokker 70	RR-Tay MK620	10	23.1 (11.5 – 46.7)	19	32.33 ± 3.05	24.10	10	0.3 (bdl–4.0)	21	1.74 ± 0.23	2.5
CRJ100LR	GE CF 34-3A	1	39.2 (38.0 – 40.5)	6	25.87 ± 12.1	42.60	1	1.00	—	—	3.82
CRJ100LR	GE CF 34-3A1	2	36.7 (31.0 – 46.6)	—	—	42.60	2	1.0 (0.8–1.2)	6	1.40 ± 0.27	3.82
CRJ200LR	GE CF 34-3B	8	38.9 (10.2 – 57.3)	5	30.83 ± 4.88	42.60	8	1.0 (0.8–1.4)	11	1.60 ± 0.56	3.82
DHC-8-300Q	PW 123B	8	9.5 (2.2 – 27.0)	8	18.00 ± 3.58	—	8	1.7 (bdl–14.5)	8	1.78 ± 0.28	—
DHC-8-400Q	PW 150A	5	8.3 (3.3 – 17.0)	7	15.88 ± 3.2	—	5	0.8 (bdl–2.2)	3	2.12 ± 0.94	—

bdl: below dedection limits, FTIR: Fourier transform infrared spectrometry; DOAS: differential optical absorption spectrometer; 'No.' denotes the measured single different engines. Modified from Schafer et al. (2003).

Table 3. Bag analysis results and fuel flow during tests for the JT8D-219 and APU engines

Engine	Power setting	Fuel flow rate (kg s⁻¹)	Total HC[a] (mg C m⁻³)	NO$_x$ (mg NO$_2$ m⁻³)	NO$_2$[b] (mg m⁻³)	CO$_2$[c] (g m⁻³)
JT8D-219	Idle	0.13	2	3.8	0	0.55
	Idle	0.13	2.3	3.8	0	0.54
	Idle	0.13	2.1	3.8	0	0.56
	Avg.	0.13	2.1	3.8	0	0.55
JT8D-219	Max.	1.4	4	31	17	1.3
	Max.	1.4	2.9	33	19	1.3
	Max.	1.4	3.9	33	17	1.4
	Avg.	1.4	3.6	32	18	1.3
APU	Normal	0.013	8	48	40	13.8
	Normal	0.013	6.3	48	40	13.3
	Avg.	0.013	7.2	48	40	13.5

[a] Measured using infrared photoacustic spectroscopy (Bruel and Kjær Gas Monitor 1302). [b] Measured using chemiluminescense (NO$_x$-monitor CLD 700 EI). [c] Measured in vol. % using non-dispersive infrared light absorption (Monitor Labs CO/CO$_2$ monitor) and recalculated to g carbon in CO$_2$ emissions per m³ (g CCO$_2$ m⁻³). Reproduced by permission from Winther et al. (2006).

Table 4. Distance flown and normalized fuel consumption and emissions for various aircraft for 2002

Aircraft	Distance flown (10^9 km)	Normalized fuel use (g km^{-1})	Normalized emissions (g km^{-1})					
			CO_2	H_2O	CO	NO_x	HC	Soot
A306	0.50	6,893	21,722	8,533	20	98	1.6	0.2
A310	0.37	5,645	17,786	6,987	16	72	1.2	0.1
A319	0.89	3,519	11,080	4,357	16	44	2.6	0.0
A320	2.11	3,531	11,122	4,371	14	48	2.1	0.0
A321	0.44	4,043	12,734	5,004	15	64	1.8	0.1
A330	0.63	6,646	20,958	8,228	12	116	3.0	0.1
A340	0.80	7,448	23,486	9,219	12	120	1.0	0.1
A34R	0.06	8,207	25,666	10,151	138	57	26.3	0.5
AT72	0.43	1,458	4,594	1,805	7	16	0.0	0.2
B703	0.19	7,662	23,936	9,487	155	48	30.3	0.5
B712	0.13	3,402	10,714	4,212	15	38	0.1	0.0
B722	0.69	6,077	19,154	7,523	15	58	2.9	0.3
B732	0.57	3,851	12,133	4,767	12	37	2.1	0.2
B734	3.19	3,710	11,677	4,592	19	39	1.0	0.0
B736	1.35	3,363	10,599	4,164	10	43	0.9	0.1
B738	1.35	3,560	11,222	4,407	8	49	0.8	0.1
B742	0.78	11,766	37,114	14,566	12	193	1.7	0.2
B744	2.06	11,289	35,589	13,975	24	142	1.7	0.2
B752	2.13	4,270	13,455	5,286	13	58	1.0	0.1
B763	2.30	6,087	19,188	7,536	15	81	1.1	0.1
B772	1.26	8,188	25,828	10,137	8	178	0.8	0.1
BA46	0.39	3,343	10,511	4,137	24	25	2.8	0.1
C130	0.06	3,672	11,573	4,554	16	25	4.9	0.2
C550	1.13	874	2,752	1,082	4	7	0.5	0.0
DC9	0.63	5,253	16,535	6,503	27	35	4.1	0.3
E145	2.04	1,561	4,908	1,933	12	11	1.4	0.1
F100	0.30	3,469	10,904	4,296	29	28	3.5	0.1
F2TH	0.43	1,089	3,425	1,350	9	14	0.8	0.0
F50	0.74	1,668	5,252	2,065	8	20	0.0	0.2
F70	0.20	2,886	9,076	3,569	17	26	3.0	0.1
F900	0.20	1,350	4,246	1,674	14	15	1.2	0.0
GLF4	0.30	1,961	6,166	2,426	12	16	2.2	0.1
L101	0.41	8,877	27,928	10,991	56	162	11.0	0.2
L188	0.02	4,212	13,229	5,184	13	22	1.1	0.2
MD11	0.56	8,063	25,421	9,984	19	98	1.3	0.2
MD80	1.81	4,511	14,218	5,585	11	47	2.1	0.2
MD90	0.13	4,428	13,939	5,477	19	60	2.9	0.1
SF34	1.57	938	2,951	1,161	5	8	3.4	0.2
YK42	0.04	5,508	17,360	6,803	12	81	1.3	0.2

Based on data reported by Eyers et al. (2004).

Table 5. Annual flights, distance flown, fuel use, emissions and emission indices for 2000-2004

	2000	2001	2002	2003	2004
Flights	29,706,287	27,673,927	28,477,399	28,780,037	30,378,593
Distance flown (million km)	33,300	31,900	32,600	34,400	37,000
Fuel burn (million tonne)	181	170	171	176	188
Fuel burn rate (km kg^{-1})	5.43	5.33	5.25	5.12	5.08
Emission (million tonnes)					
NO_x	2.51 (13.8)	2.35 (13.8)	2.41 (14.1)	2.49 (14.1)	2.69 (14.1)
CO	0.541 (2.98)	0.464 (2.73)	0.48 (2.81)	0.486 (2.76)	0.511 (2.71)
HC	0.0757 (0.417)	0.063 (0.371)	0.0639 (0.374)	0.0617 (0.350)	0.0625 (0.332)
CO_2	572	536	539	557	594
H_2O	224	210	211	218	233
SO_x	0.145	0.136	0.137	0.141	0.151

The value given in parentheses denote emission index of pollutant in (g kg^{-1}) Emission indices for CO_2, H_2O and SO_x are constant as 3155 g kg^{-1}, 1237 g kg^{-1} and 0.8 g kg^{-1}, respectively. Modified from Kim et al. (2005).

Table 6. Annual statistics of distance flown, fuel burn and global NO_x emission indices for common commercial aircraft and engine combinations by 12 months period between mid 1991 and mid 1992

	Distance flown (1000 km)	Fuel consumed (t)	NO_x emitted (t)	EI NO_x (g kg^{-1})	Emission rate (g km^{-1})	Percentage of km travelled	Percentage of fuel used	Percentage of NO_x emitted
B747JT9	3,356,166	36,758,491	831,814	22.6	247.8	14.5	24.7	33.3
DC10C50	1,341,397	13,286,359	240,952	18.1	179.6	5.8	8.9	9.7
B737JT8	3,069,913	13,018,281	168,085	12.9	54.8	13.3	8.8	6.7
B727JT8	2,073,623	12,034,858	146,596	12.2	70.7	9.0	8.1	5.9
MD80JT8	1,797,929	8,209,041	109,654	13.4	61.0	7.8	5.5	4.4
B767JT9	1,107,033	5,789,590	123,472	21.3	111.5	4.8	3.9	5.0
B747C80	539,905	5,541,228	80,704	14.6	149.5	2.3	3.7	3.2
DC09JT8	1,239,842	5,263,853	68,912	13.1	55.6	5.4	3.5	2.8
JETMNOR	542,699	4,748,545	44,878	9.5	82.7	2.4	3.2	1.8
LIIIR22	494,638	4,741,356	112,420	23.7	227.3	2.1	3.2	4.5
B737C56	1,418,369	4,308,316	38,380	8.9	27.1	6.1	2.9	1.5
B747C50	317,437	3,856,096	70,509	18.3	222.1	1.4	2.6	2.8
A310C80	643,990	3,417,480	6,456	13.6	72.1	2.8	2.3	1.9
B767CF8	571,540	3,103,782	45,670	14.7	79.9	2.5	2.1	1.8
A300C50	470,995	3,088,348	55,008	17.8	116.8	2.0	2.1	2.2
TU54NK8	510,807	2,712,038	22,000	8.1	43.1	2.2	1.8	0.9
B757P20	520,221	2,385,052	38,261	16.0	73.5	2.3	1.6	1.5

continued on following page

Table 6. Continued

	Distance flown (1000 km)	Fuel consumed (t)	NO_x emitted (t)	EI NO_x (g kg^{-1})	Emission rate (g km^{-1})	Percentage of km travelled	Percentage of fuel used	Percentage of NO_x emitted
A320C56	552,237	2,232,071	31,893	14.3	57.8	2.4	1.5	1.3
B757R21	462,843	2,160,736	57,183	26.5	123.5	2.0	1.5	2.3
DC08JT3	259,943	2,058,322	15,593	7.6	60.0	1.1	1.4	0.6
B707JT3	205,287	1,893,589	16,077	8.5	78.3	0.9	1.3	0.6
JETHN86	206,090	1,532,743	11,280	7.4	54.7	0.9	1.0	0.5
A300C80	251,684	1,510,743	21,754	14.4	86.4	1.1	1.0	0.9
MDI1PW4	136,689	1,320,528	24,683	18.7	180.6	0.6	0.9	1.0
LIlR524	129,125	1,123,335	35,531	31.6	275.2	0.6	0.8	1.4
F28SPEY	213,165	668,552	9,309	13.9	43.7	0.9	0.5	0.4
BA46ALF	219,006	523,115	4,387	8.4	20.0	1.0	0.4	0.2
A310JT9	85,637	465,131	11,288	24.3	131.8	0.4	0.3	0.5
F100TAY	105,166	384,754	5,460	14.2	51.9	0.5	0.3	0.2
CONCORDE	28,475	374,071	7,689	20.6	270.0	0.1	0.3	0.3
JETMBJ	173,874	86,696	541	6.2	3.1	0.8	0.1	0.0
JETLBJ	21,879	14,135	60	4.3	2.8	0.1	0.0	0.0
JETLNOR	17,399	13,186	60	4.6	3.4	0.1	0.0	0.0
Total	**23,085,003**	**148,624,420**	**2,496,559**					

Columns are ranked to fuel consumed Reproduced by permission from Gardner et al. (1997)

Table 7. Summary of departure statistics for May 1999

Aircraft type	Daily departures	% of global departure	Distance (km day^{-1})	% of global distance	Average route distance (km)
Turboprops	21,296	30.56	6,608,584	9.43	310
B737-300/400/500	10,224	14.67	9,147,802	13.05	895
MD-80	5,397	7.74	5,619,233	8.02	1,041
B737-100/200	4,013	5.76	3,176,590	4.53	792
DC-9	3,346	4.80	2,379,550	3.39	711
Regional Jets	3,198	4.59	2,096,416	2.99	656
A320	3,071	4.41	3,818,451	5.45	1,243
B757-200	2,741	3.93	4,828,701	6.89	1,762
B727-200	2,353	3.38	2,532,550	3.61	1,077
Fokker 100	1,697	2.43	1,079,091	1.54	636
B767-300	1,533	2.20	4,043,356	5.77	2,638
B747-400	1,006	1.44	5,664,264	8.08	5,632
BAE 146	993	1.43	630,398	0.90	635

continued on following page

Table 7. Continued

Aircraft type	Daily departures	% of global departure	Distance (km day⁻¹)	% of global distance	Average route distance (km)
A300-600	825	1.18	1,012,579	1.44	1,228
B737-600/700/800	771	1.11	1,047,151	1.49	1,357
Russian Aircraft	701	1.01	1,266,310	1.81	1,806
Fokker 28	626	0.90	358,254	0.51	572
B747-100/200/300	570	0.82	2,573,174	3.67	4,517
A319	537	0.77	692,169	0.99	1,289
B767-200	492	0.71	1,417,564	2.02	2,884
B777-200	473	0.68	1,583,564	2.26	3,345
A310	464	0.67	1,044,357	1.49	2,251
A321	448	0.64	361,687	0.52	807
MD-90	442	0.63	331,624	0.47	750
DC-10	379	0.54	1,523,344	2.17	4,022
MD-11	308	0.44	1,541,979	2.20	5,006
DC-8	266	0.38	451,733	0.64	1,699
B727-100	261	0.37	205,915	0.29	789
A330-300	250	0.36	510,219	0.73	2,044
A340-300	224	0.32	1,250,423	1.78	5,589
A300-B2/B4/F4	199	0.29	273,690	0.39	1,377
Fokker 70	198	0.28	149,699	0.21	756
Lockheed L-1011	140	0.20	288,761	0.41	2,058
B777-300	82	0.12	131,672	0.19	1,614
BAC 111	57	0.08	45,221	0.06	797
B707	41	0.06	105,766	0.15	2,607
A330-200	39	0.06	138,796	0.20	3,572
A340-200	20	0.03	140,216	0.20	6,961
Concorde	6	0.01	33,890	0.05	5,648
Miscellaneous	5	0.01	3,013	0.00	659
Total	69,690		70,107,755		

Used by permission of NASA from Sutkus et al. (2001). Permission by NASA does not constitute an official endorsement, either expressed or implied, by the National Aeronautics and Space Administration.

Table 8. Summary of fuel burn and global emission indices for commercial aircraft for May 1999

Aircraft type	Fuel burn (t day⁻¹)	% of global scheduled traffic fuel burned	Emission index (g kg⁻¹)					
			1-9 km altitude band			9-13 km altitude band		
			EI NO$_x$	EI CO	EI HC	EI NO$_x$	EI CO	EI HC
B 747-400	59,837	17.22	25.3	8.1	1.9	13.3	1.0	0.4
B737-300/400/500	30,765	8.85	13.2	11.5	0.9	9.6	3.5	0.2
B747-100/200/300	30,638	8.82	27.5	15.4	10.2	15.2	2.2	1.1

continued on following page

Table 8. Continued

Aircraft type	Fuel burn (t day⁻¹)	% of global scheduled traffic fuel burned	Emission index (g kg⁻¹)					
			1-9 km altitude band			9-13 km altitude band		
			EI NO$_x$	EI CO	EI HC	EI NO$_x$	EI CO	EI HC
MD-80	22,367	6.44	16.0	4.2	1.2	10.6	4.4	1.6
B767-300	22,153	6.37	21.3	7.0	1.4	12.5	1.2	0.3
B757-200	19,717	5.67	18.6	8.4	0.5	11.0	1.7	0.1
B727-200	14,334	4.12	11.9	9.6	3.1	8.3	5.4	1.0
DC-10	12,679	3.65	24.2	7.5	2.8	14.9	2.0	0.9
B737-100/200	12,223	3.52	11.2	10.0	3.3	7.1	6.8	1.3
MD-11	11,952	3.44	19.0	5.9	0.5	12.9	1.2	0.1
A320	11,884	3.42	17.5	5.6	0.5	12.0	2.0	0.4
B777-200	11,260	3.24	25.2	5.4	6.0	16.8	0.6	0.3
DC-9	9,130	2.63	11.0	11.8	4.2	7.6	6.7	1.0
Turboprops	8,788	2.53	11.9	3.8	0.2			
A340-300	8,242	2.37	23.0	11.3	4.7	13.7	1.7	0.2
RussianAircraft	7,138	2.05	12.3	15.5	9.0	9.2	8.4	1.5
B767-200	7,110	2.05	22.6	6.6	1.6	11.1	2.0	0.3
A300-600	6,397	1.84	17.8	10.1	2.0	12.2	1.7	0.3
A310	5,298	1.52	18.5	15.3	4.7	11.3	2.2	0.6
RegionalJets	4,479	1.29	11.6	9.4	1.4	9.1	0.6	0.1
Fokker100	3,444	0.99	11.1	21.0	2.0	6.4	7.0	1.0
A330-300	3,402	0.98	23.0	7.0	1.5	14.5	1.3	0.5
B737-600/700/800	3,219	0.93	16.3	6.4	0.9	11.8	1.8	0.3
DC-8	2,883	0.83	11.2	16.3	11.2	8.6	7.2	1.4
LockheedL-1011	2,415	0.70	18.7	19.4	13.6	14.4	9.0	2.2
BAE146	2,330	0.67	9.1	5.0	0.5	7.8	1.6	0.1
A319	2,048	0.59	14.6	5.7	0.7	10.9	2.5	0.3
A300-B2/B4/F4	2,004	0.58	22.2	13.1	5.2	14.5	1.9	1.2
A321	1,405	0.40	17.5	6.4	0.6	13.3	1.7	0.2
MD-90	1,243	0.36	16.4	5.2	0.1	11.9	1.8	0.1
Fokker28	1,211	0.35	10.5	13.5	7.8	7.4	7.2	2.7
B777-300	1,131	0.33	24.8	4.4	10.3	15.7	0.8	0.5
B727-100	1,094	0.31	10.8	14.8	5.7	7.1	10.2	2.1
A340-200	910	0.26	23.1	11.2	4.6	13.7	1.8	0.1
A330-200	848	0.24	24.3	6.5	1.0	16.3	1.6	0.3
B707	607	0.17	8.4	31.6	39.4	5.4	17.9	8.5
Fokker70	453	0.13	10.3	5.3	1.2	7.1	2.7	1.0
Concorde	351	0.10	11.0	18.5	1.3	10.0	26.1	1.8
BAC111	158	0.05	14.4	25.5	15.0	10.1	14.7	6.0
Miscellaneous	5	0.00	8.6	15.8	2.3	7.3	0.8	0.2

Used by permission of NASA from Sutkus et al. (2001). Permission by NASA does not constitute an official endorsement, either expressed or implied, by the National Aeronautics and Space Administration.

Table 9. NO$_x$ comparison with ICAO emission database for various types of engines and aircraft

Engine type	Aircraft type	Airline	Mean peak height (ppb), 95% CI[a]	Sampled plumes	ICAO NO$_x$ emission rate (g s^{-1})[b]
RR Trent 892	B777-200		115 (109-122)	152	179
	B777-300		147 (127-170)	25	
IAE V2533-A5	A320		45 (42-48)	229	52
	A321		53 (50-56)	374	
PW 4056	B747-400	Air India	61 (53-68)	30	72
	B747-400	Singapore Airlines	73 (63-81)	29	
	B747-400	Malaysia Airlines	75 (69-82)	26	
RR RB211-524G-T	B747-400	British Airways	67 (63-72)	163	74
	B747-400	Quantas	83 (76-89)	32	
	B747-400	Cathay Pacific Airways	81 (63-95)	14	
GE CF6-80C2B1F	B747-400	Japan airlines	85 (72-103)	15	60-66
	B747-400	Thai Airways Int.	61 (51-69)	18	
	B747-400	Virgin Atlantic Airways	50 (44-56)	34	

[a] Upper and lower 95th percentile confidence intervals (CIs) calculated through bootstrap simulations. [b] Assuming 100% thrust. The ranges given in parentheses denote highest and lowest measurements. Reproduced by permission from Carslaw et al. (2008).

Figure 18. Measured mean values of emission indices for CO, NO and NO$_x$ for the APU of various aircraft by FTIR emission method. Modified from Schafer et al. (2003).

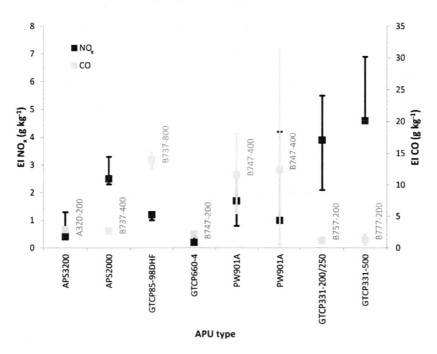

Figure 19. Measured HC emission indices and fuel flow rates for various engine types. Modified from Winther et al. (2006).

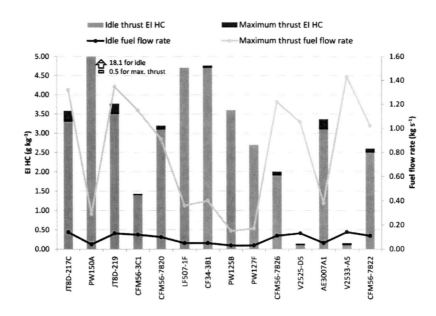

Figure 20. Flight parameters for domestic and international flights in Australian airspace during a 6-month study period. Modified from Pham et al. (in press).

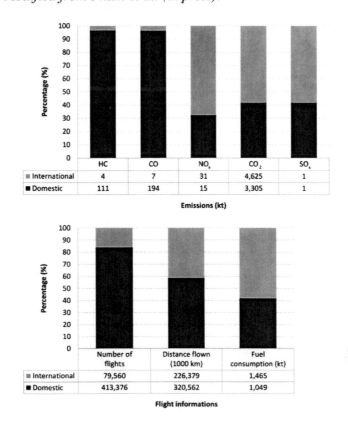

Figure 21. Fuel use and emissions for civil and military aviation in 2002. Modified from Eyers et al. (2004).

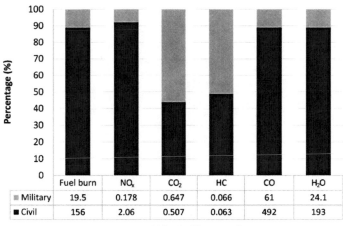

Figure 22. Distribution of fuel burn and emissions by aircraft category for 2002. (a) Breakdown of large jets category and other aircraft categories; (b) breakdown of "other" aircraft category from part (a). Modified from Eyers et al. (2004).

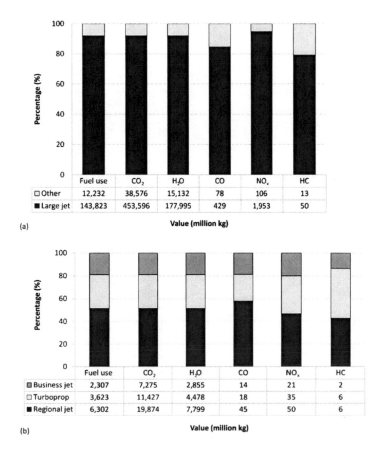

Figure 23. Distribution between flight phases of annual fuel burn and NO$_x$ emissions for 2000-2004. Modified from Kim et al. (2005).

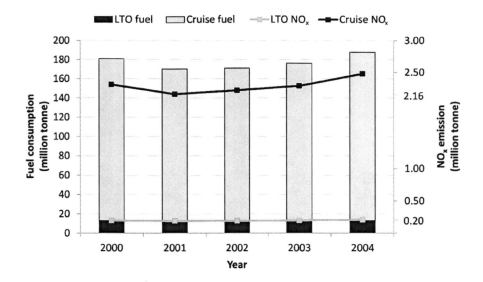

Figure 24. Vertical distribution of fuel consumption and NO$_x$ emission for civil aircraft. October and January are the months with the highest and the lowest values, respectively. Modified from Gardner et al. (1997).

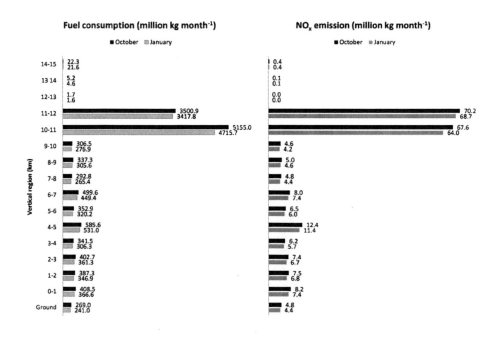

Figure 25. Aircraft daily emissions by month for 1992. The vertical range between two horizontal grid corresponds to an amount of one million kg day⁻¹. Modified from Baughcum et al. (1996).

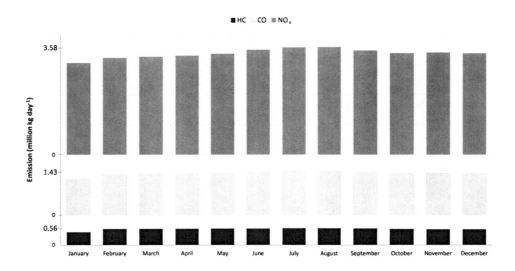

Figure 26. Aircraft daily fuel burn and NO_x emission by month for 1999. Modified from Sutkus et al. (2001).

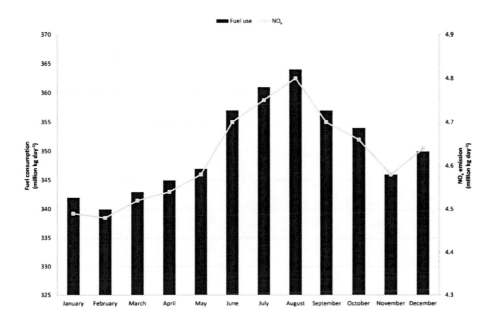

Figure 27. Breakdown (in %) of global fuel burn and NO$_x$ emissions by region for May 1999. Modified from Sutkus et al. (2001).

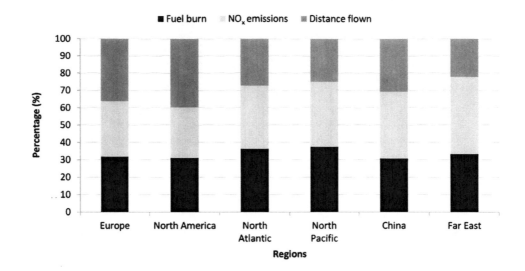

Chapter 3

Aircraft Development and Design:
Enhancing Product Safety through Effective Human Factors Engineering Design Solutions

Dujuan B. Sevillian
Large Aircraft Manufacturer, USA

ABSTRACT

Effective Human Factors Engineering (HFE) has provided the aerospace industry with design considerations that promote aviation safety in the development of complex aircraft systems, as well as the operators and maintainers that utilize those systems. HFE is an integral aspect within the systems engineering process. Measuring the effectiveness of Human Systems Integration (HSI) in the research & development stage is critical for the design of new and modified systems. This chapter focuses on how providing effective HFE design solutions enhances product design and system safety. Providing the customer with safe and reliable products augments mission capabilities throughout the product lifecycle.

INTRODUCTION

Multi-modal transportation such as aviation, marine, rail, transit, cycle, pedestrian, and motor vehicles all provide the general public with choices for determining which mode they prefer to utilize. The cost of travel, safety, and trans-portation reliability are all factors the public uses to determine their travel choices. From a safety perspective, federal agencies have provided the general public with safety statistics for all modes of transportation.

The National Highway Traffic Safety Administration (NHTSA) released data on fatal car crashes in 2008. According to the Fatality Analysis Report-

DOI: 10.4018/978-1-60960-887-3.ch003

ing System (FARS) of the NHTSA, there were more than 34,000 people killed in vehicle crashes in the U.S. (Administration, 2008). The Bureau of Transportation Statistics (BTS)-Transportation Statistics Annual Report-2008 report indicated that there were 4,654 fatalities as a result of pedestrians struck by highway vehicles, and 16,520 passenger car occupant fatalities in 2007 (Statistics, 2008). The public continues to manage their personal risks regarding travel choices and determining which mode of transportation is the safest.

Studies conducted by the National Transportation Safety Board (NTSB), a U.S. independent federal agency that investigates transportation accidents, have shown that commercial aviation transportation is a vital asset to the U.S. economy and is one of the safest modes of transportation. According to former NTSB chairman Marion C. Blakey, "The U.S has long depended on aviation, not just as a means of point-to-point travel, but as the hinge on which the door of commerce depends to stay open". "Literally, aviation keeps America open for business" (Blakey, 2007, p.1). Aviation accidents declined between 2008 and 2009. The NTSB conducted a study that demonstrated a reduction of aviation accidents in 2009—where "The total number of U.S. civil aviation accidents decreased from 1,658 in 2008 to 1,551 in 2009". "Total fatalities also showed a decrease from 566 to 534". "On-demand Part 135 operations reported 47 accidents in 2009, a decrease from 58 in 2008". "The accident rate decreased to 1.63 per 100,000 flight hours in 2009 from 1.81 in 2008". The NTSB have also been able to pinpoint that the "the majority of these fatalities occurred in general aviation and scheduled part 121 operations" (Board, 2010, p.1).

Even though commercial aviation transportation is considered a safer way of traveling, incidents and accidents continue to occur within the industry. From an international aviation perspective, high accident rates in Africa and Indonesia continue to be a critical factor in the general public's perception of air transportation safety.

Consequently, federal agencies across the globe have made substantial progress in determining influential factors that cause aviation incidents and accidents. In particular, the goal has been to reduce the rate of occurrences and provide the public with a positive image of aviation safety.

The military and civilian air transportation sectors have developed robust strategies of aviation safety management. Federal government officials, manufacturers, airlines, and independent consultant agencies have implemented aviation safety management methods for the continuous surveillance of safety in flight operations and maintenance. These methods were created to reduce the rate of incidents and accidents through various systematic processes. These systematic processes were developed to manage the 'common threads' in incident and accident causation.

After the crash of ValuJet flight 592 on May 11, 1996, the U.S. implemented the Air Transportation Oversight System (ATOS). The ATOS is an inspection system that was designed to systematically manage safety in flight operations and maintenance in the airline industry. Other reliable methods have all provided the airline industry with structured systematic ways of managing aviation safety. Some of these methods are as follows: Aviation Safety Action Program (ASAP), Maintenance Safety Action Program (MSAP), Dispatch Safety Action Program (DSAP), Safety Management Systems (SMS), and the Aviation Safety Information Analysis and Sharing System (ASIAS). Accident investigation, empirical, and theoretical research continues to provide data useful for determining the causal factors of incidents and accidents.

Aircraft design, weather, and human factors have been a constant trend of incident and accident causation throughout the history of the air transportation system. On February 12, 2009, Colgan Air flight 3407 crashed outside of Buffalo, NY with no survivors. The probable cause of this crash was "the captain's inappropriate response to the activation of the stick shaker, which led to an aerodynamic stall from which the airplane did not recover" (Board, Loss of Control on Approach, Colgan Air, Inc., Operating as Continental Connection Fight 3407, 2010, p.1). On August 27, 2006, a Comair Regional jet crashed in Lexington, KY with only one survivor. The probable cause of this accident was "the flight crewmembers' failure to use available visual cues and aids to identify the airplane's location on the airport surface during taxi and their failure to cross-check and verify that the airplane was on the correct runway before takeoff" (Board, Attempted Takeoff From Wrong Runway; Comair Flight 5191, 2007, p.1). The NTSB investigations revealed that the ValuJet, Colgan Air and Comair crashes were considered to be related to the concept most often referred to as 'Human Factors'. According to the Human Factors Accident Classification System (HFACS) Analysis of Military and Civilian Aviation Accidents: A North American Comparison; "Between 70-80% of aviation accidents are due, at least in part, to human error" (Wiegmann, 2004 p. 2).

Historically, Human Factors related deficiencies in air transportation have influenced incidents and accidents across the global airline industry. In December 2003, a ramp worker lost his life while preparing to push back an aircraft. According to the report, the tractor operator's foot accidentally hit the accelerator and "the tug jumped forward striking the radome of the aircraft fatally pinning the driver". The investigation revealed that the ramp agent did not set the parking brake (Matthews, 2004 p.6).

The constant trend of human factors related incidents and accidents in the aviation industry appear to be a parallel theme in other modes of transportation such as the railroad, chemical, highway, and marine industries. On June 11, 1997, a bus collision with pedestrians in Normandy, Missouri occurred. The causal factor in this crash was, "The driver trainee's misapplication of the accelerator resulting in the bus's override of the curb and travel onto the occupied pedestrian platform" (Board, Bus Collision with Pedestrians, 1997). On October 12, 2003, a Northeast Illinois Regional Commuter Railroad train derailed. The probable cause of the derailment was "the locomotive engineer's loss of situational awareness minutes before the derailment because of his preoccupation with certain aspects of train operations that led to his failure to observe and comply with signal indications" (Board, Derailment of Northeast Illinois Regional Commuter Railroad Train, 2003). Even in the design of Unmanned Air Vehicles (UAVs), "the Air Force Scientific Advisory Board (AFSAB) identified the human system interface as the greatest deficiency in current UAV designs" (Williams, 2006 p.1).

The term Human Factors is frequently utilized in the aviation industry. Human Factors can be operationally defined in this chapter as factors influencing the behavior of the human based on several dynamic critical interfaces. Generally speaking, the interfaces that affect human performance include: environmental conditions, software/hardware factors, and interactions with other humans. HFE is the application of human capabilities and limitations as they relate to the design of systems. HSI is the systematic process of integrating the human with the system so that the human has the ability to operate, maintain, and support personnel needs within the applicable environment. HSI is needed to reduce possibilities of human performance related errors and to promote aviation safety. These definitions will be further explained throughout the chapter. Additionally, HFE and HSIs relationship to overall aircraft systems design will be discussed.

Reducing the possibilities of human error can be a rigorous task. The industry has developed several methods to aid in the reduction of human performance based errors. Some methods such as: HFACS, Maintenance Error Decision Aid (MEDA), and the Ramp Error Decision Aid (REDA) are internationally accepted methods designed for human error reduction. Essentially, these methods are utilized to help manage the risk of human error and to better prepare the human for the tasks and environmental conditions they will experience.

Several years ago, the term 'pilot error' was commonly used by air safety investigators to describe an individual(s) flying an aircraft that was involved in an aircraft incident or accident. It is paramount to understand that 'pilot error' is not synonymous with the term human factors. The refined definition of human factors requires a complete understanding of the 'human element' as a system interfacing with a series of systems. The process of integrating the 'human element' within a system is not a parallel process within the systems engineering discipline. Therefore, understanding the dynamics of the human system interface continues to be a challenging and critical aspect in the development and design of complex aircraft systems. Nevertheless, one might ask the question, 'How is a system designed without human error?' It is impossible to design a system without the possibilities of human error. In fact, as long as there is human interface with a system, there will be the potential for errors. As engineers, investigators, scientists, and specialists, we must determine the impact of the system design on human performance, while minimizing the potential risk associated with the hazards.

The goal of this chapter is to provide the reader with a clear understanding of the need for human factors considerations at the design stage, and to examine the factors that influence safety throughout the product lifecycle. This chapter will provide case scenarios related to human factors and system design; to further demonstrate the relationship between the human and its interfaces. Furthermore, there will be a discussion of human factors issues. Finally, solutions, industry recommendations, and future research directions will be provided. An outline of topics that will be covered throughout the chapter are as follows:

- General understanding of system engineering and design architecture
- Human integration vs. system design
  - Cognitive interfaces on system design
  - Physical interfaces on system design
- Enhancing product integrity through promoting effective system safety management
  - Hazard Identification
  - Hazard Analysis
  - Hazard Mitigation
  - Hazard Monitoring

BACKGROUND

Since the Wright Brothers flight on December 17, 1903, the aviation industry has learned from, scientists, pilots, engineers, and other aviation professionals, the need to develop safe technologically advanced aircraft systems. Since World War II, the design of aircraft systems, and the apparent human factors influence on these systems has increased the ability for research and development in the human performance domain.

The aerospace industry has learned how to design aircraft systems that are reliable, safe, cost effective, and efficient for the crewmembers that utilize those systems. Recent developments in aircraft design and test, such as the Boeing 787 Dreamliner and the Boeing 747 Large Cargo Freighter (LCF)-Dreamlifter further re-iterates the continuous strategies and perspectives of safety through constant innovative new technologies. However, the design of complex aircraft systems can be a challenging effort. An abundance of engineers are required to produce quality products that are safe, human compatible, reliable, and meet the customer mission requirements. As the complication and interdependence in systems grow, the certification and testing of those systems continues to be a crucial factor in aircraft system development (Abbott et al., 1999).

It is inevitable that when engineers design a system, there will be some type of human interface with the product. Whether the system is a manned or an unmanned crew station, human system interface continues to be a persistent issue when designing any aircraft system. Software and hardware technological advances have provided the industry with the opportunity to design systems that have components that are interconnected. However, these systems have the potential to fail (Abbott et al., 1999). These failures can be considered systematic failures, those related to the aircraft system functions and human system interface failures, and those related to the integration of the human with the system.

Failures related to the system reflect deficiencies in the way a system was executed. These insufficiencies are due to lack of training, personnel selection, operating policies, maintenance and organizational support issues. Human system interface failures are design errors which include hardware, software, environment, and training factors. These design errors are the result of insufficient integration of the human with the system. In particular, Graphical User Interface (GUI), and other ergonomic equipment design related issues are some of the factors that influence the user interface (McSweeney and J.Pray, 2009).

It is important to understand that having a technologically advanced system does not constitute a safe and user-compatible product. The human utilizing the system has to be able to effectively operate and maintain the system. If the operators and maintainers cannot effectively utilize the system, then the potential of human error exists and the design becomes incompatible for the user. In order to reduce the possibilities of design human incompatibility that could lead to induced human error, an understanding of the system design architecture becomes a crucial aspect at the design inception stage. During this stage, both systems engineers and design engineers work together to understand the system fundamentals.

A system is defined as an interactive integrated composition of architecture that work together to achieve a specific mission. According to Chapanis (1996), there are three views of a system: the physical view, functional view, and the operational view. The physical view is based on the population of people that the system will contain. Next, the functional view focuses on how the system should work from an operational perspective. Last, the operational view is based on how the users interface with the system. Systems engineering is the iterative integrated process of managing the development of complex systems throughout the product design lifecycle. Without systems engineering, it would be difficult to manage the development of any aircraft system because systems engineering provides the framework for managing the design.

MAIN FOCUS OF THE CHAPTER

"The Iterative Process of Aircraft Development & Design"

When a customer solicits participation in an aircraft design contract, they want a contractor that can provide customer service that will support their mission needs. Whether these customer mission needs are to fly passengers from point A to B, or to fly a secret mission, it is paramount when designing any aircraft system to implement a 'systems thinking' tactic to design. A 'systems thinking' approach to product design is a disciplined scheme to problem solving. Typically, this process can be defined as how certain systems influence the aircraft as a whole. The approach also focuses on how the human interfaces with a system and its components. Utilization of 'systems thinking' should occur when determining the purpose of the system, functional capabilities, and operational capabilities. Finally, this approach requires knowledge of five general integrated areas related to systems engineering. The five taxonomies are as follows:

- Product definition
- System and mission requirements
- Functional Analysis/Functional Allocation
- System Design
- System Test

When designing an aircraft system, it is important to understand the conceptual design stage, which is most commonly referred to as the product definition. At this stage, the forecasting of engineering strategies, business planning development, and proposal planning are considered and executed. Reputable competition strategies are a key asset within the product definition stage. While competition strategies are key to the development and design of aircraft systems, system engineers create program plans to effectively collaborate and manage the process of design. The HSI Program Plan, Human Engineering Program Plan (HEPP), System Safety Program Plan (SSPP), Reliability/Maintainability Program Plan, and System Engineering Management Plan (SEMP) are just some of the program plans that should be developed at this stage. These collaborative program plan approaches to engineering design should be reviewed with the customer. These plans are needed so that the customer has a clear understanding of how the engineering process will be managed.

The contractor Statement of Work (SOW) provides the foundation for understanding baseline level customer requirements. From a system engineering perspective, a requirement defines a necessary element or characteristic that must be satisfied in order to achieve operational capabilities. A requirement should be further emphasized in order to determine how it applies to the system and the environment.

System and mission requirements are essential components within the systems engineering process. System requirements describe 'what' a system shall do from a hardware and software perspective. Mission requirements are performance specifications and they relate to 'how' the system shall perform in a particular environment. For example, a system requirement may be, 'the software shall be compatible for the display user interfaces'. A mission requirement may be, 'the satellite shall be capable of differentiating between various types of surface matter'. The impact of these requirements must be further explained and

reviewed by the customer and contractor. Trade studies help systems and design engineers understand how the system and mission requirements impact the user operational and environmental characteristics. It is imperative that these requirements are thoroughly examined in order to ensure consistency in the application of the requirements in system design; this is needed to ensure the requirements are attainable, verifiable and valid.

The functionality of a system drives certain requirements. Whether the system is electrical, hydro-electrical, mechanical, or pneumatic, a complete understanding of how the system functions relative to the requirements assists the designer with developing the system. Furthermore, the function of the system assists the designer with how the system will perform in an operational environment. Examples of system functionality in an operational environment are the aircraft phases of flight. Typically, the aircraft phases of flight are as follows: pushback, taxi, takeoff, climb, cruise, decent, and landing.

The function of an Aircraft Communications Addressing Reporting System (ACARS) is to provide the aircrew with a digital data link utilized as an interface between dispatch and crewmembers. Some of the characteristics of the ACARS system are the capability to send information such as aircraft system performance data characteristics, generation of weather reports, flight plan management, and maintenance message communications. The data generated from the ACARS ease communication with airline System Operational Control Centers (SOCCs). SOCCs are the communication command centers at the airline base of operations that ensure the aircraft and crew members are operating adequately throughout the phases of flight. To ensure that the crew is operating satisfactorily, maintenance control, dispatch, and other related airline functions are communicated through the use of the ACARS. The ACARS system is utilized in all phases of flight.

Another system that is utilized in flight as a power generator is the Auxiliary Power Unit

(APU). The function of the APU is to provide the aircraft with an additional power source. This power may be used for transferring air to the cabin for air-conditioning purposes or electrical power distribution. The APU also assists the crew in an emergency situation (engine loss, or failed power generator) by providing the crew with an alternative means of power. Moreover, the function of the system must be evaluated so that the system engineer understands how to meet customer requirements and the impact of system operations.

Typically, after the functions of the system have been defined and analyzed, the allocations of the functions are evaluated. Functional allocation is "a procedure, for assigning each system function, action, and decision to hardware, software, human operators, or maintainers or some combination of them" (Chapanis, 1996, p.101). Functional allocation is an important concept because it assists the designer with an understanding of the viability of technology related to software and hardware interfaces. Additionally, the software and hardware interfaces should be reviewed by HFE to determine the practicability in meeting the requirements. The data evaluated at this stage of the process are previous experiences with systems, technologically advanced performance capabilities or mechanical systems and software systems related to human limitations. Functional flow diagrams help facilitate and understanding of the functional allocation process (Chapanis, 1996).

The system and mission requirements are reviewed by the contractor and customer. These requirements are derived and further delineated through a series of customer reviews. These reviews help the customer and contractor understand the feasibility of the design for future product developments. Depending on the mission, the customer may choose how they want certain requirements implemented in particular stages of the system design; this is normally based on current and future design strategies.

An Operational Readiness Review (ORR) determines the feasibility of proceeding to the concept exploration. The inputs to this review are the operational need analysis and the operational concept analysis (Chapanis, 1996). The contractor develops the concept document to determine if customer needs are satisfied.

A System Requirements Review (SRR) is then conducted with the customer to determine how the requirements will be utilized to design the system. Depending on the program, there may be a series of SRRs developed to further delineate the requirements through incremental stages of the contract. The SRR also examines the various methods of compliance to the requirements and the methods the contractor has chosen to perform analyses to meet the requirements.

Normally, the System Design Review (SDR) is conducted after the system specifications have been reviewed. This review is required to determine if the integrated system will meet the requirements (Chapanis, 1996). The human factors requirements that are ordinarily reviewed (as they relate to the overall system) are as follows: Human Computer Interaction (HCI), ergonomics, safety, operations, maintenance, personnel requirements, and training.

During the Preliminary Design Review (PDR), the customer, program management, and various engineers are assembled to review the current design philosophy. The PDR includes the approach to various design configurations such as hardware component specification, software specifications, and detailed requirements for future test evaluations. Specialty engineers provide support to the PDR approach and they include: Reliability Maintainability & Testability (RM&T), Electrical, HFE, and system safety engineering.

After the PDR, the Critical Design Review (CDR) is coordinated. This review is the final review before the production stage of the system. The purpose of this review is to compile all of the detailed engineering data related to software and hardware configurations. At this stage, engineering drawings, mockups, HFE, safety, reliability & maintainability analyses are reviewed. Engineering data is provided to the logistics team in order to update or develop new training methods and procedures to support crewmembers in the operational environment.

Last, the Test Readiness Review (TRR) is conducted to review the design integration as it relates to the capability of the system to operate effectively in the operational environment. The design is tested in regards to the interface between the human and the system to ensure user compatibility and safety; various test plans are developed to test system effectiveness. Test reviews are performed to ensure that the contractor has met all of the customer requirements and that the system is suitable for the operational environment.

"The Human Integration Strategy"

Historically, the systems engineering process has been portrayed as a complex integrated approach to product design. The process has worked for several years in the development of aircraft systems. Each stage of the system engineering process conveys a common theme. This theme is related to the continual need to ensure 'system compatibility' so that the system can function in the operational environment. It is pertinent for design engineers to develop systems utilizing their knowledge of engineering design. However, a more important aspect is the design of the human into the overall system design.

Most of the literature related to system design covers a wealth of information regarding overall system design effectiveness and operational support. However, it is argued that there is a consistent issue with the value, understanding, and the application of implementing a HSI approach to engineering design. In addition, HSI is often considered an 'excess cost agent' to program management and to the customer. How should a system be designed to encompass program and customer requirements, and cost dynamics, while also providing the customer with a safe and effective integration of human system interfaces? This

is the primary question that most human systems integration engineers, human factors engineers and specialists ask themselves when participating in any program.

The U.S Air Force, Federal Aviation Administration (FAA), U.K. Ministry of Defense and the Canadian armed forces have all incorporated HSI within their Human Factors commercial and defense programs. HSI is the comprehensive management strategy for effectively implementing the human with the system, in order to fully support the human interfaces considering various conditions.

This strategy provides, tools, techniques, to help facilitate a more robust effort for human integration within systems. Previous approaches only focused on hardware and software integration issues. This strategy focuses on the end user and the relative operational dynamics (Reinach, 2007). Qualitative and quantitative realistic approaches should be utilized to evaluate the risk and to measure the need to utilize a HSI approach in the program development stage. In particular, trade studies should be used to provide evidence to the program regarding the need to implement a HSI program.

Customer program management, contractor program management, design engineers, and system engineers should all interface with HSI. According to the Defense Acquisition Guidebook (DAG), there are seven domains applicable to supporting the integration of the human with the system. The seven domains are:

- Manpower
- Personnel
- Training
- Environment, Safety, and Occupational Health
- Survivability
- Habitability
- HFE

"Manpower can be a major determinant of program cost and affordability" (Guidebook, 2010, p.2). Manpower focuses on operations and maintenance tasks and conditions in the operational environment that are associated with workload. Moreover, the manpower perspective focuses on the need to support military and civilian manpower. Manpower considerations are important to the customer. Program management should consider developing case studies and engineering analyses to optimize manpower and reduce costs to affordable levels. Effectively utilizing manpower considerations in system design can reduce the cost associated with design and operations. One factor that reduces the cost of manpower is an understanding of user tasks as they relate to the system functional allocation processes. If the engineer understands the user tasks, then only relevant tasks will need to be evaluated. These tasks are unique to the system design, and are related to cognitive, physical, environmental, and other related conditions. Evaluating these parameters helps the customer determine proper manpower considerations for their program.

"Our uniqueness as individuals--the way we were raised and educated, our work experiences, and genetic compositions--affect the way we perceive the world and act upon it" (Strauch, 2005, p. 47). Personnel considerations are developed based on the Knowledge, Skills, and Abilities attributes (KSAs) of individuals. The KSA attributes are related to the cognitive and physical demands of the system versus the demands for personnel needed to perform the jobs (Guidebook, 2010). These factors are used to determine the experience levels that operators and maintainers need to perform their job task. KSAs help engineers understand human capabilities and limitations as they relate to system design and personnel qualities. Lastly, personnel selection helps engineers design tasks properly in order to reduce the possibilities of human error.

"Training is the learning process by which personnel individually or collectively acquire or enhance pre-determined job relevant KSAs by developing their cognitive, physical, sensory, and team dynamic abilities" (Guidebook, 2010, p1). When the customer and program management have completed the CDR and the customer is satisfied with the design, training factors need to be discussed with the customer as they relate to the new/modified design. These factors include simulation training, On-the-Job-Training (OJT), or Computer Based Training (CBT). After these factors have been addressed, it is paramount that when personnel are trained, they understand the new or modified system functions and their components. Ongoing training should assist personnel with the new/modified operating procedures and new or modified system capabilities (Strauch, 2005). Essentially, documents such as Aircraft Maintenance Manuals (AMMs), Flight Crew Operations Manuals (FCOMs), and Technical Orders (T.O.s) are just a few of the documents that may need to be updated as a result of the new/modified system design. These updates are needed to ensure that the new/modified equipment design aligns with the nucleus of the training program curriculum.

Environment, Safety, and Occupational Health (ESOH), are essential components to the HSI process. Understanding the factors that influence the surroundings of the human when operating and maintaining a system can provide designers with an understanding of how to accommodate the user with the system. The environment includes "the conditions in and around the system and the operational context within which the system can be operated and supported" (Guidebook, 2010 p.1). In particular, certain environmental issues should be evaluated by engineering. Some of these issues include acoustical noise, vibration, and electric shock. Engineers need to determine the safety impact on the human and the mission as a result of environmental conditions. Safety related issues such as workspace design, Personal Protective Equipment (PPE), hazardous waste, and radiation need to be evaluated to determine the impact on the human and the mission. Warnings and cautions that are depicted on systems or within workspace areas should be clear and concise. These placards should provide the user with enough information to understand the warnings or cautions and provide meaningful information for the user to make the most practical decisions. Warnings and cautions should be effectively communicated in training manuals, test procedures, and OJT. While understanding the impact of safety on the human and mission is practical, it is also important to realize the effects of unsafe designs on the human in a particular environment. The occupational and health perspective is incorporated within the design to explain the potential risk of the design on the human body that could lead to musculoskeletal issues, injury, illness, or disability.

Providing the program and customer with an assessment of the occupational hazards as well as a, relative cost benefit analysis helps the customer with evaluating the risk. This evaluation can be used to determine if a design is suitable for the user population. Data presented by occupational safety specialists at the CDR provide the customer with an early indication of the hazards and the applicable mitigation processes.

The survivability domain is most often related to the military or defense environment. Survivability factors are related to design features that reduce the risk of man-made hostile environmental conditions. Survivability design and operational techniques should be outlined in the program requirements. All requirements should be further defined to provide the engineer with an understanding of the type of environment the human will be susceptible to versus the system design. Certain clothing and survivability gear may be utilized to prevent the possibilities of death or the abortion of a mission. An example of survivability gear is the Nuclear, Biological, and Chemical (NBC) gear. This gear should be designed so that the user can perform the mission operations while minimizing the potential risk of injury or fatigue. In particular, some factors to consider when designing NBC gear are egress capabilities, and extreme environ-

ments. It is important that the crewmember has the ability to exit the aircraft immediately and survive conditions on the land or sea. HFE assists with making decisions that may affect the design of the user interfaces.

Habitability focuses on the living and working conditions for the user population, in order to effectively accomplish a mission. Depending on the mission, the customer may require only certain aspects of habitability. Some characteristics of habitability are as follows: ambient lighting, food servicing, lavatory service, and medical support. HFE and other related specialists should determine the effect of the design on the user, and develop methods for evaluation to optimize user performance in the habitability environment.

HFE is the application of human capabilities and limitations as they relate to the design of systems. HFE should be implemented within a program to reduce manpower and provide effective crew training. Previously stated, when the system functional allocation is performed, the user interfaces should be identified and evaluated by HFE. HFE should be applied during the development stage of equipment design; this application provides the customer with an understanding of the human machine interfaces. Any related tools (modeling, prototyping, simulation) or techniques (Crew Station Working Groups (CSWGs), workload analysis, strategic processes) should be outlined in the HEPP to address how the user interfaces will be managed throughout the product design stages. Generally speaking, user interfaces include: display design, GUI, maintainability, training systems development, training manual development, workload analysis, and Ground Service Equipment (GSE) support. All interfaces should be reviewed with the customer and program management to understand the cost, and design impacts. A meticulous technology strategy review should be utilized when determining the best method to reduce cost associated with modeling and simulation because these cost drivers could affect the program budget. Understanding the customer requirements for HFE prior to development can reduce the cost of design by eliminating extraneous methods of analysis and evaluation. HFE should be cost effective for the program while enhancing user interface capabilities. If a HSI program is developed, a complete understanding of the HSI domains helps with system development and design, thus providing the customer and the program with data for managing the factors influencing system design.

HFE: "Indispensable Solutions in Design"

So far, the issues relative to transportation safety, system design, and the interfaces related to HSI have been acknowledged. Within the confines of HSI, the remainder of this chapter will focus primarily on the HFE domain of HSI. Also included will be a discussion regarding product safety as it relates to managing system design.

HFE can be separated in to two distinct integrated categories: cognitive engineering, and hardware/human physical ergonomics engineering. To understand the context of cognitive engineering, one must first understand the definition of cognition. Cognition is the study of how the human thinks, perceives and remembers information in a certain environment. Cognitive engineering is a multidisciplinary field focused on the design and evaluation of systems technology as it relates to human interfaces and cognitive processes (Gersh, 2005). Since the Three-Mile Island nuclear power plant incident in 1979, a focus on how the human interfaces with workplace design and socio-technical systems continues to be evaluated by scientists and engineers. In addition, human interface with system controls and displays has been on the vanguard of cognitive scientific research. Cognitive engineering may focus on HCI as it relates to GUI or other cognitive demands on the human in a particular environment. Cognitive engineering also focuses on workload and human performance in specific environments. Hardware/

physical ergonomics engineering is the application of human dimensions, physics and biomechanics on human system design integration in a specific environment.

From a cognitive perspective, when functional allocation decisions are considered, HFE should assist the program with making those decisions. It is imperative that once the functional allocation decisions have been made, that HFE outlines how those functions interface with the existing requirements. HFE should provide the program with an overview of how the system functional allocation may affect the human capabilities and limitations. For example, if the requirement is for the human to be able to control system functions, then HFE has to determine the impact of those functions and how they influence the ability of the human to perform certain tasks. These tasks should be based on data obtained from a system functional analysis. The data should be utilized to determine if the functions of the system interface are compatible with the user prospective environments.

After determining the impact on the user and system functional capabilities, the HFE should perform a task analysis. A task analysis is utilized to identify tasks and sub-tasks related to the user environment, and comparing those tasks to the operator and maintainer cognitive processes in order to achieve a goal. Essentially, the task analysis is utilized to: reduce potential hazards and errors, increase human reliability in system interface, evaluating user-system processes, and analyzing personnel needed to perform certain tasks such as operator related tasks or maintainer related tasks. The data generated from task analyses provides information to logistics for developing training documents and other related OJT. The task analysis should describe the human system interactions as they relate to human capabilities and limitations. The results from a task analysis drive new personnel and manpower requirements; these requirements may indicate a need for additional crewmembers or a reduction in crewmembers. A task analysis would be practical if the HFE is evaluating HCI on an operator aircraft display at a particular crew station. The HFE should evaluate the ability of the user to process information on the display screen while performing a task. Whether the task is evaluating the integration of overlays on a display or determining the color and text size, it is important to evaluate how the human processes this information. In order to perform the task analysis, the HFE should have a clear understanding of the Human Information Processing System (HIPS) theory.

According to Wickens (1998), there are three stages in the HIPS theory. The perceptual stage focuses on how the human perceives stimuli and compares it to memory and knowledge to give it meaning. The cognitive stage focuses on how we relate new information to memory or goals to solve problems or make inferences. The action stage is the final stage of the HIPS theory that focuses on how the central processing unit (brain) responds and coordinates utilizing certain motor signals. Once the HFE understands the HIPS theory, the HFE should utilize the Information Processing Model (IPM) to examine all human system interactions to clearly understand how the

human processes information related to the system design in a particular environment. The IPM is a model that portrays how the human interprets a stimulus and responds to certain system inputs in a particular environment. Figure 1 illustrates the IPM; this is a generic model used for human information processing.

Hardware/human physical ergonomics engineering focuses on human anthropometry as it relates to the system interface. Generally speaking, HFE integration should support at least the 5th percentile (typically who are female) and at most the 95th percentile (typically who are male) as the critical human dimensions. Critical human dimensions such as: stature, weight, chest breadth and depth are utilized when integrating the human with the system. Designing systems for the 5th and 95th percentiles provides a baseline approach for accommodation of the user population. However, depending on the user population (Army, Navy, Air-Force, Civilian), certain case scenarios may need to be further evaluated to ensure user-system accommodation. These case scenarios are related to certain body dimensions that may be on the outside boundary of the targeted user

Figure 1. Information processing model (Wickens, 1998)

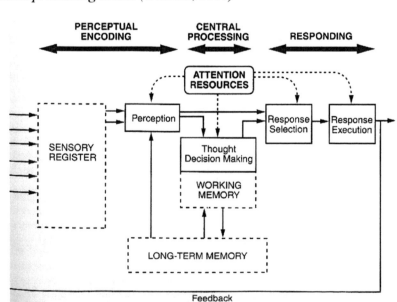

84

population for a particular environment such as chest breadth and stature. Figure 2 illustrates the 5th and 95th percentiles and Figure 3 illustrates critical body dimensions.

Locations of controls and displays, workspace design, and system maintainability are just some of the general areas that must be analyzed by HFE to determine the best method to optimize the user interface. HFE industry standards assist HFE with

determining the best method for optimizing usability.

The Department of Defense (DOD) and the FAA publish numerous design standards, and design approaches to provide HFE best practices for accommodating the user with the system. If utilized early in the engineering design process, these standards can reduce the cost associated with the development and design of hardware and software systems. These standards and approaches

Figure 2. CATIA V5 5th and 95th percentile Manikins (National Institute of Aviation Research (NIAR) Laboratories)

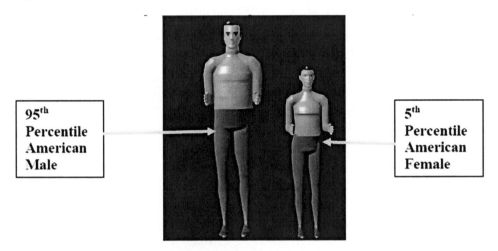

Figure 3. CATIA V5 Manikins critical body dimensions (National Institute of Aviation Research (NIAR) Laboratories)

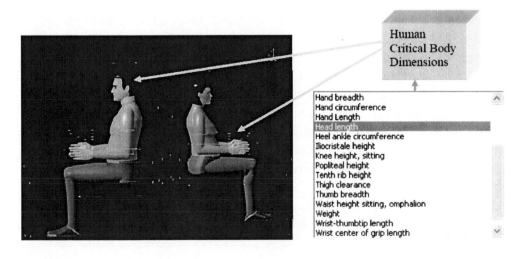

should be outlined in the HEPP to provide the program and customer with and understanding of how the standards and approaches relate to the system requirements. Furthermore, these standards and approaches provide objective data that can be utilized to substantiate issues related to engineering drawings, computer models and the overall systems integration.

MIL-STD 1472 (DOD Human Engineering Design Criteria Standard) is one of the most common utilized standards by HFE for hardware design. Many of the supplemental HFE industry standards are derived from MIL-STD 1472. From a hardware perspective, this standard provides an abundance of data related to the design of certain hardware such as: knobs, dials, pushbuttons, joysticks, pedals, toggle switches, and commonality in systems, just to name a few. The standard also provides information regarding safety, human visual fields, system color coding, and lighting. MIL-HDBK 759 (Human Engineering Design Guidelines) is utilized for determining various anthropometric variables that can be used for the design of systems. Variables such as arm length and chest breadth assist the design engineer with understanding the impact of human dimensions on system spatial integration.

HFE often utilizes design approach documents to help facilitate robust methods for human design integration. Two of the most common design approach documents are the Human Engineering Design Approach Documents-Operator and Maintainer (HEDAD-O/M). Both documents outline HFE methods to assist the HFE with understanding how to evaluate the human interface with the equipment. These documents provide information on how to meet HFE requirements and HFE design criteria. The approaches should be communicated with the customer during SRR and PDR. Communicating these approaches gives the customer a better understanding of how the operator and maintainer systems will be evaluated.

MIL-STD 1472, MIL-HDBK 759, and other applicable standards should be utilized when

creating computerized engineering models. These standards provide the HFE with data to support his/her modeling activities. One of the most common systems utilized by HFE for hardware human integration is the Computer Aided Three-Dimensional Interactive Application (CATIA). CATIA is an iterative process application, utilized to develop human engineering models that predict human performance in system design. Rather than having physical prototypes, human engineering models reduce the cost of human factors analysis by providing the customer with an early indication of possible human factors related integration issues prior to CDR. CATIA software provides the HFE with various workbenches to support HFE analysis. Two of the most common workbenches related to HFE analysis are ergonomics and kinematics. These workbenches aid the HFE with developing digital human static models (manikins) and motion activated models that support maintainability and human crew station compatibility.

Digital human static models are utilized to provide multiple case scenarios related to maintainability and crew station compatibility. These models can be manipulated to show human performance issues in system design. The motion activated models provide the customer with an analysis that conveys the relationship between engineering models (system models) and how they are put in to motion (kinematics) relative to the manikin simulated motion. A Digital Maintainability Analysis (DMA) via CATIA would be practical if there is a need to analyze the human access abilities for Removal and Replacement (R&R) of Line Replaceable Units (LRUs). It is important to ensure that the user can access, inspect, and opt for the R&R of LRUs. This analysis is also needed to meet applicable maintainability requirements. Depending on how long it takes maintenance personnel to R&R, access, or inspect a LRU, this analysis could determine the need to allocate certain personnel or manpower for the task. A tooling analysis provides an evaluation of the types of tools needed to access the equipment and the ability to R&R the system and its components; virtual tools (wrenches, screwdrivers, hammers, etc.) are utilized to determine system accessibility.

The force required to remove a LRU is computed utilizing a Push-Pull Analysis; this helps the HFE understand the effort needed to remove an LRU. A carry analysis is used to ensure the user can carry an LRU with respect to system weight and system lifting characteristics. Variability in system weight should be outlined in the system description document or other applicable equipment detail documents. It is imperative that a thorough evaluation of the effects of system weight on human performance is analyzed. User population and system weight characteristics should be identified and defined prior to PDR to ensure these factors are addressed. Depending on the type of environment, cold weather gear may need to be evaluated to determine user compatibility in maintenance operations. Arctic gloves/mittens may be tailored for a certain user population by simulating certain measurements to the hand. Applying specific measurements to the hand utilizing CATIA helps accommodate the design of arctic gloves/mittens.

A Crew Station Compatibility Analysis (CSCA) via CATIA is utilized to ensure crewmembers are able to operate the systems effectively and that they possess the ability to work in their environment. A CSCA should be utilized when determining the user's ability for grasping controls, reaching display screens, or any other applicable hardware at the crew station. For example, if an automated warning, caution, or advisory display message requires crew input to retrieve information regarding a display message, it is important that crew's ability to access the information is optimized. In other words, this means that the display needs to be designed so that the user has access to the display and has the ability to operate the controls effectively. Hand size dimensions should be evaluated with respect to control system

design to ensure that the human has the ability to grasp or press certain buttons.

Line of sight and visual field evaluations are utilized to determine human system visual compatibility. Certain equipment such as Helmet Mounted Displays (HMD) may be evaluated to determine operator spatial orientation and the ability to recognize objects in the operator Field of View (FOV). Essentially, the HMD analysis should be evaluated to ensure the user's ability to effectively perform a particular mission. Additionally, the customer mission requirements may dictate a need to accommodate a certain user population in regards to the design of the HMD.

A human posture analysis should be utilized to determine the need for the operator to maintain a certain posture while operating a system at a crew station. In regards to pneumatic systems, such as crew oxygen, it is important to evaluate the crew's ability to reach oxygen masks while maintaining a certain posture. Furthermore, variability in seat designs such as the ability of the seat to be elevated or lowered is critical to understanding how the seat design affects the user's ability to reach the oxygen mask. In particular, the stature of the human could affect the ability to operate certain controls at the crew station. The analysis should provide the designer with information to develop or modify a seat configuration so that it is compatible for the user.

Table 1 illustrates the types of analysis that HFE should conduct utilizing CATIA in order to measure the efficiency of hardware human system interfaces related to human performance.

CATIA models should be utilized to communicate with the program and the customer during the PDR and CDR. Providing these analyses assists the program with understanding the impact of HFE on the operator and maintainer interfaces. Lastly, CATIA models help facilitate the integration of system design, while providing assistance with determining the safety impact on system design.

Table 1. CATIA/human system interface design considerations

DMA	CSCA
Human/LRU Access Ability	CSCA Display Reach Analysis
Tooling Analysis	CSCA Controls Grasp Analysis
Push-Pull Analysis	Visual Field Envelope Analysis
Carry Analysis	Human Posture Analysis
Cold Weather Gear Analysis	HMD Analysis

Product Safety: 'A Management Perspective'

Thus far, various meaningful methods for accommodating the user from a cognitive and hardware engineering perspective have been mentioned. These methods are paramount for evaluating user interfaces. However, the need to evaluate user and system compatibility are not the only methods for ensuring product user efficiency.

Product safety management is the ability to manage safety related risks by providing reliable safety analysis suitable for the operational environment. Product safety management is imperative when certifying any aircraft or system. Safety risks are evaluated to determine if an aircraft and system can meet certification requirements to be certified. Various methods of reducing the risks associated with system hazards prior to system design have provided the industry with the ability to engineer-out or to provide other solutions to mitigate certain hazards. These hazards could have a direct influence on the aircraft, system/subsystem and the user. Hazards associated with the aircraft, or system/subsystems could affect mission capabilities, and hazards associated with the system/subsystem user could affect personnel life and work environments.

The most common way of reducing risks associated with a system/subsystem and the user is through the application of system safety engineering. System safety engineering provides the program and the customer with an early identification of risks. A risk is defined as the probability and severity associated with a particular hazard. System safety engineering identifies latent and active failures to reduce the potential of a hazard. Latent failures are related to system/subsystem deficiencies which may be undetected. An active failure is the result of failing to detect system/subsystem deficiencies in which case, the possibilities of an incident or accident are likely. Although latent and active failures occur in system and subsystems, the result of the failure could have an effect on the human. Thorough analysis techniques should be utilized to disclose the risk associated system and subsystem failures and their possible effects on the human. Through the use of fault tree analysis, Failure Modes and Effects Analysis (FMEA), and other related system safety analysis techniques and philosophies, the program and customer have a realistic view of potential risks on the system, subsystem and the human.

During the product definition phase, a SSPP should be utilized to outline the system safety organization plan on managing safety related program risks. The customer should provide the safety requirements to the contractor so that the contractor has an understanding of how the safety requirements may affect the system design. The system safety engineer should determine the best methods to meet those requirements. During the PDR, the system safety engineer provides a preliminary evaluation of the system/subsystem risks during certain phases of flight; risks associated with system human integration are also identified. It is essential that the system safety engineer analyzes the risk from the system component and aircraft level, in order to understand the impact of the risk for the proposed design configuration. From a military perspective, MIL-STD 882 (DOD Standard Practice for System Safety) is used to determine the levels of risk associated with certain designs and operations. ARP-4761 (Guidelines and Methods for Conducting the Safety Assessment Process on Civil Airborne Systems and Equipment) is normally utilized for commercial aircraft or systems. The techniques utilized for identifying risks may be different, but the philosophy of risk identification and mitigation should be transparent and support customer and program goals.

Classically, a preliminary risk assessment starts with the identification and description of the risk. This provides the program and customer with an understanding of the issue. For example, the risk description may be characterized as: 'Possible insufficient installation of LRUs' (Risk Identification #1). All risk descriptions in the preliminary risk assessment should be related to the proposed design configuration and must be clear and concise. The system safety engineer must have a clear understanding of the system description and the system function in order to proceed to the next step of the preliminary risk assessment. The next step of the assessment is to identify the causal factors and the effects of the risk. In this case, the causal factors could be related to the design of the LRUs and the integration in to the aircraft. The effect may be the LRU falling off an equipment support rack, leading to maintainer injury. As a result of the LRU falling off the rack and

depending on the LRU location on the aircraft, the potential for inhibiting egress operations for the operator and passengers may exist. Next, the design mitigation strategy and any residual risks are identified. In this case, the design mitigation may be to re-design or relocate the LRUs to ensure the LRU installation configuration is optimized. To ensure the design is optimized, engineering drawings and CATIA models will need to be modified. Customer and program management must understand the engineering change and the potential effects on the mission. Any residual risks (risks left after design mitigation) should be identified and communicated with the customer at PDR. Finally, a preliminary hazard risk index should be provided to identify the probability and severity associated with the hazard. Figure 4 illustrates a typical preliminary risk assessment table outline.

All new or modified risks should be identified as the design develops. As the design matures, certain risks may need to be further communicated with the contractor, program management and the customer prior to CDR. For example, certain risks that could affect mission capabilities or relatively high risks such as radiation hazards should be effectively communicated. It is the responsibility of program management to adhere to recommendations provided by system safety engineering and incorporate certain design features to reduce the potential risk of a hazard. As the program transitions in to test and operations, a thorough evaluation of the hazards should be communicated through a series of safety assess-

Figure 4. Typical preliminary risk assessment

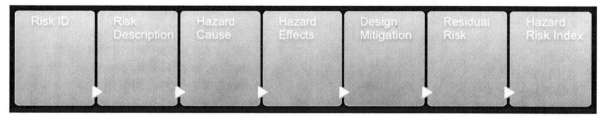

ment reports. The reports generated should be inclusive and provide conclusive evidence for the reporting of potential risks. Any analysis relative to potential system level, aircraft level or personnel hazards should be documented and monitored through a tracking system. It is up to the contractor to effectively monitor and provide updates to the customer regarding new or modified hazards.

The overall goal of safety management is to ensure that incidents and accidents do not occur within the lifecycle of the product. By implementing a product management safety perspective on system design, the user is provided with the necessary mitigation techniques in order to maintain safety in maintenance and flight operations.

Throughout this chapter, the framework for understanding the need for Human Factors design integration perspectives in the early stages of system design have been provided. There has been the discussion of the need to provide safety management perspectives in the early stages of design in order to reduce the potential for hazards in the maintenance and flight operations environment.

Theoretical perspectives in human factors and safety are important facets in understanding the core of user-system compatibility. However, theoretical perspectives are not the only methods of communicating issues in the industry relative to human factors and safety. Illustrating practical solutions provide the industry with a clearer understanding of the issues pertaining to product system safety and system usability. Three case scenarios related to human factors and system safety engineering are provided below. These case studies will provide the reader with realistic approaches to integrating the human with the system design effectively while providing safety considerations needed to optimize human performance. The case studies are organized as follows:

- Issue
- Effect
- Engineering Analysis
- Recommendations

Figure 5. Crew Station Prototype, National Institute of Aviation Research (NIAR) Laboratories

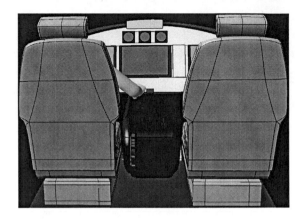

CASE STUDIES

Case Study# 1

Issues

- The flight crew's inability to reach and read display due to seat configuration (95th percentile male is 1.17 inches away from display and the 5th percentile female is 4.68 inches away).
- Crew's inability to respond to warnings and cautions due to the seat configuration. Program management is concerned with the ability to meet the aircraft transport category airworthiness standards in 14 CFR 25.1322 (flight crew alerting-warnings, cautions, and advisory lights).
- Need to ensure that the design eye angle position is compatible with user display interface in order to effectively respond to the warning and caution lights. This is also needed to meet the requirements set forth by 14 CFR 25.1321 (arrangement and visibility)
- The display characters and color scheme needs to be evaluated to ensure user system interface compatibility.
  ○ Need to evaluate the flashing rate of the warnings and cautions on the display

Effects

- May affect the crew's response in the critical phases of flight
- Mission critical and safety critical
- Could affect flight deck certification requirements
- Increased workload

Engineering Analysis

- Determine applicable methods to integrate the human with the system and to meet safety/flight deck certification requirements (Figures 6 and 7).

Case Study #1 Recommendations

- A physical hardware engineering analysis is needed to ensure user system interface compatibility. Change seat position or crew station to accommodate the 5th and 95th percentile population. The seat position change should provide the ability for both percentiles to reach and read the display. If this analysis is not sufficient, HFE many need to provide an analysis based on operator population boundary cases.

- Change seat geometry to accommodate the 5th and 95th percentile population posture. The seat change is needed to ensure the crew members can respond to warnings and cautions annunciations. This change will reduce the system safety risk and increase the likelihood of meeting flight deck certification and airworthiness requirements.

- A cognitive engineering analysis is needed for the Flight Management System (FMS). The display needs to be re-designed to ensure crew members can read the information and interpret the data effectively to ensure mission success. If a software change is not immediately available during flight test operations, a temporary bulletin will need to be implemented in to the FCOM. This is needed to assist the crew members with an alternate means of understanding how to set up the flight plan via the FMS during flight test.

- Need to relocate the FMS system from the throttle quadrant area to an optimal viewing angle to accommodate the human vertical and horizontal visual fields. This anal-

Figure 6. Case Study 1: Sevillian, Dujuan (Ph.D. candidate-Cranfield University, UK) Crew Station Human Modeling [Vision Evaluation]-NIAR Laboratories

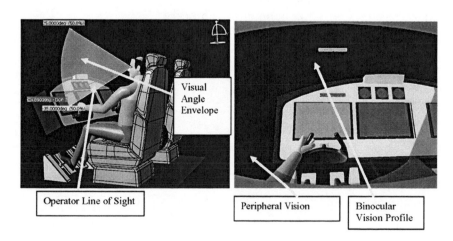

Figure 7.

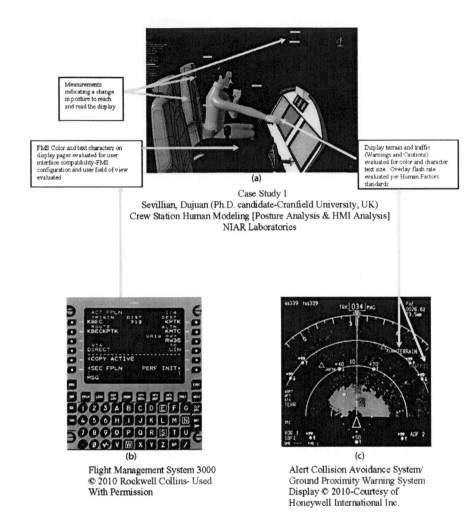

Case Study 1
Sevillian, Dujuan (Ph.D. candidate-Cranfield University, UK)
Crew Station Human Modeling [Posture Analysis & HMI Analysis]
NIAR Laboratories

(b)
Flight Management System 3000
© 2010 Rockwell Collins- Used
With Permission

(c)
Alert Collision Avoidance System/
Ground Proximity Warning System
Display © 2010-Courtesy of
Honeywell International Inc.

ysis is needed to ensure compatibility with user system view angle and to ensure the user can see the FMS system at the optimal viewing distance.

- HFE needs to provide software engineering with standards for display character size and color logic. Also, provide software engineering with display overlay flash rate standards. Last, HFE needs to evaluate the feasibility of integrating aural warnings with the textual terrain and traffic visual annunciations to provide the user with an increased situational awareness.

Case Study #2

Issues

- Need to ensure the design eye position is evaluated with new seat design, also need to ensure the operator can still see the display
- Need to ensure the applied weight/cushion compression on the seat does not affect the eye position or the seat reference point

Effects

- Inability to meet safety/certification requirements. Crew members may not be able to respond to an emergency effectively
- Safety and mission critical issue
- Could affect the program schedule

Engineering Analysis

- Determine most effective solution to comply with certification and safety requirements (Figures 8 and 9).

Case Study #2 Recommendations

- Ensure seat length and width corresponds with the human physical dimensions. This is needed to ensure compliance with MIL-STD 850.
- Ensure that the change in seat geometry does not affect the eye position or the ability to read and interpret information on the aircrew display.

- Ensure seat geometry change does not affect the ability to meet certification and airworthiness standards.
- Perform a simulation study to determine human body movement limitations within the seat configuration.

Case Study #3

Issues

- Flight crew posture and stature inadequate for flight test operations
- Human must remain strapped in lap belt in order to effectively reach rudder pedals but cannot reach pedals - (5th and 95th percentiles issue)

Effects

- Inability to reach rudder pedals
- Possible knee contact/strike with crew station
- Safety and mission critical issue

Figure 8. Case Study 2: Seat Schematic (MIL-STD 850-Aircrew Station Geometry for Military Aircraft)

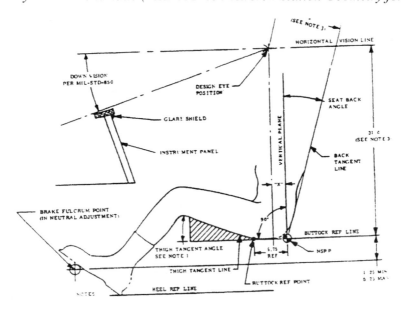

Figure 9. Case Study 2: Sevillian, Dujuan (Ph.D. candidate-Cranfield University, UK) Crew Station Human Modeling [Seat Analysis-Wireframe Configuration] NIAR Laboratories

Engineering Analysis

- Determine most effective way to accommodate crew and meet MIL-STD 1472 and the FAA Human Factors Design Standards (HFDS). Need to utilize MIL-HDBK 759C for anthropometric dimensions (Figures 10 and 11).

Case Study #3 Recommendations

- Ensure that the dimensions of the human leg and foot integrate with the design of the rudder pedal assembly.
- Ensure the posture of the human is not affected due to the change in rudder pedal configuration.
- Ensure the ability to move seat lateral and vertical to reach rudder pedals. This is needed to accommodate and optimize human abilities to operate the systems within the flight deck.
- Ensure the design configuration does not induce the possibilities of crew member's knees striking the flight deck control panel due to the new engineering design.

- Reduce the system safety risk to an acceptable level for the program and customer.

SOLUTIONS AND RECOMMENDATIONS

This chapter discussed several methods on human integration philosophies and strategies related to system design. Ensuring effective product safety management is an integral aspect in system design. Risk management is a systematic and collaborative effort between program management and the customer which can lead to safe maintenance and flight operations. Implementing effective design solutions provides a practical approach to integrating the human with the system. Understanding program requirements is essential to program success. The program must understand the effects of the requirements on system design. These requirements affect the system design and ultimately the user in their prospective environments. Processes and procedures help facilitate robust strategies for system design management. However, this is not an easy task and it requires a complete understanding of the systems engineering process and how the process interfaces with HFE and safety. Whether the program goal is to

Figure 10. Case Study 3: Sevillian, Dujuan (Ph.D. candidate- Cranfield University, UK) Crew Station Human Modeling [Rudder/Knee Strike Analysis, Vision, Seat Analysis] NIAR Laboratories

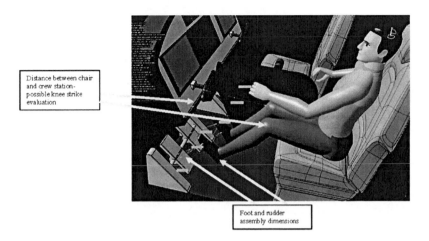

Figure 11. Case Study 3: Sevillian, Dujuan (Ph.D. candidate-Cranfield University, UK) Crew Station Human Modeling [Rudder and foot dimensions analysis/knee strike analysis, seat analysis] NIAR Laboratories

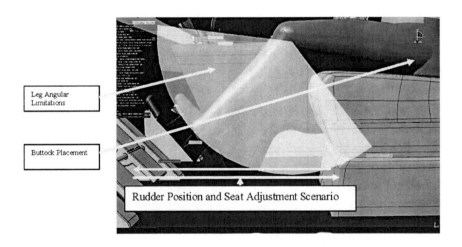

implement a HSI program or a HFE program, the program scope and cost dynamics must be further researched to determine the most effective way to integrate the human with the system. In view of the fact that the overall goal of HFE is to reduce the possibilities of human error in design, strategies must focus on influencing system design concepts with a desire to optimize human system interfaces at the beginning of the program acquisition. In

theory, there are seven general recommendations for effective human integration in system design that are recommended. The seven recommendations are as follows:

- Understand the design concept
- Understand the requirements
- Understand the system functions
- Identify the user system interfaces

- Analyze the user system interfaces
- Provide effective human engineering design solutions
- Understand the impact of system design on the user operational environment

FUTURE RESEARCH DIRECTIONS

There is a perpetual need to discover new and more effective methods for integrating the human with the machine and the environment. Research in the field of cognitive engineering appears to be a primary focus within the field of Human Factors Research and Development (HF R&D). New aircraft (unmanned and manned) are continuously being developed and the need to ensure user system compatibility is still a prominent focus.

Pivotal changes in software technology development, such as automation continue to affect human performance in aviation. There have been an abundance of airline and military incidents/accidents related to automation in the flight deck. On June 30, 1994 an A330-300 crashed in Toulouse, France killing seven people. One of the causal factors was the issue related to the user interface with the automation within the flight deck. Automation technology has been a factor influencing the ability for crewmembers to effectively communicate issues within the flight deck. There is a constant need to understand how new automation technology affects Situational Awareness (SA). In particular, CRM in flight deck operations continues to be an issue as more technology is developed for the user community. Questions are asked often by operators, engineers, and scientists regarding automation technology, and the impact on the user interface. 'Does more automation distract the operator'? 'Does more automation provide the operator with the ability to monitor the system better'? Dr. Micah Endsley provides theoretical approaches that can be utilized to understand the impact of automation on the operator interface. According to Endsley, (2000),

there are three levels of SA. The three levels are Perception (Level 1 SA), Comprehension (Level 2 SA), and Projection of the Future Situation (Level 3 SA). Understanding these levels provide the framework for determining how to design certain systems that are compatible for the user while increasing SA. There is a constant need to research the impact of new software automation technology on the operator interface during the development and design phases of aircraft systems.

Another focus in HF R&D has been Virtual Reality (VR) and Augmented Reality (AR) applications. Human Factors research in VR & AR environments has provided techniques and strategies to increase human performance. For example, the ability to convey HCI influences on human performance in virtual environments provides the HFE with an understanding of the factors that should be addressed prior to product concept. According to Stanney (1995), an understanding of human sensory and motor physiology is the foundation for understanding the compatibilities and incompatibilities of the human in virtual environments. Human Factors issues such as: Haptic, visual, and auditory perception impact system design in virtual environments (Stanney, 1995). These factors are continuously investigated and are used to determine the impact on human performance.

The book "Technology Engineering and Management in Aviation: Advancements and Discoveries" provides the blueprint for understanding 'how' aviation and aerospace challenges should be managed and executed. Essentially, this chapter illustrates the constant need to effectively manage the human system interface while providing indispensible solutions to development and design.

CONCLUSION

It is common knowledge that air transportation safety is a legitimate reason to design aircraft systems that do not impinge on the user's abilities to perform effectively in particular environments. Designing aircraft systems with the 'Human Element' in perspective provides the foundation for safe maintenance and flight operations. Whether performing maintenance on a landing gear assembly or flying from New York to Japan, it is important that mechanics and pilots have a suitable user interface that enables them to have the ability to effectively communicate and perform their job duties with the highest degree of safety. Therefore, aircraft systems must be developed to optimize the user interface. Expectations of incidents and accidents are likely to occur throughout the lifecycle of a product. However, system designs that optimize human performance can lead to a reduction of human error. Early detection of human performance errors in aircraft design leads to safe and effective maintenance and flight operations.

REFERENCES

Abbott, D. W., & Wise, M. A. (1999). Underpinnings of system evaluation. In Kantowitz, B. H., Daniel, J. A., & Garland, D. J. (Eds.), *Handbook of aviation human factors* (pp. 51–52). Mahwah, NJ & London, UK: Lawrence Erlbaum Associates, Publishers.

Blakey, M. C. (2007, February 15). *Testimony – Statement of Marion C. Blakey.* Retrieved July 19, 2010, from http://www.faa.gov/news/testimony/news_story.cfm? newsId=8184

Chapanis, A. (1996). *Human factors in systems engineering.* New York, NY: John Wiley & Sons.

Defense Acquisition University. (2010, May 17). *Community browser.* Retrieved July 19, 2010, from www.acc.dau.mil/communitybrowser.aspx? id=314774& lang=en-US

Endsley, M. (1999). Situation awareness in aviation systems. In Kantowitz, B. H., Daniel, J. A., & Garland, D. J. (Eds.), *Handbook of aviation human factors* (pp. 257–259). Mahwah, NJ: Lawrence Erlbaum Associates, Publishers.

Gersh, J. R. (2005). *Cognitive engineering: Understanding human interaction with complex systems.* Baltimore, MD: John Hopkins APL.

Matthews, R. (2004, November). Ramp accidents and incidents constitute a significant safety issue. *ICAO Journal*, 4–25.

McSweeney, J., & Pray, B. C. (2009). *Integration of human factors engineering in to design–An applied approach.* Houston, TX: ABS.

National Highway Traffic Safety Administration. (2008). *Home.* Retrieved July 19, 2010, from http://www-fars.nhtsa.dot.gov/Main/index.aspx

National Transportation Safety Board. (1997). *Bus collision with pedestrians.* Washington, DC: National Transportation Safety Board.

National Transportation Safety Board. (2003). *Derailment of Northeast Illinois regional commuter railroad train.* Washington, DC: National Transportation Safety Board.

National Transportation Safety Board. (2007). *Attempted takeoff from wrong runway: Comair Flight 5191.* Washington, DC: National Transportation Safety Board.

National Transportation Safety Board. (2010). *Loss of control on approach, Colgan Air, Inc., operating as Continental connection flight 3407.* Washington, DC: National Transportation Safety Board.

National Transportation Safety Board. (2010, April 8th). *Press release.* Retrieved 19 July, 2010, from http://www.ntsb.gov/pressrel/2010/100408.html

Reinach, S. (2007). *An introduction to human systems integration (HSI) in the railroad industry.* Washington, DC: Federal Railroad Adminstration.

Sheridan, T. B. (1993). My anxieties about virtual environments. *Presence (Cambridge, Mass.), 2*(2), 141–142.

Shneiderman, B. (1992). *Designing the user interface* (2nd ed.). Reading, MA: Addison-Wesley.

Stanney, K. (1995). *Realizing the full potential of virtual reality: Human factors issues that could stand in the way* (pp. 28–33). Orlando, FL: IEEE.

Strauch, B. (2002). *Investigating human error: Incidents, accidents, and complex systems.* Brookfield, VT: Ashgate Publishing.

Wickens, C. D., Lu, L., & Gordon, S. (1998). *An introduction to human factors engineering.* Prentice Hall.

Wiegmann, D. S. (2004). *HFACS analysis of military and civilian aviation accidents: A North American comparison.* Oklahoma City, OK: ISASI.

Williams, K. W. (2006). *Human factors implications of unmanned aircraft accidents: Flight control problems.* Oklahoma City, OK: FAA Civil Aerospace Medical Institute.

ADDITIONAL READING

Air Traffic control Communications Simulation and Aircrew Training. *Proceedings of the Royal Aeronautical Society, November 5-6, 2003.* London: The Royal Aeronautical Society.

Alexander, A. L., & Wickens, C. D. (2005). *3D navigation and integrated hazard display in advanced avionics: performance, situational awareness, and workload (AHFD-05-10/NASA-05-2).* Savoy, IL: University of Illinois, Aviation Human Factors Division.

Caldwell, J. A., & Caldwell, J. L. (2003). *Fatigue in aviation.* London: Ashgate Publishing.

Chapanis, A., & Budurka, W. J. (1990). Specifying human-computer interface requirements. *Behaviour & Information Technology, 9,* 479–492. doi:10.1080/01449299008924261

(1999, June). Does CRM Training Improve Teamwork Sills in the Cockpit?: Two Evaluation Studies. *Human Factors, 41*(2), 326–343. doi:10.1518/001872099779591169

Endsley, M. R. (2000). Theoretical underpinnings of situational awareness: A critical review. In Endsley, M. R., & Garland, D. J. (Eds.), *Situational Awareness: Analysis and measurement* (pp. 3–32). Mahwah, New Jersey: Lawrence Erlbaum Associates.

Flight Simulation. (2005). *Virtual Environments in Aviation.* Ashgate Publishing, Ltd.

Grady, J. O. (1993). *Systems Requirements Analysis.* New York: McGraw-Hill.

Grose, V. (1987). *Managing risk: Systematic loss prevention for executives.* Upper Saddle River, NJ: Prentice Hall.

Harris, D, & Muir, H. (n.d). *Human factors and aerospace safety: An international journal* [multiple volume and dates]. London: Ashgate Publishing.

Hawkins, F. (1992). *Human factors in flight.* Brookfield, VT: Ashgate Publishing.

Human Error in ATC System Operations. (1980). *Human Factors,* 535–545.

Jensen, R. (1989). *Aviation psychology.* Brookfield, VT: Gower Publishing.

Johnston, N., Fuller, R., & McDonald, N. (1995). *Aviation psychology: Training and selection.* Brookfield, VT: Ashgate Publishing.

Johnston, N., McDonald, N., & Fuller, R. (Eds.). (1994). *Aviation psychology in practice*. Brookfield, VT: Ashgate Publishing.

Lee, J., & Moray, N. (1992). Trust, control strategies and allocation of function in human-machine systems. *Ergonomics, 35*, 1243–1270. doi:10.1080/00140139208967392

Maurino, D., Reason, J., Johnston, N., & Lee, R. (1995). *Beyond aviation human factor*. Brookfield, VT: Ashgate Publishing.

McDonald, N., Johnston, N., & Fuller, R. (1995). *Applications of psychology to the aviation system*. Brookfield, VT: Ashgate Publishing.

National Research Council Committee on Vision. (1983). *Video Displays, Work and Vision*. Washington, DC: National Academy Press.

Norman, S. D., & Orlady, H. W. (Eds.). (1989). *Flight deck automation: Promises and realities. (NASA conference Publication 10036)*. Moffett Field, CA: NASA Ames Research Center.

Nunes, A. & Wickens. C.D. (2005) Clutter effects in CDTI. Unpublished manuscript Podczerwinski, E., Wickens, C.D., & Alexander, A.L. (2002). Exploring the "out of sight, out of mind" phenomenon in dynamic settings across electronic map displays (ARL-01-8/NASA-01-4). Savoy, IL: University of Illinois Aviation Res. Lab.

O'Hare, D., & Roscoe, S. (1990). *Flightdeck performance: The human factor*. Ames: Iowa State University Press.

Proctor, R., & VanZandt, T. (1994). *Human factors in simple and complex systems*. Needham Heights, MA: Allyn and Bacon.

Reason, J. T. (1997). *Managing the risks of organizational accidents*. London: Ashgate Publishing.

Roscoe, S. (1980). *Aviation psychology*. Ames: Iowa State University Press.

Salas, E., Bowers, C., & Edens, E. (2001). *Improving teamwork in organizations*. Lawrence Mahwah, NJ: Earlbaum Associates.

Salas, E., & Maurino, D. (Eds.). (2010). *Human factors in aviation*. New York, NY: Academic Press.

Salvendy, G. (1987). *Handbook of human factors*. New York, NY: Wiley and Sons.

Sanders, M., & McCormick, E. (1993). *Human factors in engineering and design*. New York, NY: McGraw-Hill.

Shappell, S. A., & Wiegmann, D. A. (1997). A human error approach to accident investigation: The taxonomy of unsafe operations. *The International Journal of Aviation Psychology, 7*, 269–292. doi:10.1207/s15327108ijap0704_2

Simonds, R. H., & Grimaldi, J. V. (1989). *Safety management: Accident cost and control* (4th ed.). New York: McGraw-Hill Professional.

Smith, S. L. (1986). Standards versus guidelines for designing user interface software. *Behaviour & Information Technology, 5*, 47–61. doi:10.1080/01449298608914498

Strauch, B. (2002). *Investigating human error: Incidents, accidents, and complex systems*. Brookfield, VT: Ashgate Publishing.

Thygerson, A. L. (1992). *Safety (Rev)*. Boston: Jones & Bartlett Publishers.

Wagenaar, W. A., & Groeneweg, J. (1987). Accidents at sea: Multiple causes and impossible consequences. *International Journal of Man-Machine Studies, 27*(5-6), 587–598. doi:10.1016/S0020-7373(87)80017-2

Wickens, C., & Hollands, J. (2000). *Engineering psychology and human performance* (3rd ed.). Upper Saddle River, NJ: Prentice-Hall.

Wickens, C. D. (1992). *Engineering Psychology and Human Performance* (2nd ed.). New York: Harper Collins.

Wickens, C. D., Gempler, K., & Morphew, E. (2000). Workload and reliability of predictor displays in aircraft traffic avoidance. *Transportation Human Factors.*, *2*(2), 99–126. doi:10.1207/STHF0202_01

Wickens, C. D., Merwin, D. H., & Lin, E. L. (1994). Implications of graphics enhancements for the visualization of scientific data: dimensional integrality, stereopsis motion, and mesh. *Human Factors*, *36*(1), 44–61.

Wiener, E., & Nagel, D. (Eds.). (1988). *Human factors in aviation*. San Diego, CA: Academic Press.

Wiener, E. L. (1989). *Human Factors of Advanced Technology ("Glass Cockpit") Transport Aircraft. (NASA contractor Report 177528)*. Moffett Field, CA: NASA Ames Research Center.

Woodson, W. E., Tillman, B., & Tillman, P. (1992). *Human Factors Design Handbook* (2nd ed.). New York: McGraw-Hill.

Yacavone, D. (1993). Mishap trends and cause factors in Naval aviation: A review of Naval Safety center data, 1986-1990. *Aviation, Space, and Environmental Medicine, 64*, 392–395.

KEY TERMS AND DEFINITIONS

CDR: The Critical Design Review (CDR) is the final review before the production stage of the system. The purpose of this review is to compile all of the details regarding software and hardware configurations.

FAA: Federal Aviation Administration (FAA)-U.S. Government agency charged with developing regulatory standards providing regulatory oversight to the aviation aerospace industry.

HFE: Human Factors Engineering is the application of human capabilities and limitations as it relates to the design of systems.

HIS: Human Systems Integration is the systematic process of integrating the human with the system so that the human has the ability to operate, maintain, and support personnel needs within the applicable environment.

Human Factors: Human Factors are the factors influencing the behavior of the human based on several dynamic critical interfaces.

IPM: The Information Processing Model (IPM) is a model that portrays how the human interprets a stimulus and responds to certain system inputs in a particular environment.

LRU: Line Replaceable Unit- A system designed to be removed and replaced expeditiously.

NTSB: National Transportation Safety Board (NTSB)-Agency charged by U.S. Congress to investigate all U.S. transportation related accidents.

ORR: The ORR is a review that determines the feasibility in proceeding to the concept exploration.

PDR: The Preliminary Design Review (PDR) is a review with the customer, program management, and various engineers to review the current design philosophy. The PDR includes the approach to various design configurations such as hardware component specification, and detailed requirements for future test evaluations.

SRR: The System Requirements Review (SRR) is conducted with the customer to understand how the requirements will be utilized to design the system.

SDR: The System Design Review (SDR) is a review developed to determine if the integrated system will meet the requirements.

TRR: The Test Readiness Review (TRR) is conducted to review the design integration as it relates to the ability for the system to operate effectively in the operational environment.

Chapter 4
Managing Human Factors in the Development of Fighter Aircraft

Jens Alfredson
Saab Aeronautics, Sweden

Rikard Andersson
Saab Aeronautics, Sweden

ABSTRACT

The purpose of this chapter is to describe important aspects of how to manage human factors in the development of fighter aircraft, by presenting prominent features of the domain, corresponding design concepts to cope with what is special, and guidance in the approach to managing human factors in the development of fighter aircraft.

The chapter includes a description of the domain of fighter aircraft development, followed by a section on how to design for what is special about a fighter pilot, a fighter aircraft, and the flight environment. Examples of design concepts from the domain, such as HOTAS and Basic-T are discussed in light of domain specific demands. Later, human factor considerations, such as evaluation of human factors and usable systems, are discussed together with how to manage human factors in development.

INTRODUCTION

This chapter is about how to manage human factors in the development of fighter aircraft. Due to natural limits in education and training or other ways of changing the user's characteristics, the long time evolution of technology to support the fighter pilot is central for future use of the fighter aircraft developed today, to become significantly better. A fighter aircraft has state of the art technology and subsystems that are designed to provide the pilot with large amounts of information and are ready to execute the pilot's commands swiftly and precisely. In addition, the pilot is highly capable, selected among the best and educated and trained over a long period of time. Frequently, there is a

DOI: 10.4018/978-1-60960-887-3.ch004

bottleneck between the competent system and the competent user, when large amounts of important information should be sent between the two in fast interaction. The question of how to manage human factors issues is therefore central to the development of fighter aircraft.

The objective of this chapter is to highlight important aspects of human factors and how they could be managed in the domain of fighter aircraft development. First, the domain of fighter aircraft development is introduced to readers who are not familiar with its characteristics, followed by a section describing what is special about a fighter pilot, a fighter aircraft, and the flight environment and some design concepts used in the domain. The latter part of the chapter concerns human factors and how human factors can be managed during development. Here, methods and techniques for human factors, as well as development, are discussed in the domain context of fighter aircraft development.

THE DOMAIN OF FIGHTER AIRCRAFT DEVELOPMENT

The domain of fighter aircraft development is special in many ways. For instance, it is extremely important to design for performance. As for many other domains, there is a professional user, not a consumer using the product for pleasure. As many other professional users, pilots have defined purposes regarding their use of a system. The context of a military system is that it will be used by professional military in extreme situations where failure could be devastating. For military pilots, it is extremely important to perform better than opponents and to be able to make fast decisions and execute them accurately. Therefore, the most important parameter for successful fighter aircraft design is performance, together with safety. Only a highly performing system, which is safe for the user, can meet the demands from the fighter aircraft domain. It is less important, for instance,

for the system to be inspiring or fun to use, even if motivation is also important for support in a fighter aircraft.

As stated above, safety is also very important. Safety critical systems in a fighter aircraft have to be robust and work all of the time. Despite redundancy being built in when possible, it has to be extremely rare for a subsystem to malfunction. Additionally, tactical systems that are not directly linked to flight safety have to work because of the potential hostile and threatening situation that the fighter aircraft has to perform in.

User involvement is important in the development of many systems, but perhaps even more important in the development of fighter aircraft than in many other systems. One reason for this is that the cockpit, its use, and corresponding experiences are not very easily available compared to other domains. The designer of a car is probably also a driver of a similar car to and from work, or the designer of a telephone probably makes some calls occasionally on a similar phone, etc. Since it takes a long time to become an experienced pilot and a long time to become an experienced designer with technical and methodological skills, it is not always possible to find these competences combined in one designer. When that is the case, it quite often the own fighter aircraft experience is based on yesterday's systems rather than the systems of today. Therefore it is often wise to include current users in all phases of the design of a fighter aircraft.

In addition to user involvement, it is desirable to have continuous broader customer involvement. For a consumer product it is common that the customer and the user are the same person, but in the domain of fighter aircraft there are actually various kinds of users and customers (often different roles and individuals within large organisations). It is important that the human factors development manages this wide spectrum of needs and expectations.

Due to the length of time a fighter aircraft is going to be in service, and partly due to the long development loops, it is very important to study not just current conditions regarding state-of-the-art technology, tactics and use of today etc., but rather future conditions, even if they are not fully known now, since they are often more valid conditions than the current operational ones. When doing so, the developer will encounter the envisioned world problem and try to figure out what is relevant for the future (Dekker, 1996; Dekker & Woods, 1999). This is an important consideration for managing human factors in the development of fighter aircraft.

There has been a long-term trend in the domain of fighter aircraft development, spanning several generations of fighter aircraft, towards changing the pilot's role in the aircraft. The trend has been for the pilot to spend less time and effort on basic piloting and more time and effort as an advanced tactical decision maker. Here, the development of automatic systems such as autopilots and advanced flight control systems has played a role in lowering those workload demands, opening up for other tasks. At the same time, the sensors, weapons systems and other tactical systems have evolved to be more competent, resulting in more tactical information for the pilot to handle, more than enough to keep the pilot mental workload extremely high at the most intense moments.

A long-term trend has been that more and more technology has been added between the pilot and what is controlled. Due to better performance of sensors that can detect at a longer range, and weapons that reach further, the pilot is physically separated more from the things that are controlled. However, it is not so much the physical distance in itself (even though the fact that the pilot, for instance, has to look at displays to perceive the position of other aircraft so far away that the pilot's eyes can not detect them directly) but rather the mental distance that has changed the working conditions for a pilot. The pilot has to use new mental models to cope with the new situation. In fact, the mental distance that the modern fighter aircraft system has created calls for design con-

siderations, for instance, supporting the creation of pilot mental model, mental anchoring, and decision support.

How to help the pilot create a useful mental model? Note, that the mental model does not have to be "correct", or it may not even be desired for the created mental model to have a one-to-one correlation with the real world since the most important criterion for the mental model is to help the pilot interact efficiently for better performance. A simpler and less detailed model may be more useful, and a consistent mental model may be easier to use. For instance, if there are delays in systems that are calculating things to be displayed alongside each other, or together with the real word, such as for head up displays and helmet mounted displays, it might be easier to regard the events as synchronised by the pilot. Another example could be a directional cue for the pilot to pull up, which might be easier to use if it directs the pilot to the shortest time to get away from the ground, instead of the actual gravitational up.

Also, it is important to be consistent in the support of mental models for the pilot so that the pilot does not have to execute unnecessary changes between different mental models. For instance, if some information (in for instance, a visual display) supports the mental model that the pilot (aircraft) is fixed and the surrounding world is moving, sometimes referred to as "inside-out", it is easier for the pilot to interact with other information presented according to the same principle than if some information supports the mental model that the pilot (aircraft) is moving in a fixed world ("outside-in").

When the system is complex, the pilot also needs assistance not just to understand and control the environment, but also assistance to understand and control the technical system itself. Here, it is important to help the pilot navigate his mind through the system model quickly and safely.

Moreover, the pilot's interaction can be aided by filtering information in order to reduce information overload. Another useful approach is to aggregate information and tasks to be able to handle more at less mental cost, such as the use of high level commands and fused and clustered information.

The trends described above impose greater need for support, in order for the pilot to be able to make good decisions out of the plentiful information that is available. Luckily, a simultaneous trend is taking place towards more and better supportive available technology. Here, the human factors' challenge is to combine needs and opportunities into a useful decision support system. Not only does the decision support system have to perform its task properly, it is also important that the right tasks are being performed. Therefore, it is essential to achieve a well balanced division of tasks between human and technology.

For a long time, the aviation domain has sorted out which tasks are best performed by man and which are best performed by machine, and lists of suitable tasks for the two parts have been around for a long time (Fitts, 1951). Function allocation, based on what humans are best at and what automation is best at, are a necessary but not sufficient condition for effective decision making in the complex context of fighter aircraft scenario and use. On top of that, the developer has to regard the cooperation between the decision support system and the deciding pilot, so that they can complement each other in the decisions making, resulting in the pilot and decision support working together effectively, referred to by Hollnagel (1999) as function congruence, when shifting focus from substitution to cooperation. A useful way of thinking about this is the joint cognitive systems approach (Hollnagel, 2007; Hollnagel & Woods, 1983; 2005; Woltjer, 2009; Woods & Hollnagel, 2006).

DESIGNING FOR WHAT IS SPECIAL

To successfully manage human factors in the development of fighter aircraft, it is important to understand how design criteria are formed.

The design criteria have to regard what is special about the fighter pilot, the fighter aircraft and the flight environment. Several design criteria are then merged into design concepts taking into account technical prerequisites, intended use, as well as user needs.

There is a design space where available technology can be formed into usable systems in a design process. What technology is currently available, or available post-development, is thereby one important consideration. Technical prerequisites are formed by the state-of-the-art in technology and the availability of that technology including costs and practical aspects.

Another important consideration is the intended use of the system. All components are created with a purpose, more or less explicitly expressed in, for instance, formal descriptions of concepts of operation, requirement structures etc. It is desirable if the design supports the aircraft and its subsystems, in order to become capable of performing well for its intended use.

Apart from the technical prerequisites and the intended use of the system, it is important to regard user needs for the design to be satisfactory. The user has to be able to effectively interact with the system to be able to efficiently observe, orient, decide, and act, sometimes referred to as the OODA-loop after a model developed by Colonel John Boyd, fighter pilot in the U.S. air force. The pilot in a fighter aircraft often has to cope with a spectrum of high workload demands simultaneously, for instance, perceptual, cognitive, motor skills etc. and needs to have a system that regards the user's needs in order to perform optimally. For the aviation domain, the concepts of mental workload and situation awareness have been successful in describing aspects of the pilot's decision-making process. Wickens (2002) describes how the two concepts of mental workload and SA are intertwined and that, for instance, task management is related to both situation awareness, through task awareness, and to mental workload, since task management makes use of the limited mental resources available.

Good situation awareness is important for effective decision making (Endsley, 1995). Situation awareness can be regarded as a concept with three levels (Endsley, 1988; 1995). Endsley, Bolstad, Jones, and Riley (2003) and Endsley, Bolté, and Jones (2003) describe situation awareness oriented design as comprised of three components: situation awareness requirements analysis, situation awareness oriented design principles, and situation awareness measurement and validation, and provided design recommendations based to the designer. Jones, Endsley, Bolstad and Estes (2004) even developed a designer's situation awareness toolkit (DeSAT) to provide support to the designer through tutorials and application specific tools. However, more important than general design advice, is to possess a deep understanding of conflicting design criteria for a specific design task. It is important to know what is special.

It is not only technical aspects that are important for design. Just as an architect designs a building, an organisation could be designed as well (Simons, 2005). Even the design process itself can be designed. The technological design (e.g. component, relations, etc.) and the institutional design (e.g. responsibilities, allocation of costs, etc.) are both created by the higher level process design (i.e. designing the design process) (Koppenjan & Groenewegen, 2005).

Design criteria formed by technical prerequisites, intended use and user needs are then to be merged into design concepts resulting in a design through a design process. There are similarities in the design and development of various complex systems. General design knowledge can guide the design but a good design for fighter aircraft also has to regard what is special about this domain. When you know what is special, this will help to form relevant design concepts from the design criteria. For the domain of fighter aircraft development, it is important to be aware of what is special about a fighter pilot, a fighter aircraft, and the flight environment.

What is Special about a Fighter Pilot?

One striking thing about fighter pilots is that they are similar to each other in many respects. Fighter pilots are a relatively homogeneous group compared to users of, for instance, consumer products such as cars or commercial computer programs. There are no old users, nor are there very young ones. They have been carefully selected and all have good physical and mental capabilities, while they also belong to a population with certain anthropometrical characteristics. None is colour blind, and all had good visual performance when recruited. On top of this, they have undergone the same lengthy and demanding training that prepares them for life as a pilot, which also contributes to them demonstrating similar behaviour in a flight situation. Pilots tend to acquire skills in training that, although behaviour and performance between pilots may vary, often a consistent behaviour for each pilot is more of a rule.

From a design point of view, a homogeneous group of users is easier to design for. However, the extreme demands of a fighter pilot still make it very challenging to manage human factors for fighter pilots. In addition, although the group of fighter pilots is relatively homogeneous compared to other user groups, individual differences as well as differences between novices and expert fighter pilots, for instance, are important to regard in design.

What is Special about a Fighter Aircraft?

Below we highlight some factors that are special for a fighter aircraft that are of major importance for the management of human factors in fighter aircraft development. A modern fighter aircraft is special in itself, and there are numerous aspects to consider when designing. Important aspects to consider are:

- High G-loads
- Ejection
- Multi-role/Swing-role
- cost-COTS (Commercial Off-The-Shelf)
- Long lifecycle, including upgrades
- Graceful degradation
- Abundant information
- Autopilot and other automation

In some fighter aircraft scenarios, high G-loads occur frequently and limit the pilot's interaction possibilities regarding both display and control. It is important to consider this when developing hardware for display and control in cockpits for fighter aircraft. However, it is also of importance for software development since the interaction that a pilot has to perform under G-loads has to be adapted to the pilot's abilities at that time. For instance, a pilot subjected to high G-loads may experience changes in colour vision and general visual limitations, such as degraded peripheral vision etc. More importantly, however, is the pilot's limited ability to move the head, limbs, and body during high G-loads. The design concept of HOTAS (Hands On Throttle And Stick), which is described further in a later section of this chapter, gives the pilot high access functions at the fingertips, which is helpful during high G-loads.

Also, it is important to design hardware in the event of ejection, when the pilot has to leave the aircraft during an emergency. Here it is not the interaction at the time of the ejection that is the challenge, since interacting with the system is not a relevant concern for a fighter pilot ejecting. However, the possibility of an ejection influences the design by criteria for maximum weight on a helmet and its distribution, for instance, as well as the possibility for the pilot to separate from the aircraft without cables, displays or other hardware injuring the pilot in a possible ejection situation, which affects the available design space.

It is a challenge for many modern fighter aircraft to be designed for multiple roles and to be able to switch between roles between or during

missions. This is also a challenge for the design of the interaction with the pilot since the changes between roles have to be supported, but also since pilots with various experiences between roles have to be supported by the design.

The cost aspect for fighter aircraft has not been a key aspect historically compared to, for instance, consumer cars, where the cost of a small detail could heavily influence total profits. For fighter aircraft it has been more important to have a higher performing system than the opponent. However, the trend is towards lower life-cycle costs and towards the integration of COTS (Commercial Off-The-Shelf) into fighter aircraft to achieve cost effective solutions, potentially generating more "bang for the buck". For human factors' development, it is important to handle the possible limitations that a cost-COTS focus will provide.

Fighter aircraft have long lifecycles, including upgrades during that period of time. Human factors' considerations therefore have to cope with that. For instance, it is important that design solutions are not only relevant for today's technical solutions, but also comply with general interaction principles that are more likely to be valid in a future context. Also, the idea of resilient design is central for human factors' considerations from a long-term perspective.

A special concern for human factors' considerations is the design concept of graceful degradation, where it is still possible to perform, but at a quantitatively lower or quantitatively different level than intended. An example of this could be when a weapon system fails but the aircraft can still contribute by providing sensor information to members of the tactical air unit.

A fighter aircraft, with all its subsystems (sensors, weapons etc.) and communication abilities, provides the pilot with abundant information. Information overload is a key human factors' concern for the development of fighter aircraft.

As an aid for the pilot, there are several automated systems such as an autopilot and other automation. There are human factors' issues concerning the interaction between the pilot and the automated system, as well as important human factors' considerations concerning which automations shall be further developed and which new tasks are to be automated.

What is Special about the Flight Environment?

The flight environment of a fighter aircraft is extreme in several ways, as various types of contextual constraints together form extreme conditions. Three examples of important aspects in the flight environment of a fighter aircraft that are of importance for human factors considerations are:

- sudden and drastic light conditions and high visual demands
- demands on rapid decision-making in a battle of life and death

Sudden and drastic light conditions occur and high visual demands are salient for operating in a flight environment. A fighter aircraft is flown quickly in any direction and the position relative to the sun can change in split seconds. The pilot can get sunlight in his eyes directly, as well as through reflections in displays. The presented display information therefore has to be designed for a wide spectra of light conditions (e.g. night and day), as well as for sudden and drastic changes. Moreover, critical information is displayed in visual displays of fighter aircraft, so it is important that the pilot can see what is there.

In the flight environment, there are high demands on rapid decision-making. For a fighter pilot, many decisions must be made under extreme time pressure, in a hyper-dynamic setting where the stakes are high. These are all characteristics of naturalistic decision making (Klein, Orasanu, Calderwood, & Zsambok, 1993) that call for professional handling by a skilled user. However, the decision-making of a fighter pilot also includes emotional elements, since the fighter pilot can find themselves in a battle of life and death. It is important that human factors are managed so that all the important aspects of the pilot's flight environment are considered.

Design Concepts

When you have determined what is special, this will help you in forming relevant design concepts from the design criteria based on what is special, as discussed above. There are design concepts that are special for the domain, but others that have been adopted by other domains. For instance, there are special concepts for the use of a head-up display or head mounted display that are related to flight specific concerns, but other concepts can be transferred, for instance, to the automotive domain such as technology and visualisation design principles for head-up displays and night vision systems.

There has also been a continuous adaptation of design concepts to cope with technology evolution, such as special concepts for fly-by-wire and glass-cockpits. A review of usability design principles in the Swedish swing-role aircraft Gripen was presented by Wikforss (2008). Examples of design concepts used in fighter aircraft in general are:

- HOTAS
- Basic-T
- Don't need – Don't show

HOTAS is used to enable the pilot to have fast access to high access controls (for instance, important controls, frequent controls, or controls that are time critical for urgent control). Without having to move the hands from the throttle and the stick, high access functions can be controlled. However, in a fighter aircraft, because there are so many things you can do, some of the lower level access functions have often been placed in areas other than the throttle and stick, such as buttons on panels and displays, where the pilot could reach with a fingertip. There is an advantage of having high access controls in the same place, and to have that place on the throttle and stick, which in themselves are significantly important controls for a fighter pilot, is even better. The pilot may, at any time, quickly shift between tasks, and for instance take direct control of the aircraft when that is the top priority.

Since the mid-1900s, flight instruments in a cockpit have been placed in a basic-T arrangement which has been the standard arrangement for a long time. In the basic-T speed is found at the top left and altitude at the top right with attitude and heading in the centre. This arrangement of information has been retained even when traditional instruments were abandoned for modern displays. There is a huge advantage in having the same relative positions for flight information presented in the various aircraft that a pilot will use during training and independent of which aircraft type he[1] will encounter in his life. Additionally, eye-movement studies have been conducted for a long time to study how pilots look at flight instruments (e.g. Fitts, Jones, & Milton, 1951). Even though pilots have their own personal scanning behaviour, they are all scanning the basic-T.

Don't need – Don't show is a principle that has been used in Swedish fighter aircraft since electronic displays were introduced. Car manufacturers have subsequently adopted similar principles under the names "black panel" or "night panel" for driving in the dark hours. By adapting what is displayed to other external conditions, such as proximity to the ground or the amount of information available (risk for information overload), decluttering and filtering may help to lower the mental workload in critical situations. However, when doing this, it is extremely important to have conducted an extensive analysis, including a cognitive task analysis, to ensure that information intended to be less available, really ought to have lower priority in actual situations.

In addition to adapting what is displayed, you can also adapt how it is displayed, for instance forming high level representations (e.g. clusters or aggregated information) that can correspond to high level commands for control. Also, the adaptation not only has to depend on external conditions, such as the direction to the ground, the sun or an opponent, but could also be adapted to the pilot and his current state. Here, information about the pilot is useful for determining the current state, such as assessed mental workload or situation awareness of the pilot. Such information could be gained by various types of pilot monitoring systems that could include psychophysical measures of, for instance, heart rate, respiration, eye-movement etc.

Consequently, there are different levels of adaptation based on various kinds of information that could be included in the design. The first level of adaptation, which is often not even considered as an adaptation, is adapting the interaction to human abilities and limitations, which is central for human factors. This first level of adaptation is less controversial than the higher levels. An example of a higher level is the adaptation to an individual. Some functions are natural to adjust to personal specifics such as adjusting the sitting position, whilst others are less obvious, such as the layout of the displays, colour settings etc. Here, you have to balance the gain of the personal adjustment with the risk of, for instance, less efficient communication among a team when referring to the displays or considerations for homogeneous education and training. An even higher level of adaptation is not only to adapt to a general user, or a specific user, but to the current mental state of that user, such as the level of mental workload. In addition to the considerations for adaptation to a specific user, this level of adaptation also has to deal with the risk of misperceiving the current mental state, since the technology and methods of assessing online mental workload or situation awareness, for example, are not very mature. Furthermore, changing the behaviour or appearance of displayed information or control principles may cause inconsistency in the interaction that has to be balanced with the potential benefits of the adjustment to the actual mental state. It can also be challenging to identify applications that are better performed manually if the workload is low and better automated when the workload is high. If an automated function that is designed as an aid during high workload situations is well designed, this will often also be of help in low workload situations.

HUMAN FACTORS

Human factors for cockpit design are fundamental, since a good cockpit design is important for limiting human errors. Human error has been reported to be a major contributor to accidents and incidents (Nagel, 1988), even though it is important to be sceptical to error-counting in general, as explained by, for instance, Dekker (2003; 2005; 2007). Since aircraft accidents are costly in more than economic terms, it is important to minimise the likelihood of an accident, by striving for good cockpit design. Good human factors' management resulting in superior man-machine interaction minimises the risks for an accident. However, human factors are very important not only for minimising the risk for accidents, but also for maximising performance.

The concept of human factors has its tradition and roots in the aviation domain. However, modern fighter aircraft do not have that much in common with mid-1900 aircraft when it comes to interaction technology, contextual constraints etc. Therefore, the domain of human factors has to co-evolve together with the domain of fighter aircraft development to best support future fighter pilots in demanding situations. There are several definitions of the term 'human factors' contained in literature and this article is not intended to go deeper in that issue, since that is not the focus of the chapter. However, it is important for the reader of this chapter to realise that human factors issues are important for the pilot in a fighter aircraft in order to be able to best contribute to the overall system performance.

Usable Systems

Important roles in the design process for designing usable systems are the role of the usability designer and the user, in this case a fighter pilot.

Users are often regarded as a subsystem of the overall system by a designer, and these are sometimes modelled through design principles or elements of a scenario, depending on the types of meetings that are held throughout the design process (Darses & Wolff, 2006). The design outcome is affected not just by the kind of meetings, but also the design perspectives adopted by the designers in the design process (Hult, Irestig, & Lundberg, 2006). It is important to include pilots continuously in the design process. Singer (2002) has reported benefits from the use of part-task simulation early in a design process (for instance, for validating the cockpit design), when designing commercial aircraft cockpit systems.

Although some researchers claim that design ability is possessed by everyone (Cross, 1995) it is important to create a design process that integrates the roles of the user and the designer in such a way that their respective knowledge and experiences contribute best to the final design.

Instead of simply delegating the design decisions to the user or to ask the user and implement the explicit user needs, it is preferable to involve the designer and user together. It is important not to cut the designer out of the design loop and to make the designers care about what user remarks mean and put these into a future work context, as pointed out by Dekker and Nyce (2004). For instance, Smallman and StJohn (2005) reported a case of naïve realism, where users prefer analogue (photorealistic) symbols instead of artificial symbols, although experimental results demonstrated better performance when using abstract symbols.

To secure the development of usable systems, human factors for design integration have to be continuously evolved, taking research and own experience into account.

Human Factors for Design Integration

When designing a fighter aircraft cockpit, the designer, as described earlier, often needs support to gather experience in the actual use of the system, for various scenarios. To cope with this, the designer can find support in:

- User participatory design
- Designer's domain knowledge
- Standards

For user participatory design, experienced users may be actively involved throughout the design process. It is essential that this involvement includes participation during the design generation and is by no means limited to expert reviews of design suggestions. There are several methods that have proven effective when making the user part of the development, e.g. task analysis, cognitive walkthroughs, workshops, etc.

Since the designer is often prevented from using the real fighter aircraft during their time as a designer, the designer's domain knowledge must be obtained using other sources. One very effective source is the use of simulators, where the designer can get a feel of how the system is used. If the simulator is an advanced, high-fidelity simulator with a tactical environment, it can add complexity to the situation. The need for domain knowledge also stresses the fact that human resources must be managed with a long-term view, in other words, the designer must be allowed to work in the same domain for several years to be able to achieve the level of knowledge needed. One practical solution to this equation is to use advanced development simulators in the design process, both for generating good designs, but also for supporting the designer's evolution from novice to expert in domain knowledge.

Moreover, standards are important for managing human factors in the development of fighter aircraft. One important aspect of the use of stan-

dards in the design process is that the designer has less freedom, and although this is one of the reasons behind standardization, there may be backlashes regarding the ability to create designs that are innovative and groundbreaking. There is also the important issue of user expectations and de facto standards, which to a high degree influence the available design solutions. Design management therefore benefits from a focus on the rationale behind the standard, and is thereby prepared to question it.

When developing a product as complex as a fighter aircraft, there are always several parallel development activities taking place that affect the pilot's user environment. Therefore, it is important to ensure a total system design that is perceived, and functions, in a clear and consistent manner with effective and harmonised design.

Examples of available approaches useful for human factors' management of fighter aircraft development are:

- Style guide
- Prototyping
- Man-in-the-loop evaluations

A style guide can be used as a decision log where general design is noted and distributed throughout the organisation. Style guides should not be seen as a cook book that guarantees a usable system, but as a help to the designer and a complement to other usability activities.

For prototyping, concept and detailed design can be prototyped and evaluated. It is important to stress the fact that the prototyping activity must be performed across all the ongoing design activities, and it is considered as great risk to perform prototyping and evaluation without taking the entire system into account.

In order to achieve good design integration, it is of crucial importance to perform man-in-the-loop evaluations with actual pilots and using a model of the entire system in a realistic tactical scenario.

Evaluation of Human Factors

To successfully manage human factors, it is important to be aware of the human factors status of the system being developed. Just as a pilot has to observe, orient, decide, and act in an interaction loop with the system he is controlling it is important for the person developing a system to observe, orient, decide, and act on the design of the system. The interaction loop of a designer is much longer than the corresponding loop of a flying pilot, and the context in which decisions are to be made is fundamentally different. However, a designer still has to be situationally aware of the design situation. One important type of information for someone developing a fighter aircraft comes from various evaluations of human factors. When provided with information from human factors' evaluations, better design decisions can be made. Various methods and techniques are useful to varying degrees depending on which design decisions are to be made. Below, we highlight some important aspects of human factors' evaluation for the development of fighter aircraft.

As discussed previously, performance is a key concept for fighter aircraft. Due to the high importance of high performance, it is important to include performance measures in human factors' evaluations for fighter aircraft developers. This can be flight performance data, tactical performance data or other data. It is assuring for a designer to receive indications that human factors' measures will contribute to a higher performance. Design solutions that contribute to high performance could thereby be favoured in the design process. Given that in the evaluation context valid performance measures can be found, it is often wise to include them in an evaluation of human factors, especially if are easy to implement and analyze, which they frequently are. Examples of such performance indicators could be "kill ratio" in an air combat scenario, where the total system performance is measured, or performance measures for the pilot, such as "reaction time".

However, it is often not enough to measure performance. Firstly, performance measures are often limited to provide quantitative information rather than qualitative information. On other words, a performance measure often helps determine which design solution, among a few tested, could be proffered from a performance point of view, but seldom guides the designer to new design ideas nor provides much help in modifying the design in any particular direction.

Secondly, performance measures often lack in resolution. In other words, many performance measures are crude responding if there was a hit or a miss for example. However, it is often important to have high resolution data in order to distinguish between the effects of similar, but not identical, design solutions. Furthermore, in a complex scenario there are often factors other than the studied phenomenon that contribute to the performance, such as a long chain of more or less controlled events that ultimately lead to failure or success. The evaluation may therefore be more effective also if it includes a more sensitive and higher resolution measure, since the practical limitations of evaluations may restrict the number of data points available, thereby making it hard to draw conclusions from crude measures.

Thirdly, even if performance measures are "direct" measures of performance, they may be indirect measures of the studied design question. For instance, if the design goal is to increase spatial awareness for a pilot or to reduce workload, it would be a more direct approach to measure spatial awareness or workload, even if the intended benefits of better spatial awareness and lower workload could, indirectly, be noted in a performance measure.

For the reasons given above it is often wise, for the evaluation of human factors in fighter aircraft development, to include assessments of situation awareness and mental workload as complements of performance assessments. The concepts of mental workload and situation awareness not only complement the concept of performance. There are also numerous methods and techniques ready to use for mental workload and situation awareness assessment. The relationships between the concepts are well explored for the fighter aircraft domain (Castor 2009; Svensson, Angelborg-Thanderz, & Sjöberg, 1993; Svensson, Angelborg-Thanderz, Sjöberg & Olsson, 1997; Svensson & Wilson, 2002).

Some measures of situation awareness not only provide quantitative information but also qualitative information to guide the developer. The lack of situation awareness often occurs before (in time) there is a lack of performance, helping the developer link the measured dip to a relevant sequence of interaction activities, for example.

An increase in mental workload can sometimes be detected even earlier than a lack of situation awareness. Since many fighter aircraft scenarios induce high levels of mental workload on a pilot, it is often wise to strive for lower levels of mental workload, even if a level of mental workload that is too low is not optimal. There are several useful methods for evaluating mental workload, suitable for various purposes (Castor et al., 2003).

To collect the data needed for human factors' evaluation it is recommended that several methods and techniques are used simultaneously in order to "triangulate" and utilise the benefits of each method together. Subjective measures can thereby be used together with objective measures, combining the benefits of the two.

MANAGING HUMAN FACTORS IN DEVELOPMENT

Managing human factors in development is more art than it is science. Rather often, a developer comes across challenging design problems that are so specific that it is not easy to take a handbook from the bookshelf to find a solution. At first you might think that the work of the developer is therefore best performed by people who can find unique ways to handle every unique problem,

but in fact it is the other way around. It is very powerful to have structured tools for problem solving, sophisticated development processes and guidance from standards and style guides to ensure consistency in the design.

Alfredson, Oskarsson, Castor, and Svensson (2003) presented a meta-instrument for the evaluation of man-system interaction in development to support the process of defence materiel acquisition. The meta-instrument included a structured view of methods, to aid the dialogue in the process of materiel acquisition. However, no solution to a human factors' design problem is independent of other solutions. To manage human factors for a structured design, structured design approaches are needed. Here we want to highlight three approaches that are useful for managing human factors in the development of fighter aircraft. These are simulator based design, representation design, and cognitive budget.

Simulator based design has been used by the aviation industry since the 1970s for more cost-efficient designs, safer designs and for obtaining answers early in the design process, and has been well described in Alm (2007). Simulator based design is based on virtual prototyping and simulation and for human factors' issues, human-in-the-loop simulations, which can provide a flexible environment where design ideas can be tested quickly and at low cost. Regarding simulations, it is important not to confuse high fidelity with validity, but rather to strive for selected fidelity, and try to build a cost effective environment with a sufficient level of fidelity. It is also important that the activities support design decisions, for instance by comparing alternative designs relatively rather than striving for an absolute design goal. When conducting comparative experimental studies, it is important that the different designs are evaluated under the same conditions and the same simulator validity (Alm, Ohlsson, & Kovordanyi, 2005). Moreover, a strictly experimental approach, comparing two (or more) competitive design alternatives also has drawbacks. Dekker and Nyce (2004) address the difficulties presented by the fact that experimental steps into the future are often narrow, risking low generalisability, and the mapping between experiment and future reality can risk missing important aspects. It is important to be aware of this. The ideas of simulator based design are also applicable for other domains, such as the automotive area (Alm, Alfredson, & Ohlsson, 2008).

Representation design is a term introduced by Woods (1995). Woods (1995) describes how representation design has the potential to become the practice of designing marks in a medium to accurately portray relevant properties in the world. This way of thinking is very useful for a designer. Noting what is relevant rather than what is visually present in a situation is a practical consideration that is worth stressing in development (Albinsson & Alfredson, 2002; Howard, 2002). For a designer it is often useful to reflect on applicable frames of reference and forms of reference, to put data into context and to highlight significance.

Cognitive budget is an approach for optimising human factors' related efforts in development. When designing a fighter aircraft, there are several possible ways to achieve an optimised system that is superior to business competitors as well as enemies on the battlefield. Since this optimisation has to be done within economical constraints, the different solutions must be weighed against each other, i.e. should we integrate a new sensor or should we integrate a new long-range weapon in order to achieve the desired operational efficiency? etc.

The system components (sensors, weapons, propulsion, etc.) that constitute a fighter aircraft are, in most cases, possible to describe with objective parameters such as range, thrust, etc. The important component of human-machine interaction is more difficult to describe objectively however, and it is therefore hard, although extremely valuable, to evaluate the contribution of the human-machine integration with the total system's efficiency when optimising the system.

One approach to this is to assign human-machine integration design solutions with shares of a cognitive budget. The exact definition of the term and the different measurements, qualitative and quantitative, associated with it are still under practical construction and research. Examples of the assessments are time to decision and amount of mental workload. The use of cognitive budget is one important aspect when managing human factors' aspects in a fighter aircraft.

CONCLUSION

This chapter has highlighted key aspects for managing human factors in the development of fighter aircraft. The specifics of the domain of fighter aircraft development has been described with examples of what is special and design concepts to handle it. Specifically useful human factors' considerations were highlighted and recommended approaches for managing human factors in development of fighter aircraft were described.

To conclude, managing human factors in the development of fighter aircraft is a war being fought on many fronts. It is absolutely necessary to know the terrain of the battle, in other words the domain of fighter aircraft development, and to be armed with suitable methods, techniques and developmental processes.

REFERENCES

Albinsson, P.-A., & Alfredson, J. (2002). *Reflections on practical representation design (FOI-R-0716-SE)*. Linköping, Sweden: Swedish Defence Research Agency.

Alfredson, J., Oskarsson, P.-A., Castor, M., & Svensson, J. (2003). Development of a meta instrument for evaluation of man-system interaction in systems engineering. In G. L. Rafnsdóttir, H. Gunnarsdóttir, & Þ. Sveinsdóttir (Eds.), *Proceedings of the 35th Annual Congress of the Nordic Ergonomics Society* (pp.77-79). Reykjavík, Island: Svansprent.

Alm, T. (2007). *Simulator-based design: Methodology and vehicle display applications.* Doctoral dissertation (No. 1078). Linköping, Sweden: Linköping University.

Alm, T., Alfredson, J., & Ohlsson, K. (2008). Business process reengineering in the automotive area by simulator-based design. In El Sheikh, A., Al Ajeeli, A. T., & Abu-Taieh, E. M. (Eds.), *Simulation and modeling: Current technologies and applications* (pp. 337–358). Hershey, PA: IGI Global.

Alm, T., Ohlsson, K., & Kovordanyi, R. (2005). Glass cockpit simulators: Tools for IT-based car systems design and evaluation. In *Proceedings of Driving Simulator Conference – North America 2005*, Orlando, FL.

Castor, M. (2009). *The use of structural equation modeling to describe the effect of operator functional state on air-to-air engagement outcomes.* Doctoral dissertation (No. 1251). Linköping, Sweden: Linköping University.

Castor, M., Hanson, E., Svensson, E., Nählinder, S., Le Blaye, P., Macleod, I., et al. Ohlsson, K. (2003). *GARTEUR handbook of mental workload measurement* (GARTEUR TP 145). The Group for Aeronautical Research and Technology in Europe.

Cross, N. (1995). Discovering design ability. In Buchanan, R., & Margolin, V. (Eds.), *Discovering design* (pp. 105–120). Chicago, IL: The University of Chicago Press.

Darses, F., & Wolff, M. (2006). How do designers represent to themselves the users' needs? *Applied Ergonomics*, *37*(6), 757–764. doi:10.1016/j.apergo.2005.11.004

Dekker, S. W. A. (1996). Cognitive complexity in management by exception: Deriving early human factors requirements for an envisioned air traffic management world. In D. Harris (Ed.), *Engineering psychology and cognitive ergonomics, volume I: Transportation systems* (pp. 201-210). Aldershot, England: Ashgate.

Dekker, S. W. A. (2003). Illusions of explanation: A critical essay on error classification. *The International Journal of Aviation Psychology*, *13*(2), 95–106. doi:10.1207/S15327108IJAP1302_01

Dekker, S. W. A. (2005). *Ten questions about human error: A new view of human factors and systems safety.* Mahwah, NJ: Erlbaum.

Dekker, S. W. A. (2007). Doctors are more dangerous than gun owners: A rejoinder to error counting. *Human Factors*, *49*(2), 177–184. doi:10.1518/001872007X312423

Dekker, S. W. A., & Nyce, J. M. (2004). How can ergonomics influence design? Moving from research findings to future systems. *Ergonomics*, *47*(15), 1624–1639. doi:10.1080/00140130412 31290853

Dekker, S. W. A., & Woods, D. D. (1999). Extracting data from the future: Assessment and certification of envisioned systems. In Dekker, S., & Hollnagel, E. (Eds.), *Coping with computers in the cockpit* (pp. 131–143). Aldershot, England: Ashgate.

Endsley, M. R. (1988). Situation awareness global assessment technique (SAGAT). In *Proceedings of the IEEE National Aerospace and Electronics Conference* (pp. 789-795). New York, NY: IEEE.

Endsley, M. R. (1995b). Toward a theory of situation awareness in dynamic systems. *Human Factors*, *37*(1), 32–64. doi:10.1518/001872095779049543

Endsley, M. R., Bolstad, C. A., Jones, D. G., & Riley, J. M. (2003). Situation awareness oriented design: From user's cognitive requirements to creating effective supporting technologies. In *Proceedings of the Human Factors and Ergonomics Society 47th Annual Meeting* (pp. 268-272). Santa Monica, CA: Human Factors and Ergonomics Society.

Endsley, M. R., Bolté, B., & Jones, D. G. (2003). *Designing for situation awareness: An approach to user-centered design.* London, UK: Taylor & Francis.

Fitts, P. M. (1951). *Human engineering for an effective air navigation and traffic control system.* Ohio State University Research Foundation Report, Columbus, OH.

Fitts, P. M., Jones, R. E., & Milton, J. L. (1950). Eye movements of aircraft pilots during instrument-landing approaches. *Aeronautical Engineering Review, 9*(2), 24–29.

Hollnagel, E. (1999). From function allocation to function congruence. In Dekker, S., & Hollnagel, E. (Eds.), *Coping with computers in the cockpit* (pp. 29–53). Aldershot, England: Ashgate.

Hollnagel, E. (2007). Flight decks and free flight: Where are the system boundaries? *Applied Ergonomics, 38*(4), 409–416. doi:10.1016/j.apergo.2007.01.010

Hollnagel, E., & Woods, D. D. (1983). Cognitive systems engineering: New wine in new bottles. *International Journal of Man-Machine Studies, 18*(6), 583–600. doi:10.1016/S0020-7373(83)80034-0

Hollnagel, E., & Woods, D. D. (2005). *Joint cognitive systems: Foundations of cognitive systems engineering.* Boca Raton, FL: Taylor & Francis. doi:10.1201/9781420038194

Howard, M. (2002). *Usefulness in representation design.* Doctoral dissertation (No. 753). Linköping, Sweden: Linköping University.

Hult, L., Irestig, M., & Lundberg, J. (2006). Design perspectives. *Human-Computer Interaction, 21*(1), 5–48. doi:10.1207/s15327051hci2101_2

Jones, D. G., Endsley, M. R., Bolstad, M., & Estes, G. (2004). The designer's situation awareness toolkit: Support for user-centered design. In *Proceedings of the Human Factors Society 48th Annual Meeting* (pp. 653-657). Santa Monica, CA: Human Factors Society.

Klein, G., Orasanu, J., Calderwood, R., & Zsambok, C. (Eds.). (1993). *Decision making in action: Models and methods.* Norwood, NJ: Ablex.

Koppenjan, J., & Groenewegen, J. (2005). Institutional design for complex technological systems. *International Journal of Technology. Policy and Management, 5*(3), 40–257.

Nagel, D. C. (1988). Human error in aviation operations. In Weiner, E. L., & Nagel, D. C. (Eds.), *Human factors in aviation* (pp. 263–303). San Diego, CA: Academic Press.

Papantonopoulos, S. (2004). How system designers think: A study of design thinking in human factors engineering. *Ergonomics, 47*(14), 1528–1548. doi:10.1080/00140130412331290916

Simons, R. (2005). *Levers of organization design: How managers use accountability systems for greater performance and commitment.* Boston, MA: Harvard Business School Press.

Singer, G. (2002). *Methods for validating cockpit design: The best tool for the task.* Doctoral dissertation. Stockholm, Sweden: Royal Institute of Technology.

Smallman, H. S., & St.John, M. (2005). Naïve realism: Misplaced faith in realistic displays. *Ergonomics in Design, 13*(3), 6–13. doi:10.1177/106480460501300303

Svensson, E., Angelborg-Thanderz, M., & Sjöberg, L. (1993). Mission challenge, mental workload and performance in military aviation. *Aviation, Space, and Environmental Medicine*, *64*(11), 985–991.

Svensson, E., Angelborg-Thanderz, M., Sjöberg, L., & Olsson, S. (1997). Information complexity: Mental workload and performance in combat aircraft. *Ergonomics*, *40*(3), 362–380. doi:10.1080/001401397188206

Svensson, E., & Wilson, G. F. (2002). Psychological and psychophysiological models of pilot performance for systems development and mission evaluation. *The International Journal of Aviation Psychology*, *12*(1), 95–110. doi:10.1207/S15327108IJAP1201_8

Wickens, C. D. (2002). Situation awareness and workload in aviation. *Current Directions in Psychological Science*, *11*(4), 128–133. doi:10.1111/1467-8721.00184

Wikforss, M. (2008). *Usability design principles in JAS39 Gripen*. Master Thesis. Stockholm: Royal Institute of Technology. In Swedish.

Woltjer, R. (2009). *Functional modeling of constraint management in aviation safety and command and control*. Doctoral dissertation (No. 1249). Linköping, Sweden: Linköping University.

Woods, D. D. (1995). Toward a theoretical base for representation design in the computer medium: Ecological perception and aiding human cognition. In Flach, J. M., Hancock, P. A., Caird, J., & Vicente, K. (Eds.), *Global perspectives on the ecology of human-machine systems* (pp. 157–188). Hillsdale, NJ: Erlbaum.

Woods, D. D., & Hollnagel, E. (2006). *Joint cognitive systems: Patterns in cognitive systems engineering*. Boca Raton, FL: Taylor & Francis. doi:10.1201/9781420005684

ENDNOTE

[1] In this chapter the fighter pilot will be referred to as he (and not he/she) even though small numbers of female fighter pilots also exist.

Chapter 5
State-of-the-Art Real-Time Jet Fuel Quality Measurement

Jed B. Stevens
Velcon Filters, LLC, USA

Greg Sprenger
Velcon Filters, LLC, USA

Miles Austin
Velcon Filters, LLC, USA

ABSTRACT

This chapter discusses the historical approaches to monitoring aviation fuel quality, and the current industry movement towards continuous monitoring electronic measurement systems. The application of mature technology from other industries is reviewed and found to be inadequate. A new type of sensing system designed specifically for the needs of aviation fuel quality is introduced, showing advanced user features and proven to be far more accurate than any other method current available. This article also discusses a typical problem in real world applications where a combination of contaminant is encountered, and how only the new type of sensing system can properly measure the contamination to aviation fuel quality specifications.

INTRODUCTION

History: Ensuring Jet Fuel Quality

From the beginning of Sir Frank Whittle's jet engine, fuel quality has been very critical to safe flight. The kerosene-based fuel used in the evolving jet engines has also evolved. From the original illuminating kerosene Whittle used, to the various modern fuels of today, quality assessment has changed as well.

The current commercial fuels are largely covered by two major specifications: ASTM D1655 and DEFSTAN 91-91 (ASTM. *Standard Specification for Aviation Turbine Fuels*. D1655-05, 2007), (Ministry of Defence (UK). *Turbine Fuel, Aviation Kerosine Type, Jet A-1*. Defence Standard 91-91, Issue 6, 2008). These specifications cover an extensive variety of requirements for these fuels, from bulk properties, such as boiling point range,

DOI: 10.4018/978-1-60960-887-3.ch005

to trace properties, such as flash point. In addition, contamination is considered in properties such as surfactant (surface active agent) levels, dissolved metals, particle content and water content and more recently issues such as cross-contamination with other petroleum products.

Once the fuel leaves the refinery, the two main contaminants of concern are solid particles and free water. This contamination ingression takes place primarily during transportation of the fuel and also during fuel storage, before the fuels are burned. The presence of these two physical contaminants is detrimental to aircraft fuel systems and engines, and compromises flight safety. Acceptable levels have been specified; however current methods of assessment in the field have not evolved to keep up with aircraft and engine improvements.

In fact, the most accepted measure of fuel quality at the aircraft is still the "Clear and Bright" test, which is a subjective, visual examination of the fuel taken into a white bucket. The tester examines the bottom of the bucket for any contamination that settles (IATA. *Guidance material for aviation turbine fuels specifications,* Part III *Cleanliness and handling,* 5th edition, 2004). Obviously, the vision of the tester and the lighting conditions have a large effect on the capability of actually finding contaminants.

Current Quantifiable Fuel Cleanliness Requirements: Solid Particles and Free Water

Current requirements for cleanliness of jet fuel can be typified by the IATA Guidelines, which state the maximum levels of free water and solid contaminant at delivery points into aircraft. For both refueler (tanker) trucks as well as hydrant equipment, the maximum level of free water allowable is 30 parts per million (ppm or mg/l) and the maximum level of solid contaminants is 1.0 milligrams per liter (mg/l) (IATA. *Guidance material for aviation turbine fuels specifications,* Part III *Cleanliness and handling,* 5th edition, 2004).

The key issue here is the large difference in allowable levels of free water versus solid particles. The allowable limit for free water is 30X greater than the allowable limit for solid contaminants. As such, different measurement methods of have been devised to check for the presence of free water and solids in jet fuel.

In some cases, a simple manual check is done to verify that the maximum amount of contaminant has not been reached. For these cases, there are chemical solutions such as Shell Water Detector and the Velcon Hydrokit (Shell Oil Company (2009, November). *Shell Water Detector.* Retrieved November 4, 2009, from Shell website: http://www.shell.com/home/content/aviation/products_and_services/products/shell_water_detector/), (Velcon Filters, LLC. (2009, November) *HYDROKIT,* 1752-R9 11/08, Retrieved November 4, 2009, from Velcon Website: http://www.velcon.com/aviation/hydrokit.html)

These are chemical tests that take a small sample of fuel and chemically test for the presence of up to 30 ppm of free water. The color of the water-sensing membrane or powder is then judged by the eye of the operator against a standard. Since these are 'snapshots,' from a small fuel sample, their use for overall fuel quality assessment is only as good as the sample taken if it is representative of the fuel batch. These tests will generally give accuracy of +/- 5 ppm of free water, and are somewhat subjective.

Another more accurate manual method to test the fuel supply for free water does exist.

In Aqua-Glo® testing, per ASTM D3240, a fluorescent pad is placed inline of a sample of fuel flowing from the main fuel supply, typically by way of a ¼" sample tube inserted into a main flow pipe. A specific amount of fuel is then run through the pad, with any free water sticking to the pad. When the flow of fuel has ended, the pad is removed and viewed under a UV light, the pad fluoresces, and the light intensity is judged against a standard pad, and the amount of water, in ppm, is determined. One inherent flaw in this system is its large error, at 32%, that gives widely varying results under constant test conditions, making it a controlled, but still inadequate metric for judging the free water content of fuel (ASTM. *Standard Test Method for Undissolved Water in Aviation Turbine.* Fuels. D3240-05, 2007).

To determine the level of solids in the fuel, a membrane test is performed per ASTM D2276. A pre-weighed pad is used much in the same manner as the Aqua-Glo® method, but instead of collecting free water on a fluorescing pad, the membrane collects solids from the fuel. In this test, a predetermined amount of fuel is flowed through this membrane, typically 2 to 4 liters. The membrane can be analyzed in two ways; for color or weight change (ASTM. *Standard Test Method for Particulate Contamination in Aviation Fuel by Line Sampling.* D2276-05, 2007).

For a color determination, the color of the membrane is compared to standards, and reported. This provides some indication of the type of particle and amount of contaminant; however, it doesn't correlate with the limit of 1.0 mg/l of solids in the fuel. This method does not require specific laboratory equipment, and can be done effectively in the field.

For the weight change the membrane is dried, to remove the weight of the fuel and any accumulated water, then reweighed. The increase in weight of the pre-weighed pad is then attributed to collected solids. The amount of collected solids in milligrams (mg) is then divided by the volume of fuel run through the pad to determine a weight by volume measurement of parts per million of solid contaminant to fuel, usually displayed as mg/l. Of the major problems with this measurement method are the introductions of error by the multiple weight measurements, requiring very low milligram accuracy, which introduce a significant error to the overall result. Another problem with this measurement method, for jet fuel applications specifically, is that a weight per volume of solid contaminants is not a very good judge of the severity of the contaminant. The density difference between iron-type particles and silica particles is quite large. In addition, it would take fewer large particles of rust (iron-based particles) compared with a large amount of smaller sand particles (silica) to register the same Millipore reading. Operationally speaking, a large amount of 0.5 micron sand particles (silica) do not pose much of a risk to the operation of equipment with tight tolerances. Conversely, a few large particles (>10 microns) with higher densities would pose a large problem for the equipment with tight tolerances. Both contaminants may have the same membrane readings and would be treated as identical solid contaminant, although their effects downstream on the equipment would be drastically different. This real issue strongly suggests that quality monitoring equipment should have capability to provide particle size information.

Most of the above test procedures rely on dedicated human interaction with the fluid medium and are therefore highly prone to further contamination ingression, sampling errors and reduced repeatability.

ASTM Specification D 3240-05 covers Aqua-Glo® testing for free water and has an error of +/- 32%. ASTM Specification D 2276-05 covers membrane color and weight solids testing with an error of +/- 17.5%.

The Application of Electronics for Monitoring Fuel Quality

In recent decades, there have been moves to automate aviation fuel quality analysis using electronic devices. In this effort, there have been several types of devices have been employed, but the main method has been particle counting, a mature technology in the field of hydraulic oil and system quality. Turbidimeter style sensors have also been widely used for industrial applications.

Particle counters function using the principle of light obscuration. Light obscuration is the light blockage by a particle onto a sensor; see Figure 1. This light blockage registers a drop in sensor

Figure 1. Photodiode voltage variation resulting from passage of spherical particle (Fitch, E.C. (1988). Fluid Contamination Control. Oklahoma State University, Stillwater, OK: FES)

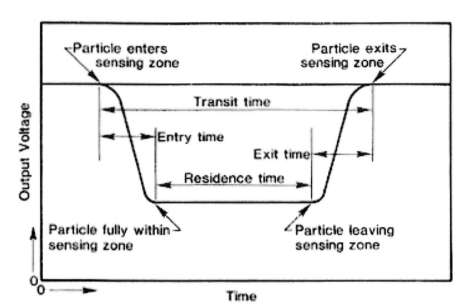

voltage, and that sensor voltage is then empirically correlated to a particle size (Fitch, E.C. (1988). *Fluid Contamination Control*. Oklahoma State University, Stillwater, OK: FES). These devices will typically amass data that will give not only a size distribution of the contaminants in the hydraulic oil, but will also give a count of how many particles of each size exist in the fluid stream at a given time. This technology was developed for the hydraulic industry several decades ago, where the primary concern was amount and size of particles, as that information had a direct bearing on wear and component service life.

Turbidimeters function using light obscuration principles, light scattering, or a combination of both. Light obscuration was already discussed above. Light scattering occurs when a particle of contamination interacts with a light source, and a small amount of light is scattered at various angles by either reflection or diffraction. In these light scattering devices, the intensity of light is measured (resulting in an inverse of Figure 1), and usually correlated into a unit of either scatter, or sometimes directly into a reading of contami-

nation. Since turbidimeters are simple sensors, and generally do not have any signal conditioning or signal processing, they are very susceptible to errors caused by different fuel colors, grades, densities and clarities. Also, since there is no higher logic, they also lack any ability to differentiate between water and solid contaminant. They have also historically had operational issues related to window fouling.

In recent years, this mature technology from the hydraulics and industrial arenas have been applied to jet fuel, and unfortunately, for the reasons listed above they are inadequate at judging aviation fuel quality (Bessee, G. (2008). *Research into Electronic Sensors for "Guaranteed Clean Fuel" [24]*. ExxonMobil Aviation Technical Meeting, October, 2008), (Kitson-Smith, A & Hughes, V. (2007, October) *The Use of Electronic Sensors in Field Measurements of Aviation Jet Fuel Cleanliness*. Paper presented at the 10th meeting of the International Conference on Stability, Handling and Use of Liquid Fuels, Tucson, AZ). Recall from earlier discussion, that the primary contaminants of concern in aviation fuel quality are the amount of free water and solid contaminant present in the fuel. Also, understanding how particle counters and turbidimeters function, it may seem a logical conclusion to suppose that these technologies would be a good fit for the aviation fuel industry. However, an analysis of the operation of particle counters versus the needs of the aviation fuel industry will prove that is not the case.

The primary problem with adopting these mature technologies to aviation fuel quality is the difference of acceptable values for solids and free water being 30x apart. This large difference in allowable concentrations is compounded by the fact that particle counters (light obscuration technology) cannot differentiate between droplets of water and particles of solids. There is no electromechanical capability present in a particle counter using light obscuration technology that can differentiate between the character of a water droplet (slightly donut shaped shadow) and a solid particle (darker shadow.) A single light sensor is used reading only a voltage difference from the baseline clean liquid. A particle counter can sense the contamination, but because it cannot differentiate between types of contaminant, it cannot alert the user to a quality problem at 1 ppm (for solids) instead of 30 ppm (for water) (ISO. *Hydraulic fluid power -- Fluids -- Method for coding the level of contamination by solid particles*. 4406:1999, 2008).

Other Shortcomings with Existing Electronic Technology

Due to the analysis of very small particles, particle counters have small sensing zones and require low sensor flow rates. This requires a small slipstream sample be taken from the main flow. This creates two troublesome issues: isokinetic sampling and the lag time of the output caused by the time required for the sample to reach the sensor itself, along with the time needed to report the result.

The sampling of a main line has to be done very carefully to ensure that there is isokinetic sampling, or representative sampling, through the sample line. As is normal operation in aviation fueling, the flow rates will vary throughout a fueling, so it is important that the flow velocity through the sample port match the flow velocity in the main line. A control system of this kind could be very expensive, prohibitively so, and in practice, little to no attention has been paid to ensuring isokinetic sampling. The result is that the flow through the sample line may not be isokinetic, and therefore may not be representative of the condition inside the main line, see Figure 2 (Hinds, W.C. (1982). *Aerosol Technology*, New York, NY: Wiley-Interscience.). Particle size and distribution can be affected. ASTM D4177 discusses this issue (ASTM. *Standard Practice for Automatic Sampling of Petroleum and Petroleum Products*. D4177-95 (2005), 2007). This would invalidate any actual readings from the particle counter, as they cannot be relied upon to accurately reflect the condition inside of the main line.

Side stream sampling (slip stream or side sampling) consists of branching out of the main flow of product through a ¼" sample line, and then further sampling of the sample stream which is actually measured by the counter. A lag time is inherent in the process of sampling from the main flow line. In addition to the 30-45 seconds of flow that a particle counter requires to 'build' a sample dataset, the actual flow from main pipe to the counter itself can take another 30-45 seconds, to begin a building a sample. This means that a particle counter can take 60 seconds or more to actually report cleanliness information.

This shortcoming can create a dangerous situation in extreme fuel quality issues like a water slug. For example, if fuel moving in the main pipe is flowing at 300 gallons per minute, and the particle counter takes a full minute to register a problem with the fuel, then at the first opportunity of the sensor to report the problem, at least 300 gallons of compromised fuel (in extreme cases, almost pure water) has reached the fuel tanks of the aircraft.

Figure 2. (a) represents perfect isokinetic sampling, where the sample flow velocity and main flow velocity are equal; (b) and (c) represent non-isokinetic sampling under different variants of alignment and flow velocities. (Hinds, W.C. (1982). Aerosol Technology, New York, NY: Wiley-Interscience.)

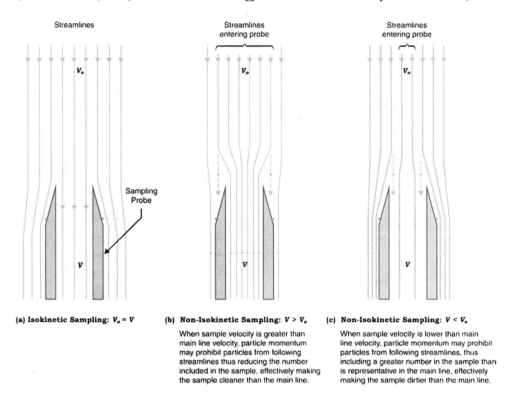

(a) **Isokinetic Sampling:** $V_o = V$

(b) **Non-Isokinetic Sampling:** $V > V_o$

When sample velocity is greater than main line velocity, particle momentum may prohibit particles from following streamlines thus reducing the number included in the sample, effectively making the sample cleaner than the main line.

(c) **Non-Isokinetic Sampling:** $V < V_o$

When sample velocity is lower than main line velocity, particle momentum may prohibit particles from following streamlines, thus including a greater number in the sample than is representative in the main line, effectively making the sample dirtier than the main line.

Development of New Technology: To Distinguish and Quantify Solids and Free Water

For the reasons stated above, particle counting and turbidimeters have not been confirmed as acceptable measurement devices for into-plane aviation fuel quality by any major group or authority. As a result of the inadequacies inherent in the above mentioned technologies, a group, working under the guidance of a US Navy Small Business Innovation Research (SBIR) program set out to develop an electronic, real time aviation fuel quality measurement device that has the ability to accurately distinguish between water and solid particles and address these mature technology shortcomings.

In June of 2003, Physical Sciences Inc. (PSI), a renowned research and development firm, was awarded a multiphase Navy SBIR contract to develop, specifically for aviation turbine fuel, a contamination sensor that had the ability to quantify contamination and distinguish between water droplets and solid particles. Also included in the design specification were the requirements that the sensor would be inline and full flow, and that the sensor be robust enough to handle the daily rigors of on-board naval duty (US Department of Defense, (2003, June) *News No. 411-03*, Retrieved October 28, 2009, from US Department of Defense website: http://www.defenselink.mil/contracts/contract.aspx?contractid=2533).

The culmination of four years of research and development, including three years of operational readiness testing onboard the active super carrier USS Ronald Reagan (CVN 76) gave rise to the advanced contamination sensor code named: "AutoGrape", aptly named since naval seamen who are responsible for fuel handling and cleanliness all wear purple jumpsuits.

Application of Technology to Commercial Needs

In light of this new advanced contamination sensor by PSI, Velcon Filters, Inc. (now Velcon Filters, LLC) entered into a joint-venture partnership to commercialize the "AutoGrape" sensor. The trade name VCA® was chosen for the commercial sensor, which stands for Velcon Contaminant Analyzer.

Instead of taking a current, mature technology that exists in some other industry, the VCA® was designed to meet the specific needs of aviation fuel quality. The basic function of the sensor uses a combination of light scatter, light obscuration, signal conditioning and computer processing in order to quantify and discriminate water droplets and solid contaminants.

A laser beam traverses the main pipe section and interacts with two different sensors in order to determine water and solid contamination in the fuel, see Figure 3. In order to sense water particles, Mie Scattering theory predicts that contamination in the 0-10 micron range scatter at very near angles, typically less than 5 degrees.

Also, predicted by Mie Scatter theory, as the contamination get larger, the angle of forward scatter continues to decrease. This property is measured using a multi-dimensional CCD light sensor that records the location and intensity of light in the near scatter region (White, M., Templeman, C., Frish, M., Nebolsine, P. (2008). U.S. Patent No. 7,450,234. Washington, D.C.: U.S. Patent and Trademark Office). This data can be directly correlated to contaminant concentration and size. Using this advanced light sensor, a proprietary process is employed so that the one CCD can be used to sense light scatter data, as well as baseline intensity of the laser beam. This provides the VCA® with the versatility to be used in fuels of all different grades, colors and even to a certain degree, clarities, by providing a normalizing reference for the scatter light. This specific portion is regarded as light obscuration, but used in a novel way to provide a more robust sensor, capable of a wide application in fuel quality monitoring.

Again, as pointed out above, one sensor is not sufficient, however complicated, to provide all of the metrics required to differentiate between water and solids. In order to sense the presence of solid particles, and differentiate between water droplets and solid particles, the VCA® uses a second sensor at 90 degrees (perpendicular) from

Figure 3. Interior section of the VCA®, showing laser light path, optical sensors, auto-calibration standard and flow meter

the laser path. This 90 degree sensor (or side sensor) collects light that is reflected off the solid surface of particles, and not scattered at similar levels by water droplets. In this way, the side sensor provides another independent means of correlating fuel quality based on the intensity of scattered (reflected) light that is collected in the 90 degree orientation.

Both types of signals are then fed into the on-board computer processor and analyzed using a proprietary algorithm developed by PSI. The signal conditioning from each type of sensor also happens in near-real time at the processor so that the system can react to conditions inside of the main line in milliseconds.

The VCA® uses these several optical principles and sensors in concert with computer technology developed by PhD scientists. The result is a high level of accuracy for both water and solids measurements (typically less than +/- 5%) even in consideration of the 30X order of magnitude difference in allowable concentration of water and solid contamination. As a result, the VCA® can sense water accurately between 0-30 ppm, with an average error of about +/- 0.5 ppm, and can simultaneously sense solids accurately between 0-2 mg/l with an average error of about +/ -0.05 mg/l, as seen in Figure 4. These metrics exactly match those used by the commercial fuel quality authorities mentioned earlier in the article.

Both the US Navy and the US Air Force have accepted the proven system as capable of providing accurate and reliable aviation fuel quality assessment. To date, no other instruments have been approved by any US military branch (SBIR Success. (2009). *Navy Transitions*, 6, p. 6-7.),

Figure 4. Table and graph showing the typical calibration data and curves of a VCA®, highlighting both linearity as well as very low error

CALIBRATION VALUES - Separate Water and Solids Injections						_		
INJECTED WATER (PPM)	0.0	5.0	10.0	15.0	20.0	AVG. ER: ppm, mg/l		
MEASURED WATER (ppm)	0.0	6.3	11.0	15.1	20.0	WATER	**0.6**	
ERROR - WATER	**0.0%**	**6.6%**	**5.1%**	**0.4%**	**0.2%**	SOLIDS	**0.06**	
INJECTED SOLIDS (mg/l)	0.00	0.25	0.50	1.00	2.00	**AVERAGE ERROR**		
MEASURED SOLIDS (mg/l)	0.00	0.15	0.46	1.04	1.95	WATER	**3.0%**	
ERROR - SOLIDS	**0.0%**	**5.1%**	**1.8%**	**2.1%**	**2.3%**	SOLIDS	**2.3%**	

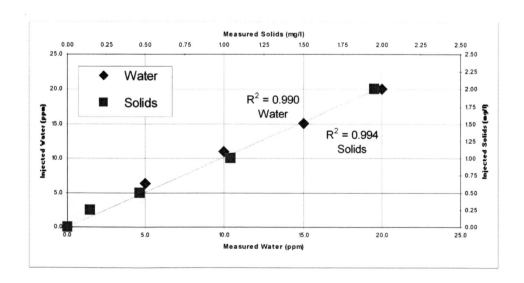

(Wilkes, Rick. (2008). *U.S. Air Force VEMSO Product Evaluation Program Final Report – Fuel Containment Analyzer Project Nr. E07-02*: Air Force Petroleum Office (AFPET)).

Designed-In Benefits to Users

In addition to the striking accuracy of the new sensor, it has several inherent benefits not shared by particle counting technology. Since the VCA® is an in-line and full flow device, there is no sample error associated with sampling. The VCA® mounts directly into the pipeline and so can make direct readings without the need to disrupt the flow or sample fuel away from the source.

Another benefit of in-line and full flow measurement is near-real time information, with a lag of only about 1-2 seconds. In contrast to the example used above with a particle counter and a water slug, the VCA® will report a fuel quality problem in approximately 3 seconds, instead of 1 minute. This translates into approximately 5-10 gallons of unfit fuel passing the sensor before action is taken compared with about 300 gallons at the same flow rate with a particle counter. This amount of fuel is unlikely to actually reach the aircraft through the fueling hose.

Several important operational features have been added to the VCA® that enhance its appeal to operational users and offer automation of continuous fuel quality monitoring on the overall process of the fuel handling. The VCA® has been designed and configured to operate automatically, with data telemetry or storage options, auto-fuel-shutdown capabilities, as well as auto-on and auto-off features built into the standard unit. Also available is an optional auto-calibration feature, using an internal calibration standard, that allows the unit to be calibrated while in place, inside the pipe, making the routine disassembly of the piping system unnecessary and guaranteeing continuous calibration (Sprenger, G., Stevens, J., White, M., Hillis, R., Lavenberg, J. & Templeman, C. (2009).

U.S. Patent No 7,518,719 B2. Washington, D.C.: U.S. Patent and Trademark Office).

In addition to the native features of the VCA®, the data management system that has been devised can alert the end user in any number of ways, from data telemetry (email alerts, SMS Text messages, etc.) to direct fuel shutdown of the equipment. The server that houses the fueling data can be used to catalog historical data for trend analysis, as well as near-real time viewing of current fuelings. All of this information can help the fuel handling managers better manage their operations.

The VCA® has also recently been integrated into a control valve, which offers easy retrofit into existing fueling systems, and built-in flow control, see Figure 5.

State-of-the-Art Custom Fuel Quality Assessment is Achieved

Fuel quality requirements are chosen based on the operational requirements of the equipment in use, and may be compromised by the measurement methods and other practical limitations.

Now that new sensing technology exists, has been proven, and is starting to be adopted by major fuel handlers, it may be a good time to re-examine the reported metrics of fuel quality. Based on the work done on the VCA® project, it is now possible to provide contamination data in ppm (or mg/l) for both water and solid particles, and also give size distribution of particles of each type of contamination. These values can still be correlated to a standard 'Water PPM' and 'Solids mg/l' in order to provide easy reference points for manual checks. In terms of operational data, the number of particles is already implied in the "solids ppm" data. However, having stated that, the size distribution of solids is important because of the physical clearances of engine components. Also, the size distribution of water droplets would give good operational feedback about filtration systems, surfactant levels in fuel, etc.

Figure 5. Photo of the VCA-CV, integrated solution to flow/pressure control and contamination analysis in one package

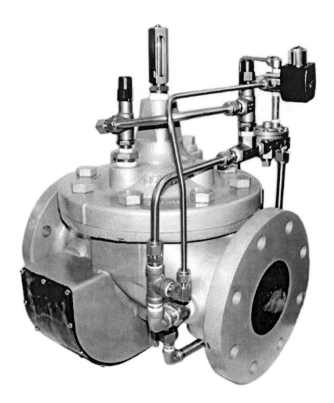

The proposed format of water and solid contaminant reporting moving forward is as follows: Water values could be reported as a correlated ppm measurement, along with a particle size distribution in an ISO 4406 Code style format; solids could be reported as a correlated ppm measurement, along with a particle size distribution in an ISO-Code style format. It would look like the format in Table 1.

In this way, the complete fuel quality profile is transmitted to the user in as simple a form as possible, but also gives a clear view of quantity and type of contamination. This will also help operators more thoroughly and efficiently diagnose problems in their fueling systems.

Table 1. Proposed contamination analysis reporting format

Water:	8.0 ppm; >4 um: 85%, >6 um: 8%, >14 um: 6%, >21 um: <1%
Solids:	0.20 ppm; >4 um: 95%, >6 um: 3%, >14 um: 2%, >21 um: <1%

CONCLUSION

The 5 ppm Dilemma Solved

Based on all the capabilities of this new sensing technology, there is no problem handling a situation where, for example, 5 ppm of a contaminant is sensed. With other electronic sensors, which do not distinguish between water and solids, the user is forced to make a choice. Does he decide to stop the fuel flow, presuming the contaminant is solids?

Or continue presuming it is water at an acceptable level, and risk loading highly contaminated fuel, containing 5 mg/l of solids, onto an aircraft?

It is clear that the situation will be handled properly in each circumstance with the new VCA® technology. If the reading is 5 ppm of water, then the VCA® reading will report 5 ppm water, 0 ppm of solids and the fueling will continue per the specifications. If the reading is 5 ppm of solids, the sensor will report within 5 seconds and the user can take appropriate action, or the automatic shutdown will end the fueling. If it is a combination of contamination, say 4.5 ppm water and 0.5 mg/l solids, then the fueling will be allowed to continue, per specification. In each case, the VCA® will provide the proper handling of the contamination profile, because it was designed specifically for the very application of aviation fuel quality analysis and assurance.

Moving Forward

Current commercial specifications are being written to govern the use of electronic monitors of fuel quality (API/EI. *Considerations for electronic sensors to monitor free water and/or particulate matter in aviation fuel.* 1598 *Draft Standard,* 2007). These specifications should strive to provide proper evaluations of technology for fuel quality assessment, and also uncover shortcomings in inadequate technology. The VCA® technology provides a critical capability; to distinguish and quantify the levels of both solids and water independently. The use of inferior technology is not in the best interests of fuel handlers or the aviation industry at large.

REFERENCES

API/EI. (2007). *Considerations for electronic sensors to monitor free water and/or particulate matter in aviation fuel.* 1598 Draft Standard, 2007.

ASTM. (2005). *Standard practice for automatic sampling of petroleum and petroleum products.* D4177-95 (2005), 2007.

ASTM. (2007). *Standard specification for aviation turbine fuels.* D1655-05, 2007.

ASTM. (2007). *Standard test method for particulate contamination in aviation fuel by line sampling.* D2276-05, 2007.

ASTM. (2007). *Standard test method for undissolved water in aviation turbine fuels.* D3240-05, 2007.

Bessee, G. (2008). *Research into electronic sensors for "guaranteed clean fuel."* ExxonMobil Aviation Technical Meeting, October, 2008.

Faudi Aviation. (2009, November). *AFGuard.* Retrieved November 10, 2009, from http://www.faudi-aviation.com/ images/ stories/ ds_af-guard_en.pdf

Fitch, E. C. (1988). *Fluid contamination control. Oklahoma State University.* Stillwater, OK: FES.

Hinds, W. C. (1982). *Aerosol technology.* New York, NY: Wiley-Interscience.

IATA. (2004). *Guidance material for aviation turbine fuels specifications,* part III: *Cleanliness and handling,* 5th edition, 2004.

ISO. (2008). *Hydraulic fluid power-Fluids-Method for coding the level of contamination by solid particles.* 4406:1999, 2008.

Kitson-Smith, A., & Hughes, V. (2007, October). *The use of electronic sensors in field measurements of aviation jet fuel cleanliness.* Paper presented at the 10th Meeting of the International Conference on Stability, Handling and Use of Liquid Fuels, Tucson, AZ.

Ministry of Defence (UK). (2008). *Turbine fuel, aviation kerosene type, jet A-1.* Defence Standard 91-91, Issue 6, 2008.

SBIR. (2009). Success. *Navy Transitions, 6*, 6–7.

Shell Oil Company. (2009, November). *Shell water detector*. Retrieved November 4, 2009, from http://www.shell.com/ home/ content/ aviation/ products_and_services/ products/ shell_water_detector/

Sprenger, G., Stevens, J., White, M., Hillis, R., Lavenberg, J., & Templeman, C. (2009). *U.S. patent no 7,518,719 B2*. Washington, DC: U.S. Patent and Trademark Office.

US Department of Defense. (2003, June). *News no. 411-03*. Retrieved October 28, 2009, from http://www.defenselink.mil/ contracts/ contract. aspx? contractid=2533

Velcon Filters, L. L. C. (2009, November) *HYDROKIT*, 1752-R9 11/08. Retrieved November 4, 2009, from http://www.velcon.com/ aviation/ hydrokit.html

White, M., Templeman, C., Frish, M., & Nebolsine, P. (2008). *U.S. Patent No. 7,450,234*. Washington, DC: U.S. Patent and Trademark Office.

Wilkes, R. (2008). *U.S. Air Force VEMSO product evaluation program final report – Fuel containment analyzer project no. E07-02. Air Force Petroleum Office*. AFPET.

Chapter 6
Design and Optimization of Defense Hole System for Uniaxially Loaded Laminates

Salih N. Akour
Sultan Qaboos University, Oman

Mohammad Al-Husban
Civil Aviation Regulatory Commission, Jordan

Musa O. Abdalla
The University of Jordan, Jordan

ABSTRACT

Reducing stress and weight of structures are most important to structural designers. Most engineering structures are an assembly of different parts. On most of these structures, parts are assembled by bolts, rivets, et cetera. This riveted and bolted structure is highly used in aerospace industry. These holes that are drilled for bolts and rivets work as stress raisers. Defense hole theory deals with introducing auxiliary holes beside the main hole to reduce the stress concentration by smoothing the stress trajectories in the vicinity of the main hole. These holes are introduced in the areas of low stresses that appear near the main circular hole. Defense hole system under uniaxial loading is investigated. The optimum defense hole system parameters for a circular hole in an infinite laminated composite plate are unveiled. This study is conducted using finite element method by utilizing commercial software package. The optimum design of the defense hole system is reached by utilizing the redesign optimization technique and univariate search optimization technique. The finite element model is verified experimentally using RGB-Photoelasticity and by reproducing some well known cases that are available in the literature. Digital Image Processing is utilized to analyze the photoelastic images.

Stress concentrations associated with circular holes in pure Uniaxial-loaded laminates can be reduced by up to 24.64%. This significant reduction is made possible by introducing elliptical auxiliary holes along the principal stress direction. The best reduction is achieved when four elliptical defense holes are introduced in the vicinity of the main hole. The effect of the fiber orientation, as well as the stiffness of both the fiber and the matrix are investigated.

DOI: 10.4018/978-1-60960-887-3.ch006

INTRODUCTION

A composite material consists of two or more materials mixed together to give a material with good properties. A typical composite material consists of a material with high mechanical strength and stiffness (reinforcement), for example unidirectional or woven fibers, embedded in a material with lower mechanical strength and stiffness (matrix). To tailor the properties of the composite material, a laminate is formed by stacking on top of each other layers of reinforcement oriented in different directions.

Composite materials, if properly used, offer many advantages over metals. Examples of such advantages are: high strength and high stiffness-to-weight ratio, good fatigue strength, corrosion resistance and low thermal expansion. Nevertheless, conventional composites made of pre-impregnated tape or fabric also have some disadvantages, such as poor transverse properties, inability to yield and sensitivity to moisture and high temperatures, which must be accounted for in the design.

Composite materials such as glass fiber, aramide fiber, boron fiber and carbon-fiber-reinforced plastics have been used for a few decades, especially in the aircraft industry. Aircraft structures also include a large number of open holes and cut-outs e.g. holes for electric wires and hydraulic pipes or holes required for assembly or maintenance where a laminate containing open holes is subjected to different types of loading.

Reducing stresses in structures and optimizing their weight are the main goals for designers and engineers which improve the structural efficiency, performance and durability. Most engineering structures are assembly of different parts. Parts and components are assembled to the main structure by bolts, rivets, etc. Joining by mechanical fasteners is one of the common practices in the assembly of structural components. Among the most important elements in aircraft structures in general and in composite structures in particular are mechanically fastened joints. Improper design of the joints may lead to structural problems or conservative design leading indirectly to overweight structures and high life-cycle cost of the aircraft. Typical examples of mechanically fastened joints in composite aircraft structures are: the skin-to-spar/rib connections, a wing structure, the wing-to-fuselage connection and the attachment of fittings etc. Since the failure of the joints can lead to the catastrophic failure of the structures, an accurate design methodology is essential for an adequate design of the joints. Because of the complex failure modes of composite materials, the mechanical joining of structures made of composite materials demands much more rigorous design knowledge and techniques than those currently available to the traditional methodology for metallic joints. The holes that are needed for such joints induce stress concentration. These high stress concentration spots are likely places for crack initiation. There have been number of incidents of aircraft fuselage failure resulting from crack initiation in the vicinity of riveted holes. The best known example of this was the Aloha airlines incident in 1988 where it was found that many short cracks exist in the row of the riveted lap splice joints. Cracks propagate under fatigue loading and can eventually link up and cause failure for the whole structure. To prevent such scenario from taking place stress relief system usually introduced within the vicinity of these riveted hole to reduce the stress i.e., increasing the load carrying capacity of the structure and reducing the weight. Up to this moment, most Airframe structures (on commercial level) are made of aluminum and some other alloys except the new Boeing Aircraft B787. This new aircraft is made of laminated composite material with many layers of different thicknesses.

Defense Hole System

Introducing auxiliary holes in the neighborhood of a main hole to reduce the stress concentration is called defense hole theory which has been known since the early years of last century. These holes are usually introduced in the low stress spots in the vicinity of the main hole. These new auxiliary holes smooth the stress trajectories around the main hole. In this research, design and optimization of stress relief system for laminate composite plate is investigated to unveil the optimum design parameters of the defense hole system.

Photoelasticity

Since there are always substantial difference between the actual stresses in a machine element and those predicted by analytical or computer models, it may be necessary to test parts to substantiate the design calculations. One testing method is Photoelasticity.

Light consists of electromagnetic radiation. It can be thought as a harmonic oscillation with the time axis along the direction of the light beam and the amplitude perpendicular to it in all directions (see Figure 1). There are some transparent materials which allow light to be transmitted on a specific plane (see Figure 2). Such materials are called polarizing filters and the resulting light is polarized. No light will pass through two polarizing filters that have their polarizing planes perpendicular to one-another (see Figure 3).

Some transparent materials have the property that they rotate the plane of polarized light by an angle that is proportional to the difference in the principal stresses:

$$\Delta\phi = \lambda\,(\sigma_1 - \sigma_2)\,h \qquad (1)$$

These materials are called photoelastic materials (birefringent material). In equation one, λ is a constant of the particular material, h is the plate thickness and $\sigma 1$, $\sigma 2$ are the principal stresses. The product $c = \lambda h$ is a constant of the particular plate.

If $\Delta\phi = \pi/2,\ 3\pi/2,\ 5\pi/2,...$ all light will pass through and fringes of full light emittance will appear on the other side of the material. If a

Figure 1. Light waves

Figure 2. Polarized light

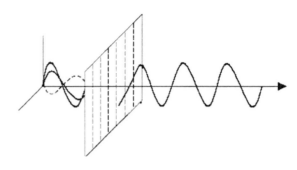

Figure 3. Perpendicular polarizing planes

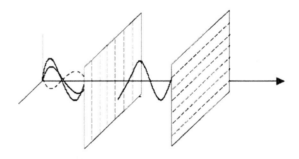

single color light source is used, different light intensities between fringes represent values of stresses. For a white light source, different color fringes will appear as it can be seen in Figure 4. The latter is due to the dependence of the angle of rotation on the frequency of the electromagnetic radiation. When a load is applied, different stress magnitudes will form at different points in the plate and different fringes will appear. Each fringe corresponds to a multiple of the stress that will cause one full cycle $\Delta\phi = \pi$). Therefore, every fringe corresponds to a stress

$$\sigma_1 - \sigma_2 = N\pi/\lambda h = Nd \qquad (2)$$

Figure 4. Fringe patterns on a bar of photoelastic material subject to four point bending

If the constant $\delta = \pi/\lambda\eta$ is known (d is a property of the plate) and the multiplicity number N is determined from the number of fringes, the difference in principal stresses $\sigma1 - \sigma2$ at every point in the plate can be determined.

The constant d is given by the plate supplier or can be measured. The multiplicity number N can be found in two ways:

a.   Start from an area which is black (zero stress, $0 = \sigma1 - \sigma2$) and count the number of fringes between this point and the point at which the stresses are to be determined.
b.   While slowly loading the element, observe the changes in color at the point of interest. The number of full color changes is N.

Therefore, in principle, the difference in the principal stresses can be found anywhere, at least to a multiple of the stress that causes one full cycle. But the difference in principal stresses in a plate is of no practical value. However, failure usually occurs near the boundary (edge) of the plate, where the perpendicular component of stress is zero. In the direction along the boundary, the stress is a principal stress because there is no shear at the boundary. Therefore, if the difference of the principal stresses is known, the stress along the boundary can be calculated:

$$\sigma_b = N\pi / \lambda h = Nd \qquad (3)$$

Since the composite material is not transparent (in general) and does not have the birefringent (doubly refracting) characteristics, reflection photo-elasticity is going to be adopted. This method gives the researcher the opportunity to apply birefringent coating to the composite material while the polariscope in the reflection-polariscope arrangement as illustrated in Figure 5.

Literature Review

Numerical and experimental studies for reducing stress levels in structures by introducing other geometric discontinuities are very few. Most of the work that has been done so far deals with isotropic material.

Erickson and Riley (1978) investigated the effect of the defense holes on the stress concentration around the original hole using two-dimensional photoelasticity. Jindal (1983) examined the reduction of stress concentration around circular and oblong holes using the Finite Element Method (FEM) and photo-elasticity analysis. Meguid (1986 & 1989) studied the reduction of stress concentration in a uniaxialy-loaded plate with two coaxial holes using the Finite Element Analysis (FEA). Rajaih and Naik (1984) investigated hole – shape optimization in a finite plate in the presence of auxiliary holes using the two dimensional photoelastic methods. Ulrich and Moslehy (1995) used boundary element methods to reduce stress concentration in plates by introducing optimal auxiliary holes. Akour et al. (2003 and 2010)

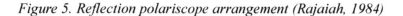

Figure 5. Reflection polariscope arrangement (Rajaiah, 1984)

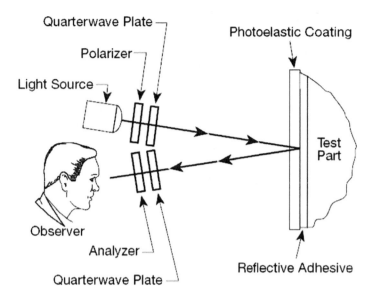

studied the defense hole system for shear biaxial loading and mixed loading (shear and tension). The study achieved maximum reduction of 13.5% and 18% respectively by introducing the defense hole system along the principal stress direction. These reductions are achieved by introducing four elliptical defense holes in the vicinity of the main hole.

Durelli (1978) investigated the optimization of geometric discontinuities in stress field under uniaxial loading. Dhir (1981) studied the hole shape optimization in plate structure under tension and shear loading.

High stress concentration points in the structure have the tendency to initiate a crack leading to failure to the structure. Many people have studied radial cracks emanating from circular holes. Previous efforts studied crack propagation by quantifying applicable stress intensity factors. Kamel et al. (1991) obtained the stress intensity factors for cracks emanating from a fastener hole using the principle of superposition and Green's function for a point force applied in an anisotropic sheet with an elliptical hole. He investigated various loading cases, such as a point load, uniform pressure applied on an arc, and a cosine distribution pressure. Wanlin (1993) has studied analytically the stress intensity factor for corner cracks emanating from fastener holes in finite plates subjected to biaxial load and three types of pin load, uniform, concentrated and cosine distribution. Gungor (1995) used optical Photoelasticity to measure stress intensity factors in corner cracks emanating from holes in sheet structures with bolted stiffeners (L-shape stringer). Dong Shan and Xing (1995) studied analytically a finite plate with a cracked hole stiffened by a ring with riveted joints. From a practical point of view the sheet structure is always attached to a stiff frame, like an aircraft, and the skin is riveted and bolted to the frame. Figure 6 and 7 show the crack formation and the damage that will occur when these cracks initiate and propagate.

William L.K, (1985) investigated the stress behavior around a small circular hole in the HiMAT (Highly Maneuverable Aircraft Technology) composite plate. Yoshiaki Yasui and Kiyoshi Tsukamura (1987) have studied stress concentra-

Figure 6. A view of cracks emanate from a row of rivet holes (Kanninen et al., 1995)

Figure 7. Schematic view of fracture due to cracks propagates from holes (Kanninen et al., 1995)

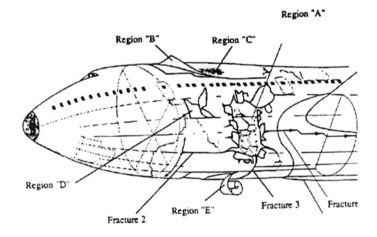

tion for FRP (Fiber Reinforced Plate) plates with a circular hole under biaxial tension. Moham-madi et al. (2006) studied the influence of eccentric circular cutouts on the prebuckling and post buckling stiffness, and effective widths of compression. Mittal and Jain (2008) investigated the effect of fiber orientation on stress concentration factor in a laminated composite plate with central hole under in-plane static loading.

Objective

Most of the previous work in defense hole design has been done for sheet metal plates (isotropic material). Some attempts are made for composite plate under uniaxial loading. In the current research, design and optimization of stress relief system for composite laminate is investigated and unveiled the optimum design parameters of the defense hole system. Moreover, it aims to obtain efficient, economical, reliable and optimum design of stress relief system for laminate composite plate that yields to significant reduction of maximum stress in structure.

The research has investigated the optimum design of the defense hole system by utilizing univariate search optimization technique (Chapra and Canal, 2006). The investigation has covered

Figure 8. Schematic drawing of defense hole system

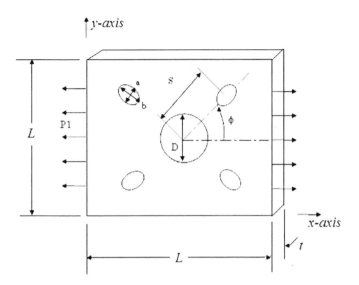

the defense hole system under in-plane uniaxial loading. The optimum size, shape and position of the defense holes, furthermore experimental verification for selected cases have been conducted. The Experimental verification for selected cases has been achieved by using RGB photoelasticity techniques.

PHYSICAL MODEL

This section presents the physical model of the composite plate, which includes geometry, boundary conditions as well as the materials used in the investigation.

Composite Plate Geometry

The plate is square in shape with circular hole in the middle. The side length of the plate is too large compared to the thickness and the diameter of the circular hole (L/D=16, L/t=100). So the plate can be considered as infinite. Two models are investigated, one with two elliptical holes and the other with four elliptical holes as defense

hole system. The four elliptical holes model is shown in Figure 8. To reduce the effect of the plate side length and edges effect, the DHS is rotated through an angle equal to the principal angle. The plate is large enough to be considered as infinite plate. Moreover the ratio (D/L) of the main-hole-diameter D to the plate-side-length L is 0.0625 (Fuchs and Stephens 1980). Table 1 presents a summary of the geometric parameters involved in the investigation.

All the geometric parameters are dimensionlized with respect to the main circular hole diameter. This has been done for generalizing the results produced by this research.

Assumptions

The following assumptions are considered:

1. The bonding between the plies of the laminate is considered perfectly bonded and the fibers are unbroken.
2. No delaminating occur between layers
3. Linear static loading

Table 1. Geometry parameters values

Parameter	Dimension	Note
Length L	400mm	constant
Width W	400mm	constant
Main Hole Diameter D	25mm	constant
Thickness t	4mm, 8mm, and16mm	variable
Design Hole Diameter (circular) d	3mm-25mm	variable
Design Hole (elliptical) diameters (a, b)	3mm-20mm	variable
Distance between two holes centers s	15mm-50mm	variable
Defense Hole System Orientation φ	0° - 90 °	variable
Material direction (Fiber Direction) FD	0° - 90 °	variable

Table 2. Study parameters values

Study Parameters	
Materials	E1/E2=1.0(Woven Glass), 1.72(Boron Aluminum), 10, 15, 20, 30, 43.94(Graphite Epoxy), 50, 63.77(Carbon Epoxy).
Design Hole	Two and Four circular holes, Two and Four elliptical holes
Distance s	Varies from 25 mm to 150 mm
Angle φ	Varies from 0° to 90°
Fiber Directions	Varies from 0° to 90°
Loading ratio	Uniaxial Tension
Layers	4 layers
Thickness	4 mm

4. The plate is large enough to be considered as infinite plate.

Study Parameters

The main parameters that have influence on the stress reduction ratio of laminate plate are; first is the material i.e. the ratio of modulus of elasticity for the fiber to the matrix, nine different materials are investigated each has different mechanical properties (modulus of elasticity, density, fiber epoxy ratio and others. Second is the dimensions of defense hole diameter d in case of circular design hole or minor diameter a and major diameter b in case of elliptical defense hole. Third is the distance s between main hole center and defense hole center. Fourth is the angle φ the defense hole system orientation angle measured between s line and the x-axis of the plate (see Figure 8). Fifth is the material orientation i.e. fiber direction. Table 2 shows all study parameters.

Material Mechanical Properties

Material properties for the composite materials such as woven Glass, boron Aluminum, Graphite Epoxy and Carbon epoxy are taken from (material handbook, 1991) where as the other materials are assumed theoretically to cover the study range for E1/E2 ratios from 1.0 to 63.77 so we have equally period's coverage. This gives designer the opportunity to fit any material by using dimensionless E1/E2 ratio in stress reductions figures by using interpolation or extrapolation. Table 3 shows the

Table 3. Material Mechanical Properties (www.matweb.com)

Material		Young's modulus (GPA)	Poisson's Ratio	Shear modulus (GPA)	E1/E2 Ratio
Woven Glass	Fiber	29.7	0.3	11.4	1.0
	Epoxy	29.7	0.3	11.42	
Boron Aluminum	Fiber	235	0.3	90.38	1.72
	Epoxy	137	0.3	52.69	
Material 10	Fiber	250	0.3	96.15	10
	Epoxy	25	0.3	9.62	
Material 15	Fiber	250	0.3	96.15	15
	Epoxy	16.67	0.3	6.41	
Material 20	Fiber	250	0.3	96.15	20
	Epoxy	12.5	0.3	4.81	
Material 30	Fiber	250	0.3	96.15	30
	Epoxy	8.33	0.3	3.21	
Graphite Epoxy	Fiber	294	0.3	113.08	45.94
	Epoxy	6.4	0.3	2.64	
Material 50	Fiber	250	0.3	96.15	50
	Epoxy	5.0	0.3	1.92	
Carbon Epoxy	Fiber	220	0.3	84.62	63.77
	Epoxy	3.45	0.3	1.33	

mechanical properties of different composite material. The composite plate is assumed to remain elastic at all times. Therefore only elastic material properties are required for the plates and they are presented in Table 3.

Applications

Composites are increasingly being used in the aerospace industry because of their bending stiffness-to-weight ratio. Floorboards, composite wing, horizontal stabilizer, composite rudder, landing gear door, speed brake, flap segments, aircraft interior and wingspans are typically made of sandwich composites. Also composites are ideally suited for the marine industries most advanced designs.

High strength-to-weight ratios of composites offer great advantages to the transportation industry. The insulating, sound damping properties and low cost properties make them the choice materials for the constructions of walls, panels and roofs.

FINITE ELEMENT MODEL

This section presents the development of finite element models for composite plate. Detailed descriptions of the boundary conditions, element types, and the loading are presented in this section. The finite element software used in the development of the finite models is I-DEAS Master Series 10. The relatively robust and user-friendly solid modeling and finite element meshing interface are the main advantages of this solid modeling and finite element software. A finite element model is the complete idealization of the entire structure problem, including node locations, elements, physical and material properties, loads and boundary conditions.

Construction of Composite Plate

A plate of size 400 mm x 400 mm with a hole at the middle of 25 mm diameter is modeled. The thickness of the composite plates is 4mm. So this size is adopted in the FEM. This laminate is built by number of plies; each ply material consists of fiber and epoxy which have been fed into I-DEAS program according to Table 3 so the ply can be created.

Model Description

Circular and Elliptical defense hole systems with different geometric parameters for a plate with main circular hole under uniaxial tension load are investigated. A defense hole system is introduced in the vicinity of the main hole. An infinite plate with circular hole of diameter D is clamped on one side (left side is free to move in y-axis). The right side is loaded by total force of 1000 N.

Laminate plate has different material properties, Fiber modulus of elasticity (E1), modulus of rigidity (G1) and Poisson's ratio (v1). The Matrix has also modulus of elasticity (E2), modulus of rigidity (G2) and poisons ratio (v2). Dimensionless material ratio E1/E2 is used to specify different materials which make the results applicable for any composite material.

Stacking sequence of 0°, 90°, 90°, 0° of laminates is studied in most cases of this research using I-DEAS laminates procedures by creating one ply of certain material (fiber and epoxy) then stacking the plies over each other in certain directions. Another parameter which has been investigated in this study is the fiber direction in each laminate for different material ratios (E1/E2). Figure 8 represents the model that is used in the current investigation. S is the distance between the centers of the main hole and the auxiliary holes. The diameter of the main hole is denoted by D. The major and minor axes of the auxiliary system are denoted by a and b respectively. All the parameters are dimensionlized; material stiffness ratio is E1/E2, the location and size parameters are s/D, a/D, and b/D.

The software package SDRC I-DEAS is used to produce the Finite Element Model (Lawry, 1998). Due to the limitations on mesh size in the software and to reduce computation time, element sizes around the main hole and the auxiliary holes are chosen smaller than those areas where no high stresses is expected.

Meshing

The plate is meshed with a thin shell mesh on one surface to analyze the stress at the main hole. Even though the part model is 3-D solid plate model, a 2-D finite element is needed since there is no bending load applied and only in plane forces are applied. In addition to that, the load difference and the non symmetry may cause membrane effect i.e. out of plane deformation, so solid mesh and plane stress (stress in depth) are not working.

Also shells for large dimensions are considered as flat plates since the curvatures are small. In comparison with steel, composites have the same behaviors for stress concentration. A fine mesh is utilized in the areas where the stress concentration is expected to be high. Figure 9 illustrates the mesh that has been used in the analysis. Since this research investigate infinite plate, large enough plate is used (L/D = 16, L/t = 100) to give accurate results for infinite plate.

A gradual change in the element size is employed to assure convergence and to obtain accurate results, as such, smaller elements near the points of high stress concentration are used, whereas larger elements far from the high stress spots are used. Also the size of the overall elements is decreased until the change in the results is lower than 1%.

Figure 9. (a): Plate thin shell meshing, (b): Design hole Shell Meshing

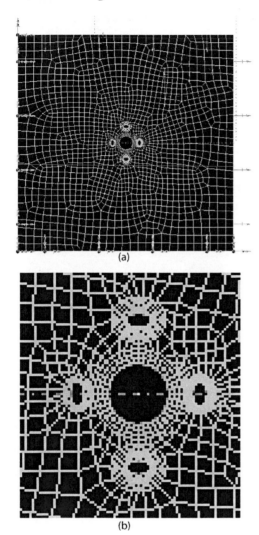

(a)

(b)

MODEL VERIFICATION

To assure validity and accuracy of FEM model comparison is carried out by two ways, first by reproducing other researches findings, second by carrying out experimental investigation for some selected cases using RGB photoelasticity. The experimental study is conducted to be more confident of the finite element model and its results. The comparison with the previous finite element analysis (FEA) and experimental findings shows very good agreement.

Redesign Optimization Technique (iterative optimization technique based on sensitivity analysis) is used in this study to find the optimum parameters of design hole system. This technique is widely used by researchers in the past three decades and has shown good results.

Comparison with Previous Research

*   William L.K, (1985) in his memorandum to National Aeronautical and Space Administration (NASA) investigated the stress behavior around a small circular hole in the HiMAT (Highly Maneuverable Aircraft Technology) composite plate. The results of our model show very good agreement against Williams' results. The stress concentration in our model appears at 54° as shown in Figure 10 whereas in Williams' study appears at an angle of 55° as shown in Figure 11. The difference is less than 2%.
*   Yoshiaki Yasui and Kiyoshi Tsukamura (1987) have studied stress concentration for FRP (Fiber Reinforced Plate) with a circular hole under biaxial tension. The calculation of stress is carried out using finite element method. The stress distribution around the hole in the plate is examined with respect to the fiber direction while the model is under biaxial load. The maximum stress concentration factor occurred at 90° for fiber direction $\theta=0°$ is 5.1 (see Figure 12) whereas the stress concentration factor obtained by our model is 4.98 (see Figure 13), the difference between our results and their results is 2.2%.

Material Mechanical Properties

Material properties of the composite materials such as woven Glass, boron Aluminum, Graphite Epoxy and Carbon epoxy are taken from (material

Figure 10. (a): Comparison with Williams' (1985) results for single ply laminate showing the stress concentration at 54°, (b): Closer image for the hole for the figure 10a that shows comparison with Williams' (1985) results for single ply laminate showing the stress concentration at 54°

(a)

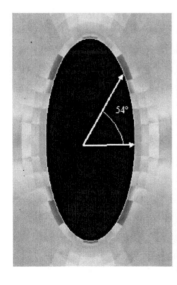

(b)

handbook, 1991) whereas the other materials are made up to cover the study range for E1/E2 ratios

from 1.0 to 63.77. This gives designer the opportunity to fit any material by using dimensionless E1/E2 ratio in stress reductions figures by using interpolation or extrapolation.

Experimental Setup

Photos of the experimental setup are presented in Figures 14 and 15. The setup consists of Light Source, Polarizer, two Quarter-wave plates, Analyzer and Specimen (that is coated by birefringent material). Figure 15 presents the machine that is utilized to load the specimens.

Experimental Procedure

1. Specimens are produced that match the numerical model.
2. Reflective adhesive is applied to the specimen.
3. Photo-elastic coating is applied.
4. Calibration specimen is produced under the same conditions that the test specimens are produced as shown in Figure 16 (Akour et al., 2003).
5. Correlation between the calibrated specimen and the test specimen is carried out and Table 4 is produced (Akour et al., 2003).

Each color represents fringe order that at the end represent strain (this could be related to the stress) level according equation 4. The equations 4 and 5 represent the relation for isotropic material. However for composite material, the relation of the fringe order with the principle stress is going to be a matrix rather than just a constant *(Ef / (1+v).* Figure 16 and table 4 represent the calibration specimen for isotropic material and the corresponding fringe order for each color intensity.

$$\varepsilon_x - \varepsilon_y = N f \qquad (4)$$

Figure 11. Williams' (1985) study of stress distribution (stress concentration) around a circular hole in a single ply composite plate

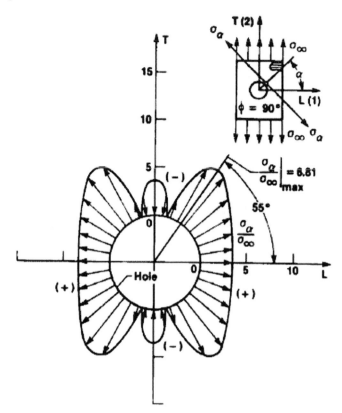

Figure 12. Stress distribution around a circular showing the maximum at 90° where Ky=Ny/Nx (Yoshiaki Yasui and Kiyoshi Tsukamura, 1987)

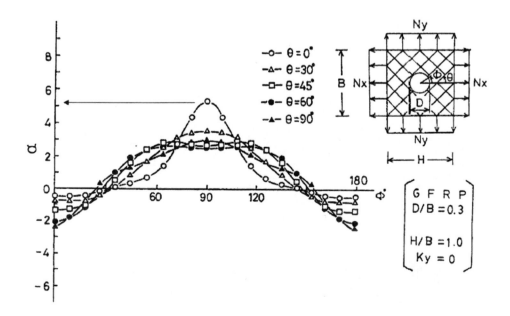

Figure 13. Illustration of the stress distribution around circular hole with similar conditions as in figure 12

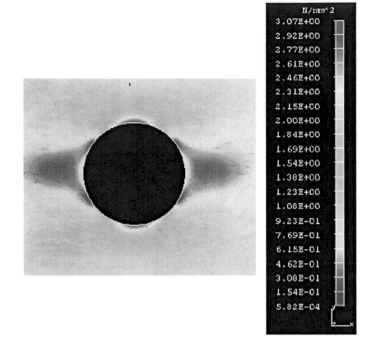

Figure 14. Photo of reflection polariscope setup that is manufactured for experimental verification

Figure 15. Photo of manufactured biaxial loading machine for carrying out the experimental verification

Where ε_x, ε_y principle strains, N = fringe order, and f = fringe value of coating

$$\sigma_x - \sigma_y = \frac{E}{1 + \nu} N \, f \qquad (5)$$

Where: $\sigma_x - \sigma_x$ = Principal stresses in test part surface

$E = E_f \upsilon_f + E_m \upsilon_m$ (Elastic modulus of test part)
$v = v_f \upsilon_f + v_m \upsilon_m$ (Poisson's ratio of test part)
Where: E_f modulus of the fiber, E_m modulus of the matrix

υ_f fiber volume fraction, υ_m matrix volume fraction

v_f Poisson's ratio of the fiber, v_m Poisson's ratio of the matrix (Reddy, 2004)

For the case of fiber reinforced composite material the calibration specimen is shown in Figure 17. Due to the anisotropic behavior of such material (carbon fiber composite), the fringe trajectories are not as smooth as those of isotropic material. The color intensities along the line shown are collected to be the reference that the specimen should be compared with, to find the corresponding fringe order. The line is selected where the variation of the stress is linear. The maximum value for the fringe order is 2.0 at the edge of the calibration specimen. Using MATLAB Image Processing toolbox the color intensity of the test specimen is

Figure 16. Calibration Specimen

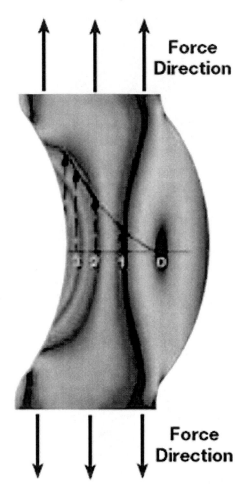

compared to those obtained from the calibration specimen. For uniaxial tension case (see Figures 17 through 21) with circular defense hole system the reduction obtained is 6.6% (± 1%) compared to the one obtained from the FEM study 5%. The experimental testing and analysis has been repeated to assure the results. The differences between the experiment and FEM results are due to many reasons. Even though the testing setup is carried in dark environment still some light has leaked into the camera and the test specimen. The loads applied by the biaxial loading machine that has been manufactured for this purpose is not 100% aligned.

Figures 17 and 18 present images of the specimens used in the investigation. Figure 17 presents the calibration specimen. The specimen without defense system is shown in figure 18. The specimen of two circular hole defense system is shown in Figure 20. Calibration error is eliminated by using the same conditions and environment; the same source of light, distance of image shooting, camera focusing, etc. Cutting operations of specimens samples are done slowly, and all samples are tested under no load to be sure that there is no any residual stresses, any specimen has residual stress is discarded of the experimental verification. High stress areas are investigated, since they are true candidate to be starting points for failure.

RESULTS AND DISCUSSION

The results of computational procedures involved numerous iterations. The optimization scheme focused on the following variables: shape of the design hole system; i.e. two circular, two elliptical, four circular or four elliptical shapes, size of the DHS i.e. diameters of the hole, the angle φ which represents the orientation of the defense hole system (refer to Figure 8). Redesign optimization module is utilized to find the optimum solution for each case (refer to I-DEAS user guide, Optimization book). The redesign optimization technique is an iterative technique that is based on sensitivity analysis in reaching the optimum design parameters. To maximize the benefit of this study the univariate search optimization technique is adopted to record the stress variation through the utilization of the parametric optimization module of I-DEAS. The parametric analysis is applied to record the routs that are leading to the optimum design.

The defense hole system theory is based on introducing the holes in the low stress spots within the vicinity of the main hole. So these holes raise the stress in these low stress spots to smooth the stress flow through the whole model.

Table 4. Correlations between the color and the fringe order for isotropic material (Akour, 2003).

	COLOR	APPROXIMATE RELATIVE RETARDATION		FRINGE ORDER *N*
		nm	in × 10⁻⁶	
	Black	0	0	0
	Pale Yellow	345	14	0.60
	Dull Red	520	20	0.90
	Red/Blue Transition	575	22.7	1.00
	Blue-Green	700	28	1.22
	Yellow	800	32	1.39
	Rose Red	1050	42	1.82
	Red/Green Transition	1150	45.4	2.00
	Green	1350	53	2.35
	Yellow	1440	57	2.50
	Red	1520	60	2.65
	Red/Green Transition	1730	68	3.00
	Green	1800	71	3.10

Figure 17. Photoelastic image of the calibration specimen of carbon fiber composite material using reflection polariscope arrangement. The black line represents the color intensities that have been extracted for correlation with test specimens.

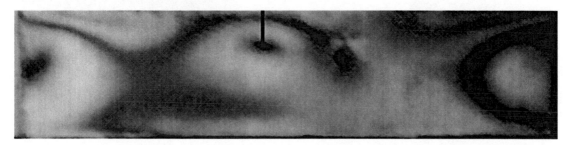

Figure 18. Photoelastic image of carbon-fiber laminate with main hole under uniaxial loading.

Figure 20. Photoelastic image of laminate under uniaxial loading with defense hole system under the same conditions of the specimen shown in Figure 18.

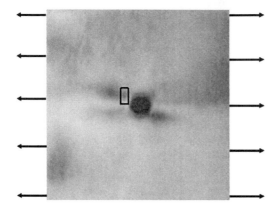

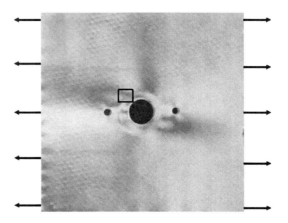

Figure 19. Illustration of the fringe order for the area enclosed in the box shown in Figure 18. Each division in the z axis represent 0.1 fringe order i.e., 15 divisions equal 1.5 fringe order. The values on the x-y axes represent the pixel coordinates.

Figure 21. Illustration of the fringe order of the area enclosed in Figure 20. Each division in the z axis represents 0.1 of the fringe order i.e. 14 divisions' equal 1.4 fringe order. The values on the x-y axes represent the pixel coordinates.

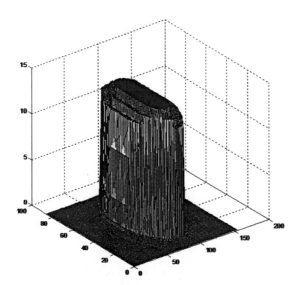

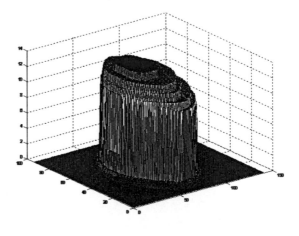

A shape study of the defense hole system for infinite plate has revealed that four Elliptical design hole system has the best stress reduction. Nine Different E1/E2 material ratios starting with Woven Glass composite material with E1/E2 ratio 1.0 to Carbon Graphite composite material with E1/E2 ratio equal to 63.77 are studied. Four of them are real materials while the other five materials are assumed to cover the range of E1/E2 ratios. So E1/E2 ratio of 1.00, 1.72, 10.0, 20.0, 30.0, 40.0, 45.94, 50.0, and 63.77 are investigated.

Results

Maximum stress reduction and optimum defense hole system parameters are achieved in a uniaxial loaded plate. For the cases under investigation, pure uniaxial load has the maximum stress reduction of 24.64%. This reduction is achieved by introducing four elliptical defense holes around the main hole. A two elliptical defense satisfy the stress reduction but less than four elliptical defense hole system. Finite element analysis is used to optimize the size and the location of the defense hole system.

Figure 22 shows the results of stress reduction versus different materials under uniaxial loading, the reduction increases as material E1/E2 ratio increases up to 45.94 ratios then starts descending.

Fiber directions are investigated for different materials under uniaxial load. The maximum reduction is obtained at zero fiber direction, while there is no any reduction at fiber direction 45 degrees for all materials as it can be seen in Figure 23.

Discussion

The optimum location and size of the defense hole system is identified for an infinite composite plate under uniaxial loading, and a baseline data for optimization is generated for different material ratios and different fiber directions. Maximum stress reduction is achieved at zero angles as shown in Figure 23, and as material ratio increases the

stress reduction increases up E1/E2 45.94 then stress reduction ratio starts decreasing.

Four elliptical defense hole system is the optimum number and shape of auxiliary holes for the case where uniaxial stress is dominant whereas introducing two elliptical defense hole system serves to reduce the maximum stress by approximately 5%. The stress for E1/E2=30 without defense hole system is 315mN/mm^2 and the stress with defense hole system is 303mN/mm^2 as it can be seen in Figure 24a, this means that the reduction is 5% whereas the E1/E2= 63.77 has a stress of 381mN/mm^2 for the case without defense hole system and 360mN/mm^2 with defense hole system as it can be seen in Figure 24b, furthermore the stress reduction is 5.5%.

CONCLUSION

The maximum reduction has been achieved for uniaxial tension loaded laminate is 24.64%. In this research, baseline data for optimizing defense hole systems under uniaxial load is introduced and achieved. This research determines the size and the placement of the defense holes that produce maximum stress reduction. The optimum location and size of the defense hole system is identified by using redesign optimization technique. The parametric study is carried out by utilizing univariate search optimization technique in an iterative manner for laminated composite material. The finite element model is verified by reproducing previous well known cases and also by carrying out experimental investigation for some selected cases using RGB photoelasticity.

The following bullets summaries the concluding remarks of this study:

- FEA study for fiber reinforced plate with and without defense hole systems are conducted. The model show good agreement along with previous work available in the literature. Also experimental verification

Figure 22. Defense hole system for a uniaxial load

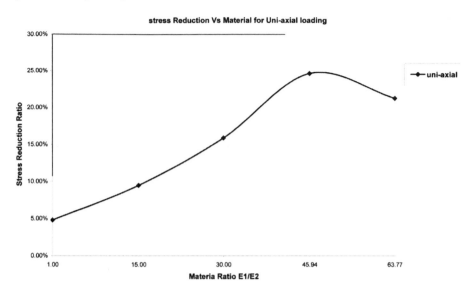

Figure 23. Stress Reduction Ratio Vs Fiber Directions for uniaxial load

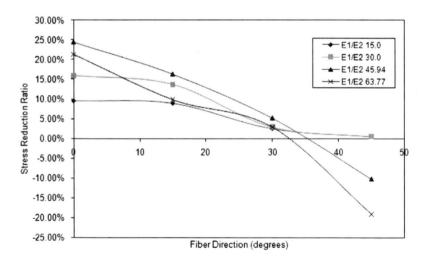

for the FEM model is carried out using RGB photelasticity. Introduction of circular defense hole system is not effective in the uniaxial case. However the elliptical defense system shows good results. The stress reduction achieved by the elliptical defense system is 24.64%.The optimum values for the defense hole system are a/D=0.24, b/D=.4, S/D=1.0

- Optimum size, shape and position of the auxiliary holes are identified for each case under certain conditions.

- Fiber direction has the largest stress reduction at zero and 90° degrees and smallest or no reduction at 45°.

- The effects of main parameters that are necessary in designing composite plates are unveiled.

Figure 24. (a): Material ratio E1/E2 30.0 under uniaxial loads (two elliptical DHS), (b): Material ratio E1/E2 63.77 under uniaxial loads (two elliptical DHS)

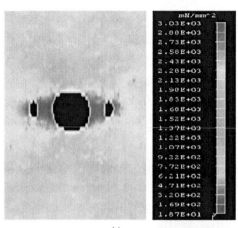

(a)

(b)

- The optimum number of defense holes is four holes of elliptical shape.
- The maximum stress reduction achieved by two-elliptical-hole or two-circular-hole defense system does not exceed 5%.
- It is obvious that the circular holes are special case of the elliptical shape when the major and the minor diameters of the ellipse are equal i.e., the circular shape is already included in this investigation.

REFERENCES

Akour, S., Nayfeh, J., & Nicholson, D. (2003). Design of defense hole system for shear loaded plate. *The Journal of Strain Analysis for Engineering Design, 38*(6), 507–518. doi:10.1243/030932403770735872

Akour, S., Nayfeh, J., & Nicholson, D. (2010). Defense hole design for shear dominant loaded plate. *International Journal of Applied Mechanics, 2*(2). doi:10.1142/S1758825110000548

Chapra, S., & Canal, R. (2006). *Numerical methods for engineers* (pp. 355–360). McGraw Hill.

Dhir, S. K. (1981). Optimization in a class of hole shapes in plate structures. *Journal of Applied Mechanics, 48*(4), 905–909. doi:10.1115/1.3157754

Dong-Shan, F., & Xing, Z. (1995). Analytical-variational method of analysis about a finite plate with a cracked hole stiffened by a ring with rivet joints. *Engineering Fracture Mechanics, 50*(3), 325–344. doi:10.1016/0013-7944(94)00205-V

Durelli, A. J., Brown, K., & Yee, P. (1978). Optimization of geometric discontinuities in stresses fields. *Experimental Mechanics, 18*(8), 303–308. doi:10.1007/BF02324161

Erickson, P. E., & Riley, W. F. (1978). Minimizing stress concentration around circular holes in uniaxially loaded plates. *Experimental Mechanics, 18*, 97–100. doi:10.1007/BF02325003

Fuchs, H. O., & Stephens, R. I. (1980). *Metal fatigue in engineering* (pp. 115–118). John Wiley & Sons.

Gungor, S., Nurse, A. D., & Patterson, E. A. (1995). Experimental determination of stress intensity factors of cracks in sheet structures with bolted stiffeners. *Engineering Fracture Mechanics, 32*(17/18), 2423–2445.

Jindal, U. C. (1983). Reduction of stress concentration around a hole in uniaxially loaded plate. *Journal of Strain Analysis, 18*, 135–141. doi:10.1243/03093247V182135

Jindal, U. C. (1983). Stress distribution around an oblong hole in a uniaxially loaded plate. *Journal of the Institution of Engineers India. Part ME, 64*, 66–70.

Kamel, B., & Liaw, M. (1991). Boundary element analysis of cracks at a fastener hole in anisotropic sheet. *International Journal of Fracture, 50*, 263–280.

Kanninen, M. F., & O'Donognue, P. E. (1995). Research challenges arising from current and potential applications of dynamic fracture mechanics to the integrity of engineering structures. *International Journal of Solids and Structures, 32*(17/18), 2423–2445. doi:10.1016/0020-7683(94)00275-2

Matweb. (n.d.). *Homepage*. Retrieved on February 12, 2008, from http://www.matweb.com

Meguid, S. A. (1986). Finite element analysis of defense hole systems for the reduction of stress concentration in a uniaxially loaded plate with two coaxial holes. *Engineering Fracture Mechanics, 25*(4), 403–413. doi:10.1016/0013-7944(86)90254-7

Meguid, S. A. (1989). *Engineering fracture mechanics*. Elsevier Science Publishers Ltd.

Mittal, N. D., & Jain, N. K. (2008). Finite element analysis for effect of fiber orientation on stress concentration factor in a laminated composite plate with central hole under in-plane static loading, *Material Science and Engineering, 498*(1-2: A), 115-124.

Mohammadi, B., Najafi, A., & Ghannadpour, S. A. M. (2006). Effective widths of compression loads of perforated cross ply laminated composite. *Science and Direct/Composite Structure Journal, 75*, 7-13.

Rajaiah, K., & Naik, N. K. (1984). Hole shape optimization in a finite plate in the presence of auxiliary holes. *Experimental Mechanics, 24*(2), 157–161. doi:10.1007/BF02324999

Wanlin, G. (1993). Stress intensity factors for corner cracks at holes subjected to biaxial and pin loads. *Engineering Fracture Mechanics, 46*(3), 473–479. doi:10.1016/0013-7944(93)90239-O

William, L. K. (1985). *NASA- Stress* concentration around a small circular hole in the HiMAT composite plate. NASA Technical Memorandum 88038, Dec. 1985.

Yoshiaki, Y., & Kiyoshi, T. (1987). Stress concentration for FRP plates with a circular hole under bi-axial tension. *Proceeding of the Faculty of Engineering of Tokyo University, 27*(2), 91-97.

Chapter 7
Design and Optimization of Defense Hole System for Hybrid Loaded Laminates

Salih N. Akour
Sultan Qaboos University, Oman

Mohammad Al-Husban
Civil Aviation Regulatory Commission, Jordan

Jamal F. Nayfeh
Prince Mohammad Bin Fahd University, Kingdom of Saudi Arabia

ABSTRACT

Stress reduction and increasing the load to weight ratio are two of the main goals of designers. Fiber reinforced composite materials are preferred by aerospace industry because of their high load to weight carrying capacity. Many parts of aircraft structures are assembled by bolts and rivets. These holes that are drilled for bolts and rivets work as stress raisers. Reducing the stress in the vicinity of these holes is a need. Defense hole theory deals with introducing auxiliary holes beside the main hole to reduce the stress concentration by smoothing the stress trajectories in the vicinity of the main hole. These holes are introduced in the areas of low stresses that appear near the main circular hole. Defense hole system under hybrid (shear and tension) loading is investigated. The optimum defense hole system parameters for a circular hole in an infinite laminated composite plate are unveiled. The study has been conducted using finite element method by utilizing commercial software package. The finite element model is verified experimentally using RGB-Photoelasticity. Digital Image Processing is utilized to analyze the photoelastic images.

Stress concentration associated with circular holes in hybrid loading (i.e., tension-compression ratios of 0.25, 0.50, and 0.75) achieved maximum reduction of 31.7%. This reduction is obtained by introducing elliptical defense holes along the principal stress direction. Finite element analysis is used to optimize the size and location for defense hole system. The effect of the stacking sequence, the fiber orientation, and the stiffness of both the fiber and the matrix are investigated.

DOI: 10.4018/978-1-60960-887-3.ch007

INTRODUCTION

Composite materials offer many advantages over metals such as: high strength and high stiffness-to-weight ratio, good fatigue strength, corrosion resistance and low thermal expansion. Composite materials such as glass fiber, aramide fiber, boron fiber and carbon-fiber-reinforced plastics have been used for a few decades, especially in the aircraft industry. Aircraft structures also include a large number of open holes and cut-outs e.g. holes for electric wires and hydraulic pipes or holes required for assembly or maintenance where a laminate containing open holes is subjected to shear loading i.e. biaxial loading case. Joining by mechanical fasteners is one of the common practices in the assembly of structural components. Among the most important elements in aircraft structures in general and in composite structures in particular are mechanically fastened joints. Improper design of the joints may lead to structural problems. Also, conservative design may lead indirectly to overweight structures and high life-cycle cost of the aircraft. Typical examples of mechanically fastened joints in composite aircraft structures are: the skin-to-spar/rib connections in e.g. the wing structure, the wing-to-fuselage connection etc. Since the failure of the joints can lead to the catastrophic failure of the structures, an accurate design methodology is essential for an adequate design of the joints.

Introducing auxiliary holes (defense holes) in the neighborhood of a main hole to reduce the stress concentration is called defense hole (DH) theory which has been known since the early years of last century. Most of the work that has been done so far deals with defense hole system (DHS) under uniaxial loading on sheet metals (isotropic material). Some efforts are done for shear loading.

Numerical and experimental studies for reducing stress levels in structures by introducing other geometric discontinuities are very few. Erickson and Riley (1978) investigated the effect of the defense holes on the stress concentration around the original hole using two-dimensional photo-elasticity. Jindal (1983) examined the reduction of stress concentration around circular and oblong holes using the finite element (FE) analysis and photo-elasticity analysis. Meguid (1986 & 1989) studied the reduction of stress concentration in a uniaxialy-loaded plate with two co-axial holes using the FE analysis. Rajaih and Naik (1986) investigated hole – shape optimization in a finite plate in the presence of auxiliary holes using the two dimensional photoelastic methods. Ulrich and Moslehy (1995) used boundary element methods to reduce stress concentration in plates by introducing optimal auxiliary holes. Durelli (1978) investigated the optimization geometric discontinuities in stress field under uniaxial loading. Dhir (1981) studied the hole shape optimization in plate structure under tension and shear loading.

Akour et. al (2003) studied the design of a DHS for a pure shear-loaded plate. Mittal and Jain (2008) investigated the effect of fiber orientation on stress concentration factor in a laminated composite plate with central hole under in-plane static loading.

Most of the previous work in DHS has been done for sheet metal plates (isotropic material). Some attempts are made to investigate the stress distribution in the vicinity of a circular hole for composite plate under uniaxial loading (William, 1985, Yoshiaki and Kiyoshi, 1987). In the current research, design and optimization of stress relief system for composite laminate is investigated and unveiled the optimum design parameters of the DHS.

The research has investigated the optimum design of the DHS by utilizing univariate and pattern search optimization technique (Chapra and Canal, 2006). This technique is multi-dimensional numerical optimization technique. DHS under hybrid in-plane loading is investigated. The optimum size, shape and position of the defense holes, the effect of laminate thickness and the stacking sequence such as cross-ply and angle ply laminate are revealed. The Experimental verification for selected cases is conducted using RGB photoelasticity techniques.

PHYSICAL MODEL

This section presents the physical model of the composite plate, which includes geometry, boundary conditions as well as the materials used in the investigation.

Composite Plate Geometry

The laminate plate under investigation is square in shape with circular hole in the middle. The side length of the plate is too large compared to the thickness and the diameter of the circular hole (L/t=100, L/D=16), so that the plate can be considered as infinite (Fuchs and Stephens 1980). Where L is the side length, t is the thickness and D is the circular hole diameter. Three models are investigated, one with two circular holes, another with four circular holes, and the last with four elliptical holes as DHS. To reduce the effect of the plate side length and edges effect, the DHS is rotated through an angle equal to the principal angle θ and a biaxial load (Tension-compression) is applied rather than the shear tension load. Figure 1 illustrates the model that is used for the current investigation. All the geometric parameters are dimensionlized with respect to the main circular hole diameter. This has been done for generalizing the date obtained by this research.

Assumptions

The following assumptions are considered:

1. The plies are perfectly bonded and the fibers are unbroken.
2. No delamination occurs between layers
3. Linear static loading
4. The plate is large enough to be considered as infinite plate.
5. Compression load is below buckling limit.

Figure 1. Schematic Drawing of DHS loaded by a) hybrid shear-tension, b) principal biaxial tension–compression

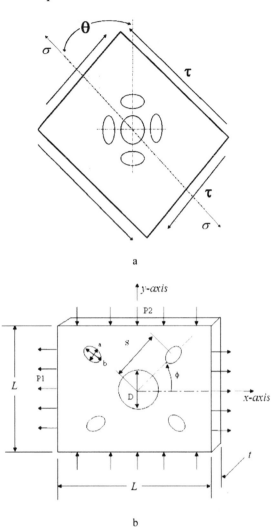

Boundary Condition

Two of the plate edges are clamped. The edge aligned to the *x*-axis is allowed to move freely in that direction. The other one which is aligned to the y-axis is allowed to move freely in that direction. So the bottom left corner is fixed. The other two edges are loaded in tension and compression.

Study Parameters

There are four main parameters that have influence on the stress reduction (calculated as the relative difference between same plate with and without DHS) of laminate plate. First is the material i.e. the ratio of modulus of elasticity for the fiber to the matrix (E1/E2=1.0 (Woven Glass), 15, 30, 43.94 (Graphite Epoxy), and 63.77(Carbon Epoxy)), five different materials are investigated each has different mechanical properties (modulus of elasticity, density, fiber epoxy ratio and others. Second is the size and the shape of the defense hole, diameter **d** in case of circular DH(arrangement of two circular hole and four circular holes) or minor diameter **a** and major diameter **b** in case of elliptical defense hole. Third is the location of DHS i.e. the distance **s** between main hole center and DH center. Fourth is the angle ϕ (varying from 0 to 90 degrees) between **s** line and the x-axis of the plate. Fifth is the fiber orientation (varying from 0 to 90 degrees). Also the effect of layers stacking sequence (symmetric, anti-symmetric), the thickness (4 mm, 8 mm and 16 mm) and the quantity of stacked plies (2 plies, 4 plies, 8 plies, and 16 plies) for some cases are highlighted.

Loading

The load is applied to the composite plate as a distributed load along two edges as shown in Figure 1. Tension is applied to one side while compression is applied to the other side. The compression side is varied from zero N to 750 N by a step of 250 N. The tension load is fixed and has a magnitude of 1000 N. The tension-compression load is applied along the principle stress directions. This loading is equivalent to the hybrid load shear-tension. The load is kept below the buckling limit of the laminate since this is beyond the scope of this study.

FINITE ELEMENT MODEL

The FE software used in the development of the FE models is I-DEAS Master Series 10, 1999. The relatively robust and user-friendly solid modeling and FE meshing interface are the main advantages of this FE software.

A plate of 400 mm x 400 mm with hole at center of 25 mm is modeled. This laminate is built by stacking of plies; each ply material consists of fiber and epoxy which have been fed into I-DEAS program.

The laminate plate has the following properties; the fiber has modulus of elasticity (E1), modulus of rigidity (G1) and Poisson's ratio (v1), the matrix has modulus of elasticity (E2), modulus of rigidity (G2) and poisons ratio (v2). Dimensionless parameter E1/E2 is introduced to define the materials. This makes the results applicable for any composite material that lies within the range of the investigated materials.

Stacking sequence of 0°, 90°, 90°, 0° of laminates is studied for most cases of this research using I-DEAS laminates module. The plies of certain material (fiber and epoxy) are created and stacked over each other in certain directions. Another parameter which has been investigated in this study is the fiber direction for each laminate used against different load ratio P2/P1 for different materials (E1/E2).

The software package SDRC I-DEAS is used to produce the Finite Element Model (Lawry, 1998). Due to the software limitations on the mesh size and to reduce computation time, element size around the main hole and the DHs are chosen to be smaller than the far away areas where no high stresses is expected. The plate is meshed using thin shell element, since there is no bending load applied and only in plane forces are applied.

Since this research investigate infinite plate, large enough plate is modeled (L/D= 16, L/t=100) to give accurate results for infinite plate. Also the size of the overall elements is decreased gradually until the change in the results is lower than 1%.

Table 1. Material mechanical properties (Www.Matweb.Com)

Material		Young's modulus (E) (GPA)	Poisson's Ratio (v1)	Shear modulus (G) (GPA)	E1/E2 Ratio
Woven Glass	Fiber	29.7	0.3	11.4	1.0
	Epoxy	29.7	0.3	11.42	
Material 15	Fiber	250	0.3	96.15	15
	Epoxy	16.67	0.3	6.41	
Material 30	Fiber	250	0.3	96.15	30
	Epoxy	8.33	0.3	3.21	
Graphite Epoxy	Fiber	294	0.3	113.08	45.94
	Epoxy	6.4	0.3	2.64	
Carbon Epoxy	Fiber	220	0.3	84.62	63.77
	Epoxy	3.45	0.3	1.33	

Material Mechanical Properties

Material properties for the composite materials such as woven Glass, boron Aluminum, Graphite Epoxy and Carbon epoxy are taken from (www.matweb.com). The other materials are made up to cover the study range for E1/E2 ratios from 1.0 to 63.77. This gives designers to obtain the optimum parameters of the DHS for any material lies within the range of E1/E2. Table 1 presents the mechanical properties of different composite material utilized in the current investigation. The composite plate is assumed to remain elastic at all times. Therefore only elastic material properties are required for the plates are presented in the table.

MODEL VERIFICATION

The model has been verified in two ways. First, some related cases in the literature are reproduced (William, 1985, Yoshiaki Yasui and Kiyoshi Tsukamura, 1987). The differences between the results of the current model and the literature are less than 2%. Second, experimental test is conducted. RGB (Red, Green and Blue Color intensities) Photoelasticity is used to verify the FE model (Akour et al 2003, Ajovalastit et al 1995). Reflec-

tion polariscope arrangement is set up. Colored images of the loaded specimens under investigation are collected through the polariscope. Using MATLAB Image Processing Toolbox, the pixel color intensity of the test specimen is correlated to those obtained from the calibration specimen. Those pixels that have similar color intensities in both the test specimen and the calibration specimen have the same stress. The test specimen and the calibration specimen are made of the same laminate. The comparison of the experimental results with the FE results shows a difference of 10% in the stress reduction. The experimental testing and analysis has been repeated to assure the results accuracy.

RESULTS AND DISCUSSION

The results of computational procedures involved numerous iterations. The optimization scheme focuses on the following variables: the shape of the DHS; i.e. circular or elliptical shape, the size of the DHS i.e. diameters of the hole, the angle ϕ between the orientation of DHS and the *x*-axis (refer to Figure 1), and the placement of the DHS. Redesign optimization module is utilized to find the optimum solution for each case (refer

Table 2. Optimum DHS parameters for $\phi=0$ and fiber angle zero

Load Ratio	E1/E2	a/D	b/D	s/D	% Reduction
0.25	1.00	0.24	0.40	1.4	4.0
	15.00	0.24	0.40	1-1.2	13.2
	30.00	0.24	0.40	1-1.2	21.5
	45.94	0.16	0.40	1.2-1.4	25.5
	63.77	0.16	0.40	1.2-1.4	20.9
0.5	1.00	0.32	0.56	1.2	10.2
	15.00	0.24	0.40	1	14.8
	30.00	0.20	0.40	1.44	22.3
	45.94	0.20	0.40	1.2	29.1
	63.77	0.24	0.40	1.2	19.9
0.75	1.00	0.36	0.60	1.2	13.5
	15.00	0.24	0.48	1.2	18.3
	30.00	0.24	0.40	1.4	20.8
	45.94	0.20	0.40	1.2	31.7
	63.77	0.20	0.48	1.2	27.4

to I-DEAS user guide, Optimization book). This technique uses the sensitivity analysis in reaching the optimum design parameters. To maximize the benefit of this study the univariate and pattern search optimization technique (multi-dimensional numerical optimization technique) is adopted to record the stress variation through the utilization of the parametric optimization module of I-DEAS. The parametric analysis is applied in a systematic iterative manner to exhaust all possible scenarios to reach the optimum design.

The DHS theory is based on introducing the holes in the low stress spots in the vicinity of the main hole. So these holes raise the stress in these low stress spots to smooth the flow of the stress trajectories in the model.

The shape study of the DHS for infinite plate has revealed that four Elliptical DHS has the best stress reduction. Five different E1/E2 material ratios starting with Woven Glass composite material with E1/E2 ratio 1.0 to Carbon Graphite composite material with E1/E2 ratio equal to 63.77 are studied. Four of them are real material while the other five materials are invented to cover the range of E1/E2 ratios.

The maximum stress reduction for loading compression to tension of 25%, 50% and 75% is 25.5%, 29.1% and 31.7 respectively for material E1/E2=45.94. This reduction depends on mechanical properties of composite material as it may be seen in Table 2. The optimum elliptical defense holes are found to be along the fiber direction and the loading direction. All dimensions are normalized by D the main hole diameter. Also all stresses are normalized by the maximum stress of the corresponding plate without DHS. Systematic approach based on the multi-dimensional numerical optimization technique, i.e. univariate and pattern search optimization technique, is utilizes (Chapra and Canal, 2006). FE analysis is used to optimize the size and location for auxiliary defense hole system.

Elliptical defense holes are introduced in the regions of low stress to reduce maximum stress. A circular two-defense-hole and four-defense-hole systems are investigated as well. The optimum shape of the DHS is elliptical. The maximum

Figure 2. (a): Percent Stress Reduction (calculated as the relative difference between same plate with and without DHS) Versus Fiber Direction (in Degrees) for different Load Ratio for Material E1/E2 = 15; (b): Percent Stress Reduction Versus Fiber Direction (in Degrees) for Different Load Ratios for Material E1/E2=30; (c): Percent Stress Reduction Versus Fiber Direction (in Degrees) for Different Load Ratios for Material E1/E2=45.94; (d): Percent Stress Reduction Versus Fiber Direction (in Degrees) for Different Load Ratios for Material E1/E2=63.77

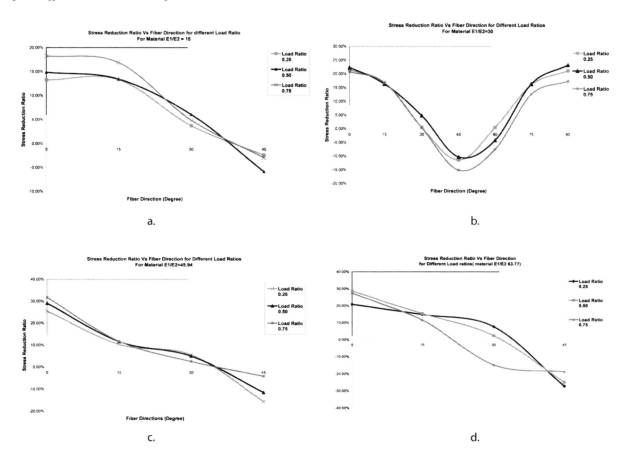

stress reduction for circular DHS that could be achieved is less than 5%. The stress reduction increases for four elliptical DHS as the compression/tension ratio increases. The maximum reduction of stress takes place always when the DHS is in alignment with the principal direction i.e. when φ is zero.

Stress reduction for different materials of the composite plate under mixed loads are presented in Table 2, this indicate that stress reduction increases as load ratio (compression to tension) increases. Also stress reduction increases as E1/E2

increases up to a certain limit around 45.94 then it starts to decrease. Maximum reduction reaches to 31.7% for Graphite epoxy at 0.75 load ratio, and minimum reduction reaches 4% for Woven Glass at 0.25 load ratio.

The criterion that has been considered in applying mixed loads in software is compression to tension force ratio. This study does not cover the load ratios 1 and 0 since they have different trend for the optimum size and location of the DHS.

An investigation of the effect of fiber direction is carried out for different materials and

Figure 3. Presentation of effect of number of plies stacked on the stress reduction for E1/E2=49.94 and load ratio of 0.5, and E1/E2=63.77 and load ratio (LDR) 0.75

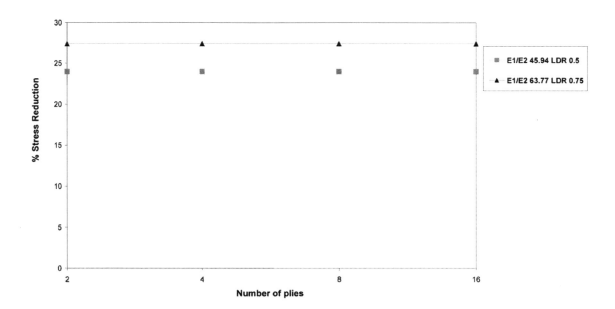

different load ratios. This investigation reveals that for stress reduction versus fiber direction is symmetric over the fiber angle of 45 degrees as it can be seen in Figure 2b. So the results on Figures 2a, c, and d show only the range of zero to 45 degrees of fiber angle.

Figure 2 show the results for each material ratio E1/E2 15, 30, 45.94 and 63.77 under three different load ratios 0.25, 0.50 and 0.75. Each material is investigated for full range of fiber angles i.e. 0°, 15°, 30°, 45°, 60°, 75° and 90° to be sure that symmetry is taking place at 45°.

Maximum stress reduction that has been obtained with respect to the fiber direction occurs at angle zero degree, this trend is observed by all materials and it is applicable for all investigated load ratios.

To find the stacking effect on the DHS, some optimum cases with 2 [0/90], 4 [0/90]$_s$, 8 [(0/90)$_2$]$_s$ and 16 [(0/90)$_4$]$_s$ layers are studied for the same laminate thickness. All of them show the same

results with differences less that 1%. It is obvious in Figure 3 that changing the number of layers does not change the amount of reduction. Anti-symmetric laminates have been considered too. However the anti-symmetric laminates experience an out of plane deformation even though the loading is in plane loading. This is due to non-zero B stiffness matrix which is responsible for the coupling of the membrane and bending effect.

Symmetry stacking sequence 0°, 90°, 90°, 0° is investigated against anti symmetry 0°, 90°, 0°, 90° with fixed and free Z movement. The results reveal two issues; first, there is a difference in the amount of stress reduction between the symmetric and the anti-symmetric when the edges of the anti-symmetric laminate are prevented to have out of plane deformation. The other issue is, the reduction in anti symmetry is less than symmetry laminate if the edges of the plate are allowed to deform freely in out of plane direction.

CONCLUSION

The optimum location and size of the DHS is identified by using redesign optimization technique and the parametric study is carried out by utilizing univariate and pattern search optimization technique, i.e. multi-dimensional numerical optimization technique, in an iterative manner for laminated composite material. The FE model is verified by reproducing some known cases in the literature and by carrying out experimental investigation for some selected cases using RGB photoelasticity. Infinite plate with mixed load (load ratios 0.25, 0.50 and 0.75) is investigated. The optimal design of the defense system is achieved.

- FE analysis study for all in plane loading types for plate with and without DHSs is conducted. The model show good agreement along with previous work available in the literature. Also experimental verification for the FE model is carried out using RGB photelasticity. The elliptical defense system shows good results. The stress reduction achieved by the elliptical defense system is 31.7%.
- Thickness of plates has no effect on stress reduction ratio and number of laminate plies as well.
- Out of plane deflection appears in the laminates that have anti-symmetric stacking due to the coupling in the stiffness matrix between membrane stiffness and bending stiffness.
- Fiber direction has the largest stress reduction at zero and 90° degrees.

REFERENCES

Ajovalastit, A., Barone, S., & Petrucci, G. (1995, September). Towards RGB photoelasticity: Full-field automated photoelasticity in white light. *Experimental Mechanics*, *35*, 193–200. doi:10.1007/BF02319657

Akour, S., Nayfeh, J., & Nicholson, D. (2003). Design of defense hole system for shear loaded plate. *The Journal of Strain Analysis for Engineering Design*, *38*(6), 507–518. doi:10.1243/030932403770735872

Chapra, S., & Canal, R. (2006). *Numerical methods for engineers* (pp. 355–360). McGraw Hill.

Dhir, S. K. (1981). Optimization in a class of hole shapes in plate structures. *Journal of Applied Mechanics*, *48*(4), 905–909. doi:10.1115/1.3157754

Durelli, A. J., Brown, K., & Yee, P. (1978). Optimization of geometric discontinuities in stresses fields. *Experimental Mechanics*, *18*(8), 303–308. doi:10.1007/BF02324161

Erickson, P. E., & Riley, W. F. (1978). Minimizing stress concentration around circular holes in uniaxially loaded plates. *Experimental Mechanics*, *18*, 97–100. doi:10.1007/BF02325003

Husban, M. (2009). *Design and optimization of stress relief system for laminate composite plate*. Unpublished doctoral dissertation, University of Jordan, Amman, Jordan.

Jindal, U. C. (1983). Reduction of stress concentration around a hole in uniaxially loaded plate. *Journal Strain Analysis*, *18*, 135–141. doi:10.1243/03093247V182135

Jindal, U. C. (1983). Stress distribution around an oblong hole in a uniaxially loaded plate. *Journal of Industrial Engineers India. Part ME*, *64*, 66–70.

Meguid, S. A. (1986). Finite element analysis of defense hole systems for the reduction of stress concentration in a uniaxially loaded plate with two coaxial holes. *Engineering Fracture Mechanics*, *25*(4), 403–413. doi:10.1016/0013-7944(86)90254-7

Meguid, S. A. (1989). *Engineering fracture mechanics*. Elsevier Science Publishers Ltd.

Mittal, N. D., & Jain, N. K. (2008). Finite element analysis for effect of fiber orientation on stress concentration factor in a laminated composite plate with central hole under in-plane static loading. *Material Science and Engineering, 498*(1-2: A), 115-124.

Rajaiah, K., & Naik, N. K. (1984). Hole shape optimization in a finite plate in the presence of auxiliary holes. *Experimental Mechanics*, *24*(2), 157–161. doi:10.1007/BF02324999

Ulrich, T. W., & Moslehy, F. A. (1995). Boundary element method for stress reduction by optimal auxiliary holes. *Engineering Analysis with Boundary Elements*, *15*(3), 219–223. doi:10.1016/0955-7997(95)00025-J

William, L. K. (1985). *Stress* concentration around a small circular hole in the HiMAT composite plate. NASA Technical Memorandum 88038, Dec. 1985.

Yoshiaki, Y., & Kiyoshi, T. (1987). Stress concentration for FRP plates with a circular hole under bi-axial tension. *Proceeding of the Faculty of Engineering of Tokyo University, 27*(2), 91-97.

Chapter 8
Design and Optimization of Defense Hole System for Shear Loaded Laminates

Mohammad Al-Husban
Civil Aviation Regulatory Commission, Jordan

Salih N. Akour
Sultan Qaboos University, Oman

Jamal F. Nayfeh
Prince Mohammad Bin Fahd University, Kingdom of Saudi Arabia

ABSTRACT

Reducing stress and weight of structures are most important to structural designers. Most engineering structures are an assembly of different parts. On most of these structures, parts are assembled by bolts, rivets, et cetera; this riveted and bolted structure is highly used in aerospace industry. Defense hole theory deals with introducing auxiliary holes beside the main hole to reduce the stress concentration by smoothing the stress trajectories in the vicinity of the main hole. These holes are introduced in the areas of low stresses that appear near the main circular hole. Defense hole system under shear loading is investigated. The optimum defense hole system parameters for a circular hole in an infinite laminated composite plate are unveiled. This study is conducted using finite element method by utilizing commercial software package. The finite element model is verified experimentally using RGB-Photoelasticity. Digital Image Processing is utilized to analyze the photoelastic images.

Stress concentrations associated with circular holes in pure biaxial shear-loaded laminates can be reduced by up to 20.56%. This significant reduction is made possible by introducing elliptical auxiliary holes along the principal stress direction. The effect of the stacking sequence, the fiber orientation, and the stiffness of both the fiber and the matrix are investigated.

DOI: 10.4018/978-1-60960-887-3.ch008

INTRODUCTION

Composite materials offer many advantages over metals such as: high strength and high stiffness-to-weight ratio, good fatigue strength, corrosion resistance and low thermal expansion. Composite materials such as glass fiber, aramide fiber, boron fiber and carbon-fiber-reinforced plastics have been used for a few decades, especially in aircraft industry. Aircraft structures include large number of open holes and cut-outs e.g. holes for electric wires and hydraulic pipes or holes required for assembly or maintenance where a laminate containing open holes is subjected to shear loading i.e. biaxial loading case. Joining by mechanical fasteners is one of the common practices in the assembly of structural components. Improper design of the joints may lead to structural problems. Conservative design may lead indirectly to overweight structures and high life-cycle cost of aircraft. Typical examples of mechanically fastened joints in composite aircraft structures are: the skin-to-spar/rib connections, the wing structure, the wing-to-fuselage connection etc. Since the failure of the joints can lead to the catastrophic failure of the structures, an accurate design methodology is essential for adequate design of the joints.

Introducing auxiliary holes in the neighborhood of a main hole to reduce the stress concentration is called defense hole (DH) theory which has been known since the early years of last century. Most of the work that has been done so far deals with defense hole system (DHS) under uniaxial loading on sheet metals (isotropic material). Some efforts are done for shear loading.

Numerical and experimental studies for reducing stress levels in structures by introducing other geometric discontinuities are very few. Erickson and Riley (1978) investigated the effect of the DHs on the stress concentration around the original hole using two-dimensional photoelasticity. Jindal (1983) examined the reduction of stress concentration around circular and oblong holes

using the FE Method and photo-elasticity analysis. Meguid (1986 & 1989) studied the reduction of stress concentration in a uniaxialy-loaded plate with two co axial holes using the Finite Element Analysis (FEA). Rajaih and Naik (1986) investigated hole – shape optimization in a finite plate in the presence of auxiliary holes using the two dimensional photoelastic methods. Ulrich and Moslehy (1995) used boundary element methods to reduce stress concentration in plates by introducing optimal auxiliary holes. Durelli (1978) investigated the optimization geometric discontinuities in stress field under uniaxial loading. Dhir (1981) studied the hole shape optimization in plate structure under tension and shear loading. Summary of the previous work is presented in Figure 1. Akour et. al (2003) studied the design of a DHS for pure shear-loaded plate. Mittal and Jain (2008) investigated the effect of fiber orientation on stress concentration factor in a laminated composite plate with central hole under in-plane static loading.

Most of the previous work in DH design has been done for sheet metal plates (isotropic material). Some attempts are made for composite plate under uniaxial loading. In the current research, design and optimization of stress relief system for composite laminate is investigated and unveiled the optimum design parameters of the DHS.

This research has investigated the optimum design of the DHS by utilizing univariate and pattern search optimization technique (Chapra and Canal, 2006). This technique is multi-dimensional numerical optimization technique. DHS under general in-plane biaxial shear loading is investigated. The optimum size, shape and position of the defense holes, the effect of thickness the laminate with different types of laminates such as cross-ply and angle ply laminate are obtained. The Experimental verification for selected cases is conducted using RGB photoelasticity techniques (Ajovalastit et al, 1995, Akour et al, 2003).

Figure 1. Summary of previous efforts of stress reduction by introducing geometric discontinuities for isotropic material

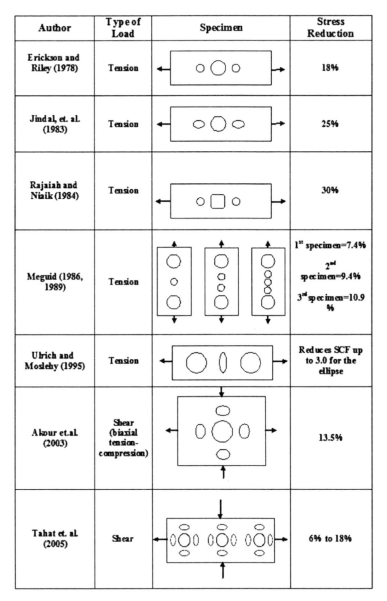

Author	Type of Load	Specimen	Stress Reduction
Erickson and Riley (1978)	Tension		18%
Jindal, et. al. (1983)	Tension		25%
Rajaiah and Naik (1984)	Tension		30%
Meguid (1986, 1989)	Tension		1st specimen=7.4% 2nd specimen=9.4% 3rd specimen=10.9%
Ulrich and Mosleby (1995)	Tension		Reduces SCF up to 3.0 for the ellipse
Akour et.al. (2003)	Shear (biaxial tension-compression)		13.5%
Tahat et. al. (2005)	Shear		6% to 18%

PHYSICAL MODEL

This section presents the physical model of the composite plate, which includes geometry, boundary conditions as well as the materials used in the investigation.

Composite Plate Geometry

The plate is square in shape with circular hole in the middle. The side length of the plate is too large compared to the thickness and the diameter of the circular hole (L/t=100, L/D=16). Where L is the side length of the plate, t is the thickness of the plate and D is the diameter of the main circular hole. So the plate can be considered as infinite. Three models are investigated, one with two circular holes, another with four circular holes, and the last with four elliptical holes as DHS. To reduce the effect of the plate side length and edges effect, the DHS is rotated through an angle equal to the principal angle θ and a biaxial load (Tension-compression) is applied rather than the shear traction. Figure 2 illustrates the model that is used for the current investigation. The plate is large enough to be considered as infinite plate. Moreover the ratio (D/L) of the main-hole-diameter D to the plate-side-length L is 0.0625 (Fuchs and Stephens 1980). All the geometric parameters are dimensionlized with respect to the main circular hole diameter. This has been done for generalizing the base line date that has been produced by this research.

Assumptions

The following assumptions are considered:

1. The bonding between the plies of the laminate is perfect and the fibers are unbroken.
2. No delaminating occur between layers
3. Linear static loading
4. The plate is large enough to be considered as infinite plate.
5. Compression load is below buckling limit (within elastic range).

Boundary Condition

Two of the plate edges are clamped. The edge aligned to the *x*-axis is allowed to move freely in that direction while the other one which is aligned to the *y*-axis is allowed to move freely in that direction. So the bottom left corner is fixed. The other two edges are loaded in tension and compression.

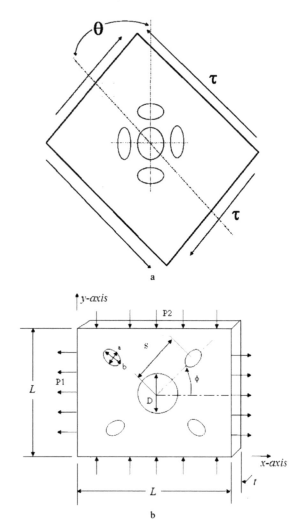

Figure 2. Schematic drawing of DHS under study a) shear loaded, b) biaxial shear loaded

Loading

The load is applied to the composite plate as a distributed load along two edges as shown in Figure 2. Tension is applied to one side while compression is applied to the other side. The load applied is 1000 N on each side. The load is kept below the buckling limit of the laminate since this is beyond the scope of this study.

Study Parameters

The main parameters that have influence on the stress reduction (calculated as the relative difference between same plate with and without DHS) of laminate plate are; first is the material i.e. the ratio of modulus of elasticity for the fiber to the matrix (E1/E2=1.0 (Woven Glass), 1.72 (Boron Aluminum), 10, 15, 20, 30, 43.94 (Graphite Epoxy), 50, 63.77(Carbon Epoxy)), nine different materials are investigated. Second is the size and the shape of the defense hole, diameter **d** in case of circular DH(arrangement of two circular hole and four circular holes) or minor diameter **a** and major diameter **b** in case of elliptical defense hole. Third is the location i.e. the distance **s** between main hole center and DH center. Fourth is the DHS orientation angle φ (varying from 0 to 90 degrees) between **s** line and the loading direction (*x*-axis of the plate). Fifth is the fiber orientation angle (varying from 0 to 90 degrees). Effect of layers stacking sequence (symmetric, anti-symmetric), thickness (4 mm, 8 mm and 16 mm) and the number of plies (2 plies, 4 plies, 8 plies, and 16 plies) for some cases are highlighted.

FINITE ELEMENT MODEL

The finite element (FE) software used in the development of the FE model is I-DEAS Master Series 10, 1999. The relatively robust and user-friendly solid modeling and finite element meshing interface are the main advantages of this solid modeling/ FE software. A FE model is the complete idealization of the entire structure problem, including the node locations, the elements, physical and material properties, loads and boundary conditions.

A plate of 400 mm x 400 mm with circular hole of 25 mm in the middle is modeled. This laminate is built by plies; each ply consists of fiber and epoxy, their material properties are fed to I-DEAS program.

The laminate plate has material properties as the following: The Fiber has modulus of elasticity (E1), modulus of rigidity (G1) and Poisson's ratio (v1), The matrix has modulus of elasticity (E2), modulus of rigidity (G2) and poisons ratio (v2). Dimensionless parameter E1/E2 is defined to designate the materials. This makes the results applicable for any composite material that falls within the studying range.

Stacking sequence of [0°/ 90°]s of laminates is studied for most cases of this research using I-DEAS laminate module. The plies of certain material (fiber and epoxy) are created and stacked over each other in certain direction.

Another parameter which has been investigated in this study is the fiber orientation with respect to the load direction for each laminate against different load ratios P2/P1 for different materials (E1/E2).

The software package SDRC I-DEAS is used to produce the Finite Element Model (Lawry, 1998). Due to the limitations on mesh size in the software and to reduce computation time, element sizes around main hole and auxiliary holes are chosen smaller than areas where no high stress is expected. The plate is meshed using linear thin shell elements. Since there is no bending load applied and only in plane forces are applied.

Since this research is intended to investigate infinite plate, large enough plate is used (L/D = 16, L/t =100) to give accurate results. Also the size of the overall elements is decreased gradually (including refinement in the high stress areas) until the change in the results is lower than 1%.

Material Mechanical Properties

Material properties for the composite materials such as woven Glass, boron Aluminum, Graphite Epoxy and Carbon epoxy are taken from website (www.matweb.com). The other materials are made up to cover the study range for E1/E2 from 1.0 to 63.77. These materials are made up by assuming the fiber stiffness 250 GPa and the epoxy stiffness is changing to produce E1/E2 =10, 15, 20, 30 40 and 50. This gives designers the opportunity to allocate any material within the range of E1/E2 ratio, i.e. interpolating the required material. The composite plate is assumed to remain elastic at all the time.

MODEL VERIFICATION

The model verification has been carried through two scenarios, the first is reproducing some related cases in the literature and the other is carrying out experimental verification by testing some selected cases experimentally using RGB photoelasticity.

William (1985) investigated the stress behavior around a small circular hole in composite plate. Some cases of William's research are reproduced. The results of our model show very good agreement against William's results i.e. the difference is less than 2%.

Yoshiaki Yasui and Kiyoshi Tsukamura (1987) have studied stress concentration for FRP (Fiber Reinforced Plate) with a circular hole under biaxial tension. Some cases of their work have been reproduced. The results obtained from our model are in very good agreement with their results i.e. the difference is 2.2%.

RGB (Red, Green and Blue Color intensities) Photoelasticity is also used to verify the FE model. Reflection polariscope arrangement is made (Akour et al 2003, Ajovalastit et al 1995). Colored images of the loaded specimens under investigation are collected through the polariscope. Using MATLAB Image Processing toolbox the color intensity of the test specimen is compared to those obtained from the calibration specimen. The comparison of the experimental results with the FE results shows a difference of 10% in the stress reduction. The experimental testing and analysis has been repeated to assure the results accuracy. The load is applied to the specimen using biaxial loading machine manufactured for

this purpose. The alignment of loading across the specimen is not 100%, this is why the difference is a little bit big. However both the FE model and the experimental model are in good agreement.

RESULTS AND DISCUSSION

The results of computational procedures involved numerous iterations. The optimization scheme focused on the following variables: shape of the DHS; i.e. two circular, four circular or four elliptical shapes, size of the DHS i.e. diameters of these holes, the DHS orientation angle ϕ i.e. the angle between the loading direction and the DHS axis (refer to Figure 2). Redesign optimization module of I-DEAS is utilized to find the optimum solution for each case (refer to I-DEAS user guide, Optimization book). The redesign optimization technique is based on sensitivity analysis in reaching the optimum design parameters. To maximize the benefit of this study the univariate and pattern search optimization technique (multi-dimensional numerical optimization technique) is adopted to record the stress variation through the utilization of the parametric optimization module of I-DEAS. The parametric analysis is applied in a systematic iterative manner to exhaust all possible scenarios to reach the optimum design.

The DHS theory is based on introducing the holes in the low stress spots within the vicinity of the main hole. So these holes raise the stress in these low stress spots to smooth the stress flow through the whole model.

A shape study of the DHS for infinite plate has revealed that four Elliptical DHS provides the best stress reduction. Nine Different E1/E2 material ratios starting with Woven Glass composite material with E1/E2 ratio 1.0 to Carbon Graphite composite material with E1/E2 ratio equal to 63.77 are studied. Four of them are real material while the other five materials are made up to cover the range of E1/E2 ratios. So E1/E2

Figure 3. Design hole system shape study, 2-CDHS and 4-CDHS: two and four circular DHS and 4-EDHS: four elliptical defense holes system

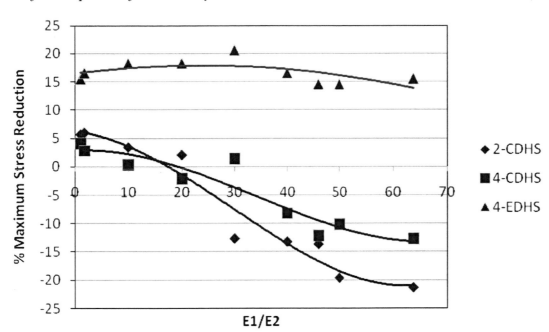

ratio of 1.00, 1.72, 10.0, 20.0, 30.0, 40.0, 45.94, 50.0, and 63.77 are investigated. Figure 3 shows the results. Maximum stress reduction for two circular design hole system reaches 5.69% for material E1/E2 equal to1.0. There is no stress reduction for E1/E2 greater than 20. Four circular DHS gives worse results and reaches to 4.07% as a maximum stress reduction ratio for E1/E2 of 1.0. The four elliptical DHS gives the best stress reduction ranging from 20.56% to 15.59% and valid for all investigated materials. It is obvious from Figure 3 that the maximum reduction occurs at E1/E2 of 30 and decreases by move above and below that value.

Concerning the fiber directions contribution in pure shear loaded composite plate; Figure 4 illustrates these results. It can be seen that the best reduction occurs at zero angle of fibers direction for all materials i.e. the maximum reduction occurs when the fiber direction is aligned with the loading direction. Due to the symmetry of the

curve around the fiber angle 45 only half of it is illustrated in Figure 4.

It is observed that the maximum reduction for all cases occurs when the DHS placed at ϕ equal to zero. Figure 4 shows that the maximum reduction occurs when the fiber direction angle is zero degree for all materials i.e. when both the fiber and the DHS are aligned with the load direction (the principle direction). Since the load is aligned with the fiber direction, the tension load will be carried totally by the fibers.

Summary of hundreds of data points showing only the optimum cases is presented in Table 1. It shows the optimum size and location of the DHS. Since the variation in the size and location of the optimum values is small, only the ratios E1/E2 of approximately step 15 are presented in the table. These optimum cases are taking place when the fiber direction and the axes of the DHS are aligned with load i.e. aligned with the principle direction.

Figure 4. Stress Reduction Vs Fiber Direction for pure shear load

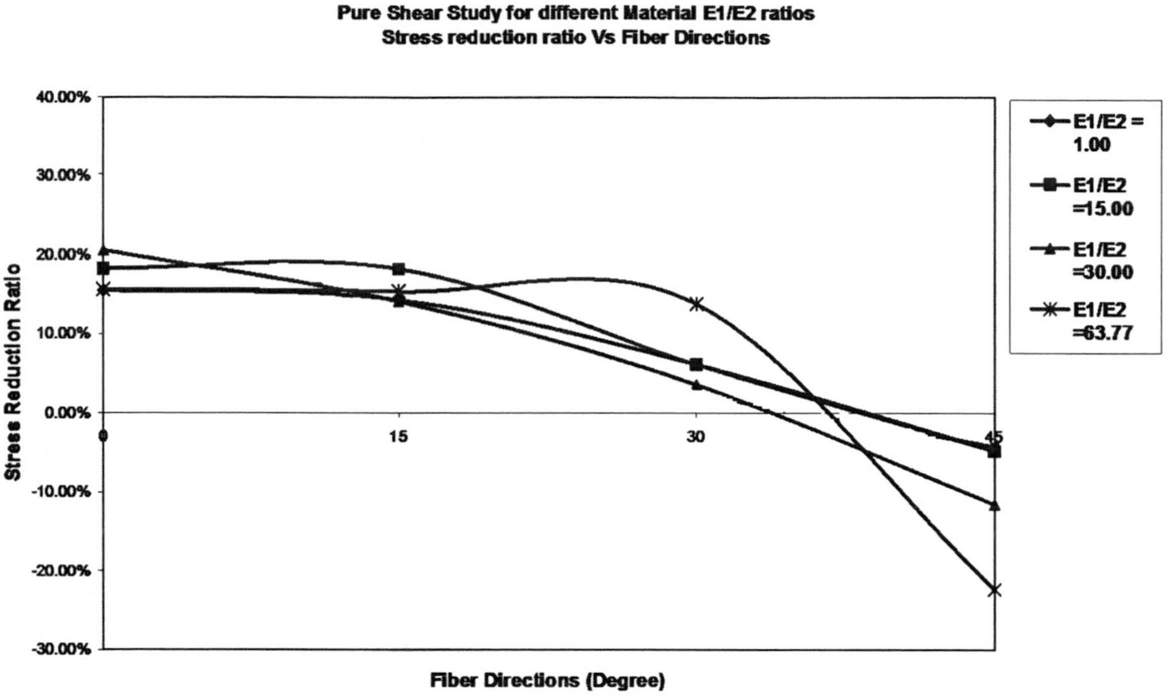

Pure Shear Study for different Material E1/E2 ratios
Stress reduction ratio Vs Fiber Directions

Table 1. Summary of the optimum cases

E1/E2	DHS a/D	DHS b/D	Dist. S/D	% Reduction
1	0.4	0.6	1.24	15.4
15	0.24	0.4	1	18.2
30	0.24	0.4	1.2	20.6
45.94	0.24	0.4	2	17.6
63.77	0.24	0.4	2	15.6

The thickness of the laminate and the number of plies are varied however the optimum cases do not change i.e. the stress reduction is the same. However the antisymmetric cases have started to experience out of plan deformation. This is due to the coupling in the stiffness matrix (B stiffness matrix is not zero, this matrix is responsible for the coupling of the bending and membrane effect) and this is beyond the scope of this investigation (Al-Husban, 2009).

CONCLUSION

Most of the previous work in DHS optimization has been performed for isotropic plates. In this research, optimum DHSs for infinite (large enough to be considered infinite) under shear load is achieved. The optimum size and placement for the defense holes that produce the maximum reduction of stress concentration are obtained.

The optimum location and size of the DHS are identified by using redesign optimization technique. The parametric study is carried out by utilizing univariate and pattern search optimization technique in an iterative manner for laminated composite plate. The FE model is verified by reproducing some previous cases from the literature as well as by carrying out experimental investigation for some selected cases using RGB photoelasticity. The model show good agreement with previous work available in the literature and with experimental verification.

FE analysis study for shear loaded plate with and without DHS is conducted. Introduction of circular DHS (2 and 4 circular holes) is not beneficial in the pure shear case. However the elliptical DHS shows good results. The stress reduction achieved by the elliptical defense system is 20.6% for shear loaded plates. Optimum size, shape and location of the defense holes are identified.

The following points summarize the main concluding remarks:

- The maximum stress reduction occurs at material E1/E2=30 while the amount of the reduction decreases below an above this ratio.
- The optimum DHS is aligned with the loading direction and the fiber direction.
- The optimum DHS is located in the low stress regions that appear in the vicinity of the main hole before the introduction of the DHS.
- The thickness of the laminate does not affect the optimum DHS as long as the stacking sequence and the fiber orientation are the same.
- The number of plies used in building the laminate does not affect the optimum DHS as long as the fiber orientation and the stacking sequence are same.
- Symmetric stacking sequence shows in plane deformation only while the ant-symmetric stacking sequence shows both in plane and out of plane deformation. This is due to the coupling of the bending and

membrane stiffness in the non symmetric case.
- The maximum stress reduction is symmetric around the 45 degrees angle, where the maximum reduction occurs at 0 degree angle, i.e. when the fiber and the DHS are aligned with loading direction.

ACKNOWLEDGMENT

The authors would like to express their gratitude to 'Seabird Aviation Jordan", Marka Airport, Amman, Jordan and "Jordan Aerospace Industries", Amman International Airport, Amman, Jordan, for their support.

REFERENCES

Ajovalastit, A., Barone, S., & Petrucci, G. (1995, September). Towards RGB photoelasticity: Full-field automated photoelasticity in white light. *Experimental Mechanics*, *35*, 193–200. doi:10.1007/BF02319657

Akour, S., Nayfeh, J., & Nicholson, D. (2003). Design of defense hole system for shear loaded plate. *The Journal of Strain Analysis for Engineering Design*, *38*(6), 507–518. doi:10.1243/030932403770735872

Chapra, S., & Canal, R. (2006). *Numerical methods for engineers* (pp. 355–360). McGraw Hill.

Dhir, S. K. (1981). Optimization in a class of hole shapes in plate structures. *Journal of Applied Mechanics*, *48*(4), 905–909. doi:10.1115/1.3157754

Durelli, A. J., Brown, K., & Yee, P. (1978). Optimization of geometric discontinuities in stresses fields. *Experimental Mechanics*, *18*(8), 303–308. doi:10.1007/BF02324161

Erickson, P. E., & Riley, W. F. (1978). Minimizing stress concentration around circular holes in uniaxially loaded plates. *Experimental Mechanics, 18*, 97–100. doi:10.1007/BF02325003

Husban, M. (2009). *Design and optimization of stress relief system for laminate composite plate.* Unpublished doctoral dissertation, University of Jordan, Amman, Jordan.

Jindal, U. C. (1983). Reduction of stress concentration around a hole in uniaxially loaded plate. *Journal of Strain Analysis, 18*, 135–141. doi:10.1243/03093247V182135

Jindal, U. C. (1983). Stress distribution around an oblong hole in a uniaxially loaded plate. *Journal of the Institute of Engineers India. Part ME, 64*, 66–70.

Meguid, S. A. (1986). Finite element analysis of defense hole systems for the reduction of stress concentration in a uniaxially loaded plate with two coaxial holes. *Engineering Fracture Mechanics, 25*(4), 403–413. doi:10.1016/0013-7944(86)90254-7

Meguid, S. A. (1989). *Engineering fracture mechanics.* Elsevier Science Publishers Ltd.

Mittal N. D., & Jain, N. K. (2008). Finite element analysis for effect of fiber orientation on stress concentration factor in a laminated composite plate with central hole under in-plane static loading. *Material Science and Engineering, 498*(1-2: A), 115-124.

Rajaiah, K., & Naik, N. K. (1984). Hole shape optimization in a finite plate in the presence of auxiliary holes. *Experimental Mechanics, 24*(2), 157–161. doi:10.1007/BF02324999

Ulrich, T. W., & Moslehy, F. A. (1995). Boundary element method for stress reduction by optimal auxiliary holes. *Engineering Analysis with Boundary Elements, 15*(3), 219–223. doi:10.1016/0955-7997(95)00025-J

William, L. K. (1985). *NASA- Stress concentration around a small circular hole in the HiMAT composite plate.* NASA Technical Memorandum 88038, Dec. 1985.

Yoshiaki, Y., & Kiyoshi, T. (1987). Stress concentration for FRP plates with a circular hole under bi-axial tension. *Proceeding of the Faculty of Engineering of Tokyo University, 27*(2), 91-97.

Chapter 9
Effect of Core Thickness on Load Carrying Capacity of Sandwich Panel Behavior Beyond Yield Limit

Salih Akour
Sultan Qaboos University, Oman

Hussein Maaitah
Royal Jordanian Air Force, Jordan

Jamal F. Nayfeh
Prince Mohammad Bin Fahd University, Kingdom of Saudi Arabia

ABSTRACT

Sandwich Panel has attracted designer's interest due to its light weight, excellent corrosion characteristics and rapid installation capabilities. It has been implemented in many industrial application such as aerospace, marine, architectural and transportation industry. Its structure consists of two face sheets and core. The core is usually made of material softer than the face sheets. The current investigation unveils the effect of core thickness on the behavior of Sandwich Panel beyond the yield limit of core material. The core thickness is investigated by utilizing univariate search optimization technique. The load is applied in quasi–static manner (in steps) till face sheets reach the yield limit. Simply supported panel from all sides is modeled using a finite element analysis package. The model is validated against numerical and experimental cases that are available in the literature. In addition, experimental investigation has been carried out to validate the finite element model and to verify some selected cases. The finite element results show very good agreement with the previous work and the experimental investigation. The study presents that the load carrying capacity of the panel increases as the core material goes beyond the yield point. Also, increasing core thickness to a certain limit delays the occurrence of core yielding and gives opportunity to face sheets to yield first.

DOI: 10.4018/978-1-60960-887-3.ch009

Figure 1. Illustration sandwich plate geometry

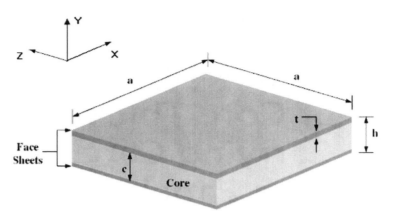

INTRODUCTION

Researchers are continuously looking for new, better and efficient construction materials. The main goal of these researches is to improve the structural efficiency, performance and durability. Light-weight, excellent corrosion characteristics and rapid installation capabilities created tremendous opportunities for sandwich panel structure in industry. Sandwich panel normally consists of a low-density core material sandwiched between two high modulus face (usually made of metal) skins to produce a lightweight panel with exceptional stiffness as shown in Figure 1. The face sheets act like the flanges of I-beam, while the core carries the shear forces. The faces are typically bonded to the core to achieve the composite action and to transfer the forces between the components.

Work on the theoretical description of sandwich structure behavior has started on the early years of the second half of the 20th century. Plantema (1966) published the first book about sandwich structures, followed by a book written by Allen (1969), and more recently a book by Zenkert (1995). Although Triantafillou and Gibson (1987) developed a method to design for minimum weight sandwich panel, and reported the failure mode map of sandwich construction, however they

didn't consider the post yield state of the sandwich structure.

Mercado and Sikarskie (1999) reported that the load carried by sandwich structures continues to increase after core yielding. Knowing that the core could not carry additional load after yield, this increasing load carrying capacity of post yield sandwich structure initiates the postulation that the additional shear load was transferred to the face sheets. To account for the above-mentioned phenomenon, Mercado et al. (2000) developed a higher order theory by including a bilinear core material module. This theory yields a fairly accurate prediction on the deflection of a foam cored sandwich beam in four point bending (Mercado et al., 2000). In addition, this theory does not take into account the core compression under localized load, or any geometric non-linearity. The classical sandwich beam theory assumes that in-plane displacements of the core through its depth are linear. In other words, it is assumed that the core thickness remains constant and cross-sections perpendicular to the neutral axis remain plane after deformation. This assumption is generally true for traditional core material such as metallic honeycomb (Frostig et al., 1992). However, this assumption is not suitable for soft, foam-based cores, especially when the sandwich structure is subjected to a concentrated load (Thomsen,

1995). With a much lower rigidity compared to metallic honeycomb, foam-based cored sandwich structures are susceptible to localized failure. Insufficient support to the face sheets due to core compression near the application points of concentrated loads can lead to failures such as face sheet/ core delamination, face sheet buckling, and face sheet yielding. This localized non-linearity is reported by many researchers such as (Thomsen, 1993, 1997), Caprino (2000) and Gdoutos et al. (2001) but the shear distribution at localized failure points was not well defined. Miers (2001) investigated the effect of localized strengthening inserts on the overall stiffness of a sandwich structure. This localized strengthening increased the rigidity of the sandwich structure, but the addition of high stiffness inserts complicates the manufacturing process of sandwich structure. The two most popular theories that include these localized effects are the superposition method (Zenkert 1997) and high order theory (Frostig et al., 1992).

To design an efficient sandwich structure, it is vital to understand the behavior of each layer in the structure. Classical sandwich theory (Zenkert 1995, Plantema 1966, Allen 1969), higher order theory (Mercado, 2000) and high order theory (Frostig et al., 1992) could predict the sandwich panel behavior fairly accurate in the linear range. However, these theories could not give an accurate prediction of the sandwich structure behavior after core yielding. Large deflection of sandwich structures due to core yielding could vary the direction of the applied load on the structure. Finite Element (FE) analysis is utilized to investigate the response of sandwich panel under distributed load. Geometric nonlinearity and material nonlinearity are considered in this investigation to unveil the behavior of sandwich panel beyond core yielding. Core thickness effect is investigated by utilizing the univariate search optimization technique (Chapra and Canal, 2006).

PHYSICAL MODEL

The sandwich panel consists of two face sheets made of metal. The thickness of each face sheet is **t**. Soft core of **c** thickness is sandwiched between those face sheets. The core material is made of foam which is soft compared to the face sheets. The panel is square in shape. The side length is designated by **a** whereas the overall thickness is designated by **h**. Figure 1 illustrates the sandwich panel geometry. The values of **a**, **t** and **c** are selected as the following: **a** is 608mm, **t** is 1.0mm and **c** is varied from 15.0mm to 50.0mm.

Assumptions

This research takes into consideration the geometric non-linearity as well as the material non-linearity. The following assumptions are made to simplify the model without losing the physics of the problem

1. Face sheets and core are perfectly bonded i.e. no delamination occur between layers.
2. Face sheets remain elastic at all time; Due to the significantly higher yield strength and modulus of elasticity of the face sheets compared to the core, face sheets are assumed to remain elastic throughout loading. The analysis stops when the face sheets start to yield.
3. The panel is simply supported from all sides.
4. Geometric non-linearity and core material non-linearity are considered.

Boundary Condition

Due to the symmetry only quarter of the sandwich panel is modeled. The loading area is square in shape; its side length is 100mm for full panel dimension. However for quarter model of the panel, the side-length of the loading area is 50mm. Figure 2 illustrates this configuration.

Figure 2. Panel span overview of quarter sandwich panel for different loading area

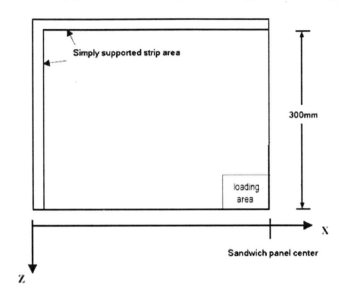

Table 1. Compression of sandwich panel material properties

Material	Property source	Young's modulus (MPa)	Poisson's ratio	Shear modulus (Mpa)	Shear strength (Mpa)	0.2% offset yield strength (Mpa)	Straign at yield popup (mm/mm)
Face sheet: Aluminum 2002 H14	Boyer and Gall 1991	69,000	0.33	25,000	120	145	Not available
Core: H100	Kuang, 2001	138.6	0.35	47.574	1.2	1.5	0.0108225

The load is applied to the sandwich top face sheet as a distributed load which is increased gradually (step by step) till face sheet stress reaches its yield strength. A distributed load is applied on the top surface of the sandwich panel. The area on which the distributed load is applied is located in the middle of the top face sheet plate (see Figure 2). The loading area in the middle of the top face of the sandwich panel is square in shape.

The core thickness plays important role in the performance of the sandwich structure. The core thickness is varied from 15mm, through 20mm, 25mm, 30mm, 40mm, to 50mm.

Table 1 presents the materials used in the current investigation. The first row shows the

mechanical properties of the face sheet material while the other row shows the core material. Figure 3 presents the stress strain curves of the core material. These materials are selected because of their wide usage in the industry.

FINITE ELEMENT MODEL

The FE package used in the development of the FE models is (I-DEAS Master Series 10 1999). The relatively robust and user-friendly solid modeling and FE meshing interface are the main advantages of this solid modeling/ FE software.

The non-linear analysis capabilities of I-DEAS are utilized in carrying the FE analysis of the model

Figure 3. Stress strain curve for material H100 (Kuang, 2001)

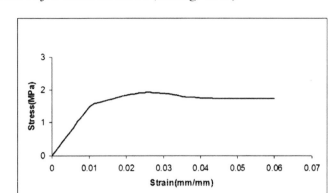

Figure 4. Sandwich panel boundary condition and loading

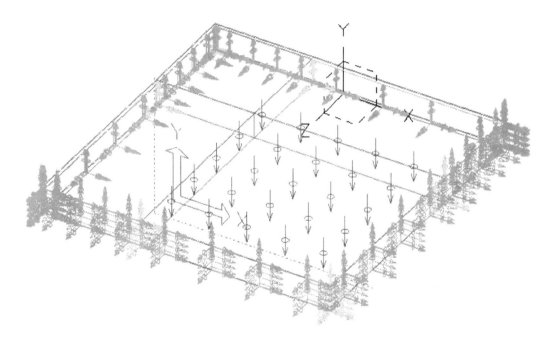

that includes geometric and material nonlinearity. Load is applied to the model in quasi-static manner by utilizing the load increment module of I-DEAS. This is similar to the actual experiment where increment loading is applied slowly to simulates the real life). Therefore, the type of analysis done for this research effort is "static, non-linear analysis".

The symmetric nature of the problem allows quarter of the whole panel to be meshed.

The boundary conditions applied are shown on Figure 4.

Symmetry boundary conditions are applied to the planes of symmetry. A simply supported boundary condition is applied to the outer sides of the quarter panel. A distributed load is applied on the top surface of the sandwich panel.

The FE software is set in such a way to solve the model at each load step. This allows all analysis of each case to be done in a single run of the FE

Figure 5. Meshed quarter sandwich plate

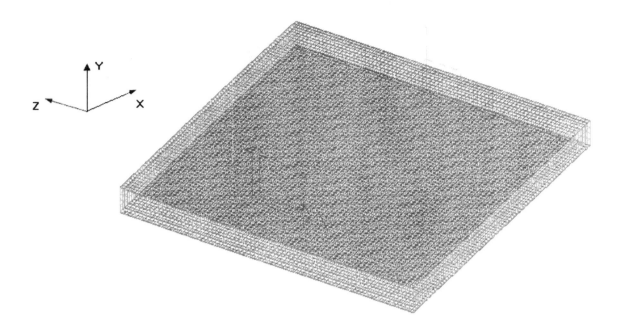

model. As a result of this, the model takes less memory space because one single solid model and FE model are used for all load steps of each case.

The numerical model utilizes the map meshing facility in I-DEAS. By controlling the number of nodes along each edge of the solid model, this function provides full control of the mesh size. The mesh is refined till the change in the result is less than 0.5%. Constant mesh density is ensured with the mapped meshing function. This is important because constant mesh density ensures that data collected from any region of the plate is of the same degree of resolution. Three-dimensional (solid) brick elements are used in this analysis. Second order (parabolic) brick elements are chosen over the first order (linear) brick elements in order to better interpolate the data between nodes. Figure 5 shows the FE mesh model of the sandwich panel.

Since the analysis involves material non-linearity, a yield function or yield criteria needs to be defined for the model. Von Mises yield criteria and its associated flow rule is used in this analysis. Isotropic hardening is also used to describe the change of the yield criterion as a result of plastic straining. Only the core elements are assigned a yield function due to the assumption that only core yielding occurs throughout the loading process. The face sheets are assumed to remain elastic at all times; hence no yield function is assigned to the face sheet elements. However the yield point of the face sheet material is fed to the software to be used as indicator for analysis termination.

Model Verification

Some cases in the literature are reproduced using the FE model. The relative difference in the results is less than 1%. To be more confident of the FE model and its results, experimental verification is carried out. The sandwich panel used in the experimental investigation is made of polyurethane foam and steel sheets. The mechanical properties are obtained experimentally for both the core and the sheets according to ASTM Designation:

Figure 6. Comparison of load versus center deflection for core thickness = 49 mm, Sheet Thickness = 0.5 mm, applied load area = 200 mm*200 mm

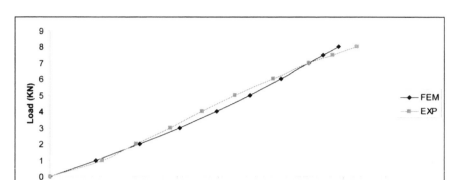

C 365 – 00 and ASTM Designation D 638 – 00 respectively.

The relation between the applied load and the deflection of the specimen center point are presented in Figure 6 for both the experimental and FE. It may be seen that the results are in very good agreement. The maximum relative error does not exceed 7%. The experimental tests are repeated more than two times and the average values are plotted in Figure 6.

RESULTS AND DISCUSSION

The thickness effect on the behavior of sandwich panel is investigated. These results are very beneficial for design engineers to obtain (or to select) the optimum parameters that fit their design. The main advantage of this investigation over the sandwich panel theory is that both geometric and material nonlinearities are considered without approximation. Usually these approximations eliminate part of the problem physics. By utilizing "I-DEAS' post processing module, stress and it is all components, strain and it is all components including the plastic strain, and deformations are obtained. The results of this work are generated according to the univariate search optimization technique (Chapra and Canal, 2006).

It is observed that the plastic deformation occurs close to the panel support (close to the area where boundary conditions are applied). This is physically true, the distributed load over the loading area becomes reaction force concentrated on the strip area on which the boundary conditions (simply supported boundary condition) are applied, and i.e. distributed load is converted to concentrated load. So the area where the boundary conditions are applied reaches the yield stress range before any other part of the panel. The criterion that is adopted by this investigation, at what load step the FE model should stop is, when anyone of the face sheets starts to yield. This criterion fulfills the designer need; in general design engineer tries to avoid panel permanent distortion. As soon as the face sheet metal starts to yield, this means that permanent deformation is taking place. So all results produced do not exceed the loading that could cause face - sheet yielding. Sample results will be presented to illustrate the behavior of the sandwich panel with respect to each parameter.

Figures 7 and 8 represent the effect of core thickness on bottom sheet, upper sheet and core maximum shear stress. For Figures 7 and 8 the

Figure 7. The variation of core maximum shear stress with load step for different values of core thickness

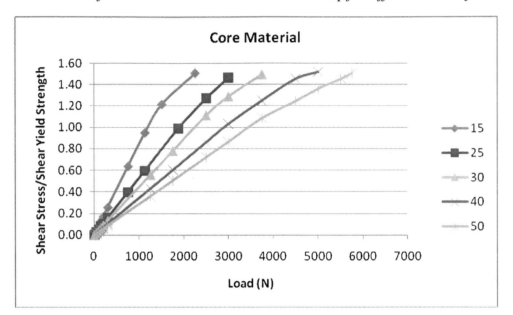

maximum shear stress is normalized by the shear yield strength of the corresponding material respectively. It may be seen from Figure 7 as the core thickness increases the load carrying capacity of the panel increases. Figure 8 presents the effect of panel - core - thickness on the bottom – face - sheet and the top – face - sheet. In all cases, it is found that the bottom - face - sheet starts to yield before the top one. Since the failure of core material is due to shear stress, all graphical results are showing shear stresses not Von Mises stress.

As the core starts to yield, the increment rate of the maximum stress decreases (see Figure 7) while the increment rate for face sheets increases (see Figure 8); i.e. the load is being transferred to the face sheets. This is the main advantage of increasing the load beyond the yield point of the core material.

In Figure 8a the yield load for each case is plotted with marker type similar to the curve that it represents. These curves are denoted by YL-thickness (where YL stands for Yield Load) for each case. These lines are plotted to show the increase in the load carrying capacity of the panel as the thickness increases. Figure 9 presents

better understanding of the gain in the load carrying capacity beyond the yield limit of the core. This percent increase is calculated based on the difference between the yield load of the sheet and the yield load of the core relative to the core yield load. It is obvious that as the thickness increases the load capacity beyond core yield increases.

CONCLUSION

Investigation of sandwich panel behavior beyond core material yield is carried out. The investigation is accomplished in sight of the core material nonlinearity and the geometric nonlinearity of the whole panel.

FE model is generated using 'I-DEAS' software. This model is validated against some analytical cases available in the literature. To assure model accuracy experimental investigation for selected cases is also carried out and compared with FE model. The model shows very good agreement with the previous work as well as the experimental one.

Figure 8. The variation of maximum shear stress to yield shear strength of the face material against the load, a) Lower face sheet, b) Upper face sheet

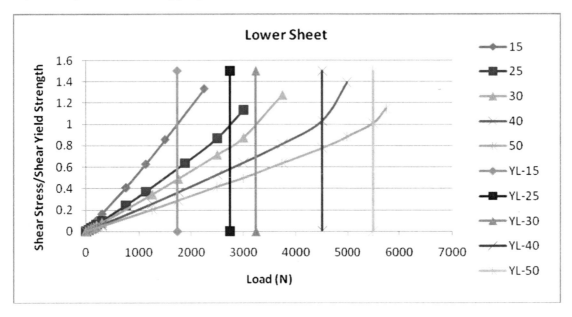

a

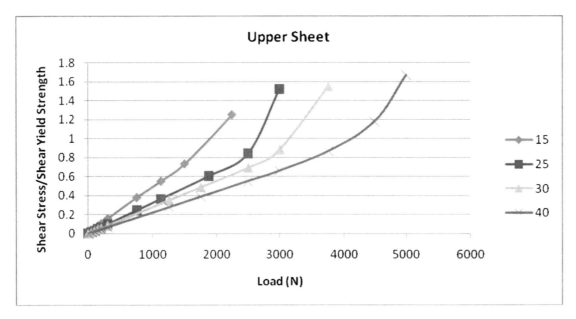

b

Figure 9. Illustration of the percent increase of the load carrying capacity beyond the core yield load plotted against the panel thickness

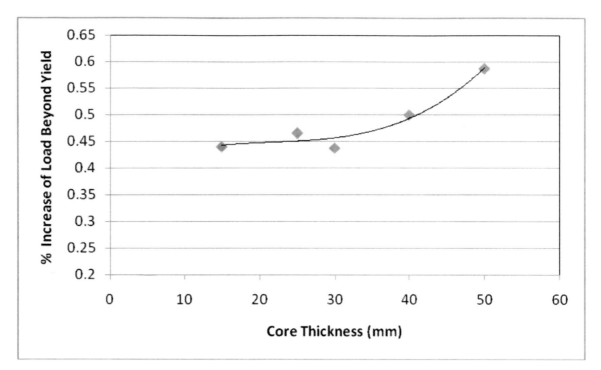

The effect of core thickness that is necessary for designing sandwich panels is unveiled. It is proved that the load carrying capacity of sandwich panel can be improved by loading the panel beyond the yield limit of the core. This load is going to be transmitted to the face sheet.

Increasing core thickness to a certain limit delays the occurrence of core yielding and gives the face sheets to yield first.

ACKNOWLEDGMENT

The Authors would like to express their gratitude to MAANI Prefab, Amman, Jordan for supporting this research.

REFERENCES

Allen, H. G. (1969). *Analysis and design of structural Sandwich Panels*. Oxford, UK: Pergamon Press.

ASTM. (1995). Standard test method for flexural properties of sandwich constructions. *Annual Book of ASTM Standards, 15*(3), C393–C394.

ASTM. (1999). Standard test method for two-dimensional flexural properties of simply supported composite Sandwich Plates subjected to a distributed load. *Annual Book of ASTM Standards, 15*(3), D6419–D6499.

Caprino, G., & Langelan, A. (2000). Study of 3 pt. bending specimen for shear characterization of sandwich cores. *Journal of Composite Materials, 34*(9), 791–814.

Chapra, S., & Canal, R. (2006). *Numerical methods for engineers* (pp. 355–360). McGraw Hill.

Frostig, Y., Baruch, M., Vilnai, O., & Sheinman, I. (1992). Higher-order theory for sandwich beam behavior with transversely flexible core. *Journal of Engineering Mechanics, 118*(5), 1026–1043. doi:10.1061/(ASCE)0733-9399(1992)118:5(1026)

Gdoutos, E. E., Daniel, I. M., Wang, K.-A., & Abot, J. L. (2001). Non-linear behavior of composite Sandwich Beams in three-point bending. *Experimental Mechanics, 41*(2), 182–189. doi:10.1007/BF02323195

Mercado, L. L., & Sikarskie, D. L. (1999). On response of a sandwich panel with a bilinear core. *Mechanics of Composite Materials and Structures, 6,* 57–67.

Mercado, L. L., Sikarskie, D. L., & Miskioglu, I. (2000). Higher order theory for sandwich beams with yielded core. *Proceedings of ICSS-5 Conference,* (pp. 141–53). Zurich.

Miers, S. A. (2001). *Analysis and design of edge inserts in Sandwich Beams*. Master of Science Thesis, Michigan Technological University.

Plantema, F. J. (1966). *Sandwich construction*. New York, NY: John Wiley and Sons.

Thomsen, O. T. (1993). Analysis of local bending effects in sandwich plates with orthotropic face layers subjected to localized loads. *Composite Structures, 25*(1-4), 511–520. doi:10.1016/0263-8223(93)90199-Z

Thomsen, O. T. (1995). Theoretical and experimental investigation of local bending effects in Sandwich Plates. *Composite Structures, 30*(1), 85–101. doi:10.1016/0263-8223(94)00029-8

Zenkert, D. (1995). *An introduction to sandwich construction, (EMAS)*. London, UK: The Chameleon Press Ltd.

Zenkert, D. (1997). *The handbook of sandwich construction*. United Kingdom: Engineering Materials Advisory Services.

Chapter 10
Next Generation Surveillance Technology for Airport and ATC Operations

Werner Langhans
ERA a.s., Czech Republic

Tim Quilter
ERA a.s., Czech Republic

ABSTRACT

Even during the economic crisis, air traffic demand has continued to increase in certain areas of the world, such as the Middle East. Other regions are on their way to recover to pre-crisis traffic demands and will shortly be back to previous growth rates. Airport operators and air traffic control service providers face the challenge to handle this traffic in an expeditious, environmentally friendly, and safe way without generating delays.

Conventional ATC concepts in many parts of the world need to be augmented with next generation surveillance technology, in order to keep pace with the required level of safety in those regions. Conventional technologies, such as primary radar and secondary radar, are not able to deliver the required cost-performance ratios for these increasing demands and need to be replaced by multilateration and ADS-B surveillance techniques. This chapter outlines the recent achievements in worldwide operational deployments in the fields of ADS-B and multilateration for airport and air traffic control applications and discusses the integration into larger aviation system applications.

DOI: 10.4018/978-1-60960-887-3.ch010

INTRODUCTION

Current air traffic control systems have severe limitations and are under economic pressure based on current and future traffic demand. In particular, existing radar-based surveillance infrastructures, which stem from the 1950s, do not have the potential to contribute to more efficient, more expeditious, more environmentally friendly and safer air transport systems. Operational concepts throughout the world (ICAO, NextGen and SESAR) have already identified the need for a next generation surveillance technique with more demanding operational requirements, due to the higher density of air traffic expected. The multilateration technique (MLAT) and Automatic Dependant Surveillance (ADS-B) have started to replace Secondary Surveillance Radars (SSR), and more and more and more air navigation service providers are using this next generation technology. Huge life-cycle cost savings and operational improvements are expected to be achieved using these next generation surveillance techniques throughout the world, which will help foster future worldwide economic growth.

THE ECONOMIC AND TECHNICAL ENVIRONMENT FOR NEXTGENERATION SURVEILLANCE

The Historical Driver for Next Generation Surveillance

In the late 1990's, air traffic control systems in the United States and Western Europe began seeing more and more delays. For the first time, EUROCONTROL reported average delays of more than 15 minutes per flight. At that time, slow turning radar technology with update rates up to 12 seconds, some of which were 15-20 years old, together with old and isolated data processing systems, were commonly used around the globe. ATC system procedures simply could not cope with the huge traffic demand created by the liberalization of the airline industry and the prosperous global economy.

The Dawn of Multilateration

Since the beginning of the last century, there have been many external forces that have greatly affected the economics of the aviation industry. We have seen the airline industry change dramatically, by mergers and bankruptcy; we have seen aircraft manufacturers merge into only 2 major companies; and we have seen air navigation service providers (ANSPs) strive for improved efficiency and increased productivity.

One major way that ANSPs have looked to balance the steadily increasing demand in air traffic with the increasing safety demands and lower costs is to look to next generation air traffic management technology. They found that technologies, such as Mode S MSSR radar with higher update rates (4-8s), multilateration for airport and wide area surveillance and multilateration for height monitoring units could serve as significant tools to help them achieve their ambitious goals.

The Pioneers of Multilateration

In the early 2000's, two ANSPs emerged as surveillance pioneers, and began replacing their radars with multilateration technology. The Czech Republic and the Austrian ANSPs were the first to deploy wide area multilateration (WAM) systems instead of radar technology. Ostrava in the Czech Republic (2003) and Innsbruck in Austria (2004) were such great successes, in terms of technical and economical performance, that a whole group of ANSPs in Europe and even the FAA started to look into this new substitute technology. Life cycle cost savings for Innsbruck have been calculated at about 13M Euro (W. Langhans, Ch. Scheiflinger, et.al., 2007), when compared to MSSR, which would have required two systems to adequately cover the extremely mountainous region of the Inn valley.

The successful introduction of these two pioneer WAM systems has triggered a sequence of actions in the industry. EUROCONTROL has established the MLAT-Taskforce, which coordinates the international standardization process and exchange of user and industry information; EUROCAE has started the WG-70 workgroup and has created the technical specification (ED-142) for WAM to support ATC separation services of 3/5NM; and ICAO has integrated MLAT technology into their Annex 10 and Doc 4444. EUROCONTROL has also published a generic safety case and a general approval guideline in order to support civil aviation authorities and ANSPs to introduce the MLAT technology in their areas of responsibility.

Over the last decade, many more ANSPs have taken WAM systems operational, such as the FAA, AENA (Spain), South African ATNS, NAV Canada, Australia, Armenia, Namibia. Even more are currently in the process of deploying them, such as UK NATS, Deutsche Flugsicherung GmbH (DFS), NAV Portugal, LVNL, ROMATSA, Tajikistan and many military organizations.

Multilateration: Best Match for Seamless ATM Concept

Many ANSPs agree that one way to achieve massive operational improvements is by implementing a seamless gate-to-gate concept with a seamless data-handover between adjacent ATM-units, as well as centralizing network management functions. In particular, the landing and departing phases of flight are vulnerable and thus a seamless surveillance in those flight phases is of paramount importance. While no radar could achieve the performance necessary to implement such a plan, multilateration is fully able. From the final approach, an MLAT system is able to precisely measure flight vector parameters in 4 dimensions with a 1s update rate. MLAT sensors on the airport surface will not only detect the aircraft, but will also detect correctly equipped vehicles performing their tasks at the airport runways, taxiways and aprons.

MLAT technology, as all other cooperative surveillance technologies, relies on an aircraft transponder. It can detect and decode all relevant information from the aircraft, which is needed for ATC-4D trajectory concepts. The 4D position information, however, is derived from a TOA (Time Of Arrival) measurement principle, which allows for a *completely independent position measurement* from the aircraft derived position.

Automatic Dependant Surveillance: Broadcast (ADS-B)

ADS-B, like MLAT, provides a good match with the seamless surveillance requirement. But unlike MLAT, ADS-B receives the aircraft position information from a data source onboard each aircraft. As such, ADS-B is *dependent* on the aircraft position information. ICAO and all relevant standardization organizations have also created operational concepts and technical standards for ADS-B based ATM procedures; however, ADS-B needs a very high integrity of its data sources in the aircraft avionics. This is in order to allow safe ATC separation services on the ground, in particular when it comes to high density and high complexity type airspaces.

Currently, ADS-B equipage for aircraft is being mandated in many parts of the world. However, the reader can expect that there will be many exceptions and exemptions granted around the globe, which will ultimately result in worldwide ADS-B based ATC-operations not fully being implemented until several years from now.

. ANSPs are realizing that when they deploy MLAT technology based on the 1090MHz Extended Squitter (i.e. the identical signal as is used for Mode S MSSR), ADS-B data reception comes along for free. All MLAT equipment is automatically capable of decoding and receiving ADS-B data. To perform proper MLAT measurements, an accurate and reliable time stamping technique needs to be employed, so the same is not true the other way around; pure ADS-B receivers cannot be used for MLAT.

The ADS-B/MLAT Symbiosis Leads to Huge Economic Savings

Because MLAT can provide independent position information, MLAT can *validate any ADS-B derived position* data and vice versa. Additionally, multiple ADS-B positions can be used to validate the overall integrity of an MLAT system. Due to this symbiotic effect, most ANSPs around the world are developing their long-term plans to rely on both ADS-B and MLAT, and thus economizing on their surveillance equipment total cost of ownership. Average numbers of 30-50% total cost of ownership savings of MLAT/ADS-B in comparison to conventional MSSR have been calculated (WHL Neven, et.al., 2005).

Further Applications of MLAT

As we have already discussed, MLAT technology can be used for:

- aircraft and vehicle detection and labeling on the ground
- aircraft detection and labeling in the air.

Due to the high accuracy and the high update rate, MLAT technology can be used in many other ATM applications as well, including:

- **Precision Runway Monitoring (PRM) for No Transgression Zone (NTZ) Monitoring:** The MLAT sensors can be placed in a manner that allows update rates and accuracy that is equal to or better than ICAO requirements for PRM. This allows ANSPs to avoid acquiring extremely expensive PRM radars, which can cost roughly 10M USD.

- **Precision Approach Radar (PAR):** When placing the sensors in a manner that will measure both horizontal and vertical calculations, a 3D measured position (based on measured height and not only barometric altitude) can be calculated and displayed just in the same way as a conventional PAR (vertical and horizontal display). However, instead of a primary response of the aircraft reflection, the Mode A/C/S transponder responses are used for this position measurement based on TOA. Using this system, ATC services like those established with PARs, can be designed.

- **Height Monitoring Units (HMU):** MLAT systems have been used for monitoring and recertifying barometric altimeters. An MLAT system can measure the 3D position of an aircraft and, with special filtering, tracking and correction for meteorological effects, a precise height measurement can be recorded with an accuracy better than 25ft rms.

NEXT GENERATION TECHNIQUES: TECHNICAL KEY CHARACTERISTICS[1]

The basic principles of Multilateration can be found in many references, such as (Xu, et.al., 2010), (WHL Neven, et.al., 2005)

Hyperbolic Positioning

Standard MLAT systems use hyperbolic positioning to locate a target and use the same aircraft interface mechanism as secondary surveillance radar. The transponder on the aircraft is interrogated and the reply from the transponder is detected at multiple ground stations. The time difference of the reply reception between each ground station pair creates a hyperboloid in space on which the target could lie. If the signal is received at four ground stations, three independent time differences can be derived and the intersection of the three resulting hyperboloids identifies the target position. It is normal for the hyperboloids to intersect in more than one place creating a false target location or ambiguity. This false position is normally easy to identify and remove when tracking a target. When the target is located on the ground or the altitude is known from its transmitted flight level, it is possible to locate a target with only three ground stations using the intersection of two hyperboloids as well as the ground or altitude plane. Reception at more than four ground stations can enhance the accuracy of the position solution.

Automatic Dependant Surveillance–Broadcast (ADS-B)

Modern avionic navigation equipment is based on Global Navigation Satellite Systems (GNSS) technology. Aircraft equipped with GNSS can transmit their location together with other useful navigational data to the ground. The most commonly used datalink for transmitting this information is the Mode S protocol via an aircraft's transponder. This form of ADS-B is referred to as 1090MHz Extended Squitter (ES). The 1090ES messages from the transponder are the same format as those decoded in response to MLAT Mode S interrogation. This means an MLAT system can easily provide both ADS-B (dependent) and MLAT (independent) data feeds, resulting in two layers of surveillance in one system.

Large Area WAM Systems

WAM systems are not constrained to their area of coverage operating from a single location like a radar system. Ground stations are positioned in a network and can cover a whole country or even continent, thus enabling lower coverage limits parallel to the curvature of the earth. The first operational WAM systems were designed

to cover a specific area, either where secondary radar was not an option or as a direct alternative to secondary radar. The true potential of WAM is to provide a seamless surveillance layer across a complete area. In order to realize these large area WAM systems, many technical challenges must be overcome. The data from many ground stations must be processed simultaneously, the optimum ground station pairs used to form the hyperboloids for a given target position must be chosen, high numbers of targets must be tracked and availability requirements must be met across the region. This requires careful design of the system layout and advanced data selection algorithms to minimize and distribute the processing.

ISSUES, CONTROVERSIES, PROBLEMS

Multipath

The 1090MHz signal used in aircraft transponders was designed many years ago and is very susceptible to multipath. In WAM systems, this multipath is not normally an issue, as the situation varies over time and the errors generally smooth out across a track. In an airport system, the presence of smooth sided structures and the proximity of aircraft can make the errors significant in localized spots. The situation with MLAT is vastly better than with Surface Movement Radar (SMR) because multiple sensors can be used to cover complex areas.

Accuracy Beyond the System Baseline

The positional errors in any MLAT system are derived from the accuracy with which the time difference between pairs of ground stations can be determined and the geometry of the system with respect to the target. When the intersection of hyperboloids is close to perpendicular, the accuracy is more than sufficient, but as the intersection becomes more tangential, the timing errors are magnified. In certain geometries, the hyperboloids come close to being parallel and the errors approach infinity. Inside the baseline of the ground stations, the error remains small, but outside it degrades rapidly. This means that a standard MLAT system based on TDOA only cannot be used to provide surveillance great distances beyond where the ground stations are located. This limits the use in coastal and border situations.

Over Interrogation

Multilateration systems typically use omni-directional antennas for interrogation. This means that the interrogation is transmitted to all aircraft within range, rather than only those within a narrow beam, like those for traditional rotating radar interrogators. For this reason, MLAT systems should perform target acquisition using the automatically transmitted squitters from the transponder and use Mode S roll call interrogations to address an interrogation to a specific aircraft. If the interrogation scheme is not well designed, there is a risk of degrading the performance of other secondary systems, including TCAS. It is on to the responsibility of the MLAT system operators and the system designers, in conjunction with the definition of the operational requirements, to carefully choose the interrogation scheme that guarantees the efficient use of the 1090MHz spectrum.

Shared Infrastructure Systems

One advantage of multilateration technology is that the same ground station can be used for surface, approach, TMA and en-route applications for both MLAT and ADS-B data. This means that ground station infrastructure can be shared between multiple systems, reducing the total cost of ownership. The infrastructure can be shared by neighboring countries either to improve coverage on either side of a border, or by an airport authority and an air navigations service provider to provide coverage both on and around an airport. The barriers to realization of these systems are not technological but institutional. While the savings can be considerable, the complexities of which organization is responsible for the integrity of the data, and which is responsible for the maintenance of a shared system, are preventing these savings from being realized.

SOLUTIONS AND RECOMMENDATIONS

Multipath

There is no simple solution for the problem of multipath in airports. The potential is limited by the physics of the signal in space. It is not normally an issue on the main movement areas, such as runways and taxiways, but accuracy and detection on stands can be seriously affected. The choice of site, the use of additional sites, the appropriate use of antennas and some signal processing can improve the situation, but the final meters up to the gate remain difficult.

Elliptic-Hyperbolic Positioning

WAM system accuracy can be enhanced by the addition of elliptical measurements to the normal hyperbolic processing. This technique is also referred to as range-aided multilateration or multi-ranging. In a basic multilateration system, the target position is calculated from the intersection of hyperboloids in space. If the target interrogation time is measured as well as the reception time then an ellipsoid is also generated, with the foci at the locations of the interrogator and receiver. This ellipsoid can be used to improve the accuracy of a pure hyperbolic solution. Elliptic-hyperbolic processing is particularly effective outside the baseline of the receivers. Here the intersections of hyperboloids are tangential rather than perpendicular and the target position accuracy is dominated by a range error. An elliptical measurement effectively contains this range error allowing an accurate solution to be calculated. This technique allows accuracy to be achieved for long ranges and is ideal for coastal and border coverage.

Adaptive Interrogation

We have seen how a poorly designed WAM system risks over-interrogation, but it is important to note that a carefully designed WAM system can actually help to reduce the congestion in the 1030 and 1090 frequency bands. One of the original aims of Mode S Radar was to reduce the 1030 and 1090 congestion by removing the need for multiple interrogations and replies to each aircraft per scan compared to SSR. However, the reality is that the Mode S messages are much longer than Mode A/C, and the increased use of enhanced surveillance, TCAS and ADS-B 1090 ES makes the 1090 band more congested than ever. WAM systems can use all these 1090 transmissions passively to position the aircraft and extract relevant data. Interrogation is only required when a specific data item (e.g. flight level) is missed or a specific performance parameter (e.g. accuracy) is not being met. These schemes are becoming increasingly sophisticated.

Shared Infrastructure Systems

The airborne domain is proactively breaking down the barriers to free fight and seamless gate-to-gate operations. Solutions are being found to the artificial problem of national boundaries, the legacy of air-lanes and the historic division between airport operations and air traffic control. The coordination and data exchange are delivering real benefits in terms of time and cost. Unfortunately this has not yet been addressed in the ground domain. If we are to realize the full potential of cost benefit that technology can offer, these institutional issues need to be resolved.

FUTURE RESEARCH DIRECTIONS

Future research for MLAT systems is going in the direction of larger scale system with more automation and self-calibration. Multipath rejections at airports and smart interrogation schemes are also fields of R&D in this market segment.

CONCLUSION

Multilateration is widely used in ATM and ANSPs as a means of replacement for conventional secondary surveillance radar or precision runway monitoring radar. MLAT is enabling future operational concepts by its fast update rates and extraordinary accuracy paired with its flexibility to create coverage performance according to the needs of the aviation industry. Many obstacles of the early days have already been overcome so that MLAT is available for wide operational use in safety to live services such as air traffic control and airport operations control. Moreover, research on the evolution and next generation of this NEXTGEN surveillance technology is welcomed in several areas, such as automation increase for calibration and deployment, further spectrum efficiency and multipath suppression.

REFERENCES

Langhans, W., Scheiflinger, C., et al. (2007). *WAM Austria-Innsbruck Eurocontrol Conference 2007*. Retrieved from http://www.eurocontrol. int/ surveillance/gallery/content/ public/documents/WAM/ 13.3.%20WAM%20Innsbruck%20 Full%20Set.pdf

Neven, W. H. L., Quilter, T. J., Weedon, R., & Hogendoorn, R. A. (2005). *Wide area multilateration. Report on EATMP TRS 131/04, Version 1.1*. Eurocontrol.

Xu, N., Cassell, R., Evers, C., Hauswald, S., & Langhans, W. (2010). Performance assessment of multilateration systems - A solution to Nextgen surveillance. *Proceedings of Integrated Communications Navigation and Surveillance Conference, 2010* (pp. D2-1-D2-8). Herdon, VA, USA. ISSN: 2155-4943

ADDITIONAL READING

Civil Air Navigation Services Organisation (CANSO). www.canso.org

Comparative Assessment of SSR vs Wide Area Multi-Lateration, Eurocontrol. Edition 1.3. September 29, 2005.

Eurocae, W. G-70: Multilateration Working Group, www.eurocae.eu

Eurocontrol Multilateration Task Force and CAS-CADE Programme. www.eurocontrol.int

Generic Safety Assessment for ATC Surveillance using Wide Area Multilateration. Helios Technology. Nick McFarlane.

International Civil Aviation Organization (ICAO). www.icao.int

International Cooperation on Airport Surveillance. www.icas-group.org

Multilateration: Radar's Replacement? Avionics. Callan James. April, 2007. Pages 30-34.

Multilateration Technology is Well Suited to a Wide Range of Applications. ICAO Journal. Number 3, 2007. Volume 62. Pages 12-14, 32-33.

(2007). Multilateration: The Challenges Ahead. *Air Traffic Management. Issue, 2*, 26–298.

Radar, L. A. (2007)... *Air Traffic Management. Issue, 2*, 18–24.

Surveillance Transition Scenario for German Airspace. Results of the DFS ADS-B Study Group. DFS Deutsche Flugsicherung GmbH. Heribert Lafferton, Dr. Roland Mallwitz. January 31, 2007.

The ATM Target Concept D3. (2007, March). Eurocontrol. SESAR Consortium. September 2007. ATM Deployment Sequence D4., Eurocontrol. SESAR Consortium. 2007. *MLAT Systems Gain Ground, Jane's AIRPORT REVIEW., 19*(Issue 2), 20.

WAM Safety Study & Surveillance Generic Safety., Eurocontrol. Bob Darby. November 9, 2007.

KEY TERMS AND DEFINITIONS

1090 MHz Extended Squitter: The main international data link for ADS-B transmissions. It uses the Mode S format from the aircrafts transponder.

Automatic Dependant Surveillance – Broadcast: A system of broadcasting navigational data from an aircraft to multiple client systems.

Elliptic-Hyperbolic Processing: The process of measuring the target position using a combination of hyperbolic processing with an ellipsoid generated from the time of interrogation.

Hyperbolic Processing: The process of determining the location of the target from the intersection of hyperboloids generated during the multilateration process

Multilateration: The process of locating a target predominantly using the time difference of arrival of signals at geographically separated locations. See also hyperbolic processing.

Multipath: The distortion of the direct received signal by the interference from reflected copies of the same signal.

Wide Area Multilateration (WAM): A MLAT system for tracking aircraft in flight, as distinct from an airport surface MLAT system. It may be used for a range of ATC applications including 3 and 5 nm separation services.

Chapter 11
The Evaluation of Wireless Communication Devices:
To Improve In-Flight Security Onboard Commercial Aircraft

Lori J. Brown
Western Michigan University, USA

Liang Dong
Western Michigan University, USA

Anthony G. Cerullo III
Western Michigan University, USA

ABSTRACT

Today, wireless technology forms the communications backbone of many industries—including aviation, transportation, government, and defense. Security breaches since 9/11 have confirmed the need for a discreet wireless communications device onboard commercial aircraft. Real time, discreet communication devices are needed to improve communication between the pilots, flight attendants, and air marshals- a concept that is essential in today's age of terrorism. Flight attendants and Federal Air Marshals (FAMs) need to be able to alert the flight deck discreetly of such dangers, and thereby pre-warn the pilots of possible attempts to enter the flightdeck or security breaches in the cabin.

This chapter will study the effectiveness of discreet, secure, hands-free, wireless communications methods for enhancing coordination during security incidents among cabin crewmembers, between the cabin and flight compartment, ground support personnel. It will identify breakthrough technologies to mitigate the likelihood of individual radical and/or violent behavior, resulting in catastrophic airline casualties, to improve communications and overall safety in-flight.

DOI: 10.4018/978-1-60960-887-3.ch011

INTRODUCTION

The United States is not the only nation with homeland security concerns nor is it the only one considering the solutions needed to address them. The International Transport Workers' Federation (ITF) recommends:

Airline operators equip cabin crewmembers with discreet, secure, hands-free wireless communications devices. Such devices would enhance communications between cabin and flight crewmembers, available law enforcement personnel including Federal Air Marshals (FAMS), and ground-based support staff, and thereby minimize the potential for a successful re-enactment of the terrorist attacks of September 11 2001 (ICAO, 2007).

To support the rapid and widespread adoption of this important terrorism prevention tool, the ITF invited industry and government security representatives to develop technical implementation plans for wireless communication devices on-board commercial aircraft.

As disclosed in a recent terror alert (Homeland Security News, 2008), Tanvir Hussain, Abdulla Ahmed Ali, and Assad Sarwar were found guilty of attempting a massive airline bomb plot- further pointing to airlines as targets for these types of terrorist attacks. The three British men were convicted of plotting to blow up flights from London to North America using bombs disguised as soft drinks. Additionally, defense experts agree that al-Qaeda is 'still plotting'. Abdulla Ahmed Ali, 28, Tanvir Hussain, 28, and Assad Sarwar, 29, were found guilty at Woolwich Crown Court after the UK's largest ever counter-terrorism operation. Their arrests in 2006 changed the face of air travel, thus exemplifying the fact that the United States does not have a monopoly on either the threats made to homeland security nor on the solutions needed to address them. This has prompted the introduction of restrictions on the carriage of liquids, pointing to the critical need for wireless alerting devices on board passenger carrying aircraft.

In China, passenger aircraft hijackings peaked in the 1990's. In 1993 alone, there were ten flights hijacked across the Taiwan Strait. In order to deal with hijacking, the governments on both sides of the Taiwan Strait collaborated and passed very strict laws again such crimes. In the years leading up to September 11, 2001, pilots, and flight attendants were taught to comply with the hijackers' demands, to get the plane to land safely, and let the security forces handle the situation. However, after the September 11, 2001 terrorist attacks, the U.S. and Chinese government, realized that this approach had to change. The crewmembers needed a more proactive operation to take control swiftly while not provoking the hijackers that could endanger themselves or other people. This global shift in thinking called for new procedures, training, and equipment.

In the United States, in a testimony before the U.S. House of Representatives (2007), Patricia Friend, International President of the Association of Flight Attendants CWA, A F L - C I O (AFA) stated:

The most basic necessity onboard a passenger aircraft is the ability to communicate quickly, efficiently and clearly between the cabin and flight deck crew. With pilots safely barricaded behind their reinforced flightdeck doors, and with instructions to limit exposure, it is crucial that a reliable and clear communication tool be provided for the aircraft crew to communicate with one another in an emergency (Friend, 2007, p. 7).

It is important to note that when various federal agencies conducted a mock terrorist attack onboard an aircraft in June of 2005, referred to as 'Operation Atlas' (Fleisher, 2005), the mock terrorists compromised and effectively disabled the interphone, thereby restricting communication between the cabin and flight deck, giving the mock terrorists ample time to kill and injure various crewmembers and passengers. Critical minutes passed before the pilots were aware that anything had happened. "While this was a mere 'mock' hijacking, such a possibility exists today" (Friend, 2007, p. 9).

Recent events have prompted legislation for a federal mandate to study the effectiveness and potential benefits of discreet, personal wireless communication systems- to be used by pilots and other personnel, (House Bill 2200, 2009, and House Bill 5590). This legislations is further supported by a global survey entitled "Pilot/Flight Attendant Communication and Training Survey"(Brown, & Rantz, 2010) reported that 87% of flight attendant and pilot respondents surveyed, from 29 countries, felt a discreet wireless communication device would enhance safety and security onboard commercial aircraft. This study also reports that 58% of the respondents felt the current interphone system does not provide discreet communications between the cabin and flight deck (Brown et al.). The new generation of aircraft, is approaching the digital world and connectivity. Digital connectivity can be used in increase coordination and communication between the flight and cabin crews in new generation electronic flight bag (EFB) equipped aircraft.

Wireless communication devices, can also utilize a "Bluetooth" style of earpiece, however, since flight attendants often have very long duty days, it was important to find out if they would be willing to wear such a device. In the survey conducted by Brown and Niehaus, (2009) 24% of the flight attendants said, they would not be willing to wear the device to achieve wireless communication in-flight.

Results from the Chinese survey (Brown & Cerullo, 2010) still show an overall willingness to use a Bluetooth-style device, although a greater number, (40%), indicated a reluctance to do so. These studies did not discriminate between different earpiece designs and weight. Some of the earpiece designs have added benefits such as hearing protection, reducing fatigue and improvement

of direct communication during meal or drinks service, and emergency evacuations. Custom molded earpiece (such as tbone Aviation™ earpieces) have the effect that the user feels 'this is a product they have introduced to make my life better', which if known, could possibly change the perception of the user (tbone Aviation, 2010).

Cabin Crew Communication Systems (CCCS) headset (tbone Aviation) would be the same design as the popular pilot headset, only with the addition of external area microphones to enhance the direct close contact communication, and provide viable hearing protection for cabin crew. The use of passive noise protection is not a practical solution for flight attendants; however, an earpiece device could provide noise protection while improving communications, protecting against headaches and reduce fatigue associated with high noise levels (tbone Aviation, 2010). In all passenger aircraft, the noise in the cabin is more than 80dB, although protective hearing equipment is not mandated.

Additional benefits of such device could be; easy to use, almost invisible, and light weight (10 grams). The proposed CCCS technology would be an integrated system, connected to the aircraft PA system and to the flightdeck interphone, with multi channel to allow for communication within work groups, such as Air Marshals. Allowing for optimal customer interaction through improved listening through external 'look-and-hear' microphones and adjustable listening volume. This could be crucial in emergencies, when aircraft power is off, such as a fire or ditching.

Some respondents of the Western Michigan University, College of Aviation study indicated that "wearing a bluetooth could make the flight attendant the target for an attack, and felt that a bluetooth style device would not be more discreet than the interphone device" (Brown, et al., p. 27). Fisher's Exact Test was used to test the differences in flight attendant and pilot responses, which were not significantly different (p =.24).

According to the Association of Flight Attendants (AFA-CWA), in the document entitled,

Discreet, Secure, Hands-Free, Wireless Communications for Flight Attendants (AFA, 2010, p. 1-2):

The AFA would like to see a device that is discreet, or as small and innocuous as possible, will allow all crewmembers to carry on their person the ability to communicate from anywhere in the aircraft at any time under any circumstance. Each personal device must have capability for encrypted, bidirectional communications to allow plain language communications during crises; this will ensure security and reduce confusion.

Security of the system is further ensured through use of dedicated hardware components that are accessible only to authorized personnel such as crewmembers and, potentially, any active law enforcement officers who may have presented credentials to the crew prior to the flight. The hands-free concept will allow crewmembers under both general emergency (e.g., medical crises, emergency evacuations) and security threat conditions to use their hands to protect themselves, the flightdeck, other crewmembers, passengers, and the aircraft while continuing to coordinate and communicate with the flightdeck, the ground, and the rest of the crew. Obviously, a device possessing such characteristics must be wireless.

Additionally, these devices will allow all emergency communications to be:

- *Recorded onto the flight recorder for future investigations (while ensuring that such communications, like flightdeck voice recordings, are protected from disclosure);*
- *Monitored by onboard law enforcement officers (if available); and*
- *Monitored by authorized outside responders for real-time information to*
- *Transportation Security Operations Center;*
- *FBI Hostage Rescue Team and local SWAT Teams;*

- *Local Airport Emergency Responders; and*
- *NORAD.*

As written in the document entitled, Discreet, Secure, Hands-Free, Wireless Communications for Flight Attendants, (AFA, 2010, p. 1-2):

Development and implementation of wireless and wired network systems for use by passengers on airplanes in-flight is being pursued by many US commercial airplane operators. If cost were the sole constraint, a wireless communications system for use by airline crewmembers might utilize such passenger-based systems. However, given the potential for security compromises inherent in shared communications hardware, AFA recommends that wireless systems for crewmembers be completely separate from passenger-accessible systems. Furthermore, to ensure system-wide conformity and harmonization, AFA recommends that development, procurement, and installation of hardware and software elements of these systems be maintained within the government. Finally, AFA recommends, "the government take responsibility for development of model operational procedures and training curricula for these systems (AFA, 2010, p. 1-2)".

With the AFA requirements in mind, we must ask what type of system can provide plain language bio-directional hands-free communications, whilst having the ability to send a discreet signal to the flight deck? Solutions may be found in digital 'wireless' systems. One example is the new generation of Airbus aircraft, like the A380 and A350. Airbus proposed a first step of "digital aircraft" through the on-board information system (OIS). This enhanced airline operations by integrating EFB and electronic maintenance within an aircraft network and introduced 'wireless' connectivity at the gate and in-flight.

Wireless Devices on Airplanes

Federal Communications Commission (FCC) rules prohibit the use of cellular phones using the 800 MHz frequency and other wireless devices on airborne aircraft. This ban was put in place because of potential interference to wireless networks on the ground.

In March 2007, the FCC terminated a proceeding that it began in late 2004 to consider potentially lifting this ban. The FCC determined that the technical information provided by interested parties in response to the proposal was insufficient to determine whether in-flight use of wireless devices on aircraft could cause harmful interference to wireless networks on the ground. Therefore, it decided at this time to make no changes in the rules prohibiting in-flight use of such devices.

In addition to the FCC's rules, the Federal Aviation Administration (FAA) prohibits in-flight use of wireless devices because of potential interference to the aircraft's navigation and communication systems. For this same reason the FAA also regulates the use of all portable electronic devices (PEDs), such as iPods and portable DVD players, during flight.

The FCC has approved rules that allow in-flight voice and data services, including broadband services using dedicated air-to-ground frequencies that were previously used for seat-back telephone service. Air-to-ground service providers are in the process of rolling out new in-flight services, such as high-speed Internet access for laptop computers. Because these services will operate in frequencies that are dedicated to air-to-ground communications and are separate from those used for wireless services on the ground, they do not pose an interference risk to wireless networks on the ground. Providers of in-flight wireless broadband and other communications services using the air-to-ground frequencies must coordinate with airlines and comply with any FAA rules in order to offer such services.

DISCREET WIRELESS COMMUNICATION TECHNOLOGIES

Wireless communication has changed the dynamics of the working environment and workforce mobility. Without being tethered to a fixed location, a wireless terminal can be concealed from passengers in a commercial aircraft. This provides a tight and discreet connection between pilots and flight attendants for aviation safety. For this purpose, the in-flight wireless communication system should be specified according to requirements in size, ease of operation, robustness, durability, fault tolerance, security, and cost effectiveness. In addition, it is important to consider end device localization and radio frequency interference with aviation navigation.

Several platforms for wireless ad hoc and sensor networks can be utilized and tailored for an in-flight discreet wireless communication system. A low-rate wireless network can be a good start towards a cost effective solution. These network standards and specifications include ZigBee over IEEE 802.15.4 for embedded sensing, medical data collection, and home automation, WirelessHART for industrial applications like process monitoring and control, IEEE 1451 for smart sensors, and EnOcean for wireless communication in building automation.

A wireless sensor network requires end node miniaturization and energy efficiency. These are also objectives of the proposed in-flight wireless system. A wireless communication device used by the flight attendant should be small enough to fit into a pocket or be disguised, for example, a collar decoration. This makes it necessary to employ current technologies of wireless system-on-chip transceivers and compact form factor antennas. To make the system robust- either the communication links should be established at all times, or a self-test should be performed regularly to ensure the pilots that the end devices are on standby and functioning properly. However, the end devices are not allowed to consume energy aggressively in order to sustain long hours of commercial flight. This makes radio front-end design and network power management challenging.

The in-flight wireless system should tolerate fault alarms caused by mis-operations and support bidirectional communications for the flight attendants to receive the pilots confirmation. It should also be secure against any intentional or unintentional system breakthrough. To achieve these, ultra-low power digital signal processors can be embedded to play an important role of implementing sophisticated coding and signal processing algorithms.

As wireless networks evolve over time, there is an increasing interest in combining location awareness with communications. The fastest growing area of location awareness applications is in wireless local and personal area networks. Localization techniques for wireless networks can be leveraged to improve the effectiveness of the in-flight wireless communication system. Upon receiving an alert from the cabin crew, the pilot should be able to locate the flight attendant who initiated the signal for appropriate response and notifying others of the troubled zone.

The wireless sensor networks, for example those under the IEEE 802.15.4/ZigBee standard, operate on unlicensed frequency bands of 868.0-868.6 MHz in Europe, 902-928 MHz in North America, and 2400-2483.5 MHz worldwide. Beyond these three bands, the IEEE802.15.4c study group is considering the newly opened 314-316 MHz, 430-434 MHz, and 779-787 MHz bands in China, while the IEEE 802.15 Task Group 4d is defining an amendment to the existing standard to support the new 950 MHz-956 MHz band in Japan. In order to minimize interference with existing aviation radio, the proposed in-flight wireless communication system operates on the 2400-2483.5 MHz industrial scientific and medical (ISM) band. This should not cause considerable interference with the current five navigation frequency bands: Very High Frequency (VHF) Omni-directional Radio Range (VOR) and Instru-

ment Landing System (ILS) Localizer, 108-118 MHz; ILS Glide Slope, 329-335 MHz; Distant Measuring Equipment (DME); Traffic Alert and Collision Avoidance System, 960-1215 MHz; and GPS, 1227.5 and 1575.42 MHz. Europe is getting ready for a decision on the final phase of the deployment of 8.33 kHz radios. Their mandatory use in all European airspace by 2018 would solve the long-standing European frequency shortage problem.

One of such technologies utilizing these networks, has been developed by STG Aerospace. The system provides the flight crew with an audible alert, coupled with a visible flightdeck annunciation signal. The signal will indicate the alert while giving a "zonal" location. The system also includes a door intercom to provide the additional audio communication between the cabin and the flight deck sides of the flightdeck door.

The purpose of this system is to provide the following functionality:

1. When a person authorized to access the flightdeck seeks entry, the existing visual identification through the flightdeck door, coupled with a new audio intercom confirmation that the door area is clear.

2. In the event of an attack on a cabin crewmember or other security breach in the cabin, the system enables the cabin crew to alert the flight crew of the emergency event, achieved by using discreet wireless "Panic Buttons" provided to the crewmembers (STG Aerospace).

This technology will allow effective communication, while keeping the cost and weight to a minimum to meet the economic constraints of U.S. Airlines, as well as other potential users such as hospitals, schools, prisons, trains, and large shipping vessels (Brown, 2010, p. 3).

Figure 1. In-flight wireless system

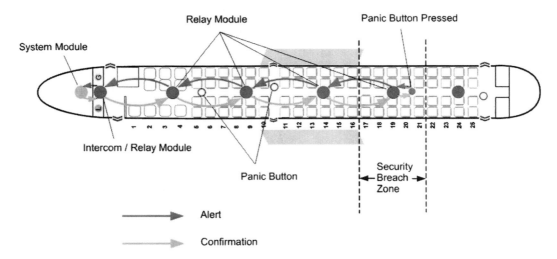

Crew Alert Monitoring Device (CAMS)

The Crew Alert Monitoring device (CAMS), a wireless device that is an ultra secure cabin alert and monitoring system, using small donut shaped alarm units held on person of each cabin crew, or FAMS, which, when activated, sends an alarm signal to the flightdeck, effectively warning them of trouble, and the expectation of escalation of that trouble to the flightdeck. The signal also tells the flightdeck where in the aircraft the alarm was triggered, and therefore an indication of the time, which may be available to them to undertake appropriate actions before attempts at intrusions to the flightdeck. The system also provides a means of voice communication between the cabin crew and pilot at the flightdeck door, and combined with the use of the door peephole provides the pilots with a good means of monitoring if anyone wishing entry to the flightdeck is under stress and possible coercion. The system is aircraft specific, extremely secure, has 'designed in' safeguards against inadvertent activation, and meets all the needs of those most closely affected - the pilots, cabin crew and passengers (Brown, 2010, p. 4).

Cost of wireless devices was cited as a reason for not mandating such devices, but CAMS costs approximately $5,000 for a B737 up to $9,000 for a B747, with minimal installation cost, providing, in combination with the existing door peephole, a discreet monitoring system for the whole aircraft (STG Aerospace, 2010). Utilizing existing wireless systems in new generation may be a cost effective means to improve on communication and integrate new generation systems.

The CAMS alarm device is completely in the control of the operator.

The 'kits' are aircraft specific, and are checked in and out of the aircraft at start and finish of every trip. The wireless signals are coded and aircraft specific, and could not compromise any entire security system – just not possible. If any alarm is lost or stolen from an aircraft then it is replaced and the whole system recoded wirelessly.

CAMS alarms stay with an aircraft and therefore only those operating an aircraft would have the device – possibly 45-50,000 individuals. It may be argued that the alarm price is small, and would not constitute an unreasonable 'burden' to the industry, in relation to the benefit of having a system that is discreet and gives an integrated audio interface (STG Aerospace).

Figure 2. Cams flow chart

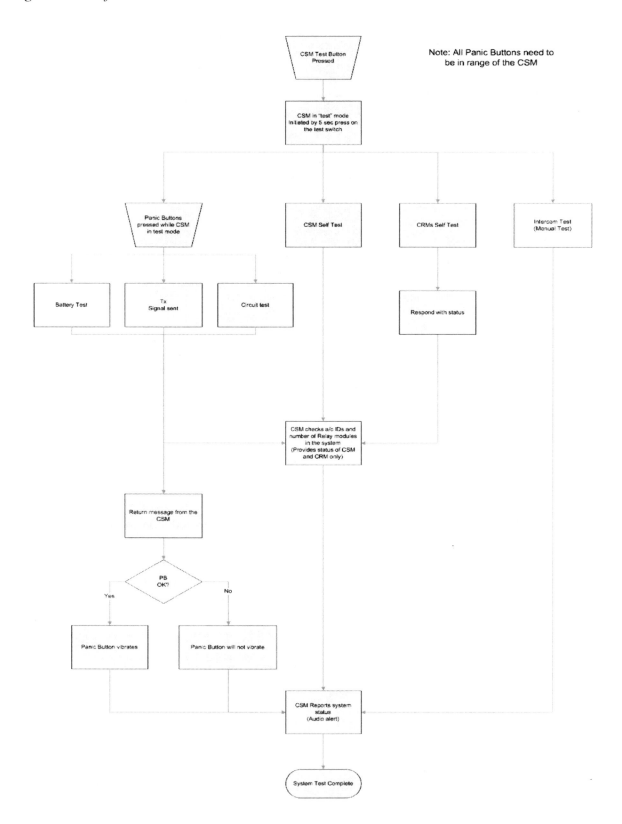

Figure 3. CAMS unit components

The system consists of four parts: Panic Button; System Module; Relay Module; Intercom.

1. A CAMS Panic Button is to be carried by crewmembers to provide discreet alert.
2. A CAMS System Module (CSM) placed in the flightdeck by the entry door, providing audio communications to the intercom along with audible and visual alerts when a panic button has been activated.
3. CAMS Relay Modules (CRM) to act as wireless transceivers for propagation of the alert signal through the aircraft. The designs of these will mimic that of the

CAMS System Module, but will be installed out of sight in the cabin ceiling area.

4. A CAMS Intercom unit is located outside the flightdeck door area, to provide audio confirmation. This is coupled with the existing visual confirmation provide by video or through the door viewing device. Since the CAMS Intercom provides audio confirmation, it can be detached from the visual confirmation device and be installed out of sight outside the flightdeck door area. Therefore, the entire CAMS system is hidden from any passenger. The CAMS system will have a specific aircraft ID to stop interference between adjacent aircraft systems on the ground. (STG Aerospace, 2010).

Analytical Approach

When building an actual wireless communication system, many specific constraints in size, power and cost affect the communication properties. While the basic principles apply, the system performance characteristics vary considerably from a regular wireless communication network to the proposed discreet wireless alert system. System test and technical performance evaluation will focus on practical issues pertinent to in-flight situations. These include delay-constrained real-time issues, network power management, network security and privacy, wireless device localization, and in-flight radio frequency spectrum measurement.

As the discreet wireless cabin alert system is developed from hardware/software platforms of wireless personal area network and sensor networks, it is imperative to make sure that it meets real-time requirements. In wireless sensor networks, many protocols either ignore real-time or simply attempt to process as fast as possible and hope that this speed is sufficient to meet deadlines of data collection. Some protocol addresses real-time but no guarantees are given, other uses feedback control to maintain an average delay but is subject to packet loss, congestion, and other problems. In the implementation of a discreet alert system, we will test and evaluate system performance under strict time constraints. For example, if a security breach happens towards the back of the cabin and a flight attendant presses her panic button, the signal may hop multiple times through several relay modules and the system module to reach the flightdeck. Again, a confirmation from the pilot may experience multiple relays. This process is required to complete within critical time constraint. Moreover, if a flight attendant moves between different cabin zones that covered by different relay modules, the handover time from one relay module to another is restricted. This guarantees that a flight attendant is ready to initiate an alert any time anywhere in the cabin.

Energy constraint of wireless sensor networks is unlikely to be solved soon due to slow progress in developing battery capacity. Most power is consumed by radio communications and therefore advanced power management of the radio is

essential for extending node longevity. Usually, energy-efficient protocols duty-cycle the radio at very low rate such that sensor nodes switch their radio according to a regular schedule. Lower duty cycles result in lower energy consumption but at the same time increase the latency in communication. Moreover, the enforced regularity rarely matches the needs of event-based applications in which an external event triggers network activity. In the in-flight discreet alert system, the system module and relay module can be powered by on-board power supplier, but the panic buttons have to rely on batteries and be active throughout the entire flight duration. Various power management schemes will be tested and their effects on quick alert response and system robustness evaluated.

Wireless network nodes are often deployed in accessible areas and interact closely with people. Security is a challenge for any wireless communication system. It is even more demanding for the in-flight discreet wireless alert system. Its special requirements pose new challenges on key establishment, secrecy and authentication, privacy, robustness to denial-of-service attacks, and secure routing. Security and privacy pervade every aspect of system design, since modules designed without security can become a point of attack. Adversaries can severely limit the effectiveness of a cabin alert. For example, in the form of denial-of-service attack, an adversary attempts to disrupt operation by broadcasting a high-energy signal. If the transmission is strong enough, the entire system could be jammed. More sophisticated attacks are also possible on the wireless links and the communication between the cabin crew and the pilot at the flightdeck door. Various security mechanisms will be tested and evaluated, in order to secure communication links against eavesdropping and tampering for a truly discreet and robust system.

In wireless sensor networks, node location information alongside data collection and distribution can enable many location-aware applications. This applies to the in-flight discreet wireless alert system, as promptly locating the security breach in the cabin can be valuable. According to its proximity to a certain relay module, coarse location information of the initiating panic button can be acquired. For more accurate localization, multiple relay modules can be involved for trilateration, or a cabin propagation map can be measured and pre-stored for fingerprinting. Many current wireless device localization solutions are developed with simplifying assumptions about the propagation environment. These solutions will be tested and evaluated in real in-flight situations with various occupancy rates of airline passengers.

An in-flight radio frequency spectrum measurement will be performed to evaluate the interference from navigation signals and portable electronics devices allowed onboard. In selected aviation-critical and personal electronics frequency bands, activity rates and average and maximum levels of received power will be monitored. Another measurement will evaluate the interference from the discreet wireless alert system to the aviation navigation and in-flight wireless access. The radio measurement tests will be carried out in two environments, a test aircraft, and the cabin simulator. The test aircraft is representative of a large passenger aircraft with operational communications and navigational aids, and will be filled with passengers for an operational test. The cabin simulator gives flexibility and longer study periods for investigating the effects of local shadowing (masking) of the radio signal as the position where the panic buttons are worn are altered.

A NEW GENERATION OF AIRCRAFT

The new generation of aircraft, like the A380, B787, and A350, is approaching the digital world and connectivity. The internet, digital documentation, and communication technologies are changing the aircraft operational environment for the benefit of airline operations, both in flight and on the ground. Now, thanks to increased com-munications capabilities and decreased costs, the introduction of the electronic flight bag (EFB) as an aircraft-installed solution can be achieved with a much more promising and efficient business case, thus bringing greater functionality from cockpit to cabin. Thanks to on-board networks, EFB facilities can be now extended to the pilots' flight "dossiers", which they commonly bring on board with them, as well as to cabin crews (Eurocontrol, 2010).

In-flight connectivity also offers a large spectrum of improvements. With the A380, Airbus is introducing and validating the new Inmarsat SBB (swift broadband) satcom service to support EFB and crew communications during flight. SBB offers a step change for crews as it securely connects aircraft in flight directly to the airline network with an Internet-like communication utility. It therefore becomes possible to send and receive pictures or updates of complex data like weather charts. In this way, numerous new and not previously envisaged services can become a reality. In-flight connectivity promises not only to change passenger entertainment, but also offer new services and facilities to the crew. This will involve IT effort and problem-solving in order to converge and integrate all these different means of communication – including the merging of aircraft communications addressing and reporting system (ACARS) with IP communications on the ground. For the on-board EFB, pilots and commercial crews must have easy access to communication equipment, such as "iPhone-like" devices, to ensure they are able to concentrate on their own priorities. This will force application developers to focus on the role of the aircraft crew and make communication access transparent. From the point of view of the aircraft, communication systems and equipment must be able to follow technology evolutions, which generally advance faster than avionics computers. They need also to bring the right level of data security (Eurocontrol, 2010). Potentially, communication equipment such as iPhone-like devices, can be used for normal and discreet communications between FAMs, flight attendants, pilots, air traffic control, and law enforcement on the ground.

CONCLUSION

Crew communications and coordination are considered "critical as they relate to the survival of all crewmembers and passengers and the overall control of the aircraft. Tactical communications experts from the military and law enforcement have advised that communication is the primary point of failure during live situational scenarios", (ICAO, 2007). With pilots barricaded behind their reinforced flightdeck doors, it is crucial that a reliable technology is provided to allow the crew on both sides of the door and relevant law enforcement agencies in the cabin and on the ground, to communicate with one another in an emergency. Appropriate wireless technology are available in the electronic security industry (i.e. STG Aerospace, Phonak Communications, tbone Aviation™, and I-Tex) to provide small transmitters in belt clips, pendants, Bluetooth-style head pieces, or wrist mounts that can communicate instantly with a receiver in the flightdeck. Such wireless-communication devices would also allow a flight attendant or Federal Air Marshal a means to notify the flight deck (discreetly) at the first sign of a security breach.

While any addition to expenditure is difficult to justify in the economic climate, which faces our industry today- the potential benefits of wireless networks used in intra-crew communications, may far outweigh the costs. Just as ACARS introduced an evolution for in-flight communications, EFBs with IP communication can bring about a major change to the way crews communicate in the future, while integrating communication with the air traffic management system. New generation wireless communication technology will force developers to focus on the role of the aircraft crew and make communication access transparent.

REFERENCES

Association of Flight Attendants. CWA, AFL - CIO. (2010). *Counterterrorism initiative, legislative update.* Retrieved August 10, 2010, from http://www.afausairways.org /index.cfm?zone=/ unionactive/ view_article.cfm& homeID=168976

Association of Flight Attendants (AFA). (2010). *Discreet, secure, hands-free, wireless communications for flight attendants.* Retrieved August 10, 2010, from http://www.afanet.org/pdf/ AFA_CabinCrewWirelessDevice.pdf

Brown, L. (2010). *In-flight security onboard commercial aircraft: Critical improvements needed.* 27th International Congress of Aeronautical Sciences. Nice, France.

Brown, L., & Niehaus, J. (2009). *Improving pilot/ flight attendant communications to enhance aviation safety.* Flight Safety Foundation, Corporate Aviation Safety Seminar. Orlando, Florida, April 2009.

Brown, L., & Rantz, W. (2010). *The efficacy of flight attendant/pilot communication, in a post 9/11 environment: Viewed from both sides of the fortress door. International Journal of Applied Aviation Studies.* IJAAS.

Eurocontrol. (2010). Communications, navigation and surveillance at the heart of the future ATM system. *Eurocontrol Skyway Magazine, (Winter, 2010).*

Fleisher, L. (2005, June 5th). Terror response is tested at Boston's Logan Airport in Operation Atlas. *Boston Globe.* Retrieved April 5, 2009, from http://www.cra-usa.net/ 1inthenews-Boston.htm

Friend, P. (2007). *Aviation security part II: A frontline perspective on the need for enhanced human resources and equipment.* Testimony of Patricia A. Friend, before the subcommittee on transportation and infrastructure. November 1, 2007, U.S. House of Representatives, Washington, D.C. Retrieved August 1, 2010, from http:// homeland.house.gov/ hearings/index.asp?ID=101

Homeland Security News. (2008, July 28). Is U.S. bio-terror attack just a matter of time? *Homeland Security News*, 2008.

ICAO. (2007). *Discreet wireless communication for civil aviation cabin crewmembers*. 36th Session of the Assembly [ICAO.]. *Agenda (Durban, South Africa)*, 15.

Report for Congress. (2006). *Issues and options for combating terrorism and counterinsurgency*. Order Code, RL32737.

STG Aerospace Ltd. (2010). *Website*. Retrieved from http://www.stgaerospace.com/

Tbone Aviation. (2010). *Website*. http://www.tboneaviation.com/

The 9/11 Commission Staff Monograph. (2004). National Commission on Terrorist Attacks upon the United States. *9/11 Commission Final Report*, p. 54.

US Congress. (2009). *HR. 2200: Transportation Security Administration Authorization Act. 111th Congress. § 234*. Washington, DC: GPO.

U.S. Congress. (2010). *H.R. 5900: Airline Safety and Federal Aviation Administration Extension Act: Counterterrorism Enhancement and Department of Homeland Security Authorization Act. 111th Congress*. Washington, DC: *GPO*.

KEY TERMS AND DEFINITIONS

In-Flight: Is the state of an object as it moves either through the air, or movement beyond earth's atmosphere (as in the case of spaceflight), by generating lift, propulsive thrust.

Radio Frequency Spectrum Measurements: Radio frequency spectrum measurement is the measuring of how the electromagnetic spectrum corresponding to radio frequencies is utilized. It measures the radio emission characteristics of individual transmitters and the extend, patterns, and amounts of individual bands or broadband in the range of about 30 kHz to 300 GHz.

Wireless Ad Hoc Networks: A wireless ad hoc network is a decentralized wireless network. The network is ad hoc because it does not rely on a preexisting infrastructure, such as routers in wired networks or access points in managed (infrastructure) wireless networks. Instead, each node participates in routing by forwarding data for other nodes, and so the determination of which nodes forward data is made dynamically based on the network connectivity.

Wireless Network Node Localization: Wireless network node localization is the process of discovering the positions of network nodes given a particular wireless network configuration. Some nodes in the network know their locations while other nodes determine their locations by measuring the distance or proximity to their neighbors.

Wireless Network Power Management: Wireless network power management refers to various power saving techniques used in wireless networking. These techniques preserve the energy consumed at each layer of a wireless networking protocol stack. They can range from the use of energy harvesting technique at the physical layer, duty-cycling the radio communications at the MAC layer, to partitioning the load of power hungry computations across multiple devices at the application layer.

Wireless Network Security and Privacy: Wireless network security is the prevention of unauthorized access or damage to systems and resources using wireless networks. Wireless network privacy is the ability of a system or a group of systems to seclude themselves or information about themselves on a wireless network and only reveal themselves selectively to authentic network users

Chapter 12
Terrorist Attacks:
A Safety Management System Training Tool for Airport and Airline Managers

William B. Rankin
University of Central Missouri, USA

ABSTRACT

This chapter examines how airport and airline managers could review their incident and command plans to enhance security counter-measures for terrorist attacks through the use of a well constructed plan-do-check-act (PDCA) tool, in the context of a Safety Management System (SMS), and incorporating a structured field survey into their emergency incident plan and command plan reviews. Thus, through the examination of actual emergency incident plan and command plan survey, airport managers are given the opportunity to work issues through the trials and tribulations of refining their incident and command plans on a recurring basis. It is suggested that a PDCA tool be implemented as a SMS model for the enhancement of these plans in the airline environment.

INTRODUCTION

According to Wells and Young, aviation illustrates the international and multinational aspect of transportation security problems. U.S. based airlines are becoming more and more involved and intertwined with foreign airlines and governments through expansion of international routes and code share arrangements. As a result, today's airports and airlines are more involved than ever with US transportation security issues, such as illegal immigration, smuggling of drugs, hazardous waste, and terrorism.[1]

These concerns arise since there is no central authority in world politics, much less the power to enforce mandates, for international and national

DOI: 10.4018/978-1-60960-887-3.ch012

security concerns. For example, in the case of Pan Am flight 103, a bomb was placed on the plane in Frankfurt through an intra-modal movement from a foreign airline originating in still another country by a terrorist group supported by yet another country's intelligence service. As a result, this bombing involved the activities of five countries.[2]

The responsibility for airport and airline security is primarily a government function at the national level in foreign countries. As a result, the variance in security systems varies from country to country in terms of organization, orientation, policies, and procedures. The US Federal Aviation Administration (FAA) has the authority to assess security at foreign airports in accordance with the International Civil Aviation Organization (ICAO) Annex 17 standards. In those cases where the FAA finds that a foreign government does not maintain the appropriate security measures, the US Secretary of Transportation may suspend service to that country. The US Department of Transportation (DOT) also issues public warning for security levels at foreign airports when their standards fall below international standards. Finally, because the Transportation Security Administration (TSA) does not control screening activities in foreign countries, the TSA may, from time to time; order US based airlines operating in foreign countries to impose additional screening procedures.[3]

BACKGROUND

The Role of the Transportation Security Administration

The Transportation Security Administration (TSA) is required to prescribe rules to protect persons and property on aircraft against acts of criminal violence and aircraft piracy, and to prescribe rules for screening passengers and property for dangerous weapons, explosives, and destructive substances. To carry out the provisions of the Aviation Transportation and Security Act of 2001

(ATSA), the TSA has adopted former FAA rules requiring airport operators, air carriers, indirect air carriers, and foreign air carriers to carry out various duties for civil aviation security. Title 49, Code of Federal Regulations (CFR), incorporates the following Transportation Security Regulations (TSRs) that concern aviation security: Part 1542 applies to certain airport operators; Part 1544 governs certain air carriers; Part 1546 applies to the operation of foreign air carriers; and Part 1548 applies to indirect air carriers such as freight forwarders, who engage indirectly in air transportation of property within the United States and sometimes operate out of GA facilities.[4]

TSR Parts 1520, 1540, 1542 and 1548

Part 1520 forbids the disclosure of information that may compromise or be harmful to the safety and security of the traveling public. Additionally, the regulation sets forth the rules that allow the federal government to withhold information from public disclosure even when requested under the Freedom of Information Act (FOIA), in litigation, or in rulemaking. Airport operators and air carriers are required to restrict the availability of information contained in security programs to those with a need-to-know. This need-to-know is defined by the Airport Security Program. Under this Part, airport management must withhold sensitive security information (SSI) from unauthorized disclosure. If SSI is released to unauthorized persons, the TSA must be notified. This permits the TSA to evaluate the risk presented by the release of the information, and to take whatever actions may be needed to mitigate that risk.[5]

New Part 1540 outlines the rules that apply to all segments of civil aviation security and includes rules that govern individuals and other persons. This regulation applies both to individuals who work at the airport and to the passengers using airports. Part 1540 also outlines definitions and terms used in Parts 1542, 1548 and others. For example, the widely used term "escort" is given a definition within this regulation. Another significant addition to the regulatory language is the inclusion of *individual accountability*. The TSA feels that "the contribution of individuals to the success of the civil aviation security program cannot be over-emphasized," and for that reason the "agency believes that holding individuals accountable for their security violations will serve as a direct and effective corrective action and may prove to be a positive deterrent".[6] The TSA feels that by incorporating a level of individual responsibility in tandem with existing airport security programs a higher level of aviation security integrity will be ensured.[7]

The TSRs contain general requirements for promoting civil aviation security. Each airport operator, air carrier, foreign air carrier, and indirect air carrier covered by these parts is required to have a security program that contains information that specifies how they will perform their regulatory and statutory responsibilities. Again, all these security programs are available only to those persons having a need-to-know

TSR Part 1542 *Airport Security* prescribes rules for airport operators servicing and facilitating U.S. certificated air carriers, foreign air carriers, and both foreign and domestic air cargo carriers. The purpose of Part 1542 is to prevent any act of unlawful interference with the safety of persons and goods in air transportation. To accomplish this goal, the TSA has extended its security regulations to airports as the first practical line of defense.[8]

The TSA's congressionally authorized area of jurisdiction and responsibility focuses on protecting persons and property in air transportation against acts of criminal violence, air piracy, and terrorism. However, to effect security of aircraft in-flight, the TSA extends security measures to the airport operator by requiring airport management to regulate the movement of persons and vehicles having access to all aircraft while on the ground and within the airport boundary. In all respects, the security of civil aviation operations begins at

an airport's perimeter fence and terminal building interface. [9]

Part 1548, indirect air carrier security program, covers security procedures for cargo that are accepted for transport on aircraft. In general, indirect air carriers are required to carry out security procedures for handling cargo that will be carried on aircraft.[10]

THE GOAL OF TERRORIST CELLS

According to, the overall goal of terrorist cells might be to effect large-scale political or ideological change; however, their immediate goals are usually designed to achieve short-term goals attached to their actions.[11] For this reason, the airport and airline terrorist training must ensure that employees know what the various types of terrorist intentions and characteristics might include. Intentions may include the following:

1. To produce widespread fear.
2. To obtain worldwide, national, or local recognition for their cause by attracting the attention of the media.
3. To harness, weaken, or embarrass government security forces so that the government overreacts and appears repressive.
4. To steal or extort money and equipment, especially weapons and ammunition.
5. To destroy facilities or disrupt lines of communication in order to create doubt that the government can provide for and protect its citizens.
6. To discourage foreign investments, tourism, or assistance programs that can affect the target country's economy and support of the government in power.
7. To influence government decisions, legislation, or other critical decisions.
8. To free prisoners.
9. To satisfy vengeance.

10. To turn the tide in a guerrilla war by foreign government security forces to concentrate their efforts in urban areas. This allows the terrorist groups to establish themselves among the local populace in rural areas.[12]

Airport and airline employees should be trained that most terrorist groups share the following characteristics:

1. They seek to intimidate by promoting fear.
2. Generally, they are militarily weaker than the government they fight (though some groups have been able to obtain advanced weaponry, e.g., tanks, in limited quantities).
3. They employ unconventional warfare tactics; terrorist may be trained in physical and mental preparation, weapons and explosives, political and religious indoctrination, combat tactics, intelligence gathering, psychological warfare, survival, and communications.
4. They do not equate tactical success with mission success. A specific terrorist attack may not achieve its desired results, but a terrorist may still view the attack as successful if it publicizes the cause.
5. They are usually urban-based and highly mobile. If urban based, terrorist have access to mass transportation (e.g., airplanes, ships, railroads, and subways). Terrorist groups with international contacts may also have access to forged passports and safe havens in countries other than their home base.
6. Generally, they organize and operate clandestinely in cells of three to five members. A cell may have contact only with another cell or only with the next up in the command of hierarchy. Therefore, the capture of one or more terrorist rarely compromises the identity or plans of the entire organization.[13]

ISSUES: TYPES OF POTENTIAL ATTACKS

Melton points out that a terrorist threat can come in many forms. He states that each features a different effect and that each type requires a different response. What is important for airport and airline managers worldwide is to be able to recognize and plan for each type of threat, as well as contingencies.[14]

Melton states that there has been much discussion about weapons of mass destruction (WMD), such as nuclear, biological, and chemical agents, but that these types of attacks have been rare. For airport and airline managers, bombs threats would be the most likely.[15]

If an airport or airline employee sees suspicious activity, the individual(s) should not be confronted. Airport and airline employees should take the following actions:

1. S – Size (jot down the number of people, gender, ages, and physical descriptions).
2. A – Activity (Describe exactly what they are doing).
3. L – Location (Provide exact location)
4. U – Uniform (Describe what they are wearing, including shoes).
5. T – Time (Provide date, time, and duration of activity).
6. E – Equipment (Describe vehicle, make, color, etc., license plate, camera, guns, etc.).[16]

THE EMERGENCY INCIDENT AND COMMAND CENTER PLAN

Airport and airline managers at all sizes of airports need to develop an emergency incident plan (EIP) and command center plan (CCP). An emergency can involve multiple aircraft, large numbers of people, and buildings and equipment. Because the consequences of an emergency can rapidly outpace the response capacity of airport and airline resources; surrounding communities and governments are often relied upon to lend assistance through mutual aid agreements. For this purpose, the development of an airport emergency incident plan and command center plan is very important.[17]

The primary purpose of an EIP and CCP is to provide for the delegation of emergency authority, the assignment of emergency responsibilities, the coordination of efforts by responding personnel, and an orderly transition from normal to emergency operations and back to normal. The EIP and CCP provide a framework upon which the various response capabilities of the airline can be identified and organized. It further provides an inventory of response options and capabilities available to the airline in the event of an incident at or near the airport. An effective EIP and CCP also provides for alternative actions to be substituted when the first choices are unavailable due to unforeseen circumstances.[18]

Developing the EIP and CCP requires that all airport managers thoroughly understand the needs of their particular operations. The EIP and CCP accommodate such factors as the environment of the facility, the airline's operational characteristics and functions, and the nature and extent of the available emergency response services in the area. Such services include those of the airline, the community, and local agencies.[19]

An EIP and CCP define, for an airport, those procedures and resources for responding to emergencies. It describes jurisdictional boundaries, the chain of command, and communication and coordination procedures. Each plan reflects the particular needs of the airline and its operations by defining general and common elements.[20]

The National Security Institute has identified the general elements of an emergency incident and command center plan in an article titled *Bomb Threats and Physical.*[21]

SECURITY PLANNING

The Emergent Incident and Command Center Plan Survey

The main purpose of this survey is to establish the operational status of an emergency incident and command center in relation to the operational recommendations of the National Security Institute.[22] The resulting survey report is comprised of three parts:

1. Fact-finding,
2. Gap Analysis, and
3. Safety assessment

The fact-finding part displays compliance or non-compliance covering the relevant proficiency in these areas. Fact-finding results will form the basis for the gap analysis and safety assessment against the elements identified by the National Security Institute.[23] For purposes of this model, the following survey is proposed for airline airport manager:

Emergency Incident Plan

1. Is there a designated a chain of command?
2. Is there an established a command center?
3. Have the airport and airlines decided what primary and alternate communications will be used?
4. Have the airport and airlines clearly established how and by whom a threat will be evaluated?
5. Have the airport and airlines decided what procedures will be followed when a threat is received or device discovered?
6. Have the airport and airlines determined to what extent the available emergency incident teams will assist and at what point the teams will respond?

7. Have the airport and airlines provided an evacuation plan with enough flexibility to avoid a suspected danger area?
8. Are there designated search teams?
9. Are there designated areas to be searched?
10. Have the airport and airlines established techniques to be utilized during search?
11. Have the airport and airlines established a procedure to report and track progress of the search and a method to lead qualified local police technicians to a suspicious package?
12. Is a contingency plan available if an explosion or other incident should occur?
13. Have the airport and airlines established a simple to follow procedure for the person receiving a threat?
14. Have the airport and airlines reviewed its physical security plan in conjunction with the development of its incident plan?

Command Center Plan

1. Is there a designated a primary location and an alternate location?
2. Are there assigned personnel with designated decision making authority?
3. Is there an established method for tracking search teams?
4. Is there a list of likely target areas?
5. Is there a blueprint of floor diagrams in the center for its operational areas?
6. Has the airport established primary and secondary methods of communication?
7. Has a plan been formulated for establishing a command center; if a threat is received after normal work hours?
8. Is a roster maintained of all necessary telephone numbers?

SOLUTIONS AND RECOMMENDATIONS

The PDCA Cycle

A promising, yet simple, quality management tool called the PDCA Cycle may provide a snapshot of potential strengths, weaknesses and security threats; thereby identifying solutions for risk factors identified using the process. The PDCA Cycle is a widely-known and very popular tool; in fact, the Airports Council Internal has endorsed this classic tool in its education and deployment strategy (see http://www.aci-safetynetwork.aero/). The PDCA Cycle was developed by Dr. Walter Shewhart, and made popular by Dr. Edwards Deming, considered the father of modern quality control. PDCA should be thought of as a continuous cycle, repeating as quickly as possible, in upward spirals that converge on the ultimate goal. The cycle accounts for limitations in knowledge and skills as well as the subsequent increase in knowledge as you work your way through the process, providing rapid improvement. The PDCA cycle is an iterative four-step quality control process. The elements include: (a) plan, (b) do, (c) check, and (d) act. In the context of a safety management system (SMS), the cycle should be thought of as a cycle within a system of cycles that makes up a security management system.[24] The power of this method is simplicity as it is very easy to comprehend as follows:

1. Plan - Establish the objectives and processes necessary to identify and deliver the stated security objectives.
2. Do - Implement the processes to carry out the plan.
3. Check - Monitor and evaluate the processes and results against the stated objectives and report the outcomes.
4. Act - Apply actions to the outcomes for necessary improvements and/or corrections. This means reviewing all steps (Plan, Do, Check,

Act) and modifying the process to improve the results before its next implementation (see Figure 1).

Using the Survey and the PDCA Cycle to Identify the Gap

Using the survey and the PDCA cycle, airport and airlines managers can perform a risk assessment and a gap analysis to identify the gap between the recommendations of the National Security Institute and the integration of these outcomes by their airport and airlines at their incident and command centers.[26] This helps provide airport and airline managers with insight into areas that have room for improvement. The gap analysis process involves determining, documenting and approving the variance between recommended National Security Institute (2004) security measures, the risk assessment and current capabilities at their incident and command centers.[27]

Risk Assessment

Airport and airline station managers can then use the risk assessment criteria of AC 150/5200-37 to assess the safety risk. The risk levels used in the FAA matrix are defined as:

1. High risk – Unacceptable level of risk: The safety measure should not be implemented or the activity continued unless hazards are further mitigated so that risk is reduced to medium or low level. Tracking and management involvement are required, and management must approve any proposed mitigating controls. Catastrophic hazards are caused by:
    a. single-point events or failures
    b. common-cause events or failures
    c. undetectable latent events in combination with single point or common cause

Figure 1. The PDCA Cycle[25]

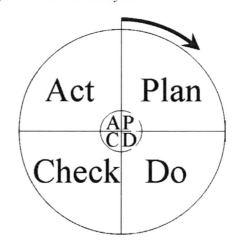

events are considered high risk, even if extremely remote

2. Medium risk – Acceptable level of risk: Minimum acceptable safety objective; the safety measure may be implemented or the activity can continue, but tracking and management are required.

3. Low risk – Target level of risk: Acceptable without restriction or limitation; the identified hazards are not required to be actively managed but are documented. Hazards are ranked according to the severity and the likelihood of their risk, which is illustrated by where they fall on the risk matrix. Hazards with high risk receive higher priority for treatment and mitigation.[28]

The risk assessments of the non-compliance are based on the risk assessment in Table 1.

SOLUTIONS AND RECOMMENDATIONS

The Gap Analysis

Gap analysis naturally flows from benchmarking and other assessments. Once the general expectations of performance in safety are understood,

Table 1. Risk Assessment Matrix

Severity / Likelihood	No Safety Effect	Minor	Major	Hazardous	Catastrophic
Frequent					
Probable					
Remote					
Extremely Remote					
Extremely Improbable					
Low Risk					
Medium Risk					
High Risk					

Note. Source: (FAA)

it is possible for airport and airline managers to compare those expectations with the level of performance at which the operational emergency incident and command centers currently function. The comparison of the risk assessment to the level of the proficiency identified by use of the survey instrument forms the gap analysis, which is the basis for airline airport or airline manager's safety recommendations at the conclusion of the emergency incident and command center reviews.[29]

Future Research and Direction

The most promising research involves on-going use of the survey, gap analysis, and risk assessment tools at airports. To date, this method of risk analysis and assessment has been used on several occasions to identify security and safety risk present at the University of Central Missouri's Skyhaven Airport. Further expansion into the air carrier airport venue would be a beneficial test on how effective the tool is to management. After implementation into the air carrier airport venue, further research involving outcomes of the SMS process is the next step in the process.

CONCLUSION

In the wake of recent events, including the September 11, 2001 terrorist attacks on America, airport and airlines managers must prepare their airports and airline systems in the event of terrorist attacks. This preparation should include the recognition of terrorist intentions and characteristics, types of potential attacks, recognizing terrorist activity, implementing emergency incident plan and command center plan reviews, and incorporating heightened anti-terrorism passenger awareness measures. This preparation is designed to prepare airport and airline managers for the potential threat of violence, including bombings. While the ideas set forth in this chapter may be applicable in a number of cases, they are intended only as a guide for further research of this SMS method of assessment in the air carrier airport venue. If there is one point that cannot be overemphasized, it is being prepared. Hopefully, the knowledge gained by use of this SMS model will be used to improve the operational safety status of airport emergency incident and command centers plans across the US and abroad.

REFERENCES

Dartmouth College. (n.d.). *The clinician's black bag of quality improvement tools.* Retrieved March 11, 2008, from www.darthmouth.edu

Federal Aviation Administration. (2007, February 28). *AC 150/5200-37.* Retrieved March 11, 2008, from www.faa.gov

Melton, H. K. (2003). *The U.S. government guide to surviving terrorism.* New York, NY: Barnes and Noble Books.

Morrison, B., & Levin, A. (2004, November 22). Flights into USA get new security. *USA Today.* Retrieved December 1, 2004, from http://www.usatoday.com/ travel/news/2004-11-22 -flights-security_x.htm

National Security Institute. (2004). *Bomb threats and physical security planning.* Retrieved December 1, 2004, from http://nsi.org/Library/Terrorism /bombthreat.html

Office of Homeland Security. (2004). *Recognizing terrorist activity.* Retrieved December 1, 2004, from http://www.national terroristalert.com/readyguide/activity.htm

Quilty, S. M. (2005). *Airport security and response to emergencies.* Alexandria, VA: American Association of Airport Executives.

Transportation Security Administration. (2008, March). *Electronic code of regulations.* Retrieved March 10, 2008, from www.tsa.gov

Wells, A. T., & Young, S. B. (2004). *Commercial aviation safety* (3rd ed.). New York, NY: McGraw-Hill.

ADDITIONAL READING

Abeyratne, R. (2010). *Aviation security law* (1 ed.). Springer.

Airport, aircraft, and airline security. (2 ed.). (1991). Butterworth-Heinemann.

Alexander, Y. (1990). *Aerial Priacy and aviation security.* Springer.

Anderson, T. (2004). Aviation security. *Security Management, 48*(6), 138(1).

Dorey, F. C. (1983). *Aviation Security.* Van Nostrand Reinhold.

Elias, B. (2009). *Airport and aviation security: US policy and strategy in the age of global terrorism.* Auerbach Publications. doi:10.1201/9781420070309

Freni, P. (2003). *Ground stop: An inside look at the Federal Aviation Administration on September 11, 2001.* iUniverse Incorporated.

Harrison, J. (2009). *International aviation terrorism: Evolving threats, evolving security. 2009.* Routledge.

Janes Special Reports. (2003). *Aviation security: Standards and technology.* US: Jane's Information Group.

National Research Council. (1993). *Detection of explosives for commercial aviation security. UC.* National Academy Press.

Nelson, N. (2004). *Airport security: One of the most deadly weapons...your tongue.* Authorhouse.

Phipps, D. (1991). *The management of aviation security.* US: Financial Times Prentice Hall.

Prasad, S. (2001). *Aviation security.* Abha Publications.

Price, J., & Forrest, J. (2008). *Practical aviation security: Predicting and preventing future threats.* Butterworth-Heinemann.

Rbeyratne, R. I. (1998). *Aviation security: Legal and regulatory aspects*. US: Ashgate Publishing.

Seidenstat, P., & Splane, F. X. (2009). *Protecting airline passengers in the age of terrorism*. Praeger.

Sweet, K. (2008). *Aviation and aviation security: Terrorism and safety concerns* (2 ed.). US: CRC Press.

Thomas, A. R. (2003). *Aviation insecurity: The new challenges of air travel*. Prometheus Books.

Thomas, A. R. (2008). *Aviation security management*. US: Praeger.

Trento, S. B., & Trento, J. J. (2007). *Unsafe at any altitude: Exposing the illusion of aviation security*. Steerforth.

Voegele, A. K. (2010). *Airport and aviation security*. US: Nova Science Publications.

Waltrip, S., Williams, C., & Waltrip, K. A. (2004). *Aircrew security: A pratical guide* (1 ed.). US: Ashgate Publishing.

Wells, A. (2004). *Commercial aviation safety* (4 ed.). US: McGraw-Hill.

Wilkinson, P., & Jenkins, B. (1999). *Aviation terrorism and security* (2 ed.). Routledge.

Zellan, J. (2003). *Aviation security: Current issues and developments*. US: Nova Science Publishers.

KEY TERMS AND DEFINITIONS

Command Center: Is any place that is used to provide centralized command for some purpose.

Gap Analysis: Is a tool that helps a company to compare its actual performance with its potential performance.

Incident Plan: Is a plan to ensure that everyone is working in concert toward the same goals set for that operational period by providing all incident supervisory personnel with direction for actions to be taken during the operational period identified in the plan.

Risk Assessment: Is the determination of quantitative or qualitative value of risk related to a concrete situation and a recognized threat (also called hazard).

Safety Management Systems (SMS): Is the term used to refer to certain regulatory and enforcement frameworks. These frameworks generally apply to transportation, but have also been explored in other industries. An SMS is the specific application of quality management to safety.

Terrorist Cell: Is an organizational structure for organizing a group in such a way that it can more effectively resist penetration by an opposing organization.

Transportation Security Administration: Is an agency that was created by the federal government in response to the September 11, 2001 terrorist attacks in the US. Prior to its creation, airline security screening was operated by private companies that had contracts with either an airline or a consortium contracted by multiple airlines that utilize a given terminal facility.

ENDNOTES

[1] Wells, A. T., & Young, S. B. (2004). *Commercial aviation safety* (3rd ed.). New York: McGraw-Hill.

[2] Ibid.

[3] Morrison, B., & Levin, A. (2004, November 22, 2004). *Flight into USA get new security.* Retrieved December 1, 2004, from http://www.usatoday.com/travel/news/2004-11-22-flights-security_x.htm

[4] Quilty, S. M. (2005). *Airport security and response to emergencies*. Alexandria, VA: American Association of Airport Executives. Transportation Security Administration (2008, March). *Electronic code of regula-*

tions. Retrieved March 10, 2008, from www.tsa.gov

5   Melton, H. K. (2003). *The U.S. government guide to surviving terrorism.* New York: Barnes and Noble Books.

6   Ibid.

7   Ibid.

8   Ibid.

9   Ibid.

10   Ibid.

11   Melton, Op. cit.

12   Melton, Op. cit. pp. 25-26.

13   Melton, Op. cit. pp. 26-27.

14   Melton, Op. cit.

15   Ibid..

16   Office of Homeland Security (2004). *Recognizing terrorist activity.* Retrieved December 1, 2004, from http://www.nationalterroristalert.com/readyguide/activity.htm

17   Melton, Op. cit.

18   Ibid.

19   Ibid.

20   Ibid.

21   National Security Institute (2004). *Bomb threats and physical security planning.* Retrieved December 1, 2004, from http://nsi.org/Library/Terrorism/bombthreat.html

22   Ibid.

23   Ibid.

24   Dartmouth College (n.d.). *The clinician's black bag of quality improvement tools.* Retrieved March 11, 2008, from www.darthmouth.edu

25   Ibid.

26   National Security Institute, Op cit.

27   Dartmouth College, Op. cit.

28   Federal Aviation Administration (2007, February 28). *AC 150/5200-37.* Retrieved March 11, 2008, from www.faa.gov

29   Ibid.

Chapter 13
ePlanAirport:
A Web-Based Tool to User-Friendly Decision-Support Systems for Airport Stakeholders and Policy-Makers

Jaime García Sáez
Ingeniería y Economía del Transporte (INECO), Spain

ABSTRACT

ePlanAirport is a Web-based tool that allows running complex studies based on airport systems. The primary goal of the tool is helping to the airport stakeholders and policy makers in the decision-support processes. Nowadays, there is a lack of tools and systems that may help the targeted users in such a process. Otherwise, this tool could guide them in the current global scenario.

So, ePlanAirport fills this gap detected allowing a non-expert user in complex tools to fulfill successfully his mission. For that purpose, a relevant set of data got in a simple and fast way via Web will be available, and they will help in the planning of airport infrastructures and operations.

For instance, the planner will know how a change in an operational procedure or a change in the fleet characteristics or a change in the current infrastructure of the airport will impact on key performance indicators such as capacity, delay, or environment.

DOI: 10.4018/978-1-60960-887-3.ch013

INTRODUCTION

Before the current economic downturn spread all over the world, Eurocontrol forecast a rising traffic above 3% in the period 2008-2014 only in Europe (Eurocontrol, 2009). Other sources say that within the European Union, air traffic demand is expected to double by the year 2025. But, taking into account the present conditions, the growing demand and the resulting mismatch between demand and supply of airport services, this will result in an increase of congestion problems at airports.

So, the simulation role becomes even more important now. There are two main characteristics in this kind of systems:

- Evaluate alternative paths that will help in the decision-support process.
- Support to research and development activities based on new concepts.

These models provide concrete answers dependent on input conditions more than a general solution.

On one hand the simulation activities guide the experimentation and sensitivity analysis in a controlled way. And, on the other hand, the simulation is able to run complex studies that otherwise would require much more effort and time until its complete calibration.

CONTEXT: AIRPORT PLANNING AND AIR TRAFFIC MANAGEMENT

Currently, there are simulation tools able to deeply analyze concrete aspects of the airport environment. Particularly, there are tools intended for capacity analysis, environment analysis, cost-benefit analysis and so on. These tools work with specific sets of data and analyze exclusively some aspects either the traffic or the airport.

The available (analytical or simulation) tools can jointly address all airport elements and flows (at the airport airside and in the airport terminal), support different levels of decision making (strategic, operational and tactical) and analyse nearly all types of performance measures (capacity, delay, level of service, third-party risk, security, environmental impacts, and cost-efficiency). However, as it was stated, each tool is only suited for a specific element, flow, decision-making level, or performance measure, so that tools have to be used in combination for conducting a total airport performance analysis.

So, this initial constraint is a drawback to extend the studies and make deeper ones, being necessary to design a new workflow for the aforementioned complex works. In this new workflow, several tools would be involved being the output of one tool the input of the next one in the workflow. This also implies that the airport planner would require either a deep knowledge of the tools involved or rely in technical experts to get specific output results. Of course, there is a background task related with the design of the correct workflows, which is not straightforward, because there is a need to measure the impact of a little change in the input conditions and check how the output is modified accordingly and in a significant way.

A first step forward in this sense was the idea behind the OPAL project that established a centralized data repository for the whole process, so all tools related in the process could read the needed data and write back the results. Moreover, that project developed the data converters for the simulation tools. So, as the tools could connect to the repository through the data converter, it would allow the final user to implement virtually all combinations of tools to solve a wide range of issues. But this first approach arose two main technical difficulties:

- Need of a high process capacity.
- High volume of information to be managed.

Figure 1. SPADE platform

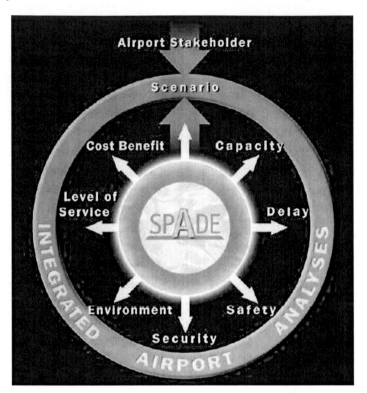

SPADE: EXAMPLE OF AN AIRPORT SIMULATION PLATFORM

The SPADE platform, seen in Figure 1, assists in seeking ways to increase airport utilization in order to accommodate the rising air traffic without undue delays, while improving safety and security, efficiency and service, and reducing the burden of operations on the environment.

SPADE platform is applicable to any airport that would like to improve its support for airport development, planning and operations, with integrated impact analyses and trade-off analyses with respect to a variety of performance measures. It also offers a user-friendly layer on top of individual simulation and analytical tools. The system makes direct interaction with the data-intensive tools obsolete, and removes the burden of data transfer and conversion between tools. Furthermore, the central repository inside the system will contribute to the consistency of the results obtained by the different tools.

In previous European research projects (like TAPE and OPAL) pre-selected tool combinations communicated and interacted in order to perform total airport performance analyses. Nevertheless, the outcomes of these projects lack a harmonised, integrated and fully-automated computing environment for executing the various tools and for presenting their integrated results. Besides that, SPADE core is around the questions that the platform should answer and not in the technical issues related with the selection, use and integration of analytical and simulation tools.

To fulfil the mentioned goal, a 'top-down' approach is followed. More specifically, SPADE first investigates what type of questions the SPADE system should support and then selects tools to address the appropriate performance measures and, finally the workflow is implemented. This approach led to the modelling concept of use cases.

A use case can be considered as a generic airport study in the form of decision-making questions supported by the system. An example of a use case is 'Examine ways of improving or expanding airport infrastructure'. Specific airport decision-making questions corresponding to this use case could be 'What is the effect of a change in runway configuration' and 'What is the effect of a new check-in'. The use case is defined then by an exhaustive list of frequently-asked questions that need to be answered.

SPADE: PLATFORM ARCHITECTURE

The SPADE software is a distributed system that, as seen in Figure 2, using an application Server, allows multiple users to access many applications through a unique interface. It performs the required operations over the data transforming them from- and to- specific tool language automatically. In any case, the final user is not aware of how the tools are really used.

Use cases are split in two categories: operational and strategic. The operational UCs fill the gap in medium and short terms as strategic UCs are intended to help planners to take decisions in the long term. Each use case has its own questions that provide a general answer to specific problems.

From the results out of the platform execution, a user evaluates the baseline scenario in terms of movements, passengers and bags, efficiency, environmental impact and cost-benefit. It also provides decision support for changes in different elements, checking the impact based on the aforementioned performance indicators.

The detailed flow is explained next:

- **First step**: The user selects the initial data previously stored in the database to prepare the study. Then, a question inside a use case is selected (for instance, what effect has a change in the traffic fleet on capacity and delay?) and the parameters are

Figure 2. SPADE: platform architecture

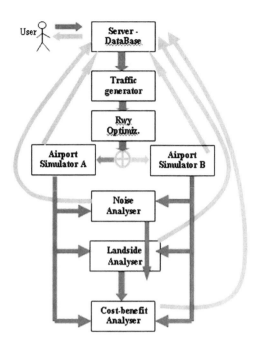

filled by the user (for instance, the traffic increase percentage).

- **Second step**: All inputs ready, then the platform will process the data in the following order without requiring further contribution of the user:

  ○ **Traffic Generation:** The initial data from the user are the baseline of the new scenario. The simulated traffic will correspond to the parameters selected by the user through the interface.

  ○ **Runway Optimization:** After the traffic generator has processed the new traffic and it is stored in the Server, the runway optimizer application is called and the output results will be post-processed and stored in the database.

  ○ **Airport Simulator:** Now the platform is ready to transform all data. To do this, there are two possible options (namely, traffic simulator A or

Figure 3. ePlanAirport software

B in the next figure). The platform will use either of the simulators based on availability and the performance indicators checked by the user. It is remarkable that a wide bunch of tools may be used in SPADE, making the results tool-independent.

- **Third step:** The output is presented as tables and graphics showing movements, delays, cost per airline and so on. One of the greatest advantages of SPADE is the work after the execution of the workflow, where information is read from files or database and transform to a more manageable and understandable format.

SPADE does not leverage the individual results of the different tools, because the results depend on the execution of the tool itself, which is not modified by the platform. Nevertheless, SPADE unifies and automates the collection of information that serves the airport planners and operators in the decision making process. Other important issue is that information sent to the platform is minimal. The simulation model is stored in the tools and they are only provided with some parameters to run the simulations, avoiding peaks of traffic in the network.

ePlanAirport: GOALS OF THE TOOL

In the context of simulation, ePlanAirport goes one step further with regards to the previously described platforms. EPlanAirport is a Web tool (Figure 3) designed to focus the airport planners in the study of the airport operations and help them in the decision making process.

This tool allows the targeted users to know the capacity, the delay or the environmental data when a change in an operational procedure or a new infrastructure is available in the airport, for instance. And, the most important thing is that this will be available from the desktop through a conventional Web browser and without any need of additional tools.

EPlanAirport takes advantage of a set of commercial tools, well-known and currently valid for experts, to run studies and workflows without the help of those experts. They will not run manual, but automatically. This automation will spread the use of the simulation tools to non-expert users which will not need to install new software in their computers, as it is only required the availability of an Internet connection and a conventional Web browser as graphic interface between the user and the simulated workflow.

Of course, the simulations will rely on specific Simulation Platforms storing the model of the airport under study. In this environment all

Figure 4. ePlanAirport architecture

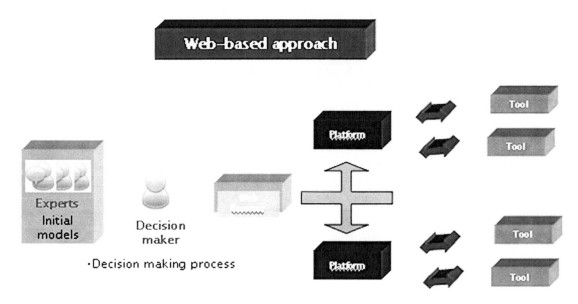

needed tools are integrated and managed in a transparent way.

The goals of ePlanAirport are:

- Develop a Web Services Framework to evaluate and integrate different Simulation Platforms. A set of web-services that may be satisfied by a simulation platform will be defined and developed. This common set is intended to be extended to all known platforms so they can be accessed via Web.
- Provide both theoretically and practically, a sensible approach based in feasible methodologies and suitable technologies.

So, ePlanAirport as web-services framework has the potential technologies to interoperate in the Web and offer the services world-wide. The core technology is able to interchange date between different applications with the goal of providing services (see Figure 4 for the ePlanAirport architecture).

So, ePlanAirport will deploy the services as remote procedures and the users (likely, simula-tion platforms) will request them through the Web. Indeed, the services will provide standard communication mechanisms between the different applications to show dynamic information to the final user.

This goal of extensibility and interoperability between applications makes possible the running of complex operations and appeals to build a standard architecture as a reference point. In this process there are some specific technologies that enable the information flow in a standard way. By one side, there is SOAP (Simple Object Access Protocol). It is a protocol based in XML that allows the communication between devices and has the ability to send and receive complex messages. Data can be transmitted via HTTP, SMTP and so on.

In the evolution of the needs of the companies based on Web applications, new mechanisms have been developed to enrich the descriptions of operations by means of semantic annotations and tags that define the behaviour of the services. This would allow a refine search of the most relevant services attached to the desirable goals. Moreover, in part due to the increase complexity of the processes in great corporations, there is a technology that helps in the definitions of such processes composing new services based on individual and smaller Web services. This technology is called choreography. EPlanAirport studies the state-of-the-art of the Web Services Paradigm from past and current RTD projects funded by EC in the last years (CORDIS News, 2009), and applies it to a new test field which is the simulation platforms in airports, doing some research concerning the market-driven opportunities detected in airport managerial processes.

ePlanAirport: USER INTERFACE DESCRIPTION

EPlanAirport provides an interface to the user that, generically, will have the following behaviour:

- The user will fill in some panels through a webpage, making some choices in different combos and text boxes.
- Through the Net, the corresponding services will be transparently invoked. This is independent of where the server and the supporting platform are hosted.
- The execution will run as if the platform is directly connected to the user system. So, the server will guide the whole process, handling the inputs and outputs of the different tools.
- Tools may be installed in the same server that is invoked from the client or they may be hosted in other server and connected to the simulation platform through specialized tool controllers.
- The user will finally receive the outputs of the whole process and show them in another webpage.

Overall, the application is a software module hosted in an application Server that will give service to multiple users connected to it by simply using a Web browser and the web services technology. This module will have access to a database management system containing the necessary data to run the different studies. Also, the module will provide connections to the platforms to be tested. Its main goal is the control of the execution workflows, which is itself the real core of the application. On the other hand, both the execution and the control of the simulation tools are up to the platform which the user connects to.

EPlanAirport is developed under the Web Service Paradigm, within J2EE platform and deployed as a Web platform, whose functions are accessible through web services. So, the user only needs a Web browser to access and use the underlying platforms, such as:

- SPADE2 (SPADE Consortium, 2006): Sixth Framework Programme project to develop a supporting platform for airport decision-making and efficiency analysis. The objective is a user-friendly decision-support system for airport stakeholders and policy-makers. This system will provide support in airport development, planning and operations, allowing both integrated impact analysis and trade-off analyses with respect to a variety of performance measures (e.g., capacity, delay, level of service, safety, security, environmental impacts, and cost-benefits).
- PITOT (AENA SNA development, 2006): Under development platform in AENA that integrates different simulation tools of the Air Navigation System. It is a process integrated platform for optimal use of analysis techniques.
- PACS (Desart, 2007): ATM Data Repository Service offered by Eurocontrol.

The wrapping of the access layer to the platforms will put its emphasis in crucial aspects of the Web Paradigm such as security, throughput, and efficiency and so on.

SUMMARY AND CONCLUSION

This paper concerns the important issue of integrated solutions to air transport and ATM problems. Across these sectors, even within a reasonably specific domain such as disruption management, there are very often a number of disparate applications, often running on different platforms and using dissimilar data structures, which present real challenges to integration and high-level solution management and scenario investigation.

This paper firstly displays a degree of technical knowledge relating to systems architecture and communications protocols and sensibly suggests that future solutions to complex problems in the aeronautical field may depend in many cases on web-based tool integration and data exchange. For this reason, it was discussed the conceptual and practical development of SPADE project as a basis to build the ePlanAirport tool.

Lessons learnt in building backend simulation platforms and research in Eurocontrol data sources provided the necessary background to design, develop and deploy the Web Services framework, which is the pillar of the tool. Decoupling the user's access to the Web from the execution of services in a remote application server was one of the toughest topics addressed in the course of the project.

Current simulation platforms are highly dependable on the tools integrated and the time they need to compute different studies. This is clearer showed in the case of analytical tools, which are desktop oriented. But, in the case of operational tools able to run in a distributed environment, the cost of the studies in a concrete airport depends on a range of factors: conversion between different protocols, import/export data, time execution of the tool, scenario modelling of an airport, license of the tool, expert assessment in an initial stage of the tests and so on. All these indicators are of great importance when it comes to designing a universal access to a range of similar tools and/ or platforms whose results may accommodate airport stakeholders' needs.

So, in the aforementioned scenario, EPlanAirport may be considered as a Web façade that decision makers would use to test the platforms in a twofold way: free access to common and open facilities and payment access to more detailed and/or complex trade-off analysis that need the platform and/or compelled tools to be further customized. Stakeholders and policy makers in an airport could access easily to complex tools, run studies and get results for further analysis, which is the core of their job.

Whilst this tool was being developed, several interviews with operations management staff in different airports in Europe were driven and preliminary results were presented to them. They interpreted the results obtained with ePlanAirport, adding new tips for the improvement of the framework and the clarity in the navigation and help system through the developed website. This made us focus in an agile and attractive front-end as well as in the use of the latest Web technologies so as to bring the user the chance of an impressive visit to the tool keeping the eye in the informative goal of the tool.

The impression of the tool was positive and they rank it helpful in their daily work. They can get fast results that would require more detailed studies under demand of the airport authorities. So, ePlanAirport results would be the first answer to correctly point the arrow of the final decision. It would need refinement so as to better assess the managerial staff questions.

EPlanAirport broadens the range of potential users as it opens the testing with costly tools and platforms, which require both investment in servers and dedicated premises and deep know-how in simulation processes, to non expert users. License of some tools is well above some thousands of euros and it is not easy to convince using those kinds of tools and/or platform from scratch without previous knowledge. It is not affordable in most cases, as it was said by some airport representatives of operations staff. Then, ePlanAirport would pave the way to adopt and spread the use of simulation platforms in more processes. The stakeholders need to see the benefits of the tools currently wrapped by ePlanAirport framework. The research strategy followed in ePlanAirport deals more with the better ways to approach the workload of decision makers in airports than with the consideration of the development of a Web framework as a beyond-standard engineering task.

The main advantages compared to other tools are:

- Simplicity, as it is not mandatory to be an expert to get results from ePlanAirport. This tool enables studies that were impractical up until now.
- Response time, as it is not necessary to build a new model each time the user needs to run a study. This tool improves the efficiency, reliability and response times of the air navigation system analysis required by ANSPs (Air Navigation Service Providers).
- Cost, as costly experts and tools are not necessary to manage a complex simulation platform. It permits a rationalised use of the analysis media, separating not only the analysis tasks from the software usage tasks (operating system usage, file, storage and data processing system handling, format preparation, etc.) but also distinguishing between the different actors involved in the decision-making processes.

A presentation was made to Eurocontrol ATM Performance Division in order to evaluate the connection between the Aeronautical Information Management (AIM) concept for a network-centric ATM environment and the data processing services of the ePlanAirport framework. Further collaboration will be stretched over the following months as significant benefits in terms of capacity, efficiency and economy can be obtained if the ATM community in Europe were to adopt Collaborative Information Sharing. Eurocontrol has the goal to use a fully secured and centralized data repository coupled to a set of web-based applications to demonstrate the stakeholders (mainly in local airports) the benefits of an effective share of information. In this context, ePlanAirport is well positioned.

Finally, SESAR (SJU, 2008) initiative supports a paradigm shift based on state-of-the-art and innovative technologies required to fully integrate the future air navigation concepts and developments.

REFERENCES

AENA SNA development. (2006). *Integrated tool PITOT for the analysis of air navigation systems.* Retrieved from http://www.aena.es/

CORDIS News. (2009). *A more realistic World Wide Web.* Retrieved from http://cordis.europa.eu/

Desart, B. (2007). *A window to airport system information sharing.* PACS Workshop. Retrieved from http://www.eurocontrol.int/aim /public/ standard_page/pacs.html

Eurocontrol. (2009). *Medium-term forecast update: IFR flight movements (2009-2015).* Eurocontrol forecast reports, September 2009. Retrieved from http://www.eurocontrol.int/

SJU. (2008). *European ATM master plan.* Retrieved from http://www.eurocontrol.int /sesar/

SPADE consortium. (2006). *Project objectives.* Retrieved from http://spade.nlr.nl/pobj.htm/

Chapter 14

Improved Airport Enterprise Service Bus with Self-Healing Architecture (IAESB-SH)

Issam Al Hadid
Isra University, Jordan

ABSTRACT

This chapter introduces the different aviation and airport Information Technology systems. Also, this chapter provides architecture based on the Service Oriented Architecture (SOA) that improves the information accessibility and sharing across the different airport departments, integrating the existing legacy systems with other applications, and improving and maximizing the system's reliability, adaptability, robustness, and availability using the self-healing agent and virtual Web service connector to guarantee the quality of service (QoS).

INTRODUCTION

Airports need to adapt new technologies to react effectively and quickly to customers' needs and to provide a better service such as the electronic ticket. In addition to the challenges of the ability to respond to the growing requirements of the automatic information interchange between the different systems to ensure safe and efficient

DOI: 10.4018/978-1-60960-887-3.ch014

airport operations. Most of the challenges in the Airport systems today lie in the ability to respond to the growing requirements of the automatic information interchange between the different departments including operational, statistical, aviation and financial information. In addition, the integration with the existing legacy systems to ensure safe and efficient airport operations. All the operations in the airport are driven by the exchanged information; Airport business units create information, transform information, distribute information, and take action on received information. Airports' systems are developed by different vendors and were not designed to be interoperable, which makes systems integration a very complicated and not easy to be implemented. Service Oriented Architecture (SOA) provides the ability to address the distributed computing requirements; protocol independent, loosely coupled, reusability and standard based (Papazoglou & Heuvel, 2007). It is based on the Web Services; distributed, loosely coupled, reusable software components that encapsulate a discrete functionality and can be accessed using standard internet and XML- based protocols (Sommerville, 2007). SOA encourages a lot of businesses to move toward the adapting the SOA architecture to enable the response to change faster and to cut the cost of replacing the legacy systems that they have and integrate with the new systems so all the information can be accessed and shared by all the systems (Keen et al., 2004; Minoli, 2008), accordingly; SOA will provide a guideline for airport Information systems architecture design, development and integration. The functionality provided by the integration platform Enterprise Service Bus (ESB) is based on the SOA that utilizes Web service standards to supports a variety of communication patterns over multiple transport protocols to connect different applications and technologies (Papazoglou & Heuvel, 2007). In addition, the features of loosely coupling and breaking up the integration logic into separate parts can be easily managed (Keen et al. 2004). ESB provides architecture based on the SOA that improves the information accessibility and sharing across the different Airport's departments. Furthermore, it provides a component interface to existing legacy system so it can be integrated with other applications, accessed over the web, and support the reusability of the legacy systems. Also, the Self-Healing Agents which will improve and maximize the system's reliability, adaptability, robustness and availability.

RELATED WORK

There are many different information technology systems that are used in the aviation industry (Abu-Taieh, 2009), shown in Table 1. Many Airports and Air Traffic Control (ATC) Units have moved toward the adapting of new open systems (Goold, n.d.). The information can be accessed, shared and flow across the different hardware systems, operating systems, networks and airport management systems, because there must be an integration framework for defining the information integration requirements, and designing the systems integration architecture to address the loosely coupled systems, standards-based interfaces, and protocol independent distributed computing. this manages information elements with a defined process and provides the ability to upgrade, replace or move systems or components without having to modify code and disrupt execution of the existing applications (Cheng, 2001). Service Oriented Architecture (SOA) provides the ability to address the distributed computing requirements; protocol independent, loosely coupled, reusability and standard based (Papazoglou & Heuvel, 2007). It is based on Web services; distributed, loosely coupled, reusable software components that encapsulate a discrete functionality and can be accessed using standard internet and XML- based protocols (Sommerville, 2007). SOA provides flexible architecture that unifies the business process by modularizing applications into services which satisfy the addressed requirements for the Airport and aviation systems integration architecture. The functionality of the integration architecture must also support a variety of communication patterns over multiple transport protocols, this requirement is addressed by the integration platform Enterprise Service Bus (ESB) which is based on the SOA, It utilizes Web service standards to support a variety of communications patterns over multiple transport protocols to connect different applications and technologies (Papazoglou & Heuvel, 2007). It

Table 1. Airport and aviation systems (Abu-Taieh, 2009)

System	Purpose
Reservations systems	Used to store and retrieve information and perform the air travel transactions.
E-Ticketing	Used to book airline tickets by passengers through a website or by telephone.
Self Service Kiosk (SSK)	Computer based guides to give directions and help airport users use the airport better.
Radio Frequency Identification (RFID)	For luggage position; used to prevent the loss of luggage or being mishandled.
Simulators and Training	used to train the pilots and the air traffic controllers
Global Positioning system	It provides reliable positioning and navigation services used to identify the aircraft route to destination and guiding the planes to a safe landing.
Security checking	Used to check the passengers' records to make sure they are not terrorists or criminals
Flight management and planning systems	It holds the flight plan and the aircraft's position, It also includes the Fuel calculation, to ensure that the aircraft can reach the destination safely, and fulfill the requirements of the air traffic control to minimize the risk of collision with another aircraft. In addition, minimizing flight cost by (route, height, and speed, and the minimum necessary fuel).
airport management system	Used to manage the airport resources efficiently and effectively; financially, flight, events, incident, operational, CRM, air traffic control and the passengers and fright information.

also, provides the integration logic between the service consumer and provider which is used to transform messages, route requests and convert transport protocols between the two parties (Keen et al., 2004; Chappellm, 2005). The configuration and orchestration of the services in unified and clearly defined processes are provided by the Service Orchestration using the Business Process Execution Language (BPEL), this allows the business operations to have the ability to respond to the underlying business needs via the different components invocations either in an event- driven or asynchronous fashion to fulfill a complex business process. As well, ESB improves the ability to upgrade, move or replace applications or services without having to modify code and disrupt existing ESB applications; due to the abstraction of the physical destination and connection information provided by the End points, they allow services to communicate using logical connection names which will be mapped to actual physical network destination at runtime (Papazoglou & Heuvel, 2007; Chappell, 2004). Beside, the Self-Healing Agent that will be added to the Web service will monitor, diagnosis and repair the detected anomalies and/or other unexpected faults. (Mehta & Medvidovic, 2002; Robertson &

Williams, 2006) it will also improve and maximize the system's reliability, adaptability, robustness and availability.

Airport management system is used to control the Airport Operations, financial data, Air traffic control and other resources. Table 2 shows the details airport management sub-system and the purpose for each sub system.

Figure 1 shows a typical airport management system. All the subsystems are connected to the central Database and the information following between the subsystems via the airport LAN or WAN.

REQUIREMENTS OF THE AIRPORT ENTERPRISE SERVICE BUS WITH SELF-HEALING ARCHITECTURE

The Airport Enterprise Service Bus with Self-Healing architecture must achieve a number of goals: reliability, maintainability, evolvability, extensibility, scalability, and interoperability. These goals help achieve the Airport's systems integration while increasing business operations efficiency between all the Airport's departments.

Table 2. Airport management and planning sub-systems (AODB, 2009; Abu-Taieh, 2009; Goold, n.d.)

Sub-System	Purpose
Aeronautical Billing	Collecting and billing the revenue of all flights and non-flights.
Resource planning and Management	Used to manage the airport resources (gates, ticket counters, baggage belts, stands).
Airport Operational Database (AODB):	The central database that contains all the essential data for airport daily operations.
Flight Information Display systems (FIDS):	Provides the flight related information throughout the airport.
Property Management:	Helps the Airport manage the rental, lease, license, services contracts, agreement, and concession related interests.
Interactive Voice Response (IVR	Used to respond to the customers inquiries and helps improve customer service.
Incident Management	Used to track all the Airport's incidents activities.
Air Traffic Control	Contains the following information: • Aircraft Registration • Runway Used • Actual Time of Landing and Departure • Number of Circuits • Number and Type of Approaches • New Estimates of Arrival and Departure • New Flight Information
Air Traffic Control (ATC) Billing	Information Includes: • Aircraft Registration • Point of Entry into Air Space • Point of Departure of Air Space • Airport Point of Departure and/or Landing • Times at the Different Points of Entry or Departure
Airline/Handling Agents:	Information includes: • Block On/Off Times • Passenger and Freight Information • Check-In Desk Opening and Closing • Departure Gate Opening and Closing • New Flight Details (If not entered by ATC) • New Estimates of Arrival and Departure Times (If not entered by ATC) • Seasonal Schedule Information
Apron Handling:	It Handles the following information: • Stand Allocation • Block On/Off • First Bag Last Bag Times • Ground Services Supplied to the Aircraft
Staff management Display	used by the staff to access the information
Aeronautical Invoice System	It's a special invoicing system designed especially for the airports to calculate the cash and the credit invoicing
Ledger systems	It contains the financial information (AR, AP, GL, payroll, etc...)

Seasonal Schedule	• Code Shared Flight Numbers
	• Multi-Sector Airports
	• Aircraft Type
	• Estimated Time Of Arrival
	• Estimated Time of Departure
	• Operational Days Of The Week
	• Exception Dates
	• Allocated Resources Including:
	  ◦ Carousels
	  ◦ Check-In Desks
	  ◦ Check-In Time
	  ◦ Departure Lounges/Gates
	  ◦ Stand/Air-Bridge

continued on the following page

Table 2. Continued

Sub-System	Purpose

Reports	• Typical reports include:
	  ◦ Timetable
	  ◦ Daily Mayfly
	  ◦ Load Factors and Income Reports: by Airline, Aircraft, Registration, Route, Operator)
	  ◦ Aircraft Movements – Hourly Activity
	  ◦ Passenger Movements – Hourly Activity
	  ◦ Traffic Distribution by Airline, Aircraft, Airport
	  ◦ This Year to Last Year Comparisons
	  ◦ Traffic Analysis Reports

Figure 1. Airport management system (Goold, n.d.)

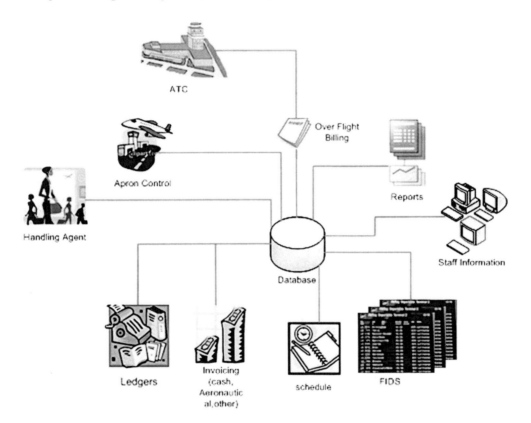

- Reliability: IEEE (IEEE, 1990) defined the Reliability as "The ability of a system or component to perform its required functions under stated conditions for a specified period of time." Airport operation is basically a sequence of business events triggered one after another; a reliable system guarantees the success of these airport event executions and delivery. If the sequence stops, the airport operations would stop. This requires a highly reliable integrated system to prevent such situations. The architecture not only needs to provide reliable integrated systems, but also ensures that a failure in one subsystem does not affect the other systems operations.

- Maintainability: IEEE (IEEE, 1990) defined the Maintainability as "The ease with which a software system or component can be modified to correct faults, improve performance or other attributes, or adapt to change environment." So the new architecture must be designed to satisfy the changing requirements; system or component upgrade, move or replace should only impact the upgraded module; so the interface must be well-defined and structured well.
- Evolvability: According to Rowe & Leaney (Rowe & Leaney, 1997) the System evolvability is a "system's ability to withstand changes in its requirements, environment and implementation technologies." It's the ability of an integrated system to adapt the continuous changes to satisfy the business requirements and technologies. When there is a change in the business needs or technology used, only related components need to be changed, the other systems and components should not be affected by the new changes.
- Extensibility: IEEE (IEEE, 1990) defined the Extensibility as "The ease with which a system or component can be modified to increase its storage or functional capacity." New business requirements might include improvement of the efficiency of the existing business operations or support new business functions which required the modification of the existing systems or components or replace it with new functionally and more efficient systems and components. The architecture of the system integration will facilitate the ease of integration for the new or upgraded systems or components functionality; thus providing a better extensibility.
- Scalability: According to Coulouris, Dollimore & Kindberg (Coulouris, Dollimore & Kindberg, 2005) "The system is described as scalable if it will re-

main effective when there is a significant increase in the number of resources and the number of users." It refers to the ability to add more systems or components to improve the system performance which is based on the loosely coupled systems and components.
- Interoperability: IEEE (IEEE, 1990) defined the Interoperability as "the ability of two or more systems or components to exchange information and use the information that has been exchanged." It's the most important motivation toward the systems integration; achieve a high level of automation. Hence high efficiency of business operations by providing necessary integration services to individual systems through well-defining those integration interfaces.

AIRPORT ENTERPRISE SERVICE BUS WITH SELF-HEALING ARCHITECTURE

The proposed Improved Airport Enterprise Service Bus with Self-Healing architecture (IAESB-SH) provides an implementation backbone for the Airport and Aviation systems to control flow and translation of the exchanged messages between systems using the supported messaging protocol.

Proposed AESB-SH provides the following capabilities:

- The ability to route a request to a particular service provider on deterministic or variable routing criteria.
- Message Enhancement via the protocol transformation by accepting the message formatted as a type of a protocol from the consumer as input (i.e., JMS) and convert it to a different protocol, and send it to the service provider (i.e., SOAP).
- The ability to translate the business processes into a business services that can be

arranged, managed and coordinated to fulfill a business services request that manage a complex business process.

- Message synchronization; manage the request and the messages' state.
- Security; prevent any unauthorized access.
- Self-Healing Agent capabilities; Adaptability, Traceability, Awareness, Autonomy, Robustness, Distribution-ability, Dynamicity, and Mobility.

IAESB-SH is designed to provide interoperability between the different airport systems with different technologies and components via standard-based adapters and interfaces using Web Service technology. IAESB-SH uses the Java Message Service (JMS) to integrate JE22, C# client to integrate.NET, MQ application to interface the legacy systems and Web services for data sources and external applications. The End points will abstract the physical destination and connection information that offers the connected airport systems and services the ability to be upgraded, moved, or replaced without having to modify code or disrupting existing systems. IAESB-SH pulls together the Airport systems and integration components to form a complex business process which automates the systems' functions by the configuration and orchestration of the services in unified and clearly defined processes. This is provided by the Service Orchestration (usually using BPEL); because it grants the Airport operations the ability to respond to the underlying business needs via the different components' invocations either in an event- driven or asynchronous fashion. In addition, Self-Healing Agent will maximize the reliability, Adaptability, Robustness and availability by monitoring all the service layer tasks; this is done by extracting all information about the service's tasks, Diagnosis errors by examining and analyzing extracted information, and Repair the fault by executing an action according to the faulty type. Also, the Virtual Web Service Connector expands the classical service oriented

architecture; it is used as Web service interface to guarantee the Web service QoS. Orchestration Manager is used to automated arrangement, coordination, and management of services to fulfill a business process. Services send notification to the Orchestration Manager to execute the related business rules including services and events that must be invoked. Figure 2 shows the proposed Airport Enterprise Service Bus with Self-Healing architecture.

SELF-HEALING AGENT

The web service Self-Healing refers to the web services ability to automatically monitor, fault detection and diagnosis, while repairing failures by executing an action to maintain an appropriate QoS. The Self-Healing process is also called the lifecycle of self-Healing. it is used to monitor the service execution by collecting and extracting information from managed service tasks accessed during the execution by tracking the tasks' behavior and the objects accessed by the tasks, examine and analyze information extracted by the monitor process by the diagnosis process, anomalies occurred if service task does not execute as expected within specific time, or return an error value. Finally, heal the service by execute an action which specifies what action to be taken in order to recover any malfunction. If the repair action is accepted, the system will be updated accordingly. Figure 3 shows the Self-Healing process.

The Self-Healing provides the following capabilities for the Web Service (WS):

- Adaptability: enables dynamic modification of the WS (behavioral and interaction).
- Dynamicity: WS's adaptability concerns during run-time (internal state consistency).
- Awareness: WS must monitor its own performance (ex.: state, availability, behavior, correctness, reliability) and its environment (ex.: network connection state).

Figure 2. Proposed framework (airport enterprise service bus with self healing architecture)

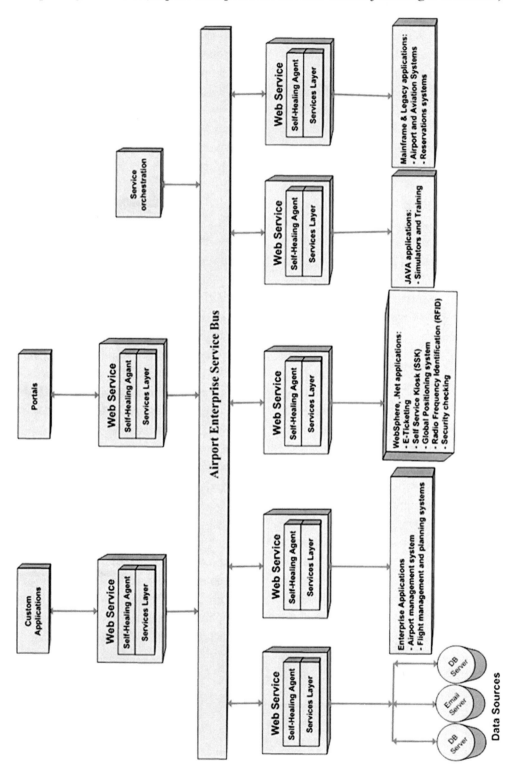

Figure 3. Self-healing process

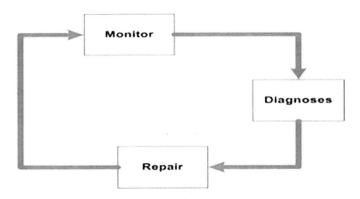

- Autonomy: the ability to detect the anomalies in the WS performance and its environment.
- Robustness: is the WS ability to effectively respond to unexpected operating conditions imposed by external environment as well as errors, faults, and failures within the WS itself.
- Distribution-ability: WS must support effective performance in the face of different distribution/deployment conditions.
- Traceability: WS must provide complete information about execution states.
- Mobility: provides the ability to dynamically change the (physical or logical) locations of WS components and resources.

Proposed Web service Self-Healing Algorithm:

```
Repeat Until All Required Tasks are
Executed
  Monitoring Manager: Call Task
    Diagnosis Manager: Check Results
      IF (Error, Fault, anomaly)
        Add Record to Log file
        Repair Manager:
          Create Repair Plan Strategy
          Execute Healing Action
          Add Record to Log File
```

```
      Else
        Notify Monitor Manager (No
Errors)
      End IF
End Repeat
```

Figure 4 shows the process of the Self-Healing Agent for the Web service execution. The Req/Resp Manger handles the sent requests to the Web service, it will check the Request parameters before proceeding in the processing to insure there is no Input Fault (IF), or invokes an unavailable operation in the actual state of the Web service. The Req/Resp Manger will then send the request to the Monitoring Manager which will start the execution of the required tasks in the service layer and extract information about each executed task and send the result to the Diagnosis Manager to examine and analyze the extracted information and to decide if there are any unexpected faults or anomalies. According to *the result of the Diagnosis Manager a notification should be sent to the Monitoring Manger to proceed in the execution or the result should be sent to the* Repair Manager to take the right action according to the error occurred, and the available solutions in the Action Knowledge. This can be used to prepare the Repair Plan; re-execute the Task, Locate backup Resources (if unavailable), restart the Web service (re-execute all tasks), Call a backup task

Figure 4. Web service self-healing model

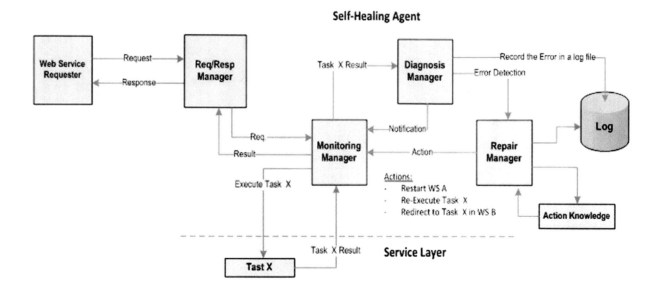

in external WS or redirect the execution to other Web services (if available).

Web service with Self-Healing Agent components:

- Monitor Manager: collects and extracts information about tasks and passive objects accessed during the execution; trace the task behavior and the passive objects accessed by tasks.
- Diagnosis Manager: detects the anomalies of the task and passive objects accessed by task based on the extracted information passed by the Monitor Manager. Task or passive objects are anomalous if they do not execute as expected within the accepted parameters. The accepted parameters:
  ○ Wrong returned values
  ○ Resources are not available
  ○ Response exceeds the specified time interval.

If no anomalies are detected, it will notify the Monitor Manager to complete the execution to execute the next task. Otherwise, it will pass the result to the Repair Manager to generate a repair plan and take action.

- Repair Manager: generates a repair plan using the Action knowledge according to the anomalies detected by the Diagnosis Manager. It also performs the healing actions to the anomalous tasks and passive objects. The healing actions might be one of the following:
- Re-Execute Task
- Locate backup Resource
- Re-Execute WS tasks
- Call backup task in external WS
- Invoke a backup WS (Send the Consumer request)
- Invoke a composed Web service to fulfill the equivalent functionality

```
** Repair plan strategy: using AND
&&, OR || to generate the repair
Plans
```

```
Example:
Action 1 || Action 2
Where Action 1: Re-Execute Task, Ac-
tion 2: locate backup Task… etc.
Or
Action 1 && Action 2
Where Action 1: locate backup re-
source (database), Action 2: Re-Exe-
cute Task.
```

- Action Knowledge: is used to determine which actions can be taken according to the diagnosis anomalies. It is used to match the recommended actions in order to heal service and ensure the action will not have a negative impact on the current process.

Figure 5 shows a detailed process in case of anomaly detected and the Self-Healing Agent response.

If Web service Results are critical, the Req/Resp Manager will send the Request to more than one WS (if available) and compare the results to make sure they are identical, as shown in Figure 6.

VIRTUAL WEB SERVICE CONNECTOR

Virtual Web Service Connector exposes Web services' interfaces to the clients, and allows the interaction between clients and Web services. Also, Virtual Web Service Connector Keeps an eye on the held requests that didn't responded, also, make sure that the response time out is not been exceeded, in addition, guarantee that the Web service will not be flooded with requests to provide the expected QoS; it will react if the Web service is hit with a higher than expected requests (flash crowed) that would slow or stop the Web service from responding to requests. As well, it will work to allocate Web service optimally and do not refuse any valid request. And finally it

will map the consumer requester parameters with substituted Web service Web Service Description Language (WSDL) input parameters (Dobson, 2006). And as a result, it will guarantee the QoS as agreed in the Service Level Agreement (SLA) for the requested users (Ludwig et al, 2005). If the Web service executed successfully, it will send the response to the Virtual Web Service Connector which will send the request results to the service consumer who invoked the Web service and a notification to the Extended Execution Engine to invoked the related Web services participating in the composed service that fulfill a specific business process. The Virtual Web Service Connector as observer between the invoked Web service and the Service Consumer, and the invoked Web service and the Execution manager to guarantee the QoS according to the Service Level Agreement (SLA). Virtual Web Service Connector is responsible for the Web service QoS based on the SLA because we don't want to dump the network with redundant information. Fault detection notification will be enough. (see Box 1)

Repair Manager Healing Actions might be one of the following actions:

- Re-Invoke Web Service
- Invoke a substituted Web Service
- Invoke a composed Web service functionality equivalent
- The repair actions are generated automatically from the WSDL specification by substituting Web service by another functionality equivalent Web service.

Figure 7 shows the Virtual Web Service Connector architecture.

Virtual Web Service Connector components:

Virtual Web Service Connector objects provide healing capabilities using monitor, diagnosis, and healing actions. The objects of the Virtual Web Service Connector are:

Figure 5. Detailed process in case of anomaly detected and the self-healing agent response

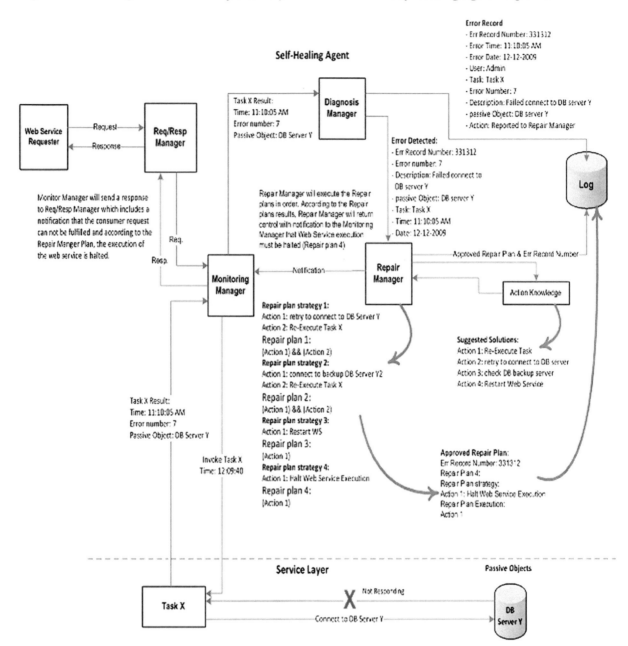

Figure 6. Web Service self-healing model for accurate and critical results

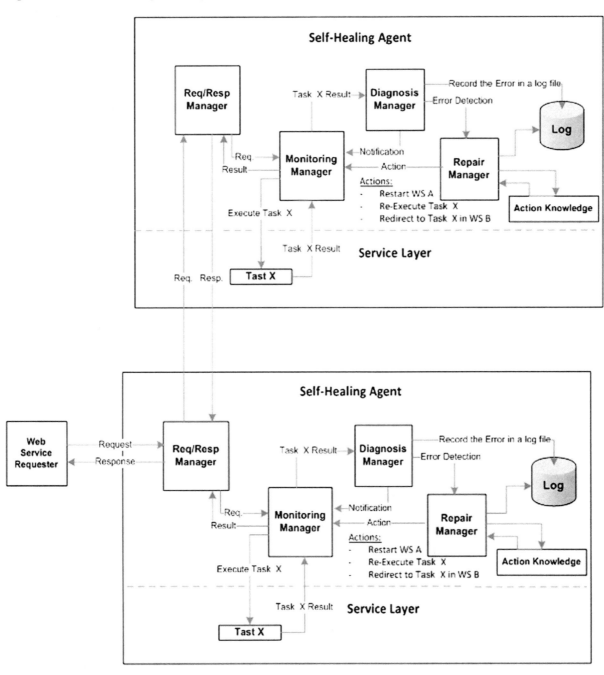

Box 1.

```
Proposed Virtual Web Service Connector Algorithm:
Applying Virtual Web Service Connector Algorithm will guarantee that Web services will not be flooded with requests to provide the
expected QoS according to the Service Level Agreement (SLA).

  Req/Resp Connector Manager: Get Clients' Requests and insert QoS Parameters (service in-
vocation time)
  WS Reconfiguration Manager: mapping of the input parameters (if necessary), and send re-
quest to the real Web service.
  Real Web service: execute and send response to Virtual Web Service Connector.
  WS Reconfiguration Manager: receive the Web service response and forward it to the Diag-
nosis Manager.
  Diagnosis Manager: analyze the Web service response:
    IF (Error, Fault, Anomaly):
      Send analyzed information to WS Healing & Reconfiguration Planner.
    Else
      Remove the QoS parameters from the Web service response and send it to the Req/Resp
Connector Manager.
      Analyze statistically the QoS parameters and the Web service QoS History and Send re-
sults to WS Healing & Reconfiguration Planner.
    End IF
  WS Healing & Reconfiguration Planner: decide the action about the current Web service ac-
cording to WS Diagnosis Manager results (anomalies detected, QoS parameters, and SLA), and
send decision to WS Reconfiguration Manager.
    Actions:
    IF (Error, Fault, Anomaly):
      Send request to the Real Web service (B)
      Send request to composed Web service.
      Re-Invoke Real Web service (A)
        Else
    IF (SLA not satisfied, history of Web Service QoS):
      Use Web service (B) for the future requests.
      Keep using Web service.
    End IF
  Req/Resp Connector Manager: map the response results (if necessary), and send response to
the Web service requester.
```

- Req/Resp Connector Manager: Used to Intercept the consumer request to check the sent request parameters number and types (Input faults or Type faults), also to insert the value of service invocation time QoS parameter, and forward it to WS Reconfiguration Manager. Also, send the response of the real Web service to the requesters and map results if necessary. In addition, prevent any unauthorized request to invoke the Web Service.
- WS Reconfiguration Manager: Responsible for sending requests and receiving responses of the real Web service which is selected according to WS Healing & Reconfiguration Planner, mapping process between the different Web service providers' WSDL input parameters. As well, react according to the WS Healing & Reconfiguration Planner notification to execute the healing action
- WS Diagnosis Manager: Analyze the real Web service response for any errors, in addition, analyze the QoS parameters' values and Web service QoS history statistically and send the analyzed information to the WS Healing & Reconfiguration Planner. Also, store the QoS parameters values in the WS QoS Database (DB). In case of no anomalies detected in the Web service

Figure 7. Virtual web service connector

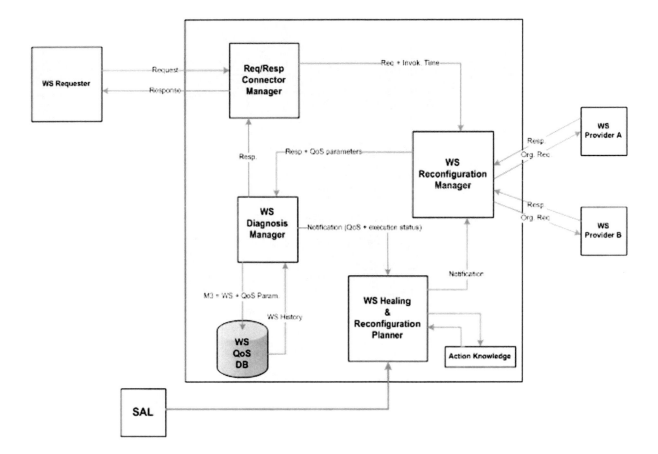

response, it will send the real Web service response to the Req/Resp Connector Manager to forward it to the Web service requester.

- WS Healing & Reconfiguration Planner: Analyze the WS Diagnosis Manager results and decide the healing action in case Web service execution failure using the action knowledge. Also, compare the analyzed QoS results and the Service Level Agreement (SLA) to decide if the current Web service provides the expected service (might ask the WS Reconfiguration Manager to leave Web service "1" and bind the request to Web service "2"), and finally send the decision to WS Reconfiguration Manager.

- WS QoS Database: Used to store the Web service QoS parameters' values.
- Action Knowledge: Used to match the recommended healing action in order to heal the request.

Figure (8) shows a detail process for the Virtual Web Service Connector and the Repair Manager healing action taken to heal the anomalies detected during the process execution.

CONCLUSION

Improved Airport Enterprise Service Bus with Self-Healing architecture (IAESB-SH) improves information accessibility and sharing across dif-

Figure 8. Virtual web service connector process

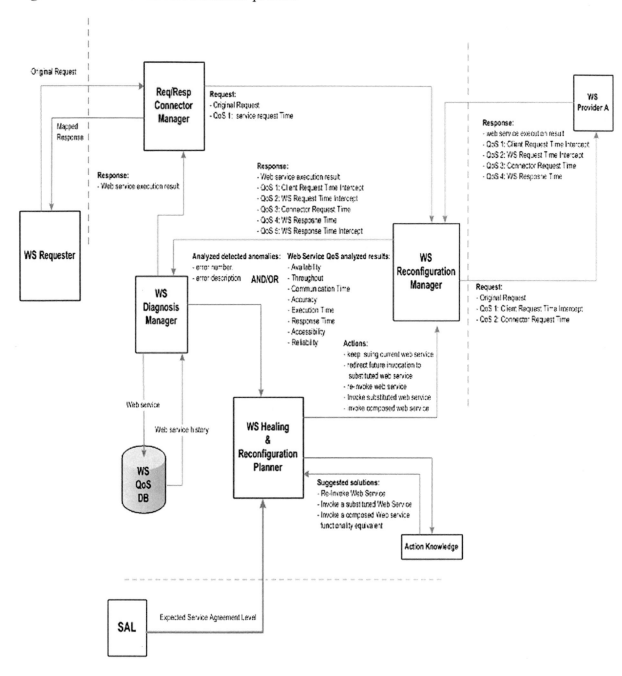

ferent Airport departments, the ability to upgrade, move, or replace applications or services can be done without having to modify code and disrupt existing ESB applications, and maximize the system's reliability, adaptability, robustness and availability using the Self- Healing Agent and Virtual Web Service Connector.

REFERENCES

Abu-Taieh, E. M. O. (2009). Information Technology and aviation industry: Marriage of convenience. In Abu-Taieh, E., El-Sheikh, A., & Abu-Tayeh, J. (Eds.), *Utilizing Information Technology systems across disciplines: Advancements in the application of computer Science* (pp. 153–164). Hershey, PA: Information Science Reference. doi:10.4018/978-1-60566-616-7.ch011

Anda, B. (2007). *Assessment of software system evolvability*. Ninth International Workshop on Principles of software evolution: In conjunction with the 6th ESEC/FSE (pp. 71–74).

AODB. (2009). *Airport Operation Data Base (AODB)*. Retrieved October 1, 2009 from http://www.aims.com.mo/upload/ brochure-AODB-final-eng-061204.pdf

Chappell, D. (2004). Introduction to the enterprise service bus. In M. Hendrickson (Ed.), *Enterprise service bus* (pp. 1-22). California: O'Reilly Media, Inc.

Chappell, D. (2005). *ESB myth busters: 10 enterprise service bus myths debunked. Clarity of definition for a growing phenomenon*. Retrieved September 1, 2009, from http://soa.sys-con.com/node/48035

Cheng, N. J. (2001). *An integration framework for airport automation systems*. Retrieved October 10, 2009, from http://www.mitre.org/work/tech_papers/tech_papers_01/cheng_integration/cheng_integration.pdf

Coulouris, G., Dollimore, J., & Kindberg, T. (2005). *Distributed systems concepts and design* (4th ed., p. 19). Harlow, UK: Addison Wesley.

Dobson, G. (2006). Using WS-BPEL to implement software fault tolerance for Web services. *Proceedings of the 32nd EUROMICRO Conference Software Engineering and Advanced Applications (SEAA 2007)*, (pp. 126-133).

Goold, R. (2009). *Airport management systems for the 21st century*. Retrieved September 25, 2009, from http://www.airport-int.com/ categories/airport-management-systems/airport-management-systems -for-the-21st-century.asp

IEEE. (1990). *IEEE std 610.12-1990, IEEE standard glossary of software engineering terminology*. The IEEE Standards Association. Retrieved October 2, 2009, from http://amutiara.staff.gunadarma.ac.id /Downloads/files/7081/IEEE+Standard+Glossary+ of+Softwar.pdf

Keen, M., Acharya, A., Bishop, S., Hopkins, A., Milinski, S., Nott, C., et al. (2004). *Patterns: Implementing an SOA using an enterprise service bus*. IBM Redbook. Retrieved September 25, 2009, from http://ck20.com/MQ/WBIMB/sg246346%20Implementing %20SOA%20using%20ESB.pdf

Ludwig, H., Gimpel, H., Dan, A., & Kearney, R. D. (2005). Template-based automated service provisioning - Supporting the agreement-driven service life-cycle. In *Proceedings International Conference on Service Oriented Computing (ICSOC'05)*, (pp. 283-295).

Mehta, N., & Medvidovic, N. (2002). Architectural style requirements for self-healing systems. *Proceedings of the first Workshop on Self-Healing systems*, (pp. 49-54).

Minoli, D. (2008). *Enterprise architecture A to Z frameworks, business process modeling, SOA, and infrastructure technology*. New York, NY: Taylor & Francis Group. doi:10.1201/9781420013702

Papazoglou, M., & Heuvel, W. V. D. (2007). Service oriented architectures: approaches, technologies and research issues. *The VLDB Journal, 16*(3), 389–415. doi:10.1007/s00778-007-0044-3

Robertson, P., & Williams, B. (2006). Automatic recovery from software failure. *Communications of the ACM, 49*(3), 41–47. doi:10.1145/1118178.1118200

Rowe, D., & Leaney, J. (1997). Evaluating evolvability of computer based systems architectures-an ontological approach. *Proceedings International Conference and Workshop on Engineering of Computer-Based Systems* (pp. 360-367).

Sommerville, I. (2007). *Software engineering* (8th ed.). London, UK: Pearson Education Limited.

Web Service MQ. (n.d.). *Providing a messaging backbone for SOA connectivity*. Retrieved October 10, 2009, from ftp://ftp.software.ibm.com/software /integration/wmq/WS_MQ_ Messaging_Backbone _for_SOA.pdf

Chapter 15

The Aviation Operational Environment:
Integrating a Decision-Making Paradigm, Flight Simulator Training and an Automated Cockpit Display for Aviation Safety

Ronald John Lofaro
Embry-Riddle Aeronautical University Worldwide, USA

Kevin M. Smith
United Airlines (ret.), & USN (ret.), USA[1]

ABSTRACT

This chapter will focus on the role of pilot/flightcrew training and performance evaluation in the identification and management of risk, especially while aloft and in changing conditions. The chapter will integrate different- but we posit interrelated, topic areas: First, a decision-making paradigm for flight crew's use in the operational environment. Second, training and performance evaluation in flight simulators (FS), as well as the design and development of FS scenarios to test decision performance. Third, Relevant Federal Aviation regulations (FAR's) and approved programs in current pilot/flightcrew training. Fourth, accident investigations; the role and use-value of accident investigation data in flying safety. Finally, the authors will present recommendations for the next steps in the development and use of new and emerging technologies for maximum pilot/flight crew decision performance and safety. This will be done via a collaborative ground-air, automated system and is what we propose to achieve our goal, increasing safety of flight.

DOI: 10.4018/978-1-60960-887-3.ch015

FOREWORD

It should be already clear that this Chapter will be somewhat different, even to format; for example, no other Chapter will have a Foreword. Yet, the authors felt constrained to add this brief Section. We will be presenting original concepts that we have worked on for 20 years, as individuals and as a team. There is little prior or current work done for comparison, argumentation and discussion. But, we have brought enough other works for valid comparison. Next, we are aware that many persons reading this Chapter may not have significant aviation expertise. With that in mind, we have attempted to present enough background and introductory information to make the aviation aspects as clear as possible. For those with aviation expertise, please bear with us. We have included a large Key Terms Section at the end of this Chapter. This Key Terms Section includes definitions of aviation-specific terminology.

Finally, what is presented in is the context of United States civil and military aviation, regulatory agencies and regulations. Today, aviation safety is an international concern with many international, cooperative efforts in that arena. It is our hope that what we present is applicable across other nations' aviation training and safety efforts and can lead to safety improvements for all of aviation. All we can ask is that you read on and form your opinions at the end of the Chapter.

INTRODUCTION

The perspective in this chapter is identifying seemingly unrelated aviation safety issues, demonstrating that they are interrelated, then describing a set of methodologies whereby these issues can be addressed and resolved. We are at a tipping point in aviation safety where there are many efforts, within the airline community and by governmental agencies, to deal with the same genre of accidents that has plagued civil aviation since its initial development. No attempts seem to be completely sufficient or totally effective. The initial definition and concepts of what was called pilot judgment has been with aviation for many, many years. We now re-define pilot judgment as the capability to make an optimum decision; a decision that is based on the identification, analysis and evaluation of the risk factors in play, with resultant action. We have called this *operational decision-making* (ODM; Lofaro, Smith or Smith, Lofaro: 2009, 2008, 2003, 2001, 2000, 1999, 1998, 1993) and define it as a cognitive process that is not solely, or in the main, the result of experience, as was previously thought. What will be presented is our effort to met the challenge of this last frontier in aviation training and evaluation: how to accurately identify and manage risk before and while in flight. We now go past ODM to what we call *decision performance*; the processes for making such decisions and implementing them to manage risk and ensure safety.

We will present two major aircarrier accidents that, literally, boggle the mind as to the decision processes...if any...in play. Some accident investigations have been less than stellar and of minimal help in what accident investigations are purported to do: prevent future accidents. In point of fact, one very recent accident is eerily similar to one only four years prior. Currently, there are tools and techniques for the training and evaluation of flightcrews that address the core problem: decision-making. Further, modern technology has given us new tools and pathways to develop on-board, automated threat displays, increasing flight safety; we need new ways to maximize such technology.

BACKGROUND

After World War II, the aviation industry was faced with major issues: the unreliability of power plants, weather, the need for a ground-based aircraft control system that used radar, as well the aerodynamics of the jet age. These have been largely resolved. Almost 60 years later, we now have extraordinarily reliable jet engines; accurate weather (Wx) forecasts made available to pilots before and during flight...even on-board cockpit weather displays; a radar-based air traffic control (ATC) system that is not only good but is entering into what the U.S. Federal Aviation Administration (FAA) terms a "NexGen" phase of improvements based on new technology. Lastly, our knowledge of the aerodynamics of sub and supersonic flight had taken quantum leaps. Yet, puzzling accidents still happen and the question always is: how and why can perfectly sound airplanes crash?

In the late 1970's and 1980's, aviation experienced accidents that were bizarre and seemingly impossible. What we now term CFIT (controlled flight into terrain, a term which is paradoxical) has replaced, to a degree, the phrase and question: how and why can perfectly sound airplanes crash? But, this question remains unsolved. There are many answers, but seemingly no effective counteractions. The aviation industry, airline manufactures, military services and federal agencies have worked for years to analyze these and other accidents and develop training and procedures to mitigate the accident rate. For the past approximately 35 years, we have heard that the reason for these and other aviation accidents is human error/human factors. This venue has been relentlessly studied and analyzed. In spite of this, the accident rate showed a 35+ year plateau which may be ending, but ending with an upswing.

Since 1978, aviation has grappled with pilot and crew decision-making. Ground breaking work was done in what was then called 'pilot judgment" by people such as Ruffell-Smith (1979) and Jensen (1982). While many realized aeronautical decision-making's (ADM) importance and worth, not much R&D money was made available for decision-making as the emphases on crew resource management (CRM) training and situation/situational awareness (SA) blocked large-scale efforts in decision-making R&D, training development and delivery.

There have been many approaches to ADM. We begin with the "common wisdom" was that pilots made good decisions easily and almost naturally, aided by (some) increase in experience. The facile assumption that additional experience will teach pilots to make better decisions has proven to be a dangerous fallacy. Experience can be a nasty teacher, often giving the test before, or without, giving the lessons and materials needed for the test. Experience can also reinforce poor decisions and behaviors that seemingly "worked" in the past (blind luck? the "not your day to die" phenomenon?). There is also the view that more information will lead, inevitably to better decisions, a view that is best shown by the ever-increasing soft/hardware placed into the cockpit displays that either aid or even make (traffic collision avoidance system, version II; TCAS II as one example) decisions for the flight crew. This will be discussed later.

There is the commonly accepted view that aeronautical decision making is but one of the components of crew resource management (CRM) training. This was, and is, dangerously inaccurate. CRM, with its emphases on communications and team function, is one enabler of good decisions. As such, it is a part of decision-making, not vice-versa. Decision-making is the primary tool to be used by the pilot/crew in their primary functions: risk identification and risk reduction. This is especially true for these pilots/crews of major carriers, with their highly complex, automated craft and their responsibility for hundreds of lives each time they fly. In short, risk management for safety. It seems that the need for operationally oriented decision-making training was and still is undeniable, as we shall later show.

Airline training and pilot/flight crew performance evaluation is another part of the problem. Current aircarrier training is built around (1) a sequence of discrete events and/or conditions and whether or not these are, separately, within legal limits and, (2) the pilot's responses to these events, often delineated in a standard operating procedures (SOP) manual. However, it seems clear to us that there are, all-too-often, interactions and interplay of seemingly discrete events and conditions. While any one, or all, of a group of events/ conditions might be within limits, the interactive resultant (cumulative effect) of them may place the aircraft and mission at risk. The pilot is then outside what is contained in the SOP and into decision-making. We are not aware of aircarrier training that uses this perspective.

In the early 1990s, air lines were allowed to revamp their pilot training, thus "escaping" the 60 year old Combined Federal Regulations (CFR), Title 14, Aeronautics and Space; Part 121; Operating Requirement: Domestic Flight and Supplemental Operations (commonly referred to as the FARs) strictures. The air lines were allowed, even encouraged to use a new flightcrew training model, the Advanced Qualification Program (AQP) and, there was enacted a special Federal Aviation Regulation (SFAR 58) for AQP.

Then, there is the issue of pilot/flightcrew performance evaluation, to include structure, metrics, methodology and environment. We will return to both AQP and flightcrew performance evaluation when we deal with Line-oriented flight training (LOFT), a realistic, real-time flight scenario to be flown in a flight simulator.

Closely allied with the above discussion is accident investigation (especially of the major ones where there are fatalities, major injuries, loss of hulls) and the ensuing reports and implementation of recommendations contained therein. We believe that we may well be experiencing a significant loss of technical expertise, in part due to retirement and both political and professionalism issues. In sum, we see all of the above factors as interrelated, seemingly making it more difficult to arrive at solutions. In point of fact, when the nexus is found and the Gordian knot severed, it actually should become easier to arrive at solutions. The objective of this Chapter is to provide a roadmap by which the above issues and concerns are addressed and a solution is delineated.

PROBLEMS, CONTROVERSERIES AND SOLUTIONS

We will be integrating the problems and solutions in this extended section. In this way, we hope to present the reader with a step-through of the issues, our solutions and rationale for them. While this represents a departure from the suggested format, we firmly believe it is the best method of presenting the material. The final Sections on future R&D in automation for aviation safety and recommendations will emphasize the "how," whereas this Section is more focused on identifying the "what" as to solutions.

CRM and Decision Making

In making decisions, the in-flight aspect of aviation operates in a time-compressed, constrained and, often unforgiving, milieu. The July of 2000 issue of *Aviation Week and Space Technology* printed four somewhat interrelated articles on pilot/crew errors and decision-making. These articles presented audit information, research and analyses along with current "fixes." Some findings include flightcrews ignoring increased evidence that the original flight plan was no longer appropriate; tactical decision-making (what we would call ODM) errors were the second most prevalent crew error in crew-involved accidents... in all-too-many case pilot misperception of risk was a key factor in safety.

CRM has not been found to be the human factors silver bullet in accident prevention, as was claimed. Captain Hal Sprogis has asked, "Is the Aviation Industry Experiencing CRM Failure?" (Sprogis, 1997). Captain Daniel Maurino has written "Crew Resource Management: A Time for Reflection" (Maurino, 1999). Both indicate that we may have expected too much from CRM; that the relationship between CRM and safety, which was, and is, the prime rationale offered for teaching CRM, has not been proven; that CRM is a process, not an outcome and certain efforts to assess outcomes (i.e. individual performance) may be misguided. American Airlines, in July of 1996, set aside much of CRM training as had been done because their flight crews had valid objections to, and concerns about, CRM: "CRM was too often viewed as a number of interpersonal issues that simply do not define the problems that we face in aviation." "CRM training will most likely always be defined and suffer in terms of the first generation of courses...which were seen as" touchy feely", "getting along", and "managing human relations or resolving personality conflicts" rather than dealing with truly important concerns (Ewell and Chidester, 1996). American's new focus is on preparing flightcrews for the daily challenges of normal operations encountered while "flying the line." Delta Airlines, in the same timeframe, revamped their "CRM for New Captains" course and now calls it "In Command." As with American, Delta is emphasizing leadership, responsibility and performance.

In 1996, we saw two of the Big 3 American flag aircarriers eschewing over-emphasis on communication and interpersonal relations in some CRM. United Airlines' prior version of CRM was called C-L-R, where the C is for Command and the L is Leadership and R is Responsibility. Yet, even United changed aspects of their CRM in 1997 to reemphasize the pilot's command and leadership functions. United returned to calling their program C-L-R. This new program also had gone past the interpersonal and on to the performance issues with an emphasis in managing the consequences of error.

We have said that CRM is an enabler of good decisions and, as such, is a part of decision making and not vice-versa. However, the problem here is that, whether you look at civil or military

flightcrew training, there is little, if any, actual training in DM. Even the latest Federal Aviation Administration (FAA) Advisory Circular on CRM (FAA AC 120-51E; 2004) while mentioning decision making and showing it as a part of communication, does not give any methods for teaching decision making or evaluating it. However, this training and evaluation is crucial to flightcrew training for flight safety.

In the early 1990's, a new concept came into play. integrated CRM. The integration, and assessment, of Crew Resource Management (CRM) and flight control skills received considerable attention—and, a fair share of concern and skepticism—in the late 1980's and early 1990's. As one response, the U.S. Air Transport Association (ATA) in 1990 formed a joint aircarrier/FAA/academe working group to deal with this, and other CRM issues; both authors were on that group. A set of flight crew CRM performance markers ("CRM behavioral markers") with behaviorally-anchored rating scales was developed. The ATA working group recognized that the CRM skills must be integrated with a corresponding set of technical skills (flight control skills) in an interactive matrix in order to fully evaluate overall crew proficiency. Further, such an "Integrated CRM" approach would serve as training tool, both in line-oriented flight training (LOFT) design and in specifying where the CRM/flight control skill linkages existed. However, the development of a mirror-image set on performance markers for flight control skills was still lacking.

This lack was resolved in 1991 when Captain Kevin Smith (United Airlines) and Jan DeMuth (FAA Flight Standards) developed an initial set of performance markers for the Technical/Flight Control skills. Both the CRM and the Technical sets of markers were used in the next step of CRM Integration. An approach to Integrated CRM, along with both human factors and flight control/technical skill evaluation scales, was partially developed during an ATA/FAA-hosted workshop in 1992. Dr. Lofaro was the designer and facilitator of this Workshop. Captain Smith, along with several training Captains from Northwest Airlines, Delta Airlines, United Airlines, the Chief Pilot for Boeing and others were the participants. (The results of that workshop are in DOT/FAA/RD-92/5, 1992)

Upon completion of the 1992 Integrated CRM Workshop, a new set of issues and concerns arose. The Integrated CRM concept and the mission performance model (MPM) were well received by the major air carrier CRM/AQP training pilots as well as Boeing's chief pilot who attended the workshop. However, many factors, such as a lack of FAA interest for follow-on efforts and a CRM "establishment" that was not open to taking CRM to either another level or in new directions, made it clear that Integrated CRM and the MPM) had become dead issues. (Aside: this is now not the case. Dr. Lofaro regularly teaches such courses as Aviation Psychology, Aviation Sensation and Perception and Memory and Cognition in graduate classes where many of the students are Instructor Pilots (IP's). These IP's come from the USN, USMC, USAF and USCG and they have modified some of the integrated CRM aspects and performance markers to make them both type of plane and mission-specific). We will return to the MPM and Integrated CRM as flightcrew training design and evaluation tools along with a related issue, jeopardy, later in the Chapter.

We would be remiss if we did not draw a clear distinction between what we call Integrated CRM and what is now referred to by the same name in the current FAA Advisory Circular (AC) on CRM, AC 120-54E. That document speaks of integrating CRM training into an AQP program. It takes the stance that, within AQP, you can do "appropriate integration of CRM and technical training." However, its view, almost a polar opposite of our integrated CRM concept, goes on to say that CRM is composed of global activities which do not fit neatly into a hierarchical set of technical activities and, *de rigueur*, cannot be evaluated in conjunction with, if at all, technical skills as CRM

skill scan be generalized to a host of a variety of situations. Therefore, when the FAA CRM AC speaks of integrated CRM, all they mean is to have CRM training integrated into flight training; no integrated evaluation schema and, in point of fact, no evaluation at all.

We are aware of a European system to evaluate the non-technical (CRM) flight crew skills: NOTECHs, standing for non-technical system for assessing pilot's CRM skills (Flin, Martin *et al.*, 2003; reprinted 2005). This system has many good points, the chief two being that both decision making and situational awareness are separated out for training and assessment. An interesting point is that NOTECHS was not designed to be a "...tool for introducing psychological jargon into the [pilot's] evaluation." (Flin, Martin, et al. 2005, p.151). All this having been said, NOTECHS does not have a methodology for integrating CRM and flight control/technical skills for evaluation and, as with CRM evaluation in the USA, the evaluation of NOTECHS skills in check ride cannot be grounds for a "failure," unless there is a corresponding technical consequence which compromises safety of flight (Flin, Martin, *et al.* 2005. p.148). However, as said, NOTECHS does not have a methodology for integrating CRM and flight control/technical skills for evaluation, which makes such a failure assessment somewhat arbitrary.

Of much more import was the authors' realization, in 1992 and 1993, that the overarching issue in safety was actually that of decision making, not communication or almost any other of the CRM components. We posit that risk management is the primary role of the captain/flight crew; that decision making is a functional aspect of risk management and good decision making skills are the primary tool in a pilot's safety arsenal. Decision making is THE key pilot activity. To that end, we developed the operation decision making paradigm (ODM) to enable flight crews to best perform risk identification, evaluation and management.

In summation: We are proponents of CRM training for flight crews, albeit a very different CRM training than now given. We propose training that places the emphasis on decision making, and situational awareness, as will be shown when the ODM model is presented in detail. We hold that ODM is the primary tool for training pilots and crew to do risk identification and management and, that risk identification and management are the primary functions of a flightcrew. The common fallacies about decision-making were discussed early in the chapter. We add this: Captain Smith functioned as a line check airman (LCA) during the late 1990's. He was checking out fairly senior pilots transitioning to a large (250 plus passenger capacity), highly automated aircraft. He was saddened, angry and frightened to see the number of times pilots either did not recognize a decision point in time to stay ahead of the power curve; did not recognize a decision and action point at all; made poor decisions, decisions which raised the risk of completing safe flight. At this point, we have explicated one problem, we now go to the groundwork for a solution.

Decision Making

Decision making can be seen as identifying and choosing alternatives. In this view, making a decision implies that there are alternative choices to be considered, and one wants to identify as many of these alternatives as possible, choosing the one that has the highest probability of success or effectiveness. Another view is that decision making is the process of sufficiently reducing uncertainty and doubt about alternatives to allow a reasonable choice to be made from the alternatives. This viewpoint stresses the information-gathering function of decision making. It should be noted here that uncertainty is *reduced* rather than eliminated. Very few decisions are made with absolute certainty, every decision involves a certain amount of risk. As will be seen when we present and explicate

ODM, we want to accurately identify the risk, evaluate the level of risk and take actions to deal with it.

Decision Making is a Recursive Process

A critical factor is that decision making is a non-linear, recursive process. That is, most decisions are made by moving back and forth between the criteria and the identification of viable alternatives.

The Decision Environment

Every decision is made within a decision environment, defined as the collection of information and alternatives *available at the time of the decision.* An ideal decision environment would include all possible information, all of it accurate, and every possible alternative. However, both information and alternatives are often constrained because decisions need to be made that are both accurate and timely. Since decisions must be made within this constrained environment, we can say that *the major challenge of decision making is uncertainty,* and a major goal of decision analysis is to reduce uncertainty. We can almost never have all information needed to make a decision with certainty, so most decisions involve an undeniable amount of risk. Since decisions must be made within a limiting decision environment, this explains why hindsight is so much more accurate and better at making decisions that foresight. As time passes, the decision environment continues to grow and expand. New information and new alternatives appear, even after the decision was made. Armed with new information after the fact, one can use hindsight to look back and make a much better decision than the original maker, *because the decision environment has continued to expand.*

Decision Streams

A common misconception about decision making is that decisions are made in isolation from each other: one gathers information, explores alternatives, and makes a choice, without regard to anything that has gone before. But, decisions are made in a context of other decisions. Previous decisions have "activated" or "made operable" certain alternatives and "deactivated" or "made inoperable" others.

A decision left unmade will often result in a decision by default or made for you. Every decision you make affects the decision stream and the collections of alternatives available to you both immediately and in the future. In other words, decisions have far reaching consequences.

In developing and refining ODM, we have used the above four paragraphs a basic guide, but extensively modified the above for the aviation environment.

Operational Decision Making (ODM)

A formal definition of ODM is the process that is the functional aspect risk management of decision making, done by the pilot/flightcrew, often under time pressure and often with little or no margin for error. ODM involves integrating SA with the recognition and assessment of those factors that are critical to safe flight in order to identify and respond to the risk level. If decisive action is not taken, risk will continue to rise to a point beyond which one experiences catastrophic mission failure. Such a point is called the critical event horizon (CEH).

Since approximately the early-mid 2000's, the concept of risk management as an integral part of safety seems finally to have emerged in aviation, most notably beginning in military aviation where it is termed organizational risk management (ORM). ORM is somewhat static in that the risk assessment is done prior to launch and somewhat top-heavy in that, if the pre-launch

risk level is too high, permission to change plans or to go-no, must come from higher up the chain of command. This having been said, ORM is a very useful tool and we will incorporate it in a later Section.

Since 1993, we have stressed mission completion and risk management; completing the mission of flying the public, because of its economic benefits, but in a way that does not place people or equipment at undue risk. These should be the goals of any airline pilot and indeed, of any pilot. By risk, we mean specific risk or danger to the aircraft, passengers and crew, and the corporation. If the level of risk rises, effective risk reduction strategies need to be employed to keep that risk within manageable limits Risk management is critical to the retention of the flying public customer base and the long-term viability of the industry. Risk management is the key operational activity that works hand in glove with mission execution skills. Central to a completed and safe mission is the pilot as risk manager, which operational responsibility undergirds our model. The ODM model is as follows:

Components of the ODM Model

The ODM model and any resultant training program begins with the concept that all operational decisions have the same decision-making structure. Integrated throughout the model is the pilot's role as risk manager. The other components are:

- The Operational Envelope (Mission Space)
- Situational Knowledge/Risk Location (Location within the Ops Envelope)
- The Critical Mission Impact Areas and the Critical Mission Factors which comprise the Impact Areas/Risk Location (Hostile Agents invading the Mission Space).
- The Rising Risk Continuum/Risk Location

- Cumulative effect (a concept whose use is embedded throughout all the above components).

Thus, ODM training, built around these components, is the process by which optimal ODM is achieved and, with it, optimal risk management and successful mission completion. *Finally, it must be strongly noted that all of these components are so intertwined that any separation or sequencing of them is artificial, since it will become apparent that to speak of any one component brings in the others. Actually, there is no sequence per se. Rather, there is a constant interplay and branching across the components; perhaps a continuously rotating spiral best describes the process the pilot is involved in.* However, for the purposes of presentation, we will treat each one separately, but we refer to and revisit others, as it is a spiral process..

Risk Management: Risk, Rising Risk and "Flying the Line"

There are three levels of risk in the paradigm. High risk is defined as the likelihood or high probability of injury, damage or death. Moderate risk, if left unchecked, could continue to rise and/or likely result in significant flight trajectory deviations. Low risk, finally, is a normal situation where routine, normal procedures are sufficient. Rising risk refers to the facts that, (1) If some, or any, problems went unnoticed and/or (2) The decisions made were not accurate, timely and appropriate, then, (3) The risk to the successful completion of the flight could rise to the point where the flight is truly endangered. We will return to these later. If the level of risk rises, effective risk reduction strategies need to be employed to keep that risk within manageable limits. Risk management is the key operational activity that works hand in glove with mission execution skills. Central to a completed, and safe, mission is the pilot as risk

manager, which operational responsibility under girds the model.

Risk Management: The Operational Envelope, Risk Location and Cumulative Effect

We will work from the center of the model outwards, then return to the inside. The first--and last-- component in this spiral-like process is a graphical representation of the boundary conditions in a mission. This is the pilot's "world" and we term it the operational envelope, as does the military. It paints "the big picture"; the ops envelope is the context in which all the ODM paradigm's components are activated. The operational envelope gives a straightforward way to grasp the pilot's task universe. The ability to locate oneself within the ops envelope results from the pilot's situation knowledge which encompasses the pilot's situational awareness/SA and ability to do accurate risk location. Civil aviation has never taken this larger view of the envelope because it has focused on discrete data points instead of on the big picture. Even today, we find that Airbus and Boeing are working on software for flight control computers that are aimed at keeping the aircraft from exceeding a predetermined flight envelope (*Aviation Week and Space Technology*, August 28, 2000). However, these efforts focus on alarms and/or software to stop overstress, possibly dangerous pitch angles and over-banking. These deal with a small, specific set of discrete flight control errors, rather than flightcrew decision-making and pilot error management.

The ops envelope can be visualized as a four-sided figure, composed of sides which intersect at right angles to each other. These sides--"boundaries"--are Adverse Conditions; Restricted Visibility; Mission Critical Alerts and Warnings; and System and Human Limitations, as shown in Figure 1. These sides embed 72 factors (see Table 1) that critically impact (can threaten) mission completion; these can also be referred to as hostile agents. Below is a flat cross section. The actual envelope is three-dimensional as the plane can not only fly any direction on the Compass Rose, but can also climb/descend. While this seems obvious, one must not lose sight of the fact that this is a linear representation of a 3-D world.

Figure 1 is the Operational Envelope and contains some sample conditions that correspond to each boundary/side:

First, at the very center, there is an area of normal flight conditions. Within this inner area of low to no risk, normal SOP's will suffice. However, a single critical factor/hostile agent such as freezing rain, or event sets composed of many hostile agents, such as contamination and strong crosswinds, encountered during flight may drive a pilot/crew out of that inner area and towards the edges/boundaries. If that happens, there may still be operational guidelines on what to do. The second risk management aspect is

that, while each critical factor may be able to be handled by itself in keeping the aircraft inside the edges/boundaries of the envelope, how does one handle the combined/cumulative effect, i.e., the resultant of multiple events and factors? If the cumulative resultant of combined factors in any event set drives the aircraft toward any corner, one first encounters what is a "gray zone" where the aircraft is still within the envelope, but close to a corner. If near a corner, in a gray zone, there exists no SOP's or procedures. Certainly, cumulative effects will effect one's position in the ops envelope and, thereby, impact risk. We will return to this later.

Risk Management: The Critical Mission Impact Areas and their Components

Critical to mission success is the ability to operate in adverse conditions while simultaneously

Figure 1. Operational envelope

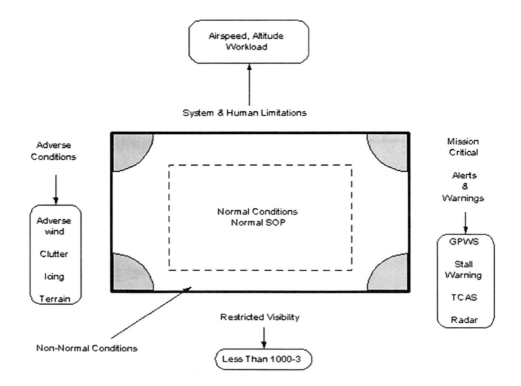

executing the mission plan and managing risk. The need to have the capability to operate under adverse conditions and thus provide reliable transportation is self-evident. But this leaves unanswered what events/factors cause a (rising) risk and what are some of the "risk drivers"? We must first look to identifying, defining and understanding the mission critical impact areas that engender risk.

A mission impact area, if encountered, denotes some level of risk and, if action is not taken, will cause the risk to rise.

The boundaries/sides of the ops envelope function as the critical mission impact areas. Table 1 shows the boundary conditions of the ops envelope, with their sub-areas and critical mission components/hostile agents. Note that that the

Table 1. Critical mission factors/hostile agents

Adverse Conditions	Restricted Visibility	System and Human Limitations
Adverse Weather	Below 1000/3	Performance limited ops
Slippery	Below circling minimums	Clutter
Clutter	Below CAT1	Weight/fuel
Contamination	Below CAT2	Flaps
Icing aloft	Below CAT3	Packs
Freezing precipitation	Takeoff below landing minimums	Delayed VR
Volcanic ash	Blowing snow	Cruise/speed altitude
Convective activity	Blowing sand	Landing
Adverse Wind	Patchy fog	Missed approach
Headwind	Night mountainous	MEL
Crosswind		Memory saturation
Tailwind	**Mission Critical Alerts and Warnings**	Task demand and overload
Severe gust	Stall Warning	Channel saturation
Non-normal Operations	Predictive windshear	Information overload
Nonprecision takeoff	Actual windshear	Fatigue
ICAO takeoff	Predictive terrain	Activity prioritization
Day only restrictions	Actual terrain	Risk awareness
Cold temperatures	TCAS RA	Multi-tasking
Extreme latitudes	Improper configuration	Task interference
Raw data	PRM breakaway	
Unfamiliar	Radar doppler return	
Engine out operations	Convective return w/hook	
Takeoff alternate	Wing contamination	
T-procedure	Any severe conditions	
V-1 cut	Upset	
V-2 cut	Wake turbulence	
EO approach		
EO landing		
EO missed approach		
ETOPS divert		

mission is complex. This complexity may have caused many to avoid a high level of specificity in decision making. But, one cannot and must not succumb to avoiding complex issues if we truly want a viable air transportation system for this millennium.

Risk Management: Situation Knowledge and Risk Location

The continuous task of the pilot is risk identification by using situation knowledge and consequently, accomplishing (ever-changing) risk location. Accurate and timely risk location is the only way to achieve both accurate and timely action response to risk. Situation knowledge and risk location are interrelated concepts. Situation knowledge is that part of the ODM structure that consists of the continually changing set of elements (knowledge bits) comprising the pilot's awareness of (1) the area of the ops envelope in which the captain believes the aircraft is located and, (2) which of the critical components/hostile agents of the ops envelope boundaries are in play. The resultant of these components "tells" the pilot what is the cumulative effect of the critical factors that are in play. By using the resultant of a cumulative effect, the pilot can re-locate the aircraft in the operational envelope. This enables the pilot to use the rising risk continuum as a decision tool for action response. As said, accurate risk location is the key, when in a (rising) risk situation, to making the optimal selection of a course of action, i.e., an action response that is an alternative to the original mission plan.

Risk Management: Rising Risk, Risk Location and Response Action

Rising risk is a teaching concept and the decision making tool that enables the crew action responses needed for optimum risk management. The teaching aspect of rising risk is in engendering the awareness that, as issues and problems arise,

the risk will rise. As risk will rise under virtually any set of adverse conditions, risk management demands that the crew execute, in a timely fashion, specific mitigation procedures to prevent "risk migration to the right." This is somewhat similar to the drift into failure concept "...a slow, incremental movement of systems operations toward the edge of their safety envelope." (Decker, 2005).

Rising risk also is the decision tool for response action(s), based on the rising risk continuum (Table 2). If risk is low, the original plan can be executed to completion. When the risk rises to a moderate level, modifications to the original plan must be implemented in order to maintain an acceptable "location" in the risk dimension. When the risk rises further to some critical threshold, the original flight plan needs to be discontinued and the mission is aborted. In sum: If the risk rises, the captain, working with dispatch and other crew members, either (1) continues with the mission as originally planned for low risk situations; (2) modifies the mission plans as needed in moderate risk situations; or (3) abandons the mission altogether in high risk situations. However, the pilot must, in order to have an accurate perception of risk location, be aware of any adverse factors that can have a cumulative effect on the airplane's trajectory. In short, the pilot's perception of the risk must be consonant with the reality of the plane's location in the ops envelope. We are now back to cumulative effects and their impact on risk location in the ops envelope.

A Distinction: Cumulative Effect Resultant versus Additive Effect Resultant

We have discussed the cumulative effect that results from the interaction of factors that come from any 2 contiguous sides of the ops envelope. In this, we have relied, albeit loosely, on vector/tensor mathematics; our cumulative effect vector is somewhat analogous to the resultant vector that can be computed from 2 or more forces at right

Table 2. Rising risk continuum

Continue with the mission	Continue with modification to the mission	Discontinue the mission
Low	Moderate	High

angles to each other. (An aside: perhaps mathematicians and engineers working with pilots could develop a formal mathematical algorithm or set of equations to quantify such vectors. Surely a worthwhile effort). The keys here are that (1) The cumulative effect resultant (CER) has a new and different direction than either/any of the forces that comprise it; (2) The CER drives the plane towards a corner and, (3) The CER changes the shape and position of the corner the plane is being driven towards. As the cumulative resultant of combined factors in the event set drives us toward a corner, we first encounter what we term a "gray zone" where we are still within the envelope, but close to a corner. If near a corner, in a gray zone, there exists no SOP's or procedures. The same is true if forced out of the envelope (See Figure 2).

As said, the CER of an event set, composed of critical mission (and boundary) factors can drive a pilot towards one of the four corners (pun intended, as you are really "cornered."). The closer to a corner, the more the risk has risen.

However, the corner of the ops envelope has also changed shape; it has become a triangle, with the (45 degree) hypotenuse of the 2 sides being the new boundary line. Thus, the corner boundary lines are replaced by a new line (the hypotenuse), termed a Special Boundary. This Special Boundary is closer in than the former corner so that before you would get to the former corner, you have actually crossed this new (special) boundary line and are outside of the ops envelope. (See Figure 3). At this point, you are in abnormal/emergency conditions and at high risk. A typical set of conditions that cause a change in the position and shape (becoming a new boundary line) of what once was a corner could be low ceilings and visibility (restricted visibility) with a slippery runway and strong crosswind (adverse conditions).

Note: As one rule of thumb--any factor encountered on a mission will cause the risk to rise; two factors generally will result in moderate risk; the interplay of three or more factors usually results in a high moderate to high-risk situation

Figure 2. Cumulative CER

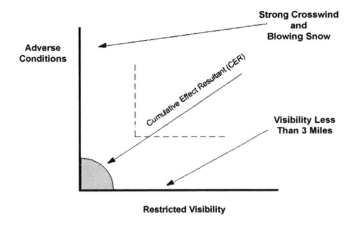

Figure 3. Formation of a special boundary

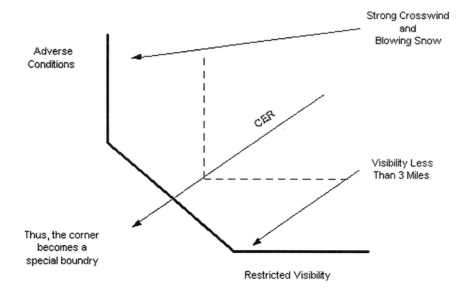

There is another effect that comprises more than one factor and that can act in a way to drive the aircraft either out of the ops envelope area of normal operations or out of the envelope entirely; we will term this the additive effect resultant (AER). The difference here is that an AER is composed entirely of factors that do not give the resultant a new direction. This is to say that all the factors in play come from one and only one particular Mission Impact Area or set of boundary factors. These factors combine, additively, to produce an effect larger than any of the factors. The AER pushes the plane towards a boundary of the envelope, rather than a corner. One simple example would be a series of Adverse Conditions such as icing aloft, clutter on the way down, slippery runways on arrival. Pilots seem to be able to recognize that these linked factors do raise the risk and, usually, take the necessary actions. Still, we are surprised at the level of stress engendered when more than one factor is encountered.

Relationship: The Operational Envelope and Rising Risk Continuum

There is no simple isomorphic relationship between the aircraft's position in the ops envelope and its position on the rising risk continuum. However, by looking again at the ops envelope, we can develop a loose mapping to take us from a position on the ops envelope to the risk continuum. What follows is somewhat rough, but seems intuitively obvious.

Clearly, if a plane is in the normal ops zone of the ops envelope, that is roughly equivalent to being at no or low risk. The non-normal ops zone is, then roughly equivalent to, at least, moderate risk. Penetration of any boundary, or special boundary, of the ops envelope puts the aircraft at high risk. The gray areas, where an aircraft is nearing a corner due to a CER, we term as high moderate, bordering on high, risk.

Putting It All Together: Risk Location and Risk Management in the Operational Envelope

We now can return to the ops envelope and risk location within it, i.e. the big picture, or in current phraseology, maximal situation awareness. As was said, all the paradigm's components are intertwined, and the pilot is using them by a constant process of mentally rotating through them; a process that, as it deals with three axes, is a spiral one. The pilot must always ensure that his risk location encompasses not only the critical mission factors but the plane's location in, or more dangerously, outside the ops envelope, and the risk factors in play due to that position as well. This is what we have termed "cumulative effect." What must be used by the pilot/flightcrew is situation knowledge to accurately develop the resultant of the combined critical factors and locate themselves (risk location) in or outside the ops envelope. Next, they translate that ops envelope location into a location on the risk continuum. Once they locate themselves accurately on the risk continuum, taking into account the cumulative effect of being in, or very near, a corner, with all the factors that are in play, they can formulate and execute optimum decisions. Again, when we encounter the phenomenon of cumulative effect and rising risk (gray zones; new, special, boundary lines), the needed decisions cannot be made by simply following procedures (SOP's).

By returning to the ops envelope focus, the pilot can accurately place the plane on the rising risk continuum, then, take the needed action. Risk location in/out of the ops envelope, translated into risk location on the rising risk continuum, followed by response actions are the keys to risk management.

Before we go further, the ODM model can be used to analyze accidents and, melded with Integrated CRM and the Mission Performance Model (MPM), used to both design line-oriented flight training (LOFT) scenarios for use in a FS and to evaluate flightcrew performance in those scenarios. All of these will be detailed in later Sections of this chapter. A quick word about the MPM. Captain Smith created the framework for a model that demonstrated that the CRM human factors skills and the technical/flight control skills are interrelated, interdependent, and often simultaneous in execution. This model is called the Mission Performance Model and, as will be shown, relates directly to LOFT design and flightcrew evaluation while "flying" a LOFT. At this point, we leave ODM and proceed to a pair of interrelated topics: flight crew training and AQP along with LOFT design and usage.

FLIGHTCREW TRAINING, USE OF FLIGHT SIMULATORS (FS), AQP AND LOFT

In this Section, we will examine flight simulators, capabilities and uses with a modicum of history about their development. Again, we do this to try and provide the non-aviation readers with a strong background. Then we will proceed to current flightcrew training and the most recent, FAA-approved flightcrew training program. In doing this, we will look at relevant flightcrew training FAR's and their context. We will end with LOFT and aircrew training needs in this millennium.

Why Flight Simulators?

The primary functions of any FS, the functional definition as it were, are 1. To present information like that which the real system would present, for the purpose of training and, 2. To provide a practice environment that facilitates and enhances the skills and knowledge of the pilot and, thus provides learning which enhances performance in the real system, the airplane.

Put into other terms, a flight simulator is a system designed to "imitate" the functions of another system (a plane) in a real operational environment; to be a realistic substitute that responds realistically to flight crew inputs. The key here is that a FS can be programmed to offer varied experiences to a flightcrew, but experiences that are safe in that if you "crash" in a simulator, there is no injury, save to your pride. Basically, a FS is a training device that is safer; less expensive; capable of quick modification; can operate in all weather and for all or any part of a 24 hour day. The FS presents accurate cockpit displays to the flightcrew and accurately (except for complete motion capability) responds to control and avionics inputs; all the time processing and storing data on the crew's control inputs and the like.

The "characteristics" of a FS are that it:

1. Stores data that can be re-played and analyzed on crew input/crew response.
2. Stores data that can be used to generate a realistic "environment"/mission or portion; control and other responses to crew inputs
3. Displays such data/info both to the flightcrew and to the FS operator.
4. Responds to crew inputs and does do accurately as to their effects on both system and environment.
5. As would the actual plane, has a both accurate and valid display of the status of on-board systems/components so vital for the pilot to see and monitor.
6. Provides 2-way training interfaces for the FS instructor and the flightcrew being trained.

Although there are several types of flight simulators, the focus in this Section will be what is often termed the "full-up" FS. Simply put, a "full-up" FS has:

1. A three-axis motion base (Pitch/Roll/Yaw) in 2 directions each (hexapod motion; six degrees of freedom). The FAA, NASA and the DOT/Volpe Transportation Center are examining the benefits and need for hexapod motion for the platform on which the airline

FS is mounted. The ultimate goals of these experiments ((Burki-Cohen, Judith; Go, Tiauw, et al., 2003 and Go, Tiauw, Burki-Cohen, J., et al., 2003) were to provide information for a possible FAA AC and to develop information for a possible FAA policy on FS motion requirements in airline pilot training and evaluation. As of 2006, the FAA has published a document that only lists relevant research in the area. A brief overview of the research findings to date shows that hexapod platform motion has a significant positive effect for flightcrew evaluation, but no significant benefit for training. Further, certain enhancements to the motion washout filters (lateral side-force and heave motion) seem to be beneficial in all cases. However, for more complete information, the reader is referred to the works cited and to the www. FAA.gov website

2. Computer-generated graphic displays for the "out the windows" visual scene. These include most types of weather and environments, to include night scenarios.
3. A complete cockpit that is the same as in the aircraft type that the FS models.
4. A vast amount of computer memory, allowing for superb capability in realistic flight simulation.

In short, the FS has ultra-realism (except a damping out as to the motion bases.)

These FS capabilities are usually subsumed under the term "fidelity"-- the closeness to which the FS mimics actual flight.

What may be of more import is the FS's veridicality. In the main, the use in training and evaluations of FS deals what is called fidelity (of the FS to the aircraft) in terms of (a) The motion bases; (b) The out–the-windscreen/window visuals and, (c) The simulator response to control inputs. All these issues can be subsumed under the more general question of "How veridical to actual flight does a FS have to be in order to insure training that fully prepares a pilot/crew for actually flying the plane?" Veridicality is the closeness of the correspondence of the knowledge structures formed by using the FS (learning and using controls/input responses/instrument responses/visual scene/motion, et al) to the information environment it represents, i.e., the actual aircraft type. Since the FS is used in training to build knowledge structures in the crew that will be used in actual flight, it is obvious that veridicality is the primary factor in FS design and use. The higher the fidelity of a FS, the more veridical the knowledge structures built in the FS become, thus making the flightcrew optimally prepared for actual flight. To be sure, we do not want to give the impression that other issues do not exist and could include incorrect control inputs; incorrect sequencing; poor or incorrect decision-making and more. These issues are beyond the scope of this chapter.

There are three obvious benefits from the use of a flight simulator in training. These are the underpinning of the now extensive use of FS in general aviation, civil aviation and military aviation. Briefly put, they are: 1. Cost reduction and increased efficiency by replacing the real system, the plane, with the FS. 2. Reduction in the hazards of training in the plane; the loss of life and injuries that result from training accidents and incidents are well-documented. 3. The ability to train skills and performances that cannot be trained in the aircraft such as malfunctions and adverse conditions and, more important, missions/tasks that may never be performed in real word operations, but are essential components of the operational mission profile of the aircraft. (*cf.* Flexman and Stark, 1987)

The extensive use of FS in the civil aviation world has resulted from both recognizing these benefits and a confluence of other factors. The first, as said, is that the FS is a safe environment, putting neither crew nor planes at risk. However, while crew and aircraft safety is foremost, looking further we see the following set of converging vectors. In the formative years, and continuing

well into the late 1980's and early 1990's, both civil aviation and the FAA relied in ex-military (mainly former active duty personnel who left after fulfilling their initial commitment, but also some retirees in their 40's and even 50's) for a source of experienced manpower. In civil aviation, these former military included pilots and aviation maintenance technicians (AMT). The same was true in the FAA, where there was, perhaps, a higher emphasis placed on the ex-military pilot, who could be placed in the flight standards, aircraft and pilot certification areas without missing a beat. The rationale for the aviation industry seeking (and welcoming) military personal is both obvious and subtle. It is obvious that the military was a source of highly trained and qualified personnel whose training was both extensive and standardized. On the subtle side, it is true that there is a "brotherhood of airman," where inclusion is highly dependent on airmen background, experiences, training and even common acquaintances. Add in a common "language," one that is technical, acronym-laden and replete with idiomatic expressions. All of the above is still active, albeit to a lesser degree, today. One resultant of the influx of military was the use of the FS as the major training tool for aircrew.

En passant, it must be remembered that the earliest FS, with replicated cockpit instrumentation, controls and most aspects of flight built by Edward A. Link in 1929, originally was a generic FS for General Aviation (GA) use. This Link trainer, whose picture we probably all have seen (it looks like a child's drawing of a plane mounted on base) soon evolved into an instrument flight trainer as a result of WWII. From there, the great advances in FS had to do with its use in military training. The history of the early FS shows the development of both capability and capacity for training military pilots/aircrew. It was natural, as many of the early aircarrier and FAA personnel had military backgrounds, that the use of the FS in training and certification became paramount.

Add to this a piece of reality that is often overlooked: the air carriers simply do not have enough planes to take significant or even small item numbers out of service for pilot training. As one example, the largest US flag carriers own/lease approximately 300 planes each, although with recent mergers, this number has gone up. Unlike the military, which has large numbers of aircraft that are dedicated only to training pilots, the air carriers must use their planes, in the main, to generate income. Considering that the aircarrier's run 365/24 schedules, it is apparent that the FS is, must be, and will continue to be the training and certification tool in civil aviation as the planes owned, or leased by the carriers are and must be used to generate revenue.

To resume: The military use of FS drove much of the early FS technology, as the aircraft become more complex, more automated and higher performance. On the civil aviation front, the aircarriers began to insist that the delivery of a FS for a new type aircraft be simultaneous, or even before, the new aircraft type was put in everyday service. Similarly, the civil side of aviation demanded more and more simulator capability (fidelity and veridicality). The major air carriers used their pilot and crew training facilities to house a growing number of FS. Companies that made FS worked closely with both the airframe manufacturers and the aircarriers to design and deliver FS that met the changing the needs of pilot training and certification. Today, we have civil aviation FS for such as the B-757/767 and 777 which cost upwards of 35 million dollars each, and cost in the thousands of dollars/hr when in operation. These operational costs include maintenance, the simulator operator and the A/C needed to maintain the temperature in the FS faculty at a level that does not impact the highly-sophisticated computers which drive the FS.

The final vector has been the technological advances in FS capability during the 1980's and 90's. The advances in the fidelity of the visual scene presented to the crew, as well the fidelity of the response of FS instrumentation (be they "glass" or "steam") to inputs by the crew are the outstanding examples of this fidelity.

In sum: Through a convergence of causal vectors, the period from approximately 1970 to today has seen the emergence of the FS as the primary and absolutely essential tool for pilot training and certification in the civil aviation area.

Flightcrew Training: Regulations and the Advanced Qualification Program (AQP)

A brief look at flightcrew training and the relevant FAR: Combined Federal Regulations (CFR), Title 14, Aeronautics and Space; Part 121; Operating Requirement: Domestic Flight and Supplemental Operation, Aircrew training for Air Carrier pilots; Part 121 (usually referred to as just Part 121). This FAR was originally written more than 70 years ago, when the DC-3, a small, 150 MPH, twin piston-engine plane, capable of carrying approximately 25 passenger was state-of-the-art. Over the years, civil aviation has progressed in leaps and bounds to the Boeing "triple seven" as one example: Two monstrous jet turbofan engines, each capable of 110,000 lb of thrust, a "glass cockpit," highly automated with seven on-board computers, capable of sustained Mach 0.84 flight, a range of over 9,000 miles, can carry over 300 passengers and can take-off with a weight of over 230 tons. Space precludes even a cursory look at the enormous advances in military aviation since the 1940's.

Training had changed and evolved to keep pace, somewhat, with the changes in aircraft. This was due to five main drivers. They are: First, Technical advancements in Aircraft Systems and Simulation Realism. Second, Engine Out operations. Third, Mission Critical Alerts and Warnings. Fourth, Adverse Condition Operations and, fifth, Human factors.

Technical advancements in aircraft systems has for obvious reasons driven pilot training programs. The features of a new system and how it should be used operationally have always been built into the curriculum. A good example of this is traffic collision avoidance system, version II (TCAS II); a hard/software system that provides the aircrew with alerts for terrain collision avoidance While TCAS was initially introduced with a part task trainer, it is now an important feature of full mission simulators and collision avoidance training is now possible. Engine out operations as a major driver of pilot training is not quite so obvious. With the advent of multi-engine aircraft, engine out training had always been important to pilot certification. However, with the advent of swept wing turbojet aircraft, this training took center stage. This was due to the unique aerodynamic properties of swept-wing aircraft, more specifically, asymmetrical thrust and axis coupling. During asymmetrical thrust operations, the swept-wing turbojet aircraft experiences pronounced axis coupling, manifested in a rapid roll off or wing drop along with an equally pronounced yaw. Increased pilot skill and specific training was and is the only counter-tactic to this potentially fatal condition.

In 1968, a DC-8 training accident involving asymmetrical thrust prompted the effort to conduct all training in simulators. Such an effort was successful, prompting, among other things, advancements in simulator realism. We will bring in flight simulators shortly

Although often overlooked, the third major pilot training and certification driver has and continues to be Mission Critical Alerts and Warnings. This area includes such things as stalls and steep turns, wind shear recovery, ground proximity warning recovery, etc. Recent additions include CAT III auto land system failure recovery maneuvers. It is important to note that all of these recovery maneuvers require the aircraft to be "hand-flown" by the pilot.

This last statement brings to the surface what we think is a major challenge facing training managers today: The tension line between increasingly sophisticated autopilot systems and the continuing and pressing need for a high degree of "stick and rudder" pilot skills. Pilots therefore need to demonstrate proficiency in (1) Adverse conditions (which include engine loss/" out" operations). (2) Low visibility operations. (3) Responding to mission critical alerts and warnings. (4) System and human limitations. We have shown that these 4 conditions form the boundaries for the pilot's worldview (the ops envelope) and how they can be incorporated into a training/operational model to manage and reduce risk: ODM.

AQP

Yet, in spite of aeronautical and technological advances, and some resultant flightcrew training changes, the training was still carried out under Part 121, which itself had grown and changed over the years, to approximately 900 sub-parts, many with sub-sub-parts and Appendices. In fact, it had grown so complex and complicated that it was becoming difficult to train under as well as difficult for the FAA to oversee and enforce. In response to the recommendations from a Joint Government-Industry Task Force and from the National Transportation Safety Board (NTSB), the FAA put forward SFAR 58, Advanced Qualification Program (AQP), in October, 1990. Please note that any Special FAR (SFAR) expires within 5 years unless extended or made into an FAR. In the case of SFAR 58, it would have expired in late 1995, but was extended until 2 October, 2005. It is interesting to note that the original AQP AC accompanying SFAR 58 was published in 1990. It was to have updated and be re-issued in its newest version in early 2004, or some 14 years since the original AC was published. In June of 2006, a revised version of the AC was published by FAA: FAA AC 120-54-AQP. However, in July of 2005, Sub-Part Y was added to FAR Part 121; it essentially recaps the FAA AC on AQP and is now part of an FAR with all that entails operationally and legally.

AQP was established to permit a greater degree of regulatory flexibility in the approval of innovative pilot training programs. Based on a documented analysis of operational requirements, an airline (FAA certificate-holder) under AQP may propose to depart from traditional training practices and requirements for pilot/crew with respect to what, how, when, and where training and testing is conducted. This is subject to FAA approval of the specific content of each proposed program. SFAR 58 requires that all departures from traditional regulatory requirements be documented and based upon an approved continuing data collection process sufficient to establish at least an equivalent level of safety. AQP provides a systematic basis for matching technology to training requirements and for approving a training program with content based on relevance to operational performance. AQP emphasizes crew-oriented training and evaluation. These training and evaluation applications are now grouped under the general term of Line Operational Simulations (LOS), including Line-Oriented Flight Training (LOFT), Special Purpose Operational Training (SPOT), and Line Operational Evaluation (LOE).

In short, AQP is a pilot training and certification path in which air carriers could design training specific to their types of operations (e.g. Winter Ops) and aircraft types flown (e.g., B-757/767, CRJ series) and avoid some of the prior complexities of Part 121.

Whether an aircarrier uses the relevant sections of Part 121 of the newer Sub-Part Y to train flight crew, the training today, as said earlier, uses a sequence of discrete events and/or conditions and whether or not these are, separately, within FAA FAR's and/or aircraft limits, along with a pilot's responses to these events. Again, interactions and interplay of seemingly discrete events and conditions are not formally taught. However, while any one condition or even set of condition may well be within limits, the interaction of them may place the aircraft and mission at risk. We are not aware of aircarrier training that uses this

perspective. Let us now examine at today's major flightcrew training arena: the flight simulator and LOFT scenarios

LOFT and Flight Simulators

We have previously explicated what a FS is and can do if flight training. Line Oriented Flight Training (LOFT) is flightcrew training in a FAA-rated Level D simulator; what we earlier referred to as a 'full-up" simulator. A LOFT uses a pre-developed and computer-presented "scenario" of about 1 to 2 hours. There is a complete crew and the scenario uses representative flight segments that contain normal, abnormal, and emergency procedures that may be expected in line operations.

LOFT emphasizes an orientation on events that could be encountered in line operations ("flying the line"). Thus mission realism...making the LOFT session correspond as closely as possible to event sets that could or would be encountered in flying one or more point A to point B legs... becomes the major driver in LOFT design. In other words, events that make up a LOFT scenario should pass the test of mission realism where it is reasonable to assume that this "could" happen in the real world.

The LOFT in a FS has become increasingly important in training and evaluating flightcrews. Aircarriers have come to rely on simulators for a large part or all of their flight training programs. The use of flight simulators for civil pilot/crew training and performance evaluation is at the point where it is the key and indispensable tool for air carriers, and will continue to be so. This is due to the already cited factors: safety; cost; simulator fidelity and fairly recent changes and additions to the Federal Aviation Regulations (FARs). One such change and a reason why FS today have assumed such an important place in aircarrier training: An aircraft type-rating can be obtained (almost) entirely in a FS. Finally, the FS of today has become the key piece in the first new pilot/crew training effort in over 40 years, the

Advanced Qualification Program (AQP), because of FA and LOFT scenarios as training tools. There is one major issue which we must mention and must deal with before we go into LOFT design using ODM and MPM

History and Context of Human Factors/CRM Training

CRM training was first developed in the late 70's/ early 1980's after a series of disastrous and fatal aircarrier accidents; accidents where perfectly functioning planes crashed. Human factors errors by pilot/crew were seen as the cause of these accidents. As a result, the FAA wanted the aircarriers to implement a new, human factors-oriented training, i.e. CRM. Rather than append existing FARs to make CRM mandatory, the FAA chose another path. (Of interest here is that, in SFAR 58, the FAA decided, after many years, to make CRM mandatory but only in AQP training). To return; the FAA in order to make CRM training costs palatable to aircarriers, offered to waive some hours of pilot recurrent training in lieu of CRM training. As training is a "big buck" item for aircarriers, this allowed the carriers to "save" (not spend additional funds) and give CRM. It also allowed the FAA to ensure the CRM training would, to a great extent, be given thus silencing some critics who, understandably, wanted new HF training to counter the rash of accidents. However, there was a sticking point; jeopardy. Somewhat simply put: as a further inducement to carriers to give CRM, the FAA and Air Line Pilots Association (ALPA) agreed that CRM would be "no jeopardy" training. When CRM skills were evaluated in a LOFT, neither pilot nor crew could fail be given a "down" which requires additional training and checking and is made part of a pilot's record. CRM in LOFT is "no-jeopardy" training, i.e., the instructor does not issue a passing or failing grade to a participating crewmember. As a LOFT scenario progresses, it is allowed to continue without interruption so crewmembers

may learn by experiencing the results of their decisions. Decisions and/or actions that produce unwanted results do not indicate a training failure, but serve as a learning experience. If the LOFT instructor identifies crewmember performance deficiencies, additional training or instruction will be provided. This training or instruction may be in any form, including additional LOFT. Before the crewmember may return to line operations, the performance deficiencies will be corrected and the instructor will document the training as satisfactorily completed. The "no-jeopardy" concept allows crewmembers to use their full resources and creativity without instructor interference. At the end of a LOFT session and after debriefing, the instructor certifies that the training has been completed. Therefore, the "evaluation" of a LOFT and CRM skills consisted of videotaping the LOFT and a critique/de-brief given to the crew upon completion of the LOFT. The videotape is then erased.

Our view is that LOFTs built around the MPM and ODM must be evaluated as a jeopardy LOFT session. The MPM has CRM behavioral markers with rating scales. It also has technical/flight control behavioral markers with rating scales; these encompass actions and skills normally evaluated in a check ride (FS or actual flight), and which can be failed. The flight control markers, as will soon be seen, include attitude management, course deviation, power management, *et al.* In short, the full panoply of flight skills that are usually evaluated in FS sessions/flights can be included, so it would seem that failure should be an option. Added to this, we would have LOFT scenarios which enable ODM and both pilot and crew must be allowed to succeed or fail.

In sum: The Air Transport Association of America (ATA), through many committees and working groups, and the FAA have tried to find the best ways to design, develop and implement LOFT and pilot/crew line-oriented evaluation (LOE) in a FS. This collaboration led to FARs and FAA Advisory Circulars (AC). The end-result is that LOFT, which is done entirely in a FS, has become over the past 20 years what could be called the crown jewel of air carrier pilot/crew training and evaluation. LOFT and LOE received initial impetus from their use in human factors/crew resource management (CRM) training. However, as said, CRM in the mid-1990's, encountered problems with one resultant being the "Big 3" (Delta, American and United Airlines) US flag air carriers almost completely revamping and re-naming their CRM programs. LOFT, always recognized as the key element in CRM, has thus become more and more recognized as the indispensable training, practice and evaluation tool for flight crew. The FAA's emphasis on air carrier's going to a new training and certification paradigm, the above-mentioned Advanced Qualification Program (AQP) as spelled out in SFAR 58, has further enhanced the role of LOFT.

MAXIMIZING LOFT CAPABILITY: USING MPM AND ODM LOFT DESIGN LOFT AND EVALUATING FLIGHTCREW PERFORMANCE

The Mission Performance Model (MPM)

As promised, we now return to the MPM. Captain Kevin Smith created the framework for a model that demonstrated that the CRM human factors skills and the technical/Flight Control skills are interrelated, interdependent, and often simultaneous in execution. Plainly put, for safe and efficient flight, some aspects of CRM are integral to flight control, and vice-versa. This model is called the Mission Performance Model (MPM). Captain Smith (UAL) worked with Captain Hamman (UAL) and others to develop exemplars of the application of the MPM to actual flight maneuvers, such as an engine out at V1 with a turn procedure required by the terrain. (this is called a V1 "cut" where V stands for velocity and V1

is the critical engine failure recognition speed or takeoff decision speed, above which the takeoff will continue even if an engine fails. The speed will vary between aircraft types. "Cut" stands for single engine failure at a specific point on take-off roll; that point is V1 and is before rotation, i.e., leaving the ground),

The model is based on these concepts: Flying is an integrated, mission-orientated activity and must be evaluated as such. The crew's performance is not adequately captured by totaling the sum of the component tasks/sub-tasks/elements. The focus must be on crew function—usually at the task and critical sub-task levels. *Flight proficiency skills/ knowledge are interwoven, interdependent, and necessarily interact with the CRM skills/knowledge differentially across tasks and conditions. These interactions can be identified/specified by a matrix-type crew mission performance model using the tasks, which comprise a mission/flight leg.* (This is what the authors term Integrated CRM). The model can capture these interactions, and can be sensitive to changes in both task and mission, *i.e.,* show that, for different tasks and conditions, the technical/flight proficiency skills, the CRM skills, and their interactions, will vary. This is an indication that the model has a measure of discriminatory power, or "sensitivity" to changes in task and conditions.

To summarize: The FAA's "Behavioral Markers" (*viz.* FAA AC 120-54E) can adequately delineate CRM skills and provide one basis for the (flightcrew) Mission Performance Model. As we said, the basis for the technical proficiency evaluation currently exist, in an analogous behavioral marker-type format with rating scales. As said, Captain Kevin Smith and Jan DeMuth developed a set of performance markers with rating scales for the Technical/Flight Control skills to be an analog to the CRM markers and capture the flight control skills needed for safe flight. These markers, "Crew Performance Markers—Technical Factors" focus on the crew as a unit and how well they discharge the technical aspects of the mission. They specifically addresses precision maneuvers across these areas:

- Flight Maneuvers and Attitude Control
- Propulsion/Lift/Drag Control
- System Operations
- Malfunction Warning and Reconfiguration
- Energy Management

Taken together, these two sets of markers form the MPM basis.

Another rationale for the MPM, and later, ODM, is that pilot/crew performance has often been seen as series of discrete tasks, where each task was further decomposed to reveal a set of subtasks combined with the requisite knowledge and skills necessary for subtask completion. For many applications, such as aircrew training, this produces a large collection of task, knowledge and skill data. In most traditional pilot or crew training programs, these are taught individually as isolated knowledge components. Consequently, the trainee is left with the responsibility of combining these isolated knowledge components into integrated wholes. However, the linear decomposition of individual tasks does not address integrated functioning nor does it reveal how tightly coupled teams (flightcrews) perform, thus an analytical process other than the traditional task analysis approach is considered necessary. The MPM, instead, uses a functional modeling approach

The Mission Performance Model has embedded within it the concept of functions. It is proposed that the model, as constructed, represents all significant functions necessary for the successful completion of an air transport mission. This model views crew performance as consisting of system level functions which represent the mechanisms that are used to perform a mission activity. The importance of a model that is founded on a set of systems level functions cannot be overstated. Moreover, the model delineates crew performance at a level of abstraction that is significantly dif-

ferent than the current descriptions of individual performance.

The MPM consists of a set of functions that can be activated by inserting an a specific instance/example; in other words, asking the function to specify/describe a particular activity or situation in the mission. If a particular function, e.g. "Workload Management," was asked to "spin out" the components of a particular mission activity, such as take off with an engine failure at V_1, then the function should be able to organize, sequence, distribute, and coordinate key crew actions so that a successful outcome could be assured.

This workload management function, then, can be viewed as a generic performance statement that:

- Can be applied to many mission activities/situations, and
- Can be activated for the application to, and specification of, any one of these activities/situations.

The Mission Performance Model specifies the components of flightcrew "effectiveness" (effective performance). That the model represents effectiveness is important to understand since, if the crew is really engaging in **the** set of functions that are both germane and linked to the problem at hand, and if these functions are the prerequisites for a successful outcome, then effectiveness had been demonstrated.

As had been said, the model is a generic one; many military IP's have, quite easily they said, have modified it for specific aircraft (from F-15's to helicopters) and specific military missions.

Similarly, the model is prescriptive as it prescribes what needs to be accomplished for the crew to perform effectively. For example, we can specify, during the LOFT design process, what are very likely to be the necessary crew actions.

In sum: In the MPM, CRM/human factors as well as technical performance clusters are specified along with the applicable descriptors ("markers") under each cluster. For example, under workload management and situational awareness, key markers include: preparation, planning, vigilance, workload distribution, and distraction avoidance. Similarly, under the cluster entitled "propulsion/lift/drag control, the key makers include: instrument interpretation, energy management, power control, lift control and drag control. For each marker, there are a set of behaviorally-anchored performance descriptors. When all these are combined into a matrix array, the MPM emerges.

Using MPM and ODM to Design LOFT

LOFT has been described from its initial development to its current form and content. We have stated that LOFT is the major pilot and aircrew training and check tool in an AQP Program. LOFT and LOE, as performed in the FS, simply put, are both the optimal training/testing environment and the "court of last resort," as it were. Upon successful completion of LOFT/LOE, the pilot/crew have earned new type-ratings/certifications or are "good to go" for another year. However, the current LOFTs and LOEs need to be strengthened, for the exact reasons cited above; they are the best, and safest methods for cutting edge, realistic training and evaluation, and they provide a final stamp of approval in an AQP—as well as a more traditional Part 121-based training program. We have set the stage to present how our earlier statements about the tremendous potential and existing use-value of LOFT can be merged and realized via the MPM and the ODM models

In designing and developing LOFT scenarios, the basic unit, as proposed in 1993 (viz., Hamman, Smith, Lofaro and Seamster) is the event set. The scenario is a set of event sets selected from real world ops reports or, made from amalgams of events and incidents as reported, put together from the experiences of the LOFT design team (aside: NTSB accident reports can also be used).

It should be clear that superior LOFT design is a team effort and the team should be carefully selected The LOFT design team must consist of senior pilots with extensive flight time in the aircraft type the LOFT scenario is being created for. These pilots should also have experience in the air carrier's training complex. It is desirable to have a person from the training department with ISD credentials, as well be shown later. A LOFT design team member should be a FS operator, to ensure that the event sets selected can be replicated in the FS. Finally, it is truly preferable if one or more LOFT design team have been line check airmen.

Having selected the team, the next step would be to layout an overview of the LOFT mission/ flight leg(s). This overview would include the basics, such as weather (Wx); time of the year ops (e.g., winter); departure and destination airports as well as alternate airports. Into this skeletal framework, the team will select the event sets for each phase of flight (take off-cruise-descent-landing) as well as any pre- take off event sets that may impact the flight leg.

The next steps are the crucial ones: carefully select the problems that you want the flight-crew to encounter such as mechanical, system malfunctions, et al. Then, plan the sequence into which you want to embed the problems. Remember that the overall goal is not to create the fabled "LOFT from hell," one which cannot be successfully flown but must result in a loss of flight control.

In both the sequencing of the event sets and the selection of the problems to be embedded in the LOFT, the MPM and the ODM are be used as the structural underpinnings. It is done in this manner:

Upon selection of the problems and the phases of flight that these problems are to occur in, the ODM is used to build a sequence that results in a rising risk. The decision points are identified (a "decision point" is a point in the flight where if no decision and resultant action is taken, or if a wrong decision is made, the risk rises from low to moderate or from moderate to high). Upon identifying the decision points, the basic sequence is modified to add the various outcomes from no decision/ wrong decision; that is to say, the sequence now contains braches which are dependent upon the decisions made/unmade. Each branch, or node will also need have any changes in conditions and systems (again, Wx, en-route or at destination, systems malfunctions, *et al.*) built in.

Next, the MPM is integrated with ODM The concept here is to make the consequences of a following a no decision/wrong decision model such that the risk continues to use until it reaches the high level and crew action must be taken in order to regain any possibility of successful flight completion. The "successful completion" may involve an air turn back (ATB) to the airfield of flight origin or diversion to the/an alternate airfield as "success" simply means landing the plane.

This integration is a two-step process. The first involves taking the selected event sets and identifying the critical tasks that are to be performed during those times. These high-level critical tasks (e.g., V1 "cut" on takeoff) are then decomposed, using the instructional system design (ISD) process, into the complete list of sub-tasks involved. The MPM is then used to further identify which of the critical tasks track to which of the both the relevant CRM and lift control functions necessary for successful task performance. The set of MPM functions will organize, sequence, distribute, and coordinate the actions key to successful performance.

Looked at another way: With the V1 "cut" at take-off exemplar, we find that the needed CRM function is workload management. The MPM, with ISD decomposition, will spin out the specifics of the critical actions and flight control skills embedded in the workload function. The flight control tasks for this example include propulsion/ lift/drag; operational integrity and altitude control- -with such sub-tasks as disconnecting the auto throttle at 400' above-ground-level (AGL), setting airspeed to xyz knots, checking flap setting, and

so on. The MPM well also spool out the crew performance markers for each subtask, both the CRM and the Flight Technical markers. Not only that, but the functions and actions of both the pilot flying (PF) and the pilot not flying (PNF; usually a first officer), will be clearly spelled out. These, as said, will be spelled out at the sub-task level. In fact, this is true CRM integration; the place where both CRM and flight control actions are presented as a unified whole. However, space and scope preclude further explication and, there is the CRM Integration document previously cited.

It seems clear that, since the necessary performances are specified, the performance markers can be used not only to track the crew's actions, but, if desired, to evaluate them. This evaluation can be done simultaneously using a FS operator and a check airman or, done post-hoc, using the videotapes that are normally part of LOFT sessions.

As a summary, we see that the LOFT design has been driven by using the ODM to do initial event set and sequence design. Then, the MPM was used to generate the task and sub-task breakout for selected events within the event sets. The MPM further sequenced the events selected (as an example, CRM and flight control integration was performed at the sub-task level, with optional evaluation procedures)

The LOFT design can be seen as embedding event sets into the LOFT scenario which can take the crew and plane into the moderate and even the high risk areas of the rising risk continuum thusly:

1. If the conditions causing the risks are either not identified or their interactions are not recognized.
2. If the decision points are either missed or result in an incorrect decision(s).
3. As a result of 1 and or 2, no actions are taken or incorrect actions are taken.

We now have shown the LOFT design process as one where the event sets, as well as initial and changing conditions, are used to generate decision points. The decision points, if missed or responded to incorrectly, cause a rise in the mission risk. The MPM is overlaid to give a level of detail whereby an analysis will determine if and where the errors were made: flight control, CRM, CRM and flight control. The MPM also offers an evaluative framework. However, LOFT design could be done in a slightly different manner.

LOFT Design: Another Approach

The initial LOFT developmental sequence described above followed this course. Events/event sets were selected that required decisions (and, actions) to prevent risk from rising; to prevent the aircraft's position in the ops envelope from approaching a corner or a boundary. As an example, the event set could include deteriorating Wx enroute or at the destination airport, perhaps with braking advisories or crosswinds on approach/landing. From there, the MPM was used to develop the flightcrew tasks and functions for the pilot flying (PF) and pilot not flying (PNF). An initial bifurcation could be made, one path of event sets following the correct identification of rising risk and attendant risk reduction actions; the second path based on non-identification of rising risk. There would also be sub-paths, for example showing the correct identification of rising risk but incorrect response(s). The branching process could be repeated as needed. Thus, the ODM is the driver and the MPM is the method used to develop functionality.

However, LOFT can be developed in a different way, still using the MPM and ODM. A series of event sets (based on incident reports, "hangar talk," experiences, etc) can be selected, linked and the PF/PNF functions identified. These event sets we will term "expanded event sets" of "fully articulated event sets." By analyzing these sets, the decisional points can be identified. In fact, "identified" is not the exact term, "selected" is more appropriate. This is because it seems clear that, in any flight, new conditions or changes in

conditions (Wx; flight system problems, *et al.*) will result in changes in the aircraft's position both in the ops envelope and on the rising risk continuum.

With the initial set of fully-articulated functions and actions developed via the MPM, changes are introduced using the ODM's boundary conditions as guidelines. That is to say, an initial set of boundary conditions will be specified and used as a basis for carefully selecting changes to them that result, if left unidentified and/or unchecked, to additive or cumulative interactions; such interactions driving the aircraft towards a corner or side in the ops envelope. Of course, this means that the risk has risen to moderate or high moderate—even to high risk.

The changes to the boundary conditions should be introduced at different points in flight so that the risk does not rise suddenly. The rationale here is that one goal of the LOFT is to keep situation awareness high by the introduction of on-going series of changes, rather than a compressed set of events that lead to an immediate abnormal ops or emergency, with limited options for the flight crew. If the changes in the boundary conditions are introduced over the first hour or so of the LOFT session, their additive and/or cumulative interactions and impact will be sequences so that the flight crew's ODM skills are tested. ODM skills are tested rather than skills at handling an overt and immediately apparent abnormal or emergency situation, which situations are often trained in other venues.

Aside: This is not to say that missed decision points as well as incorrect decisions and actions may not lead to an abnormal or emergency situation. If that occurs, then the LOFT can also demonstrate flight crew skills in the emergency arena. However, as said, the training, to include recurrent or special item training, of flightcrews does provide for certain emergency training. As one example, recently some aircarriers have instituted upset recovery training (recovering the aircraft from unusual or abnormal attitudes).

To resume: by carefully introducing boundary conditions changes into the event sets, the risk can be caused to rise from additive or cumulative interactions. As before, when we indicated how to use the ODM to MPM LOFT design methodology, there will be a branching effect, contingent on decisions (made; unmade; correct; incorrect) and resultant actions (taken; not taken). The LOFT scenario must be designed to include the various pathways, so that the FS can be pre-programmed for the contingencies.

It would seem that, optimally, the OPM to MPM and the MPM to ODM methods would operate simultaneously, or in an intertwined manner. Thus, it may be fairly said that the use of the MPM and the ODM is actually a necessary and sufficient condition of effective LOFT design

We believe that we now have presented, in sufficient detail and with examples, the ODM. The same can be said for the MPM. We have given references for the reader who wants more information and exposition of either model. We have presented the framework for developing an the ODM/MPM-based LOFT scenario(s). The evaluation of the flightcrew in the LOFT training session has been discussed.

We now take an apparent step backward from an earlier assertion. The ODM, the MPM, or any LOFT developed using them need have an evaluative aspect. If evaluation is to be an aspect of the LOFT session, it need not be a jeopardy situation. However, we still hold to our original view that LOFT should have a jeopardy component.

While we have not mentioned or emphasized the training aspects of the MPM on the ODM, it is clear that here are necessary training considerations for both. The MPM needs little, if any training terms of flightcrew, the reason being that the CRM components are already included as part of either initial or recurrent training. The flight-control maneuver components are all included in flight training/type training, and many of the flight control aspects are used in mandatory and yearly recurrent training. Additionally, these flight control tasks/subtasks are all part of the handbooks *et al.* used by pilots for each type of aircraft. Put another way, from the task through the sub-task level of flight control, pilots are familiar with and have been trained in all of it. Of more importance, the performance of these flight control tasks, and the FAA and carrier standards to which they must be performed are already known to the flight crew—they have learned and been tested in them on the ground and been evaluated on their ability to perform the standard in the air. Therefore, for any critical task decomposition used in a LOFT, the flightcrew is well aware of the sub-tasks required to perform the task. However, no matter which LOFT design schema is followed, there is training needed to those doing LOFT design in understanding and use of the MPM.

ACCIDENTS, INVESTIGATIONS AND IMPACT ON FLIGHT SAFETY

Introduction

We have dealt with three of the four issues we raised at the Chapter's onset and go on to accident investigations and their resultant findings. The stated purpose of accident investigations is the prevention of future accidents, thereby reducing the accident rate and fatalities. This can only occur if the investigation results are accurate and the causal factors identified are valid; by valid, we mean identifying the actual causal factors involved. While hesitating to say this, we will use a recent accident and United States' National Transportation Board (NTSB) report as an example of a report that missed the real causes.

Such accident findings data also must be used by aircarriers' in their training. We will look at a very recent accident that was strikingly similar to another recent accident; yet the flight crew made the same poor decisions. This gives one to wonder not only if the NTSB report on the prior accident had been folded into training, but if the report had even been read. The accidents we will use are Southwest Airlines (SW) flight 1248 and American Airlines (AA) flight 331. These two accidents mentioned will be re-examined using ODM to find the real causes. Lastly, we will show how accident investigation, and more importantly, subsequent analyses data can be folded into effective flight crew training.

Accident Causation and Technical Tools for Analysis

The current paradigm in use for finding accident causes is a linear path, often referred to as a causal chain. The major assumption here is that, if the place in the causal chain can be found where something untoward occurred, such data can be used to prevent future accidents. There are two major problems with this: The first in that any accident is a compendium of, almost entirely, one off conditions, personnel and actions. Finding out, even if accurately, what caused airplane A to crash is a snapshot in time; a snapshot that can never be replicated. Looked at another way: minimum, if any, generalizable knowledge in the accident findings inheres. Before we get bombarded with critiques, we do not deny some generalizable knowledge may occur, the operative words being "some" and "may."

The second problem we see with accident investigations refers back to the linear pathway viewpoint. We posit that an accident is more on the order of a mosaic, whose pieces are composed of personnel, actions, condition; a mosaic that can have many ways of fitting the pieces together, with no real negative consequences. But, there exists one or more ways where the "fit" results in an accident.

Put another way, as we have said, there can prior poor decisions and behaviors that seemingly "worked" in the past (blind luck? the "not your day to die" phenomenon?), but, at some point in time, with changed conditions, the fit of these

previous decisions and actions goes no longer make a viable mosaic. Result? An accident.

Today, there are a host of technology-based information sources: Cockpit voice recorders (CVR) and Flight data Recorders (FDR) capable of simultaneously recoding 128 channels of information as to control inputs, instrument readings, control surfaces settings and much more. The CVR and FDR are usually called the "black boxes." There may be radar-based records of flight trajectory and recordings of air traffic controller and pilot communication. The data needed to find out the "what" of the accident is usually present; the causation or "why" seems elusive lately. We will be seeking an analytic tool that can offer pathways to training that will apply, across many venues, can be useful to devise training that can be effective in accident prevention.

Let us go to a recent accident, NTSB findings and an examination using ODM.

Accident One

On December 8, 2005, a Southwest Airlines flight #1248, attempted a landing at Chicago's Midway International airport in adverse conditions, rolled through a blast fence, an airport perimeter fence, across a crowded adjacent roadway striking several passenger automobiles. One innocent bystander's (parked on a street across from the airport perimeter fence) life was lost, several people on-board seriously injured, and property destroyed. The conditions, at the time of the approach and attempted landing were as follows: Low visibility and falling snow, compounded by the fact that the runway was slippery and braking action advisories were in effect. Since there are only two types of advisories issued to pilots and dispatchers, (wind shear and braking action) one wonders what, at times, these are used for?

At the time of the accident, a greater than 8 knot tailwind did exist…beyond acceptable limits… and, when coupled with the poor braking, a recipe for disaster. It also seems that the SW instrument used to calculate landing distances had a default tailwind setting of 8 knots, which is the FAA legal limit for landing. Even though the crew entered the real tailwind speed, 11 knots, the computer defaulted to 8 knots. The reality for SW 1248 was that attempting to land in low visibility conditions, with breaking action advisories in effect, with a tail wind on a short runway and with no real overrun was not a viable option. Were the crew unaware of the issues? A quote from the NTSB accident report (NTSB/AAR-07/06; 02/10/07) may be instructive: Cockpit voice-recorder transcripts indicate the pilots had been concerned about the weather and, prior to landing, jokingly alluded to the movie *Airplane,* saying "I picked a bad day to stop sniffin' glue."

To return to the accident: The Captain of flight 1248 was clearly faced with a rapidly rising risk he either did not understand or was constrained in his decision by other, non-safety of flight, issues. The Captain was faced with about a 200 foot ceiling and ½ mile visibility, close to CAT 1 minimums. But he also had to deal with unfavorable wind conditions, and runway contamination reducing breaking effectiveness *on a short runway*. When the factors are combined, the Captain had an aircraft that was outside normal conditions and into non-normal/emergency operating conditions and, therefore at high risk. A landing should never have been attempted. But it was. A landing where the touch-down point was at 4,500 feet down the 6, 502 foot runway, when (with the plane's weight and speed) it would have needed 5,300 to safely stop). At touch-down, the plane was slightly over the speed for landing and, had there not been some dryer runway conditions on portions of the runway, this would have happened: because of its speed, the plane would have departed the runway, easily gone through the perimeter fence and not only crossed the street but crashed into the building there with resultant loss of lives and injuries.

The NTSB Report

It seems both amazing and puzzling that pilots and other aviation experts have not risen as a group and demanded this report, NTSB AAR-07/06, Southwest Airlines flight #1248 be revised...or, better yet, almost completely re-done in order to have congruence with the reality and facts of the accident. This has not occurred. The NTSB report "findings' seemed to turn logic on its head and were insufficient in scope and incorrect as to causation. The NTSB inexplicably determined that the probable cause was the pilot's failure to stop the airplane on the runway (under conditions that would almost guarantee that this could not even be possible). Ergo, the accident happened because the pilot did not complete the landing within the confines of the runway. What an enlightening piece of information! One supposes that, under this logic, when a CFIT happens, the cause is that the plane hit the ground. While superficially plausible to an uninformed observer, this NTSB finding is manifestly wrong. As an oblique afterthought, the NTSB made reference to the fact that perhaps a diversion to another airport was in order. These findings turn logic on its head; they imply that it is somehow perfectly acceptable if an airline sends flights from both coasts of the USA to the midwest part of the USA in wintertime, into a snowstorm, and then "hope for the best." It seems that NTSB tossed away the accumulated of knowledge of more than 50 years of flying military and civil aircraft out the window.

What of the NTSB finding that the fault was in the SW training that was either not received or not tested as to engine thrust reversers and auto brake systems? These cannot be the real culprits as with the time/distances/speed needed before they would/could be deployed, they would have had little effect. However, we do not mean to say that SW did not (does not?) have training deficiencies. The Captain did lack some needed training, both in winter ops, in his 737 model and, of more use-value, in risk identification, management and, most importantly, decision-making. These should be both SW and FAA concerns. Are they? One wonders.

Using the ODM for a Re-Examination

We will speak of the role of Dispatch in some detain and mission planning in the next Section. For, now, what of Dispatch and mission planning for this flight? Part 121 Carriers are not permitted to launch airplanes into the 'wild blue yonder" any time they feel like it, but must comply with Dispatch protocol and constraints. Some conditions make successful completion of some flights so improbable that they cannot be attempted. For example, if the forecast weather at the intended destination is below landing minimums, the flight should not and cannot be flown. Armed with the latest forecast of wind and weather, one has to ask why this flight was even attempted. Even more telling, why was there not an in-flight diversion to a designated, alternate field, or, as a last resort, a rejected landing? Why did not the captain abandon the approach, reject the landing and proceed to the alternate?

What occurred instead is a staggering series of either non-decisions or poor decisions, the penultimate being attempting to shoot the landing. The Captain/flightcrew, as per our ODM and risking risk paradigms, were at high risk. But, they were either not aware of or, not concerned with risk. They were dealing with one condition at a time and seeing if they could legally attempt a landing. But, rising risk is non-linear. While we tend to think of one thing at a time, like wind or visibility, the reality is a cumulative effect can, and often does, occur where the real impact of the conditions taken together result in a much higher risk than the conditions taken as discrete events. We assert here that the Captain should have abandoned the approach and proceed to the assigned alternate. The weather at the time of the accident was such that only the most carefully flown aircraft had even the slightest chance of

landing safely at this airport and that prospect was rapidly fading when the plane was well out of the approach phase to Midway. The conditions were such that only the most perfectly executed landing could possibly have brought the aircraft to a safe stop on the existing runway. That this was not done is obvious and the price for poor, incorrect and inaccurate decision making was death and injury.

Causation

The accurate "most probable cause" of this tragedy was the failure of the Captain, given the deteriorating weather conditions, marginal breaking action and adverse wind conditions. to make a timely and accurate decision to abandon the approach and landing and then proceed to the assigned alternate. As a last resort, he could have pulled up and declared a missed approach. That, however, has to be reported to the FAA with subsequent investigation. The set of conditions encountered and their ODM cumulative effect, show the plane to be in a rapidly rising risk spiral, where the risk had gone to the highest level. The Captain was unaware of the reality of his situation, for very obvious reasons. What occurred is that the Critical Event Horizon (CEH) had been passed through. Having said this, let us look at American Airlines 331 which is eerily similar in many aspects to SW 1248.

Accident Two

Even though this accident happened years afterward SW 1248, in December, 2009, it seems to mirror it, except that in was in summer, not winter, ops. The plane type, B 737-800, was the same as were most conditions. Sadly, the result also was the same. American Airlines (AA 331) from Miami literally broke apart after over running its runway at Norman Manley International Airport in Jamaica. The plane skidded across a road and halted at the edge of the Caribbean Sea, a scant 40 feet from the Sea. As with SW 1248, there was a seemingly miraculous lack of injuries and death. This flight did result in four serious injuries, all to passengers. A quick aside here: at the writing of this Chapter, there is no NTSB report, preliminary or final, available. The data presented comes from various news services.

Let us turn to the condition, actions and result. The flight was an hour late departing Miami International Airport. The aircraft attempted to land in Kingston, Jamaica, in heavy rain about two hours after taking off. The crew had contacted Jamaica Air Traffic Control (ATC) to request the Instrument Landing System (ILS) approach for Runway 12, the designated runway broadcast by the Automatic Terminal Information Service (ATIS) for arrivals that night. ATC, however, advised them of tailwind conditions on Runway 12 and offered a circling approach for landing on Runway 30. The crew repeated their request for Runway 12 and were subsequently cleared to land on that runway with the controller further advising the crew that the runway was "contaminated" (wet). The aircraft had a relatively heavy fuel load at the time of landing as it was carrying enough fuel for a roundtrip flight back to the US.

The aircraft landed only slightly below its permitted landing weight and with an airspeed of 148 knots (170 mph). However, because of the 14 knot tailwind, groundspeed at landing was 162 knots (186 mph). This is far above the recommended landing speed for this aircraft. The flight data recorder (FDR) showed that the aircraft touched down some 4,100 feet down the 8,910 feet long runway. Normally touchdown would be between 1,000 feet and 1,500 feet. The FDR also showed that the aircraft bounced once before settling down on the runway, which further reduced the amount of remaining runway. In terms of the ODM model, the risk trajectory had moved to high. The trajectory in the mission space had been pushed from the normal/safe part of the ops envelope to being outside of the envelope due to the hostile agents involved; the aircraft was unstabilized, in

abnormal/high risk conditions. The results below should not be unexpected.

Upon contact with the runway, the autobrakes deployed, the crew engaged reverse thrust and spoilers, and also used maximum manual braking. In spite of this, the aircraft veered to the left of the runway centerline and departed the end of the runway at a ground speed of about 63 knots (72 mph). The aircraft went through a fence and crossed a road before coming to rest on a beach about 175 feet beyond the end of the runway, and about 40 feet from the sea. The fuselage was broken into three major pieces, and the right engine, right main landing gear, and parts of the right wing separated from the aircraft; in short, damaged beyond economic repair.

As with SW 1248, the flight crew, with an over-the-limit tailwind, "decided" to landed "long and hot" on a contaminated runway with no overrun. The plane departed the runway, blasted through a fence, crossed a road and was within feet of a disaster. Fortunately, there were no mechanical or systems failure; the flight data recorder did not indicate any malfunctions or other anomalies with the brakes, spoilers, or thrust reversers, and that braking was normal given the wet runway and the autobrake setting. No mechanical problems have been found with any part of the aircraft, and ground based navigation and landing aids were operating normally.

In sum: the facts as they unfold have many similarities – both from an operational standpoint, as well as the aircraft type and runway environment – to Southwest Flight 1248 which overran its runway in December 2005. Let us look at the AA 331 conditions and decision. We immediately notice that the crew chose to land with a tailwind, even after air traffic controllers advised them that the runway was wet and offered the crew an option that would have allowed the aircraft to land with a headwind. The aircraft had sufficient fuel on board to divert and reach its alternate airport at Grand Cayman Island, and they had sufficient fuel to return to Miami. However, they decided to land on the contaminated runway rather than turn around or land on the other runway. Once again, a flightcrew did not recognize the risk and passed through their CEH. As with the SW flight, why was there not an in-flight diversion to a designated, alternate field, or, as a last resort, a rejected landing? Again, why did not the Captain abandon the approach, reject the landing and proceed to the alternate? This option was offered by ATC, as well the use of a different runway. Yet, the pilot/flighcrew persevered with another poor decision in a series of either non-decisions or poor decisions; the "decision" to ignore ATC and shoot the landing.

There was an NTSB report on the SW accident of 5 years before. Was it not somehow included in the AA training regime, especially as the plane type, B 737-800 was identical? In point of fact, was the data on the year 2000 SW overrun any part of SW training afterwards? Southwest Airlines already had an runway overrun accident, SW 1455, in the year 2000. Now, they have had another. American Airlines had a very similar disaster in 1999, at Little Rock, AK. The Captain continued an approach when severe thunderstorms were in/over the airport; the crosswind was over AA limits for a crosswind landing, the spoilers were not deployed on touchdown. Result? The plane overran the runway, crashed into light poles and ILS stanchions and eleven persons died, with score of injuries. Were any training changes made as a result at either AA or SW after four very similar accidents over a very few years? What is the type, quality and evaluation of training is being received by airline aircrews?

Recommended Solutions

We would be remiss if we did not inform the reader of a little-publicized fact. The NTSB, after completing an accident investigation, makes a list of recommendations for FAA implementation; these relate to NTSB findings and are designed to prevent future accidents. These recommendations have no teeth, i.e., they are not legally binding on the FAA to enact. The FAA has, in the past, rejected NTSB recommendations, or placed them on hold, or said that studies were need (paralysis by analysis?) and, on occasion, implemented them.

One solution is that NTSB accident recommendations go through a process which includes FAA, NTSB, Air Transport Association of America (ATA) and Air Line Pilots Association (ALPA) and, if approved, become legally binding on the FAA, with a short time-line to enact.

Next, the ODM model is applicable here and can produce accident that that does not focus on minutia, but on root cases. The data from an ODM analysis can then be rolled in training, especially AQP and most particularly LOFT. This recommended solution should not be unexpected. What follows, even if somewhat out of order, are both new problems and some attempt at solutions/recommendations.

LOFT is a useful training method because it gives crewmembers the opportunity to practice line operations (e.g., maneuvers, operating skills, systems operations, and the operator's procedures) with a full crew in a realistic environment. Crewmembers learn to handle a variety of scripted real-time scenarios, which include routine, abnormal, and emergency situations. They also learn and practice cockpit resource management skills, including crew coordination, decision-making, and communication skills. The overall objective of LOFT is to improve total flight crew performance, thereby preventing incidents and accident during operational flying A solid AQP program would provide both instruction and simulator practice for this Captain (and all other Captains) in how to exercise Captain's Authority in such a way as to make timely and accurate decisions and avoid the very accidents we are discussing.

There are problems that we are hesitant to raise; they involve AQP and the FAA. At this point, an underlying cause of these accidents may well be the failure, at the most senior levels of management, to develop, install and implement a quality, state of the art pilot training program: AQP. Further, it falls to the FAA to approve, and then oversee any such AQP an airline develops. The major feature of AQP is to provide "mission realistic" training and evaluation, concentrating not only on flight maneuvers, but on higher order skills like decision-making needed in actual line operations. Indeed, Captain's authority, workload management, and decision-making are the three underpinnings of any successful AQP. These higher order skills can and must be taught and evaluated by any airline that wants to produce and maintain quality, trained pilots. As said, accurate risk location is the key, when in a (rising) risk situation, to making the optimal selection of a course of action, i.e., an action response that is an alternative to the original mission plan

To return to the FAA: Their public statements were that they would provide support, approval and oversight if any airlines' development and implementation of an AQP. Yet, at the time of this Chapter, there are approximately 10 personnel doing this for the FAA. The number of major and even regional carries using AQP would seem to require 100 or more personnel.

Finally, we turn to what we consider the poorest alternative to airline safety; a new government law. As one result of the accidents referred to and others, such as Colgan Air 3401, crashing near Buffalo, NY and killing 50 people, the U.S. Government responded with a proposed Law to deal with such accidents. Before we go further, the NTSB report we cited, AAR-10/11, cited, among many things, a familiar theme: shortcomings in pilot training. We have spoken to and will continue to speak to pilot training. To let politicians write laws governing air safety is bizarre for the obvious reasons. Even a cursory look at the proposed legislation, HR3371, shows it to be a shot-gun compendium that, in very few pages, tries to deal with everything from pilot training to pilot fatigue to disclosure issues. It mentions many things and resolves nothing, save to call for ever-more legislation, rule-making, enforce-

ment and other supposed fixes to a problem that is extremely complex. If enacted as a law, it can well have deleterious consequences. We advise the reader to look at the website (http://www.gov-track.us/congress/billtext.xpd?bill=h111-3371) that contains the current draft and form their own conclusions as to the efficacy of the law in resolving civil aviation safety issues.

A BREAKTHROUGH: DEVELOPMENT AND USE OF AN COOPERATIVE AIR-GROUND, DECISION MAKING TOOL: THE AUTOMATED RISK MANAGEMENT DISPLAY (ARMD)

Introduction

We propose a concept for a new cockpit display: The Risk Management Display. We would point out that the modern cockpit, such as employed in the Boeing-777, already has risk-relevant displays and associated warning systems. Two such systems: the enhanced ground proximity warning system (EGPWS) and the Color Weather Radar are instructive. Both use the well-accepted Green-Yellow-Red color scheme and both employ some sort of predictive capability. Can we use this same three color representation to make visible other risk factors to the mission? We think the answer is yes.

As said, the pilot's main function and responsibility is that of risk manager and the pilot and crew's main functions are risk identification, assessment and mitigation. This recognition is true, at least, in military aviation with its Operational Risk Management (ORM); a checklist completed prior to launch, which can result in mission planning changes and even aborting the mission. However, this proactive approach does not provide for changing conditions and factors aloft that can result in a rising risk after launch. The entire purpose of risk identification, assessment and mitigation is to enable the pilot to make the most timely, and accurate decision, in real time, in a time-compressed and unforgiving environment. Having said that, ORM should be included as the first, pre-launch step, in risk assessment.

What follows is a systematic approach for flightcrew (Cockpit) and ground (Dispatch) integration in risk identification and management. This approach is based on ODM and will result in a way to strengthen the Captain and Dispatch team before and during a flight via up-linked ODM and an automated cockpit display. The end result is enabling the pilot/flightcrew to identify, manage and thereby reduce risk to the aircraft, and all onboard. What is proposed here is not more information transmitted from the Dispatcher to the pilot, but "better" information; better meaning on-going, timely info that incorporates risk identification, evaluation and management. What we envision is the Captain, supported by a team which includes the dispatcher, managing and maintaining a shared mental model that includes, *inter alia*, mission risk and mission critical factors, via a cockpit display. As we already stated, the NTSB report on the Midway accident in 2005 reported that the pilots either did not know, or recognize the potential impact of the risks that they were actually encountering, thus they continued the approach, even though they should have abandoned the approach and proceeded to the alternate field. This mistake was not only tragic--loss of life—but unnecessary. The same can be said of the recent accident at the Jamaica airport. In sum: if risk factors were employed in the decision making process and were a component of a shared mental model, we believe that these accidents could have been avoided

Our proposal then is to incorporate within the Dispatch-Cockpit system a software tool, with cockpit display, that is designed to quickly and easily communicate mission risk. Such a display tool could be uses interactively enroute to bring to the Captain's situational awareness the risk-

laden factors that could have a detrimental effect on mission success.

Ever More Information? Yes and No

In the world of cockpit automation, it has long been a tenet that if it can be automated, it should and will be. A corollary is that the more information that can be given/displayed, the better where the belief is that more data and information lead to better decisions by the flightcrew. The possible negative outcomes of using ever more cockpit automation to give the flightcrew more and more information have long been discussed, written on, researched and proven out. In spite of these works, it seems that the response has been more automation and more information pumped into the cockpit.

In an environment where not only spot-on decisions are *de rigueur*, but the time compression for such decisions can be daunting, one sometimes finds a time-consuming search for some magic piece of information to resolve a problem, rather than make a needed decision with correlate action. Many pilots/flightcrews have a tendency to seek more information when required to make a decision. It is a fact that some data/information that flightcrew receive can be incomplete or ambiguous, perhaps seemingly contradictory. Yet, interpretation of the data needs to be done in flight in order to make decisions that minimize risk and allow a successful completion of the flight/mission. When too much information is sought and/or obtained, one or more of several problems can arise. Information overload will result in a state where so much information is available that decision-making ability actually declines because the information in its entirety can no longer be managed or assessed appropriately. Information overload can and often does result in forgetfulness. When too much information is taken into memory, especially in a short period of time, some of the information (often that received early on) will be pushed out. It is pushed out by selective use of the information. That is, the decision maker will choose from among all the information available only those facts which support a preconceived solution or position...tunnel vision, or more scientifically, functional fixation and decision bias. Mental fatigue can occur resulting in slower work or poor quality work and/or decision fatigue occurs. In that instance, the decision maker tires of making decisions. Often the result is fast, careless decisions or even decision paralysis--no decisions are made at all. Another outcome for the thrust for ever-more information is the "paralysis by analysis" effect.

In working with and teaching military Instructor Pilots (IP's), as well as commercial pilots, we both found that information overload indeed has become a problem. Dr. Lofaro regularly teaches courses such as Aviation Psychology, Aviation Human Factors, Sensation and Perception and Memory and Cognition in graduate classes where many of the students are IPs. These IPs come from the USN, USMC, USAF and USCG. Captain Smith, during his 34 years at United Airlines (UAL) was a senior Captain, line-check airman (LCA) as well as an Instructor in UAL's Training Center.

The answer then is not ever more information and data pumped onto the flightcrew's plate BUT

a. Some information sifting.
b. Some information prioritization (these may be a part of Dispatch's role).
c. A model for sorting/sifting and placement of the data that are vital to operational decision making and placement in that paradigm. This model already exists, ODM. While originally developed using civil aviation as a target, has shown its robustness by the fact that, over the past 5 years, it has been modified (in academic venues) for use with specific airframes and specific missions by military pilots and IP's in the USN, USMC, USAF and USCG. While we cannot verify that

the modifications have actually been used (security issues exist), we can verify that more than anecdotal evidence and perhaps slightly less than empirical validity exists.

d. A cockpit display that integrates data both from the flightcrew and dispatch into a tool for timely and effective decision making and, most important, results in actions to minimize risk.

In sum: we believe that it is not ever-more information and displays that are needed, but one particular and fairly simple display that is collaborative with Dispatch and results in quick and quality decision making.

Dispatch and Flight Crew: The Ground and the Air

Before we go further, we will explicate what we have called "Dispatch." In US civil aviation, there are aircraft dispatchers (called Dispatch). Dispatch provides two important functions: Mission Planning and Flight Following. In mission planning, dispatch begins the process by bringing together the myriad of needed details, and constructs an early draft mission plan. This plan covers the basics, such as legal limits avoided, fuel and optimum route proposed, and, making sure all Federal Aviation Administration (FAA) regulations are followed. In collaboration with the Captain, the final mission plan is put together and signed by the Captain; his or her signature being the legal statement that the Captain agrees to the content of the plan and it does not involve undue risk. Flight following is equally important, and a good dispatcher will anticipate the informational needs of the crew and provide these needs without prompting. Dispatchers maintain contact with flight crews after they are airborne to keep them advised of weather conditions, alternate landing plans, and necessary changes in altitude. Today's dispatchers have many sources of information that are unique, such as holding times, discussions with a crew that just landed ahead of your flight, etc. There is no reason why the Dispatcher also cannot provide the Captain with risk relevant information, as well as legal parameters. (Captain Smith, from the flight deck, with some thoughts on the pilot-dispatch connection: I always appreciated a dispatcher who developed a good mission plan, I would often call up the dispatcher and discuss things by voice. A dispatcher who anticipated my needs during flight following was always held in high regard. Timely information delivery was always appreciated and helped me do a better job. Such things as breaking action advisories, tropical storms, wind shear, etc, which made flying the mission highly stressful, were helped to a considerable extent by a well trained, experienced dispatcher). We will now outline a way to make the pilot-dispatch effort collaborative and continual.

Since we propose training for dispatch personnel, we would be remiss if we did not mention the current FAA AC on Dispatch Management Training /DMC (FAA AC 121-32A; 2005). The very name is a give-away as to content. It is little more than the FAA AC on flightcrew CRM training, as mentioned, with some minor modifications, additions, deletions; in short the D replacing the C.

FUTURE RESEARCH

An Outline Protocol For R&D

1. The first and primary step in this effort to identify, manage and thereby reduce risk to the aircraft, passengers and flight crew is a shared model, mental and actual, that flight crew and dispatch both utilize. That model is the ODM which can be used in the mission planning stage and, as stated, in the en-route stage of a flight.
2. The second step is joint and interactive Dispatch and flight crew training on the model and its use.

3. Then would the development of a software-driven, cockpit display; the ARMD. ODM-trained Captains and Dispatch personnel could be used along with software engineers in a set of workshops using a knowledge engineering technique called Small Group Delphi Paradigm (Lofaro, 1992) in this effort
4. Next, usability testing involving both dispatch and flight crew (Maliko-Abraham and Lofaro, 2003) would follow.
5. The final step is implementation. The first air carrier to develop and test this new approach would both serve as a trial case and as a proof of concept. It must be noted that this dispatch/cockpit crew integration would be entirely at the discretion of each air carrier.

The Software Tool and Cockpit Display

This Section could be placed before the prior Section. However, the actions and activities described in both Sections may actually be concomitant and/or simultaneous. An ODM-based Automated Risk Management Tool (ARMD)... configuration, display placement, operational algorithms...would be developed along with protocols and procedures for the ARMD usage. This tool would be highly useful to both flight crew and Dispatch as Dispatch could load and uplink new/changed risk factors to the airborne flight crew. As a corollary: The Captain would also, as they are encountered en-route, downlink any critical mission factors to Dispatch. Thus, both the flight crews and Dispatch must identify various intervening factors as hostile agents; these would come from the aforementioned 72 critical mission factors/hostile agents into the ARMD.

An in-flight risk management display could be integrated into the lower engine indicator and crew alerting system (EICAS) of the cockpit display system and be connected to a similar display utilized by the Dispatcher. This display would show a series of three lights: green yellow and red.

- **Green Light:** Risk is Low; normal operations prevail; continue mission as planned.
- **Yellow Light:** Risk has risen to Moderate; one or more mission critical factors are in play and have, or will impact flight safety (e.g. windshear advisories at destination airport). Some modification to flight plan needed.
- **Red Light:** Risk level has risen to, or is expected to, rise to High. At least three mission critical factors are in play. Significant alterations to the mission flight plan are urgently required.

For example, the dispatcher doing flight following could uplink the illuminated light display, say the yellow light. The Captain could then query the illuminated light to see what has changed and what factors are involved. Say severe turbulence over Iceland was reported by Air France, with resultant on-board injuries. The Captain would either agree to the yellow, or red light, by pushing on the light. If a collaborative modification is developed and agreed-on, the dispatcher would then uplink a reroute away from the severe turbulence, with new fuel calculations. But, if the Captain did not agree to the color-related risk factors, he would not push any light. As a corollary: The Captain would also, as they are encountered en-route, downlink any critical mission factors to Dispatch.

As food for thought, a follow-on to the ARMD would be a full-up cockpit display that shows the current trajectory and position of the aircraft in the ops envelope. We will only briefly examine such an onboard risk management display. This could be called a threat display although threat displays for the air transport mission may sound bizarre. However, if one takes a look at a modern cockpit, we see that already have some sophisticated threat displays. As an example, adverse winds are, in our ODM model, listed as a hostile agent. i.e., a threat. In fact windshear is adverse winds and there currently is a cockpit display for it. Such an ops envelope threat display should have the following design objectives:

1. During the mission planning phase, preload and make visible all current and known mission critical factors/hostile agents. This risk data can be uploaded to the cockpit along with the flight planning forecast (FPF). 2. Update risk relevant mission critical factors/hostile agents en route. Simple communication protocols from the pilots to the dispatcher can be arranged and the dispatcher can then feed updated information into his and the cockpit's risk displays. 3. Using straightforward, pilot-dispatch communication protocol, along with smart computer aids to quickly access risk data and risk calculations, create an intuitive display environment that can quickly provide focus to "the things that will bite you" (To use some pilot lingo); this display would be the aforementioned ops envelopment and environment. This actual in-flight risk management display could be integrated into the lower EICAS of the cockpit display system and be connected to a similar display utilized by the Dispatcher, perhaps replacing the 3-light ARMD. But, we are perhaps getting too far ahead of ourselves. As said, we only bring such a display up as a next risk management and flight safety tool. Still, many of the design steps of the software would have been accomplished during the ARMD development.

SUMMARY

We have attempted to present both problems and solutions as we wrote this Chapter. Recapitulating the problems: the primary problem is that the pilot/flightcrew's prime function is risk identification, assessment and action. This occurs in an time-compressed, unforgiving, decision making environment. However, in flight crew training, they are not give a systematic approach to decision making in order to identify and assess risk, then take the required actions to preserve safety of flight. Put another way, they are not given, nor taught the use of, a model within which to embed the information they possess; a model that has a process by which to make risk identifying and reducing decisions.

The second problem is that the primary tool used today in assessing flightcrew performances, especially decision performances, is LOFT scenarios in a flight simulator. However, there really was no systematic way to either design LOFT's to focus on decision performance or to evaluate flightcrew performance across CRM and flight control skills. Then, there is the issue of the use-value and adequacy of accident report data.

We proposed an approach to all three problems that involved a decision making model(ODM) and a framework to embed the model in (MPM)... in training, for LOFT design and evaluation, for accident analysis that provides data what can be useful in LOFT and other training and in true accident prevention. We then proposed what we saw as a breakthrough step: to design and develop an automated cockpit display that is based on the ODM operational envelope, shows pilot his position in that envelop, has a 3-light display warning system that up-links with Dispatch and is embedded in a cooperative ground-air decision making process. This we see as the needed and next step to make risk identification and, with resultant and accurate decisions based on that identification and assessment enduing. Aviation safety and the flying public are the real beneficiaries.

REFERENCES

Burki-Cohen, J., Go, T. H., et al. (2003). Simulator fidelity requirements for airline pilots training and evaluation continued: An update on motion requirements research. In *Proceedings of the Twelfth Annual International Symposium on Aviation Psychology*, Dayton, OH: ISAP

Dekker, S. W. A. (2005). Why we need new accident models. In Harris, D., & Muir, H. (Eds.), *Contemporary issues in human factors and aviation safety* (pp. 181–198). Burlington, VT: Ashgate Publishing Co.

Ewell, C. D., & Chidester, T. (1996). *American Airlines converts CRM in favor of human factors and safety training. The Flightdeck, July/August, 1996. Flight Department, American Airlines.* DFW Airport.

Federal Aviation Administration. (2004, September 27). *AC 120-35C: Line operational simulations: Line oriented flight training, special purpose operational training, line operational evaluation.*

Federal Aviation Administration. (2004, January 22). *AC 120-51E: Crew resource management.*

Federal Aviation Administration. (2004, June 23). *AC 120-54A: Advanced qualification program.*

Federal Regulations (CFR). (2003). *Title 14, aeronautics and space- SFAR 58: Advanced qualification program.*

Federal Regulations (CFR). (2005). *Title 14, aeronautics and space- Part 121- Operating requirement: Domestic flight and supplemental operations, note sub-part Y.*

Flexman, R. H., & Stark, E. A. (1987). Training simulators. In Salvendy, G. (Ed.), *Handbook of human factors* (pp. 1012–1038). New York, NY: John Wiley and Sons.

Flin, R., Martin, L., et al. (2003, 2005). Development of the NOTECHS (non-technical skills) system for assessing pilots' CRM skills. In D. Harris & H. C. Muir (Eds.), *Contemporary issues in human factors and aviation safety* (pp. 133-154). Burlington, VT: Ashgate Publishing Co.

Go, T. H., Burki-Cohen, J., et al. (2003). The effects of enhanced hexapod motion on airline pilot recurrent training and evaluation. In *Proceedings of American Institute of Aeronautics and Astronautics AIAA Modeling and Simulation Technical Conference*. Dallas, TX: AIAA

Hamman, W. R., Seamster, T. L., Lofaro, R. J., & Smith, K. M. (1992). The future of LOFT scenario design and validation. *Proceedings of the Seventh International Symposium on Aviation Psychology*. Columbus, OH: ISAP

Jensen, R. S. (1982). Pilot judgment: Training and evaluation. *Human Factors, 2*.

Lofaro, R. J. (1992). A small group Delphi paradigm. *The Human Factors Society Bulletin, 35*(2).

Lofaro, R. J. (1992). *Workshop on Integrated Crew Resource Management (CRM)*. DOT/FAA/RD-95/5. Springfield, VA: National Technical Information Service.

Lofaro, R. J., Gibb, G. D., & Garland, D. (1994). *A protocol for selecting airline passenger baggage screeners. DOT/FAA/CT-94/110*. Springfield, VA: National Technical Information Service.

Lofaro, R. J., & Smith, K. M. (1993). The role of LOFT in CRM integration. In the *Proceedings of the Seventh International Symposium of Aviation Psychologists*. Columbus, OH: ISAP.

Lofaro, R. J., & Smith, K. M. (1998). *Rising risk? Rising safety? The millennium and air travel. Transportation Law Journal, 25(2)*. Denver, CO: University of Denver Press.

Lofaro, R. J., & Smith, K. M. (1999). Operational decision-making (ODM) and risk management (RM): Rising risk, the critical mission factors and training. In *Proceedings of Tenth International Symposium on Aviation Psychology*. Columbus, OH: ISAP.

Lofaro, R. J., & Smith, K. M. (2000). Operational decision-making: Integrating new concepts into the paradigm. In *Proceedings of Eleventh International Symposium on Aviation Psychology*. Columbus, OH: ISAP.

Lofaro, R. J., & Smith, K. M. (2008). Flight simulators and training. In Vincenzi, D. A., Mouloua, M., Hancock, P. A., & Wise, J. A. (Eds.), *Human factors in training and simulation* (pp. 257–286). Philadelphia, PA: Taylor and Francis.

Lofaro, R. J., & Smith, K. M. (2010). Collaborative risk identification and management between cockpit and dispatch: Operational decision-making. In *HCI-Aero 2010 Conference Proceedings*.

Maliko-Abraham, H., & Lofaro, R. J. (2003). Usability testing: Lessons learned and methodology. In *Proceedings of the 2003 International Military Testing Association (IMTA) Conference* (pp. 1-5). Retrieved from www.International IMTA.org

Maurino, D. (1999). Crew resource management: A time for reflection. In Garland, D., Wise, J., & Hopkin, V. D. (Eds.), *Handbook of aviation human factors* (pp. 215–234). Mahwah, NJ: Lawrence Erlbaum Publishers.

North, D. M. (2000). Finding common ground in envelope protection systems. *Aviation Week & Space Technology, 153*(6), 23–24.

NTSB. (2007, February 10). *AAR-07/06*.

NTSB. (2009, February 12). *AAR-10/11*.

Ruffell-Smith, H. S. (1979). *A simulator study of the interactions of pilot workload with errors, vigilance and decisions*. NASA Technical Memorandum 78482. Washington, DC: NASA Scientific and Technical Information Office

Smith, K. M., & Hastie, R. (1992). Airworthiness as a design strategy. In *Proceedings of the FSI Air Safety Symposium*, San Diego, CA.

Smith, K. M., & Lofaro, R. J. (2001). A paradigm for developing operational decision-making (ODM). In *Proceedings of 2001 SAE World Aviation Congress (WAC) Conference*. Seattle, WA: SAE WAC.

Smith, K. M., & Lofaro, R. J. (2003). The finalized paradigm for operational decision-making (ODM) paradigm: Components and placement. In *Proceedings of the 12ᵗʰ International Symposium on Aviation Psychology*. Dayton, OH: ISAP.

Smith, K. M., & Lofaro, R. J. (2009) Professionalism in airline operations…and accident investigations? In *Proceedings of the 15ᵗʰ International Symposium on Aviation Psychology*, Dayton, OH: ISAP.

Sprogis, H. (1997). Is the aviation industry expressing CRM failure? In *Proceedings of Ninth International Symposium on Aviation Psychology*. Columbus, OH: ISAP.

ADDITIONAL READING

Lofaro, R.J., Adams, R.J., and C.N. (1992). Workshop on Aeronautical Decision-Making (ADM) DOT/FAA/RD-92/14; Vol. I, II. Springfield, VA. 22161: National Technical Information Service

U.S. Federal Aviation Administration Advisory Circulars (AC): AC 120-40B: Airplane Simulator Qualification (7/29/91)

AC 120-45A: Airplane Flight Training Device Qualification (2/5/92)

Article in *Aviation Week and Space Technology,* 145(10), 15

Series of articles in *Aviation Week and Space Technology,* 172 (22), 66 ff.

Various articles in *Aviation Week and Space Technology,* 15 (3), 58-63.

Note: Smith and Lofaro also did a one-day Workshop: Operational Decision-Making in Accident Investigation and Prevention with accompanying Workbook

KEY TERMS AND DEFINITIONS

Accident Investigation (Airplane): The procedures followed by an accident investigation team to determine the most probable causal factors in an aircraft incident of accident. The investigation will result on a final report detailing all findings of the team

Aircraft Type-Rating: FAA certification that, after appropriate training and evaluation, a pilot is qualified to fly a particular aircraft; i.e., type-rated in that particular aircraft. Type-rating goes far beyond initial training and pilot licensure.

Category I,II, II: There are three categories of instrument landing System (ILS) which support similarly named categories of approach and landing operation. **Category I (CAT I)** – A precision instrument approach and landing with a decision height not lower than 200 feet above touchdown zone elevation and with either a visibility not less than 2,625 ft or a runway visual range (RVR) not less than 1,804 ft. FAA Order 8400.13D allows for special authorization of CAT I ILS approaches to a decision height of 150 ft above touchdown, and a runway visual range as low as 1400 ft. The aircraft and crew must be approved for CAT II operations, and a heads up display in CAT II or III mode must be used to the decision height. CAT II/III missed approach criteria applies. **Category II (CAT II)** – Category II operation: A precision instrument approach and landing with a decision height lower than 200 feet above touchdown zone elevation but not lower than 100 feet and a runway visual range not less than 984 ft for aircraft category A, B, C and not less than 1,148 ft for aircraft category D. **Category III (CAT III)** is further subdivided into **Category III A** – A precision instrument approach and landing with a) a decision height lower than 100 feet above touchdown zone elevation, or no decision height (alert height) and b) a runway visual range not less than 656 ft. **Category III B** – A precision instrument approach and landing with: a) a decision height lower than 50 feet above touchdown zone elevation, or no decision height (alert height) and b) a runway visual range (RVR) less than 656 ft but not less than 246 ft. Autopilot is used until taxi-speed. In the United States, FAA criteria for CAT III B runway visual range allows readings as low as 150 ft. **Category III C** – A precision instrument approach and landing with no decision height and no runway visual range limitations. Cat. IIIC approach is termed a zero-zero approach. Aside: Aircraft Categories A through D are often referred to as aircraft performance categories. Technically they're not performance categories, they're approach categories.

Cockpit Displays: All the CRT and LCD display screens, as well as dials and gauges in the instrument panel of an aircraft. Together, these provide pilot and crew with the information needed for a safe flight.

Decision Making: The process of the identification of alternatives for a course of action. this is followed by the selection of one of the alternatives and resultant action

FAA Advisory Circular (AC): A series of documents issued by the FAA which function as an aircarrier road-map, a "how to" for specific programs or addressing specific issues. They often are issued in conjunction with an FAR, attempting to show how accomplish and be in compliance with that FAR. A key point is that an AC does not have the legal status and ramifications of an FAR; i.e., the FAA cannot hold an aircarrier or other entity legally liable for non-use or non-compliance with an AC. Therefore, AC's can function as "best practices" and guidelines without the burden of legal liability.

Flight Crew Training: All the information, practice and assessment given to a flightcrew, both initially and yearly (recurrent). to ensure their performance

"Flying the Line": This is how civil aviation pilots refer to their function/mission; sometime referred to as flying from Point A to Point B.

"Glass" (vs. "Steam") Cockpit Instrumentation: What is referred to nowadays as a glass cockpit is one where CRT/LCD instrument panel screens have incorporated and integrated the data from many of the dials and gauges of early cockpits...and thus replaced many of these dials and gauges. Included in glass cockpits may also be heads-up-displays (HUD's). "Steam" refers to older cockpit instrument panels with their myriads of dial and gauges, where much of the information they presented had to be mentally integrated and interpreted by the flightcrew. As an aside, the glass cockpit has led to the two-person crew being standard in many civil aircraft, as what was ordinarily the third person, the flight engineer, was no longer needed as the pilot/first officer can now see and monitor the data that was only displayed for the flight engineer.

Hull: As airline terminology is sometimes borrowed from naval terminology, it would seem that hull refers to the airplane's fuselage But, it actually refers to the entire plane; a hull loss means that the plane is beyond any repair.

LOFT: An acronym for line-oriented flight training. It is given to flight crews in an FAA-rated level D flight simulator. It is a realistic, pre-programmed scenario of a flight and is both practice and assessment of flight crew performance

Recurrent(cy) Training for Pilots: In the world of civilian aviation, the FAA requires yearly training for pilots, called ' 'recurrency training." This training can consist of ground school as well as practicing emergency maneuvers, flying a LOFT or other training in a FS.

The Operational Environment: the 3-D airspace in which an airplane is flying, as well as the conditions in the airspace. It is the most current view of the flight trajectory and can be used for a projected trajectory.

ENDNOTE

[1]   The views and opinions expressed in this paper are solely those of the authors. They do not necessarily represent the positions or policies of any private, public or governmental organizations.

Chapter 16
Augmentation Systems:
The Use of Global Positioning System (GPS) in Aviation

Mohammad S. Sharawi
King Fahd University of Petroleum and Minerals, Saudi Arabia

ABSTRACT

The global positioning satellite system (GPS) has been utilized for commercial use after the year 2000. Since then, GPS receivers have been integrated for accurate positing of ground as well as space vehicles. Almost all aircrafts nowadays rely on GPS based system for their take off, landing, and en-route navigation. Relying on GPS alone does note provide the meter level accuracy needed to guarantee safe operation of aircrafts. Thus several augmentation systems have been deployed worldwide to enhance the accuracy of the GPS system. Several augmentation systems that serve local as well as wide coverage areas are discussed in this chapter, specifically the LAAS system, the WAAS system as well as the EGNOS system. The architecture as well the performance metrics for each of these augmentation systems are presented and discussed.

INTRODUCTION

In the year 2000, selective availability (SA) was removed from the global positioning system (GPS) signals, and thus the position estimates on earth improved from several hundred meters to tens of meters. This significant improvement opened the door for a wide variety of civilian and commercial applications that can utilize GPS for navigation, location determination and tracking.

GPS relies on signals transmitted from visible satellites (at least 4) to estimate the location and time of user equipment on earth (or close to earth). It uses a carrier frequency of 1575.42 MHZ with a bandwidth of 2 MHz. The navigation data from each satellite (also called ephemeris) is spread with a unique code for each satellite to allow the receiver to identify the satellite being used. In the receiver, the spread GPS signal is de-spread with the same spreading code used by the satellite, and the signal is reconstructed again. The difference between the starting of the spreading code in the

DOI: 10.4018/978-1-60960-887-3.ch016

receiver and the one received gives an estimate of the time traveled, and thus the distance from the satellite. This ranging estimate is called the pseudorange (PR). The PR is not precise and needs refinement to provide more accurate location estimates. In addition, the GPS signal goes through the atmosphere, and thus suffers from Ionospheric and Tropospheric delays which add to the position estimate errors.

Also, the GPS signal at the receiver might get degraded because of its multipath component that represents an attenuated and delayed version of the original signal due to its reflection from the ground close to the receiver terminal. This multipath can degrade the location estimate significantly if not taken into account, especially in applications that involve human lives, such as avionics. Most of the delays and errors in the GPS estimate due to atmospheric, receiver or multipath can be compensated for and corrected for to achieve meter level accuracies. Some of the corrections can be performed using more complicated receiver models, while others rely on augmentation systems that correct for atmospheric errors utilizing what is called differential GPS (DGPS).

In DGPS, reference stations in known locations are used to calculate the differences between the true location and the one given by a GPS receiver. The difference is a common estimation error that can be used in the vicinity of the reference station to correct for common errors. The geographical area can be up to several kilometers. Thus, if these corrections can be broadcasted to users in the vicinity, the common errors seen by the users can be eliminated, and their position estimates will improve.

Augmentation systems for GPS utilize DGPS for correction creation, and then broadcasting the correction information for location and time back to users, aircrafts or vehicles. Several GPS augmentation systems have been proposed and implemented in the USA and Europe. In the USA, the Federal Aviation Administration (FAA) have implemented the local area augmentation system (LAAS) for airports to aid in precision approach and landing, and the wide area augmentation system (WAAS) to aid in aircraft navigation as well as precision approach of aircrafts to airports all over the continental USA. While in Europe the European geostationary navigation overlay service (EGNOS) is deployed to aid aircrafts in their navigation and precision approach and landing in Europe.

In this chapter, we will go over the architecture of the three augmentation systems; LAAS, WAAS and EGNOS, as well as discuss their features and accuracies.

THE LOCAL AREA AUGMENTATION SYSTEM (LAAS)

The Local Area Augmentation System (LAAS) is an augmentation between the satellite based Global Positioning System (GPS) service with ground based stations to provide accurate correcting information for aircraft landing and approach in airports. The initiative was proposed by the Federal Aviation Administration (FAA) in late 1990's. LAAS has to meet stringent requirements that will provide accuracy, integrity, continuity and availability of service for aircrafts during their final approaches to airports which require the greatest safety and reliability. It relies on local area differential GPS (DGPS) for horizontal and vertical position fixing that is broadcasted to aircrafts in the airport vicinity to enhance positioning accuracy.

LAAS has been deployed in several airports around the continental USA, and is currently under heavy investigation to meet the stringent requirements of the three categories of precision approach and landing in airports.

Architecture

The current LAAS architecture consists of three major segments: The ground station segment,

Figure 1. LAAS architecture components

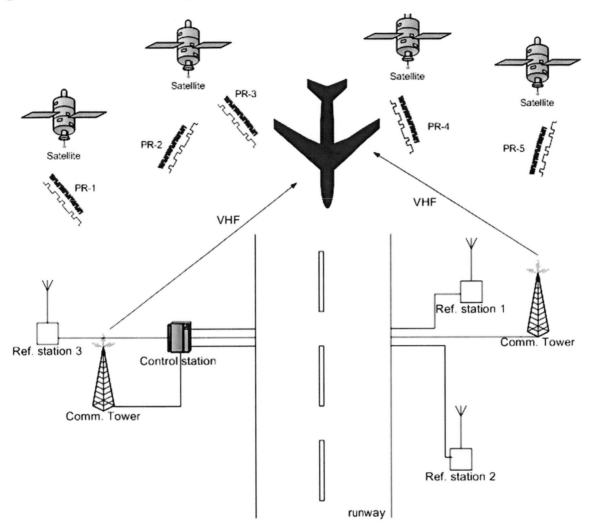

the aircraft segment and the satellite segment. The ground station segment includes the ground reference receiver stations as well as the reverse communication link via very high frequency (VHF) links. The aircraft segment includes the on board navigation systems that accept both GPS signals and corrections from ground stations via VHF links. Finally, the satellite segment includes the GPS satellite constellation transmitting in the Link 1 (L1) frequency band.

Figure 1 shows the various segments of a LAAS system in an airport environment. The visible satellites broadcast their GPS signals with their coarse/acquisition (C/A) codes. The aircraft as well as the ground station receivers can estimate their pseudoranges (PR) accordingly. The ground stations then compares the estimated ranges from the GPS signals with their actual locations and forms a differential error that is common to all signals in that local area. The ground reference stations then broadcast the error in the location estimates as corrections for the vehicles in their visible area. The approaching aircraft receives these differential GPS (DGPS) corrections via VHF links, and this will enhance their position estimates upon approach and landing.

Figure 2. LAAS ground station layout

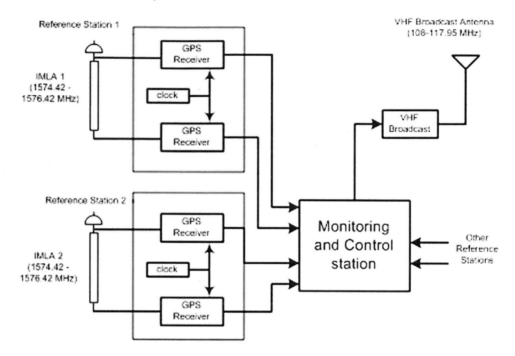

The satellite segment of LAAS in comprised of the visible satellites that broadcast the GPS ephemeris and navigation data to both the aircraft and the ground reference stations. Current GPS systems use the carrier frequency 1575.42 MHz (L1) with a navigation data period of 20 ms (i.e. 50 Hz). A spreading code that is specific to each satellite is used with a data rate of 1.023 MHz for the commercial standard positioning service (SPS). This code is called the C/A code.

The aircraft segment consists of the on aircraft GPS receiver and antennas as well as the VHF data link receiver and antennas that allow it to calculate its PR and then correct using the data received from the reference ground stations. To increase the availability of the system, the on aircraft receivers should be able to also accept signals received from airport pseudolites (APL). APLs are used in low visibility conditions (Bartone, 1997).

The ground segment is designed to meet the LAAS requirements, and can be altered to enhance the performance for meeting the Federal Aviation Administration (FAA) different approach and

landing category requirements. This segment consists of specialized antennas that will reduce the ground multipath errors that are not removed by DGPS. The specialized antenna used is called the integrated multipath limiting antenna (IMLA). This antenna covers from 0-90° in elevation angles and reduces the effect of ground multipath errors by 94% (Thornberg, 2003). It also includes several high end GPS receivers for range estimation.

Airport pseudolites are also part of the ground segment of the LAAS architecture. They are used when the number of visible satellites is not enough to obtain the PR due to bad weather conditions which might result in degradation in the system performance and position estimates. These APLs send GPS like signals to approaching aircraft but with different duty cycles and code rates compared to the satellite signals to avoid blocking them.

The ground segment also includes monitoring and control centers that perform differential corrections and send them via the VHF links back to approaching aircraft to enhance their position estimates upon runway approach. They also check signal quality by correlating received signals between adjacent stations, perform ephemeris checks, sigma monitoring and maintenance checks.

VHF links operate in the 108-117.95 MHz band that is divided into 25 KHz channels. Time division multiple access (TDMA) is used with 2 frames per second, each is divided into 8 time slots. Its message encapsulates the differential corrections along with integrity monitoring information that covers the approach and landing of aircraft.

Figure 2 shows the main components in a ground segment (station) with two reference stations connected to the control and monitoring unit. Note that the IMLA antenna consists of two antennas; the high zenith antenna and the multipath limiting antenna. More on these can be found in (Aloi, 2004).

Features and Performance Metrics

There are four major performance metrics in LAAS; Accuracy, availability, integrity and continuity. Accuracy is a measure of the difference between the estimated and actual position of the aircraft. Availability is the fraction of time for which the system is operational and is providing position corrections to incoming aircrafts with the required accuracy and integrity. While integrity and continuity characterize the system response to any failures or rare natural events, and thus address the ability of the navigation system to detect and possibly repair system failures (threats) in a timely fashion. This increases the safety measures and reliability of systems that deal with human lives.

Integrity fails if the system detects a position error that exceeds a predefined limit and the pilot or navigation guidance system is not alerted about this error in a timely manner. Continuity fails if the aircraft operation must be aborted for any unscheduled reason (Enge 1999; FRP 2001).

The FAA has setup three categories of operation for the LAAS architecture based on the aircraft height (also denoted as the decision height (DH)) and runway visual range (RVR). The three categories are called Category I (CAT-I), Category-II (CAT-II) or Category-III (CAT-III), with CAT-III being the most stringent in terms of accuracy, integrity, availability and continuity. CAT-I conditions exist when the DH is at least 200 feet and the RVR is more than 2400 feet. CAT-II conditions start when the DH is between 100 to 200 feet and the RVR is at least 1200 feet. Finally, CAT-III conditions exist when the aircraft is automatically landing using the flight guidance system.

In CAT-I precision approach, the lateral accuracy is 16 meters while the vertical accuracy is 4 meters 95% of the time. The availability of LAAS should be between 0.99 and 0.99999, but it is also airport dependent. Integrity is specified by two parameters or metrics; probability of hazardously misleading information (PHMI) and the time to alarm. In CAT-I, the PHMI is about 10^{-7} with a time to alarm of 6 seconds. For CAT-II and CAT-III, more stringent requirements are needed, and thus PHMI is 10^{-9} with time to alarm of 1 second. Table I summarizes the system performance metrics for the three LAAS operational categories (FAA 2002).

Although LAAS is based on DGPS, errors due to noise, interference and ground multipath are not correctable via differential corrections. These errors can be mitigated using special antennas and high quality GPS receivers. The variance of the noise error can be reduced using narrow correlators within the GPS receiver tracking loops. While ground multipath can be reduced using special antenna arrays such as the IMLA. The IMLA has a special radiation pattern that attenuates ground reflected waves by at least 30 dB more than the ones that arrive at positive elevation angles (i.e. with no ground reflection, direct from satellites). Innovative methods for interference mitigation for un-intentional and intentional radio frequency (RF) sources can be found in (Sharawi 2009).

DGPS can compensate for errors that are common within a coverage area of several 10's of

Table 1. LAAS performance metrics

	Vertical Accuracy (95%)	Horizontal Accuracy (95%)	integrity (PHMI)	time-to-alert	continuity risk	availability
CAT-I	6 - 4 m	16 m	2×10^{-7}	6 sec	8×10^{-6}	0.99 – 0.99999
CAT-II	2 m	6.9 m	10^{-9}	1 sec	4×10^{-6}	0.99 – 0.99999
CAT-III	2 m	6.2 m	10^{-9}	1 sec	2×10^{-6}	0.99 – 0.99999

kilometers, such as clock errors, ephemeris errors and atmospheric errors. But DGPS cannot correct for multipath or RF interference. LAAS has the following advantages over conventional DGPS:

- Higher accuracy: because LAAS equipment limits multipath errors which are considered the dominant source of errors by 94%.
- Integrity monitoring: because LAAS control stations incorporates methods for providing and monitoring the corrections sent via VHF links.
- Signal processing in the ground station relaxes the complexity of the aircraft equipment complexity.

THE WIDE AREA AUGMENTATION SYSTEM (WAAS)

The wide area augmentation system (WAAS) is an augmentation system for GPS. The system relies on geosynchronous earth orbit (GEO) satellites that broadcast GPS like signals to flying aircrafts to provide position fixes for GPS satellite locations and clocks. Also, they provide correction estimates for atmospheric delays that affect the ranging estimates from GPS signals. WAAS provides continental and national coverage, and it has been also initiated by the FAA in the USA. Currently, WAAS is already deployed in the USA (along with its counterpart that is called EGNOS in Europe, EGNOS will be discussed later in this chapter), and it is a part of the satellite based augmentation systems (SBAS).

Architecture

Figure 3 shows the architecture of WAAS. It consists of the visible GPS satellites in the coverage area (between 5-11 satellites out of 32 total), geosynchronous earth orbit satellites (GEOs), wide-area ground reference stations (WRS) and wide-area master reference stations (WMS). The true location of the satellite and indicated location signify the error in the satellite position, this is represented as a range error for the $\mathbf{x}$, $\mathbf{y}$ and $\mathbf{z}$ coordinates ($\Delta \mathbf{r}^k$) for the $\mathbf{k}^{th}$ satellite, as well as the clock error represented by $\Delta \mathbf{B}^k$.

The GPS satellites will broadcast the GPS navigation and ranging signals to the aircraft using the L1 frequency. The same ranging GPS signals will be received by the WRS. WRS are located in predefined locations and thus DGPS corrections are formed and passed to the WMS. These corrections include corrections for the satellite position as well as its clock frequency. A correction for each visible satellite is needed. The WRScontain dual frequency receivers that are used to estimate the atmospheric effects on the ranging GPS signals. More specifically, the link 2 (L2) frequency of GPS signals is utilized for this purpose. L2 has a center frequency of 1227.6 MHz. WRS generates the corrections for Ionospheric and Tropospheric delays for the service area, and these delay corrections are also sent to the WMS to be broadcasted to the GEO satellites.

The WMS forms a correction message and broadcasts it to the GEO satellites serving that area. Once the GEO satellites receive the correction information, it passes them to the aircrafts

Figure 3. WAAS architecture

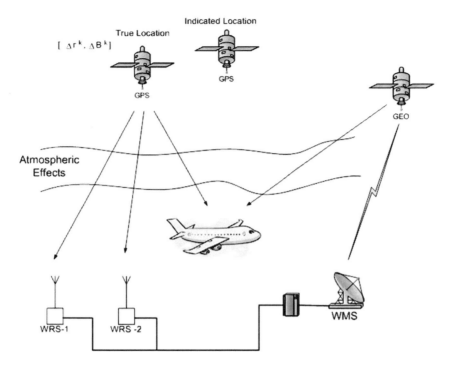

in its coverage area. These aircrafts receive the GEO signals using slightly modified GPS user equipment on board. The transmitting frequency of the GEO satellites is the same as that of the L1 GPS signal. If the WMS cannot form any of the corrections needed it does not send the correction, and it sends a "Do Not Use" flag for the problematic satellite. This is an integrity warning.

Features and Performance Metrics

WAAS will increase the accuracy over GPS only based positioning system in two main ways (Enge 1996):

1. It will minimize the range estimation error by providing differential corrections for each visible GPS satellite, thus decreasing the pseudorange error from about 30 m to about 2 m.
2. It improves the geometry by adding more ranging signals to the set of available GPS satellites.

WAAS will support both precision approach and non-precision approach modes of operation. In non-precision approach, the aircraft relies on a baro-altimeter for vertical position information and a ground based radio navigation aid for horizontal information. The later information is sent via a VHF radio from the airport facility, a distance measuring equipment, or an instrument landing system (ILS). Precision approach on the other hand, relies on GPS based services along with WAAS. The three categories defined for LAAS apply for WAAS precision approach and landing as well. But WAAS also covers aircraft oceanic en route operation and nation wide navigation.

In order for WAAS to provide the required availability, accuracy, integrity and continuity defined in the FAA specifications for CAT-I to CAT-III, complicated formulations are used to calculate differential position and clock corrections for each satellite along with atmospheric delays via dual frequency measurements. These formulations and methods will not be covered in

Table 2. WAAS Performance metrics

	Vertical Accuracy (95%)	horizontal accuracy (95%)	integrity (PHMI)	time-to-alert	continuity risk	availability
NPA	-	100 m	10^{-7}	8 sec	10^{-5}	0.99999
PA	7.6 m	7.6 m	10^{-7}	5.2 sec	-	0.999

this chapter and the reader is encouraged to check (Misra 2004).

For WAAS, there are two sets of performance metrics; one for non-precision approach (NPA) operation and the other for precision approach (PA). The later is more stringent than the former as more accuracy is needed when the aircraft is to rely on GPS and WAAS for airport approach operations. Table II summarizes the performance metrics of WAAS (FAA 1999). The availability of the NPA mode is higher than the PA mode, while the accuracy of the PA mode is more than its NPA counterpart. This is due to the fact that for precision approach to airport runways, the horizontal and vertical estimates that also translate to DH and RVR are extremely important for a safe landing operation.

THE EUROPEAN GEOSTATIONARY NAVIGATION OVERLAY SERVICE (EGNOS)

The European geostationary navigation overlay service (EGNOS) is a satellite based augmentation system (SBAS) that is intended to provide navigation support to aircrafts as well as ground users in Europe. The system is to support GPS, GLONASS as well as Galileo navigation services. EGNOS is supposed to provide vertical and horizontal accuracies of less than 7 m. EGNOS services have been available since 2009 for a user with an EGNOS enabled GPS receiver.

Architecture

The architecture of EGNOS is similar to that of WAAS presented in Figure 3. The user equipment (UE) (an aircraft or a ground user) receives ranging signals from one of the three supported global navigation satellite systems (GNSS); GPS, GLONASS or Galileo based on the receiver he/she uses. At the same time, the user receives corrections from GEO satellites that operate using the same frequencies as the receiver used with slight modifications (i.e. an EGNOS enabled GPS receiver). The UE should be capable of receiving correction information from the EGNOS system to enhance position estimation and time.

There are 3 GEO satellites used in the EGNOS system. These satellites broadcast corrected DGPS data, time and health information for the visible satellites. These corrections are created within the ground mission control stations (MCS) (4 of them distributed across Europe). The MCS receive the date from more than 30 ground ranging and integrity monitoring stations (RIMS) to create correction and health messages that are then uplinked to the GEO satellites for broadcasting to earth and space users (on ground users and aircrafts, respectively). Currently, corrections are broadcasted via the L1 frequency from GEO satellites to UE. EGNOS is also being utilized for marine time, rail road and ground vehicle traffic positioning.

Features and Performance Metrics

Unlike WAAS, EGNOS is supposed to provide several types of services to users. The open ser-

Table 3. EGNOS Performance metrics

	Vertical Accuracy (95%)	horizontal accuracy (95%)	integrity (PHMI)	time-to-alert	continuity risk	availability
APV-I	220 m	20 m	1-2 x10⁻⁷	10 sec	1-8 x10⁻⁶	0.99 - 0.99999
APV-II	16 m	8 m	1-2 x 10⁻⁷	6 sec	1-8 x10⁻⁶	0.99 - 0.99999

vice will be available for all users with EGNOS enabled UE for better position and time estimation in continental Europe. Also it will provide commercial and safety of life services. These extra services will be according to the service provider and will differ from one to another. But the fact that EGNOS will support ground users will allow it to have a huge impact on very accurate positioning and timing services that the commercial sector will benefit from.

For aircraft navigation and approach to airports, EGNOS will provide meter level accuracies in the two precision approach categories defined by the European Space Agency (ESA). Table III summarizes the features of the APV-I and APV-II precision approach categories.

It is evident that EGNOS will also have several advantages over using stand alone GPS receivers such as:

- More precise guidance for aircrafts under bad weather conditions
- Improved guidance during all phases of flight due to the availability of extra visible GEO satellites that provide both range and correcting signals.
- Improved collision avoidance with the ground due to the very accurate vertical and horizontal positioning that is not affected by barometric or weather conditions.

SUMMARY OF AUGMENTATION SYSTEM PERFORMANCE AND APPLICATIONS

In summary, LAAS is to be used in the vicinity of airports with limited coverage area compared to WAAS and EGNOS, but it will include the use of specialized antenna arrays to combat multipath that is a significant error contributor in this environment. Accuracies of LAAS range between 2 m to 7 m for vertical and horizontal accuracies for 95% of the time, while in WAAS and EGNOS, the accuracies are a little less and range between 8m and 20m. WAAS and EGNOS cover much wider areas and play an important role in both en route and precision approach modes.

The three systems have to satisfy certain availability, integrity and continuity requirements. They all have mechanisms to increase the system availability by either used airport pseudolites or use extra GEO satellites. They all convey corrections back to the aircraft either by a separate VHF radio link (LAAS), or by sending correcting information on the same carrier frequency of the GPS satellites (WAAS and EGNOS).

While LAAS and WAAS are specifically designed to operate with GPS, EGNOS is designed to also augment future navigation system constellations such as the Russian GLONASS system, or the European Galileo system. This will be part of the receiver equipment in the aircraft that has to support EGNOS functionality. Finally, EGNOS is the only augmentation system that will support both aircrafts and on ground vehicles such as marine applications, trains and regular vehicles.

WAAS and LAAS only support space vehicles and are not considered for ground applications.

FUTURE DIRECTIONS

There is a lot of work and research in the area of augmentation systems to enhance accuracy, availability, integrity and continuity of satellite based navigation. In addition, the GPS system is being modernized to include several transmissions that will enhance the position and time estimates on earth, more specifically, the introduction of the L2C (commercial L2 transmission) and the L5 (1176.45 MHz) carrier transmissions.

Also, several SBAS augmentation systems are being considered worldwide other than WAAS and EGNOS such as the Indian GPS aided Geo Augmented Navigation (GAGAN) system and the Japanese Multi-Functional Satellite Augmentation System (MSAS). Once these systems become operational, aircraft navigation and precision approach will be of greater accuracy and reliability, which should be the case in any equipment that deals with human lives.

CONCLUSION

For aviation applications, very accurate positioning and time are required for airplanes during their airport approach and landing. GPS based position and time estimates do not provide the meter level accuracies required for safe and accurate aircraft landing. Augmentation systems are systems that enhance the accuracy, integrity, availability and continuity of GPS transmissions to aircrafts. They provide position corrections and provide meter level accuracies when combined with GPS signals. This chapter touches upon the different augmentation systems being deployed worldwide with the emphasis on their architectures as well as performance parameters and metrics. Specifically, this chapter discusses the LAAS, WAAS and EGNOS augmentation systems that are currently deployed in the USA and Europe.

REFERENCES

Aloi, D., Sleity, M., & Kiran, S. (2004). Analysis of LAAS integrated multipath limiting antenna using high fidelity electromagnetic models. *Proceedings of the Institute of Navigation Annual Meeting*, (pp. 441-470).

Bartone, C., & Van Graas, F. (1997). *Airport pseudolite for precision approach applications* (pp. 1841–1850). Institute of Navigation GPS Proceedings.

Department of Defense and Department of Transportation. (2001). *Federal radionavigation plan.* Technical Specification (FRP) DOT-VNTSC-RSPA-01-3/DOD-4650.5

Enge, P. (1999). Local area augmentation of GPS for the precision approach of aircraft. *Proceedings of the IEEE, 87*(1), 111–132. doi:10.1109/5.736345

Enge, P. (1996). Wide area augmentation of the global positioning system. *Proceedings of the IEEE, 84*(8), 1063–1088. doi:10.1109/5.533954

Federal Aviation Administration (FAA). (1999). *Federal Aviation Administration specification for the wide area augmentation system* (WAAS). (DTFA01-96-C-00025, FAA-E- 2892b).

Federal Aviation Administration (FAA). (2002). *Navigation and landing transition strategy.* FAA Office of Architecture and Investment Analysis, ASD-1.

Lyon, A., Westbrook, J., & Guida, U. (2005). *Operating EGNOS. Proceedings of the 18th International Technical Meeting of the Satellite Division of the Institute of Navigation* (ION GNSS 2005), California, USA, (pp. 419-423).

Misra, P., & Enge, P. (2004). *Global positioning system: Signals, measurements and performance.* Lincoln, MA: Ganga-Jamuna Press.

Sharawi, M. S. (2009). *Interference mitigation for local area augmentation navigation systems, methods and analysis.* Saarbrucken, Germany: VDM Verlag.

Thornberg, D. B. (2003). LAAS integrated multipath limiting antenna. *Navigation: Journal of the Institute of Navigation, 50*(2), 117–130.

Chapter 17
Applying the Certification's Standards to the Simulation Study Steps

Issam Al Hadid
Isra University, Jordan

ABSTRACT

This chapter presents the certification standards applied with the simulation study steps, in addition to the Confidence Grid which is used to assets the quality (Reliability and Accuracy) of the data and the process of the simulation study step which will be the base for the validation and verification.

INTRODUCTION

Simulation certification is one of the most important tasks that should be applied to ensure the simulation Model's credibility. The Certification process is used to ensure the target behavior and the expected characteristics of the simulation Model (SM) are achieved. It should be integrated within the development process of the SM to make sure that the entities of each process are certificate and fulfills the expected target. The process entities include Input Data, Process and Output Data, in addition to the over all certification for the system as a whole.

The certification has been defined by the International Organization for Standardization (ISO) as follows (Balci and Saadi 2002):

Certification is a procedure by which a third party gives written assurance that a product, process, or service conforms to specified characteristics.

On the other hand, according to wikipedia-1 (2009)

Certification refers to the confirmation of certain characteristics of an object, person, or organization. This confirmation is often, but not always provided by some form of external review, education, or assessment.

DOI: 10.4018/978-1-60960-887-3.ch017

This definition in general can be applied anywhere to check if the product quality satisfies the requested, expected and the unexpected requirements, characteristics and behavior. Certification should be applied to all the steps; which means that a concurrent process should be run during the development of the SM steps which includes Verification and Validation V&V to make sure the product quality possesses the desired set of characteristics. In this chapter we will present the applying of the certification standards to the simulation study steps in addition to the confidence Grid as a technique to assess the quality of the processes and data based on its reliability and accuracy.

BACKGROUND

What is the Simulation?

According to Pidd (20004) simulation's definition is the development of a model which is:

an unambiguous statement of the way in which the various components of the system interact to produce the behavior of the system. Once the model has been translated into a computer program the high speed of the computer allows a simulation of, say, six months in a few moments. The simulation could also be repeated with various factors at different levels.

Chung (2004) defined the simulation as:

the process of creating and experimenting with a computerized mathematical model of a physical system.

On the other hand, Abu-Taieh (2004) has quoted El Sheikh Simulation definition as:

Simulation is the use of a model to represent over time essential characteristics of a system under study,

which is the most comprehensive and complete definition.

Why Do We Need the Simulation?

Instead of using direct experiments using real model, we can use the simulation model for the following reasons:

- Cost: real experiments might be expensive; such as materials and constriction time.
- Time: using simulation we can simulate days, weeks, months, or even years in few seconds which is not available in the direct experimentation.
- Replication (Re-Runs): using the simulation model we can repeat the experiment with different scenarios which allow testing all the aspects that might affect the behavior of the system.
- Safety and legality: according to Pidd (2004), he claimed that "one of the objectives of simulation study may estimate the effect of extreme condition, and to do this in real life may be dangerous or even illegal."

On the other hand, the mathematical models will not represent the dynamic effects and the behavior of the system. Due to that the simulation model will be the best choice because it provides a study of all dynamic effects during the operation instead of values and averages in the mathematical models. Furthermore, as Pidd (2004) claimed:

Is it possible to sample from non-standard probability distributions in a simulation model. However, queuing theory models permit only certain distributions and therefore cannot cope with many types of problems.

According to Chung (2004) the benefits of simulation can be described as follows:

- *Gaining insight into the operation of a system*
- *Developing operating or resource policies to improve system performance*
- *Testing new concepts and/or systems before implementation*
- *Gaining information without disturbing the actual system*
- *Experimentation in compressed time*
- *Reduced analytic requirements*
- *Easily demonstrated models*

What is Certification?

Certification has been defined by the International Organization for Standardization (ISO) as the following (Balci and Saadi 2002):

Certification is a procedure by which a third party gives written assurance that a product, process, or service conforms to specified characteristics.

On the other hand, according to wikipedia-1 (2009):

Certification refers to the confirmation of certain characteristics of an object, person, or organization. This confirmation is often, but not always, provided by some form of external review, education, or assessment.

What is Simulation Study?

The simulation study is the framework that manages simulation activities including process and data (data collected and data content between the processes) and the experimentation.

Figure 1. Quality of the three Ps

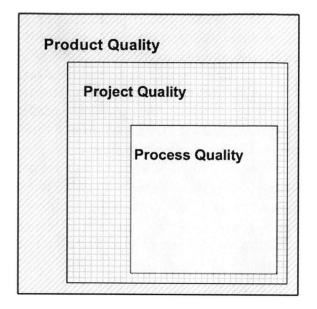

What is the Confidence Grid?

The Confidence Grid (C.G.) is an alpha technique which is utilized to describe the quality of the data and the process using a predefined acceptance level for the Reliability and Accuracy of the Data and the process which can be used as a base for the Validation and Verification processes.

Why Simulation Study and C.G.?

The simulation study (analysis, design, modeling and experimentation) is a project which needs to be managed through all the study steps (scope, communication, time, budget, change and risk. In addition to the quality assurance of the simulation study processes, input data and exchanged data between the processes.

SM CERTIFICATION

From the ISO's certification definition, the SM should be given the assurance by a third party, so the first and the second parties will be the SM

Figure 2. The evaluation process

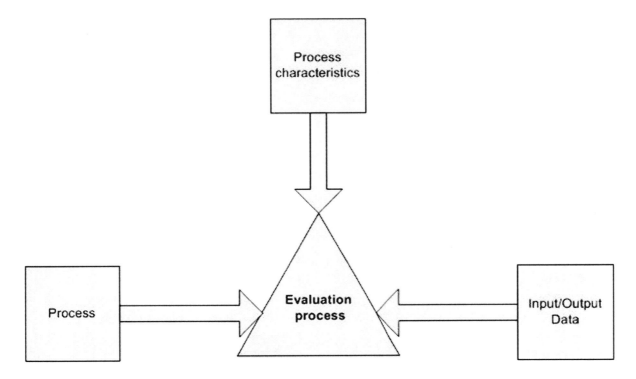

sponsor and the SM developer. Balci and Saadi (2002) have claimed that the third party should have a formal recognition from an accreditation authority. This will check that Certification agent maturity includes the process and the team who will conduct the process. It should also be fully independent (technically, managerial and financial), Balci and Saadi (2002).

A research by Balci and Saadi (2002) showed that to certify the SM we have to check the quality of the three Ps (Product, Project and Process) as shown in Figure 1.

- Process Quality: Balci and Saadi (2002) claimed that the process quality is *"the degree to which the process holds a desired set of characteristics"*.
- Project Quality: Balci and Saadi (2002) claimed that the Project quality is *"the degree to which the Project acquires a desired set of characteristics"*.

- Product Quality: Balci and Saadi (2002) claimed that the Product quality is *"the degree to which the Product possesses a desired set of characteristics"*.

As shown in Figure 2, the three Ps quality assurances depend on the V&V processes which include the evaluation for the Input/Output Data, The process itself, and the related process's characteristics.

The Verification of the SM is the process and steps we use to make sure we are building the SM right. On the other hand, the validation of the SM is the process we use to make sure we are building the right SM.

In my opinion, the certification of the SM is the process of giving the approval and the assurance by a third independent party that the SM fulfills the validation and verification for the requirements, specifications, characteristics and behavior. It satisfies the expected certification's target for each process and for the system as a whole. According to Balci (1994), the credibility assessment is situation dependent, so we need a methodology to give the certification of the SM a degree as a target to the approval and the certification.

CONFIDENCE GRADE

According to Alegre, et al. (2000), the Confidence Grade (C.G.) is a coding technique that uses the alpha numeric code to describe the reliability and accuracy of the data or process. (A1; for the best input/output data or the best process's behavior and characteristics, D6 for the worst) where the letter refers to the reliability and the number refers to its accuracy.

The technique is based on two bands to specify and describe the quality:

1- Reliability Band: according to Winschiers and Paterson (2004) the probability that process behavior and characteristics will perform the required function without failure for a predefined period of time under satisfied conditions. In addition to the dependable of the input/output data.

According the Alegre, et al. (2000) the Reliability Bands are:

A. *Highly Reliable: Data based on the best available methods.*
B. *Reliable: as in band A, but with minor shortcomings.*
C. *Unreliable: Extrapolation from limited samples.*
D. *Highly Unreliable: Unconfirmed verbal reports or cursory inspection or analysis.*

2- Accuracy Band: H. Alegre, et al. (2000) claimed that it is the "approximation between the results and the correct values". This represents the quality and degree of conformity of being close to the data actual value or process expected behavior.

Alegre, et al. (2000) suggested the Accuracy Bands based on the intervals, as follows:

[0; 1]: Better than or equal +/- 1%.
[1; 5]: Not band 1, but better than or equal +/-5%.
[5; 10]: Not band 1 or 2, but better than or equal +/-10%.
[10; 25]: Not band 1, 2 or 3, but better than or equal +/-25%.
[25; 50]: Not band 1, 2, 3 or 4, but better than or equal +/-50%.
[50; 100]: Not band 1,2,3,4 or 5, but better than or equal +/-100%.
Values are out of the valid range; such as >100%

For example, A2 is the C.G. for the input data. It means the data based on the best available methods (high reliable band A) which is estimated to be within +/- 5% (Accuracy band 2).

Figure 4 shows the C.G. Matrix to evaluate the reliability and accuracy bands suggested by the author.

Applying the C.G. schema to the SM Certification process as the target for the V&V processes would increase the degree of the reliability and accuracy for each step of the SM's study and the data that will be collected (Input/Output Data). Therefore the C.G. target should be assigned by a third party. Figure 3 shows the C.G. assessment process.

SIMULATION STUDY STEPS

J. Banks et al. (2005) have suggested in their book that the simulation study consists of 12 steps. We will avoid the 6th and 7th steps which are Verified and validate for the following reasons:

• The Verification process runs only after the model translate process to check the computer program performance, the selected input parameters and the conceptual model.
• The validation process runs only after the verification process to check the models results, behavior and characteristics. It also

Table 1. Confidence grade matrix

Confidence Grade			
	Reliability Band	**Accuracy Band**	**Result**
Case 1	✓	✓	Confidence Grade is Achieved
Case 2	✓	✗	Accuracy Band is not achieved
Case 3	✗	✓	Reliability Band is not achieved
Case4	✗	✗	Confidence Grade is not Achieved

Where:

✓ (Achieved); equal or more than the Target

✗ (Not Achieved); less than the target

Figure 3. The C.G. assessment process

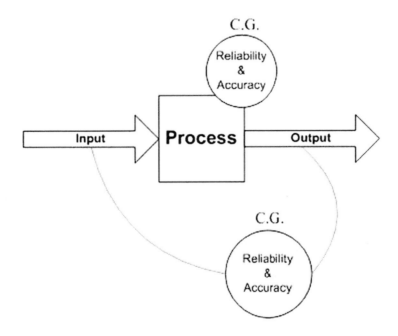

checks the achieved objectives of the project plan.

As mentioned earlier, the certification is given to the process in case the process achieved the V&V C.G.'s acceptance level (target), so we have to run he V&V processes in concurrent with each phase to check the certification for the process.

Figure 4 shows all the SM steps with the certification (V&V) methodology.

Problem Formulation

We use this process to identify and describe the real problem to make sure it's clearly understood. In this process The C.G. should be used to evaluate the reliability and accuracy of the statement of the problem.

- Verification: The C.G. target is achieved if we describe the problem right.

Figure 4. SM study steps with certification (V&V) methodology

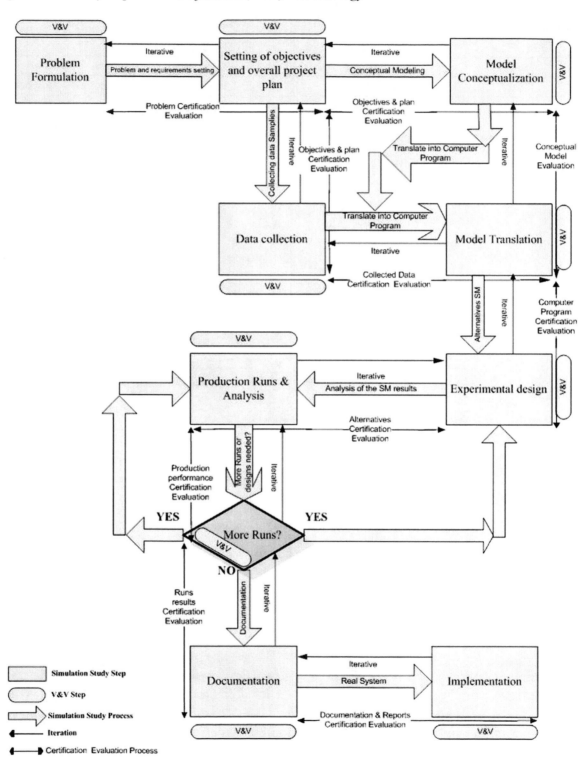

- Validation: The C.G. target is achieved if we describe the right problem.

Pressman (2005) claimed that to understand the problem we have to know:

Who has a stake in the solution to the problem? That is, who are the stakeholders? What are the unknowns? What data, functions, features, and behavior are required to properly solve the problem? Can the problem be compartmentalized? Is it possible to represent smaller problems that may be easier to understand? Can the problem be represented graphically? Can an analysis model be created?

Setting of Objectives and Overall Project Plan

Setting objectives that can be represented as questions that will be answered by simulation. At this point we should decide if the simulation is a suitable methodology that can be used based on the problem at hand and objectives that have been identified.

We use this process to specify the objectives and properties that the SM most or should satisfy, it can be mandatory, desired or optional.

Balci et al. (2000) stated that the project plan is used to:

- Manage the resources efficiently (including the V&V resources).
- Control the development process (including the V&V process).
- Identification for the responsibilities and the participant's roles.

In this process the C.G. evaluates the reliability and accuracy of the objectives, project plan, and their acceptance levels. Also, other alternative models that should be considered in the plan, and other methods for evaluating the effectiveness of these alternatives models.

- Verification: The C.G. target is achieved if we:
  - Set the objectives and project plan right.
  - Consider the alternative models and methods to evaluate its effectiveness accurately.
- Validation: The C.G. target is achieved if we:
  - Set the right objectives and project plan.
  - Consider the right alternative models and right methods to evaluate its effectiveness

Model Conceptualization

The conceptual model is a blueprint to the real system, it's used to address simulation framework, validate the requirements, and make sure it represents real world problems. According to Balci (2003), it can be used:

- As a tool to control, assess, and improve the SM requirements.
- Evaluate SM content errors.
- As a foundation for SM design
- To verify and validate the SM design.

The Development of the conceptual model is considered as the base step to develop more detailed models. Also, it's used to validate the real world problem which is represented using the conceptual model before the code translation process to make sure we are solving the right problem.

In this process we use the C.G. to evaluate the reliability and accuracy of the following process showing in Figure 5.

- Verification: The C.G. target is achieved if:
  - The conceptual model represents the real world problem right.

Figure 5. Conceptual design process

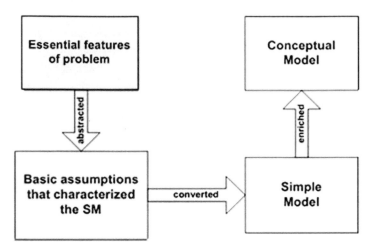

- We abstract the basic assumptions that characterized the SM from the essential features of the problem right.
- We convert basic assumptions into simple system right.
- We enrich the simple model into complex conceptual model right.
- Validation: The C.G. target is achieved if:
  - The right Conceptual model represents the real world problem.
  - The right basic assumptions that characterized the SM have been abstracted from the essential features of the problem.
  - The right simple model has been converted from basic assumptions.
  - The right complex conceptual model has been enriched from the simple model.

Data Collection

In this process the C.G. should be used to evaluate the reliability and accuracy of the collected data that will be used as the input parameters.

- Verification: The C.G. target is achieved if we:

- Select the data types right
- Select the approaches that will be used to collect the data right.
- Select the samples types right
- Identify ranges of the samples right; if random numbers will be used.
- Validation: The C.G. target is achieved if we:
  - Select the right data types.
  - Select the right approaches that will be used to collect the data.
  - Select the right samples types
  - Identify the right ranges of the samples; if a random numbers will be used.

J. Banks et al. (2005) suggested that Uniformity and independence are the desirable properties of the random number. Therefore we need a methodology to test generated random numbers if they satisfy the specifications mentioned before.

Model Translation

Figure 6 shows that the conceptual model has been translated into an executable program, a general purpose programming language can be used in the translation process (C++, JAVA,

Figure 6. Model translation

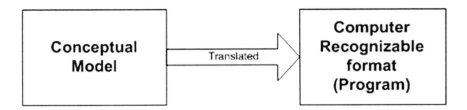

C#, VB.NET, etc…), or one of the commercial simulation packages (Simul8, Witness, Flexsim, Automod, etc…). If the problem can be solved using the existing modules or some of them; this will affect the development time by reducing the time of development and testing in addition to the extra feature that the existing modules might include which will enrich the results and the model options.

The C.G. should be used to evaluate the reliability and the accuracy of the computer program (simulation program)

- Verification: The C.G. target is achieved if the:
  - Computer program performs right.
  - Input parameters and logical structure of the model are represented in the computer program right.
- Validation: The C.G. target is achieved if the:
  - Appropriate performance of the computer program is achieved.
  - Right Input parameters and logical structure of the model are represented in the program.

J. Banks et al. (2005) claimed that if the SM is an animated model, it has to be compared with the real system to check if it satisfies the certification target which will include the conceptual model and the system requirements.

Also, J. Banks et al. (2005) said in their book that the certification process for the model translation can be conducted using Interactive Run Controller (IRC) or debugger for the following reasons:

- Observing the values of the variables, attributes, etc…
- Temporarily pause the Simulation
- Monitor the SM progress.
- Trace maybe focused on a function, procedure or even a piece of code.

Experimental Design

In this phase, we decide how long the simulation run length will take, the number of the simulation run replications, and which alternatives we need to use and how long it will be run. The C.G. should be used to evaluate the reliability and accuracy of the simulation runs, and the alternatives SM that will be used.

- Verification: The C.G. target is achieved if we:
  - Select right length of the simulation run.
  - Select the number of simulation run replications right.
  - Select the alternatives to be simulated and how long it should be run right.
- Validation: The C.G. target is achieved if we:

- Select the right period for the length of the simulation.
- Select the right number of simulation run replications.
- Select the right alternatives to be simulated and the right period for the length it should be run.

Production Runs and Analysis

The process of analyzing the SM runs results. In this process the C.G. should be used to evaluate the reliability and accuracy of the performance measurements and the analysis of the output results of the SM run.

- Verification: The C.G. target is achieved if we:
  - Estimate the performance measurements of the SM right.
  - Analyze the output results of the SM run right.
- Validation: The C.G. target is achieved if we:
  - Estimate the right performance measurements of the SM.
  - Analyze the right output results of the SM run.

Balci and Saadi (2002) suggested that the SM output runs needs to be interpreted so the stakeholders and the decision maker could understand it.

More Runs

After the analysis process for the output results, we have to decide if more experiments or other alternatives are required to run. In this process the C.G. should be used to evaluate the reliability and accuracy of the output results of the SM and if we need additional parameters or designs.

- Verification: The C.G. target is achieved if we:
  - Analyze the output results of the SM right
  - Run the SM using additional parameters or designs right.
- Validation: The C.G. target is achieved if we:
  - Analyze the right output results of the SM.
  - Run the SM using the right additional parameters or designs.

Documentation

In this process the C.G. is used to evaluate the reliability and accuracy of the documentation and reports.

- Verification: The C.G. target is achieved if the:
  - SM processes documented right.
  - Progress (History of simulation) is documented right.
- Validation: The C.G. target is achieved if the:
  - Right processes are documented.
  - Right progress (History of simulation) is documented.

J. Banks et al. (2005) mentioned in their book that the V&V for the documentation and reports phase is important and essential in the certification process. It includes the documentation of all the development life cycle phases in addition to:

- The definition of all variables and parameters.
- The description and comments about the operational model (as sections and as one unit).
- The history of the progress of the SM and the results, in addition to the trace for each/some output(s).

Figure 7. Converting the SM to the real system

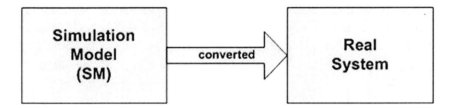

Documentation might be used by different users such as developers, testers and the end users, so it must describe the aspects of the SM from all point of views including data, code, and how it can be used.

SM's Documentation must include the following:

- Statement of the real world problem definition
- Objectives, project plan and the requirements of the SM.
- Principles used in construction of the conceptual model and the tools used to control the SM and improve it.
- C.G.'s acceptance levels (targets) for the processes and data.
- Collected data which will be used as the input of the SM including the data types, approaches used to collect it and the select samples types.
- Steps of translation process from the conceptual model into the executable program including the selected programming language or simulation package and why it has been chosen.
- Experimentation design including the run length, replications and the alternatives.
- Results of the runs including the performance measures and the output of the SM.

This phase will make the SM much simpler to understand, clarifying the SM and the phases of the development life cycle.

Implementation

Implementation is the process of converting the SM to the Real system, Figure 7.

In this process the C.G. should be used to evaluate the reliability and accuracy of the real system.

- Verification: The C.G. target is achieved if:
  ○ We convert the SM to the Real System Right.
- Validation: The C.G. target is achieved if:
  ○ We convert the SM to the Right Real System.

CONCLUSION

As we can see, the certification process for the SM is not an easy process, and it should not be performed using the ad hoc manner, it should be planned, scheduled, and run concurrently within the development life cycle for the simulation model.

However, we might face some barriers during the certification process such;

- A conflict between the Reliability and the Accuracy.
- Some processes are hard to estimate their reliability or accuracy.
- According to Balci and Saadi (2002), some developers may not in corporate or provide the information to the Certification Agent.

Using the C.G. as a unified approach to evaluate the certification of the SM will make it easy to measure the Reliability and accuracy of the Simulation study steps and the Simulation Model as a whole.

ACKNOWLEDGMENT

The author thanks Dr. Evon Abu Taieh for the assistant and support provided by her.

REFERENCES

Abu-Taieh, E. (2004). *Computer simulation with enhanced three-phase modeling using the activity cycle diagram.* Information Resources Management Association International Conference, New Orleans, Louisiana, USA, May 23-26 2004.

Alegre, H., Hirner, W., Baptista, J. M., & Parena, R. (2000). *Performance indicators for water supply services.*

Balci, O. (1994). Validation, verification, and testing techniques throughout the life cycle of a simulation study. In J. D. Tew, S. Manivannan, D. A. Sadowski & A. F. Selia (Eds.), *Proceedings of the 1994 Winter Simulation Conference.*

Balci, O. (1995). Principles and techniques of simulation validation, verification, and testing. In C. Alexopoulos, K. Kang, W. R. Lilegdon, & D. Goldsman (Eds.), *Proceedings of the 1995 Winter Simulation Conference.*

Balci, O. (2003). Verification, validation and certification of modeling and simulation application. In S. Chick, P. J. Sanchez, D. Ferrin, & D. J. Morrice (Eds.), *Proceedings of the 2003 Winter Simulation Conference.*

Balci, O., Carr, J., Ormsby, W., & Saadi, S. (2000). Planning for verification, validation, and accreditation of modeling and simulation applications. In J. A. Joines, R. R. Barton, K. Kang, & P. A. Fishwick (Eds.), *Proceedings of the 2000 Winter Simulation Conference.*

Balci, O., & Saadi, S. (2002). Proposed standard processes for certification of modeling and simulation applications. In E. Yücesan, C.-H. Chen, J. L. Snowdon, & J. M. Charnes (Eds.), *Proceedings of the 2002 Winter Simulation Conference.*

Banks, J., Carson, J. S., Nelson, B. L., & Nicol, D. M. (2005). *Discrete-event system simulation* (4th ed.). Upper Saddle River, NJ: Prentice-Hall.

Chung, C. (2004). *Simulation modeling handbook-A practical approach* (pp. 16–18). CRC Press.

Iba, T., Matsuzawa, Y., & Aoyama, N. (2004). *From conceptual models to simulation models: Model driven development of agent-based simulations.* 9th Workshop on Economics and Heterogeneous Interacting Agents. Kyoto, Japan.

Konaté, D. (2008). Mathematical modeling, simulation, visualization and e-learning. *Proceedings of an International Workshop held at Rockefeller Foundation's Bellagio Conference Center*, Milan, Italy, 2006. New York, NY: Springer. ISBN: 978-3-540-74338-5

Law, A. (2003). How to conduct a successful simulation study. In S. Chick, P. J. Sanchez, D. Ferrin, & D. J. Morrice (Eds.), Proceedings of the 2003 Winter Simulation Conference.

Pace, D. (2000). Ideas about simulation conceptual model development. *Johns Hopkins APL Technical Digest, 21*(3), 327.

Pidd, M. (2004). *Computer simulation in management science* (4th ed.). England: John Wiley and Sons.

Pressman, R. (2005). *Software engineering: A practitioner's approach* (6th ed.). London, UK: McGraw Hill.

Radice, R., & Phillips, R. (1998). *Software engineering: An industrial approach (Vol. 1)*. Englewood Cliffs, NJ: Prentice Hall.

Wikipedia. (2009). *Certification*. Retrieved February 8, 2009, from http://en.wikipedia.org/ wiki/ Certification

Winschiers, H., & Paterson, B. (2004). Sustainable software development. *SAICSIT '04: Proceedings of the 2004 Annual Research Conference of the South African Institute of Computer Scientists and Information Technologists on IT research in developing countries*.

KEY TERMS AND DEFINITIONS

Certification: It has been defined by the International Organization for Standardization (ISO) as the follows (Balci and Saadi 2002): *"Certification is a procedure by which a third party gives written assurance that a product, process, or service conforms to specified characteristics."*

Conceptual Model: Is the blueprint to the real system, and it's used to address simulation framework, validate the requirements, and make sure it represents a real world problem.

Confidence Grid: According to H. Alegre, et al. (2000), the Confidence Grade (C.G.) is a coding technique that uses the alpha numeric code to describe the reliability and accuracy of the data or process.

Simulation Study: Is the framework that manages the whole simulation of activities including process and data (data collected and data content between the processes) and the experimentation.

Simulation: In her paper, Abu-Taieh (2004) has quoted El Sheikh Simulation definition as "Simulation is the use of a model to represent over time essential characteristics of a system under study".

Validation: According to Pressman (2005) "Validation refers to a different set of activities that ensure that the software that has been built is traceable to customer requirements."

Verification: Pressman (2005) stated that "Verification refers to the set of activities that ensure the software correctly implements a specific function."

Chapter 18
Evaluating the Performance of Active Queue Management Using Discrete-Time Analytical Model

Jafar Ababneh
World Islamic Sciences & Education University (WISE), Jordan

Fadi Thabtah
Philadelphia University, Jordan

Hussein Abdel-Jaber
University of World Islamic Sciences, Jordan

Wael Hadi
Philadelphia University, Jordan

Emran Badarneh
The Arab Academy for Banking & Financial Sciences, Jordan

ABSTRACT

Congestion in networks is considered a serious problem; in order to manage and control this phenomena in early stages before it occurs, a derivation of a new discrete-time queuing network analytical model based on dynamic random early drop (DRED) algorithm is derived to present analytical expressions to calculate three performance measures: average queue length ($Q_{avg,j}$), packet-loss rate ($P_{loss,j}$), and packet dropping probability ($p_d(j)$). Many scenarios can be implemented to analyze the effectiveness and flexibility of the model. We compare between the three queue nodes of the proposed model using the derived performance measures to identify which queue node provides better performance. Results show that queue node one provides highest $Q_{avg,j}$, $P_{loss,j}$, and ($p_d j$) than queue nodes two and three, since it has the highest priority than other nodes. All the above results of performance measure are obtained only based on the queuing network setting parameters.

DOI: 10.4018/978-1-60960-887-3.ch018

INTRODUCTION

Congestion is considered as the main problems in computer networks and data communications systems, which is occurred when the required resources by the sources at a router buffer is more than the available network resources. Congestion has significant role in degrading computer network performance (Kang and Nath, 2004; Welzl, 2005) such as decrease the throughput; obtain high packet queuing delay and packet loss and unfair share of network connections (Braden et al., 1998; Richard, 1997; Welzl, 2004), so many various methods have been developed to control this problem in computer networks (Athuraliya et al., 2001; Aweya et al., 2001; Braden et al., 2001; Feng et al., 99; Feng et al., 2001; Floyd, 2001; Lapsley and Low, 1999; Wydrowski and Zukerman, 2002). The earliest queue congestion control mechanism was the end-to-end Transport Control Protocol (TCP) (Brakmo and Peterson, 1995; Richard, 1997).

When the congestion is detected just only the rate of resources is reduced (Brakmo and Peterson, 1995). Drop-Tail (DT), Random Drop on Full, and Drop Front on Full are Traditional Queue Management (TQM) algorithms, which operate at the designated node, and drop packets only when the queue is full (Braden et al., 1998; Brandauer et al., 2001). TQM have several drawbacks, including, lockout phenomenon, full queues, bias versus burst traffic, and global synchronization, and consequently they contribute in degrading the Internet performance (Feng et al., 2001; Floyd et al., 2001; Welzl, 2005).

Active Queue Management (AQM) algorithms were proposed to overcome some of TQM drawbacks (Athuraliya et al., 2001; Aweya et al., 2001; Braden et al., 2001; Feng et al., 99; Feng et al., 2001; Floyd, 2001; Lapsley and Low, 1999; Wydrowski and Zukerman, 2002). The AQM algorithms have several objectives, which presented as follows:

1. Managing and controlling congestion at routers' buffers in the network in an early stage.
2. Achieving a satisfactory performance through obtaining high throughput, low queuing delay and loss for packets.
3. Sustaining the queue length as small as possible to prevent building up the router queues.
4. Distributing a fair share of the existing resources among the network connections.

Random Early Detection (RED) (Floyd and Jacobson, 1993), Dynamic Random Early Drop (DRED) (Aweya et al., 2001), Adaptive RED (Floyd et al., 2001), BLUE (Feng et al., 1999; Feng et al., 2001), Random Exponential Marking (REM)(Athuraliya et al., 2001; Lapsley and Low, 1999), and GREEN (Wydrowski and Zukerman, 2002) are examples on the AQM algorithms. Despite the fact that AQM algorithms demonstrate better performance than TQM algorithms, but at a standstill to gratify networks users and services an elevated performance becomes inevitability. Using a multi-queue system is considered as an alternative solution, which proposed and widely used as a practical solution to the queue congestion problem at the network router.

The objectives of this chapter are: a new discrete-time queuing network analytical model based on DRED algorithm is derived, for presenting analytical expressions average queue length ($Q_{avg,j}$), packet-loss rate ($P_{loss,j}$), and packet dropping probability ($p_d(j)$) many scenarios can be implemented in this chapter to analyze the effectiveness and flexibility of the proposed model, where Ababneh (2010) derived another new discrete-time queuing network analytical model based on DRED algorithm for another Performance measures; throughput (Tj) and average queuing delay *(Dj)* (Ababneh et al., 2010). This analytical model can be utilized as a congestion control method in fixed and wireless networks. This new analytical model of DRED will applied on a network that has three-queue nodes ($j = 1, 2, 3$), Furthermore, comparing three proposed nodes in the developed model in order to determine the queue node that produces better performance. Comparison between the nodes is based on three performance metrics: Queue node average queue length ($Q_{avg,j}$), Queue node packet-loss ($P_{loss,j}$), and packet dropping probability $p_d(j)$.

The rest of this chapter is outlined as follows: background discusses tow sections first one intro-duces a DRED algorithm and discrete-time Queue Approach presented in second section. The new discrete-time queuing network analytical model is explained in Section 4. Section 5 describes a comparison between the queues nodes ($j = 1, 2, 3$) of the proposed model with reference to the performance metrics mentioned above. Finally, conclusions and suggestions for future research work are given in Section 6.

BACKGROUND

For achieving the performance of queuing net-work systems, they analyzed and modeled by discrete-time queues (woo, 05). The performance measures are; average queue length, average queu-ing delay, throughput and packet dropping prob-ability. Where this approach used many methods for congestion management and control such as AQM techniques; DRED models (Ababneh et al., 2010; Abdeljaber et al., 2008; Abdeljaber et al. 1007), BLUE models (Abdeljaber et al., 2008).

Aweya and his friends (2001) presented DRED algorithm, to control the problem of congestion in networks (Aweya et al., 2001). Unlike RED, which its average queue length Q_{avg} relies on the number of network sources, i.e., TCP connec-tions, DRED stabilizes the queue length (Q) at a predetermined level called the target level of the queue length (Q_{tar}) separately from the number of TCP connections in the network (Flo, 01; Flo, 00). Which mean; when the number of TCP connec-tions in the network is large, the calculated Q_{ave}

becomes large and also may exceed the maximum threshold. As a result, the RED router buffer drops every arriving packet, which consequently increases the number of dropped packets.

DRED depends on many parameters where each of them will explain in details:

1. First parameter is fixed time units (t) and for each t, the current Q and the error signal (E) are calculated to acquire pd.

2. The calculated error signal relies on both parameters ;the current Q and Q_{tar}, and it can be expressed as:

$$E(t) = Q(t) - Q_{tar} \qquad (1)$$

3. Depends on error signal $E(t)$, the filtered error signal $F(t)$ can be calculated as:

$$F(t) = F(t-1)(1 - wq) + wq\, E(t) \qquad (2)$$

From Equation (2) we be able to conclude that DRED algorithm used a low-pass filter in manipulative the $F(t)$ similar to RED.

4. Queue weight parameter which is control constraint used in Equation (2).

5. Packet dropping probability (Pd) constrains and it is value updated for every t using the following expression:

$$p_d(t) = \min\left\{\max\left[p_d(t-1) + \varepsilon\,\frac{F(t)}{K}, 0\right], 1\right\} \qquad (3)$$

6. K is the capacity of the DRED router buffer.
7. Control parameter (ε) used to control the feedback gain.

Aweya (2001) stated that, the *pd* parameter is updated only when the current Q is equal to or greater than the minimum (no-drop) threshold (H_{min}) for maintaining high-link utilization. That mean when $Q(t) < H_{min}$ there will be no packet

Figure 1. Calculating Dp in DRED

Table 1. DRED parameters

Recommended parameters values of DRED algorithm	
The DRED parameter	**Setting the DRED parameters**
Units of time (t)	Time for sending 10 packets or another appropriate value.
Target level for the queue length Q_{avgt}	$Q_{avgt} = k*0.5$.
Indication congestion Threshold	$H = 0.9* Q_{avgt}$
The buffer capacity	K.
Queue weight Qw	0.2%.
(ε) System control parameter	0.005%.

dropping. In order to make your mind up whether or not to drop packets DRED mechanism relies on Q(t). Hence, the congestion metric for the DRED method is the current Q. for dropping packets apparatus DRED uses randomization technique where this method is comparable to (Awe, 01). Besides that, it marks the coming packet by tow ways, first one by dropping it or by adding an explicit congestion notification (ECN) bit in its header (Ram, 01). Figure 1 showed in details how dropping packets computed due to (Che, 02).

Chen (2002), proposed how to set recommended parameters values of DRED algorithm, Table 1 illustrated this idea (Chen, 02).

Discrete-Time Queue Approach

There are many methodologies used for analyzing multi-queue system performance during congestion; mathematical (or analytical) modeling, practical experiments and simulations. However, mathematical models can help in giving appropriate solutions, not expensive, and uncomplicated to use (Robertazzi, 2000; Welsl, 2005). Nevertheless, building analytical models for queuing network systems using discrete-time queues is not easy,

and this explains the lack of research works which have done in this field (Ababneh et al., 2010; Abdel-Jaber et al., 2008; Abdel-Jaber et al., 2007; Leeuwaarden et al., 2006; Woodward, 2003;). Mathematical models for AQM algorithms can be urbanized as continuous or discrete-time models. When packets are allowed to arrive or depart at any time during the time span, it is called continuous models. While, in discrete-time models the time span is divided into slots of certain length, and packets are allowed to arrive at the beginning of time slots and depart at the end of time slots (Leeuwaarden et al., 2006; Woodward, 2003). In each slot packet arrival and/or departure can be happened. Analytical models were developed in (Ababneh et al., 2010.; Abdel-Jaber et al., 2007.; Abdel-Jaber et al., 2008) to evaluate the performance of a discrete-time AQM-based queuing systems of one and two-queue nodes.

Generally, every queuing system can be described by several notations, i.e. kendall's notations that consists of five components (Woo, 05) and described by (Ababneh J., 2009) thesis as follows:

1. The arrival process: A stochastic process that shows how customers (packets) arrive to the queuing system. The arrival process is denoted A.

2. The service process: A stochastic process that illustrates the amount of time spends by a customer (packet) in the server. The service process is denoted B.
3. The number of servers (C).
4. The system capacity: It represents the maximum number of customers (packets) inside the system including packets currently in the service. The system capacity is denoted K.
5. The customer population: It represents the limit on the total number of customers who participate in the arrival process. The customer population is denoted P.

It should be noted that there is another factor related to kendall's components called the queuing service discipline, which can be defined as the set of laws that make a decision of which customer in the queuing system should be served, i.e. first come first served (FCFS), last come first served (LCFS), etc. The service unit is distributed fairly among the customers in the queue.

K and P components can be removed where in this case we consider them infinite values. C, K and P components are positive integers, and A and B components are selected based on the set of descriptors such deterministic (D). In D, the interarrival and/or service time's distribution(s) are constant. If the descriptor is Markovian (M), the interarrival and service time's distributions are exponentially distributed in the continuous-time queues. For instance, the interarrival time is Poisson process and the service time is exponentially distributed. On the other hand, in the discrete-time queues, the interarrival and service times are geometrically distributed, and also we can use the exponential distribution in case of if the arrivals and departures are multiple (Woo, 05). Lastly, if the descriptor is generally distributed (G), then no constraints can be enforced on the distributions types. In the discrete-time queues (Woo, 05), we use basic time unit called slot, where in each slot, single or multiple events can occur. An example of a single event is the occurrence of packet arrival

or departure, whereas, both packets arrival and departure may occur in multiple events.

In discrete-time queuing systems, packets arrival take place after the starting of the slot, where else packets departure happen before the end of the slot. The number of arrivals and departures at a slot n are defined by $\{a_n, n=0, 1, 2, ...\}$ and $\{d_{n+1}, n=0,1,2,...\}$, respectively. Where $\{a_n\}$ denotes the sequence of identical, independent distribution (i.i.d) random variables with a specific distributions. $d_0=0$ since no packets are departed before they arrived. The state of the discrete-time queue with arrivals and departures at each slot are depicted in Figure 2.

We observe in Figure 2 that the process of the queue length at the slot borders $\{yn, n=0,1,2,...\}$, yn is random. Therefore, y_{n+1} result is shown in Equation (4).

$$y_{n+1} = y_n + a_n - d_{n+1} \tag{4}$$

Equation(4) denotes the process of the queue length after the arrivals happen $\{X_n =, n=0,1,2,...\}$.

$$X_n = y_n + a_n \tag{5}$$

A Discrete-Time One and Two-Queue Nodes DRED-Based Analytical Algorithm

Abdel-Jaber and his friends (2007) proposed a single queue node DRED algorithm relies on a certain threshold (H) level for determining the congestion in its router buffer as shown in Figure 3 (Abdel-Jaber et al.,2007)

The single queue node DRED algorithm described above relies on a certain threshold (H) level for determining the congestion in its router buffer. Abdel-Jaber and his friends developed and derived discrete-time analytical model based on a queuing network consisting of two queue nodes (j=1, 2) to evaluate and analyze the performance of multi-queue systems utilizing the DRED AQM

Figure 2. The state of the discrete-time queue with arrivals and departures at each slot

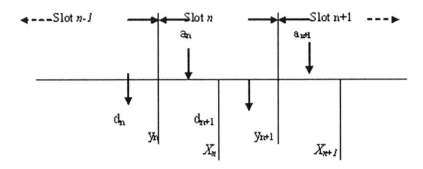

Figure 3. Uninode queuing system model

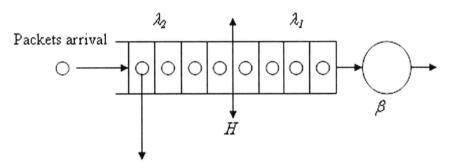

algorithm to manage congestion at the network buffers (Abdel-Jaber et al., 2008).

A Discrete-Time Three-Queue Nodes DRED-Based On Analytical Algorithm

Ababneh and his friends (2010) proposes a derivation of discrete-time queuing network analytical model based on dynamic random early drop (DRED) algorithm to manage and control congestion, which is referred to as the 3QN model and tow performance measures are calculated, namely; throughput (Tj) and average queuing delay (Dj), beside that two scenarios were performed; the variation of packets arrival probability against throughput), and average queuing delay (Ababneh, et al., 2010).

SOLUTION AND RECOMMENDATION

Discrete-Time Analytical Model Based on DRED Algorithm

The proposed queuing network system is shown in Figure 4 to overcome the congestion problems of the network routers buffer and enhance the networks performance and develop hard platform discrete-time analytical model based on AQM for using it in many respects that consist of three nodes. The analysis of the queuing network is given as:

The queue nodes shown in the above figure have a finite capacity of packets, where K_1, K_2 and K_3 correspond to the capacities for queue nodes 1, 2 and 3, respectively, including packets that are currently in service. we assume that the arrival process used is the identical indepen-

Figure 4. Three queue nodes queuing network (i=1, 2, 3) model

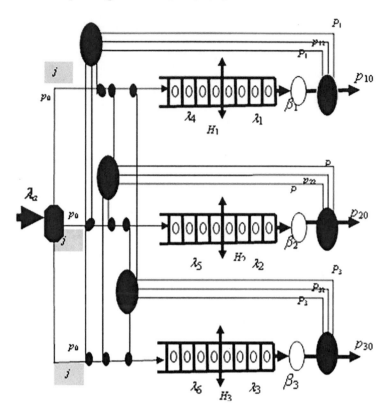

dently distributed (i.i.d) Bernoulli process a_n ε $\{0,1\}$, n= 0,1,2,3,… where a_n represents packet arrival in a slot n. Beside that the above queue node is finite and it is capacity is K packets, including packet currently in service. The thresholds H_1, H_2 and H_3 are expressed in Equations (6-8) as follows:

$$H_1 = [0.9k_1/2] \tag{6}$$

$$H_2 = [0.9k_2/2] \tag{7}$$

$$H_3 = [0.9k_3/2] \tag{8}$$

Moreover, in Figure 4, the probability of packet arrival for the router buffer of queue node 1 is λ_1 (λ_1 is defined below), only when the instantaneous $Q_{av,1}$ is below H_1 and thus there will be no packet dropping (P_{d1}=0). While, if the instantaneous $Q_{avg,1}$ reaches H_1 index, the probability of packet

arrival for queue node 1 reduces from λ_1 to λ_4 (λ_4 is defined below) in order to alleviate the congestion at the router buffer, and therefore P_{d1} increases from 0 to $(\lambda_1 - \lambda_4)/\lambda_1$. For queue node 2, the probability of packet arrival at the router buffer is λ_2 (λ_2 is defined below) in cases where the instantaneous $Q_{avg,2}$ is less than H_2 index, and therefore the dropping probability of queue node 2 (P_{d2}) becomes zero. Though, if the instantaneous $Q_{avg,2}$ is equal to or larger than H_2 index, P_{d1} increases from 0 to $(\lambda_2 - \lambda_5)/\lambda_2$, and the probability of packet arrival decreases from λ_2 to λ_5 (λ_5 is defined below) in order to control the congestion. For queue node 3, the packet arrival probability of the router buffer is λ_3 (λ_3 is defined below) in cases where the instantaneous $Q_{avg,3}$ is less than H_3 index, and therefore the dropping probability of queue node 3 (P_{d3}) becomes zero. Though, if the instantaneous $Q_{avg,3}$ is equal to or larger than H_3 index, P_{d3} increases from 0 to $(\lambda_3 - \lambda_6)/\lambda_3$, and the probability of packet arrival decreases from λ_3 to λ_6 (λ_6 is defined below) in order to control the congestion. This is since the packet arrival probabilities (λ_1, λ_2 and λ_3) are decreased from probability value to another fixed ones (λ_4, λ_5, λ_6), and the values of (P_{d1}, P_{d2}, P_{d3}) are increasing from 0 to another fixed probability.

To explain the queuing network system shown in Figure 4, we analyze each queue node separately in order to come up with the traffic equations for each one of them. We assume that λ is the probability of the packet arrival of the external packets (packets arriving from outside the network) in a slot. λ_1 and λ_4 are the probabilities of packets that arrive at node 1 before reaching the H_1 index and after reaching the H_1 index at the router buffer, respectively. λ_2 and λ_5 are the probabilities of packet arrivals to queue node 2 when the instantaneous $Q_{avg,2}$ is smaller than H_2 index and when it is greater than H_2 or equal to index, respectively. λ_3 and λ_6 are the probabilities of packet arrival to queue node 3 when the instantaneous $Q_{avg,3}$ is smaller than H_3 index and when it is greater than or equal to H_3 index, respectively. Further, β_1, β_2 and β_3 represent the probabilities of packet departure in a slot from nodes 1, 2 and 3, respectively. We also assume that the queuing network is in equilibrium, and the $Q_{avg,j}$ process of each queue node is a Markov chain with finite state spaces.

The state spaces for queue node 1 are:

$$\{0,1,2,3,...,H_1 - 1, H_1, H_1 + 1,..., K_1 - 1, K_1\}$$

The state spaces for queue node 2 are:

$$\{0,1,2,3,...,H_2 - 1, H_2, H_2 + 1,..., K_2 - 1, K_2\}$$

The state spaces for queue node 3 are:

$$\{0,1,2,3,...,H_3 - 1, H_3, H_3 + 1,..., K_3 - 1, K_3\}$$

Finally, we consider that

$\lambda_1 > \lambda_4$, $\lambda_2 > \lambda_5$, $\lambda_3 > \lambda_6$, $\beta_1 > \lambda_1$, $\beta_2 > \lambda_2$

and $\beta_3 > \lambda_3$
thus, $\beta_1 > \lambda_4$, $\beta_2 > \lambda_5$
and $\beta_3 > \lambda_6$.

Since we analyze each queue node in the queuing network system separately, the packet arrival probability for each queue node requires an independent evaluation. Equations (9-11) represent the arrival probability of packets for queue nodes 1, 2 and 3, respectively.

$$\lambda_1 = \lambda r_{01} + \lambda_1 r_{11} + \lambda_2 r_{21} + \lambda_3 r_{31} \tag{9}$$

$$\lambda_2 = \lambda r_{02} + \lambda_1 r_{12} + \lambda_2 r_{22} + \lambda_3 r_{32} \tag{10}$$

$$\lambda_3 = \lambda r_{03} + \lambda_1 r_{13} + \lambda_2 r_{23} + \lambda_3 r_{33} \tag{11}$$

After solving Equations (9, 10, and 11 recursively, we obtain the final form for λ_1, λ_2 and λ_3 as in Equations 12, 13, and 14. (see Box 1)

In Equations (12, 13, and 14, r_{ij}, $i, j = 1, 2$ and 3 represents the routing probabilities of packets between the three queue nodes in the queuing network, and (r_{i0}, r_{j0}) represent the probabilities of packets leaving the queuing network from queue nodes 1, 2 and 3, respectively. Lastly, (r_{0i}, r_{0j}) represent the routing probabilities of packets that come from outside the network to queue nodes 1, 2 and 3, respectively. The state transition diagrams of queue nodes 1, 2 and 3 are shown in Figures 5

When using the state transition diagrams in Figures 5, 6 and 7, one can derive the balance equations for queue nodes 1, 2 and 3 as shown in Equations (15-17).

We assume $j=1, 2$ and 3, where 1 represents the first queue node, 2 represent the second queue node, and 3 represents the third queue node.

$$\Pi_0 = \Pi_0\left[(1 - \lambda_j)\right] + \Pi_1\left[\beta_j(1 - \lambda_j)\right] \tag{15}$$

Box 1.

$$\lambda_1 = \left| \frac{\lambda\left[[r_{01}(1-r_{22})+r_{02}r_{21}][(1-r_{22})(1-r_{33})-r_{23}r_{32}]+[r_{31}(1-r_{22})+r_{21}r_{32}][r_{03}(1-r_{22})+r_{02}r_{23}]\right]}{[(1-r_{11})(1-r_{22})-r_{12}r_{21}][(1-r_{22})(1-r_{33})]-[r_{31}(1-r_{22})+r_{21}r_{32}][(1-r_{22})(1-r_{33})-r_{23}r_{32}]} \right| \quad (12)$$

$$\lambda_2 = \left| \frac{\lambda\left[[r_{02}(1-r_{11})+r_{01}r_{12}][(1-r_{11})(1-r_{33})-r_{13}r_{31}]+[r_{32}(1-r_{11})+r_{12}r_{31}][r_{03}(1-r_{11})+r_{01}r_{13}]\right]}{[(1-r_{11})(1-r_{22})-r_{12}r_{21}][(1-r_{11})(1-r_{33})-r_{13}r_{31}]-[r_{23}(1-r_{11})+r_{13}r_{21}][r_{32}(1-r_{11})(1-r_{33})+r_{12}r_{31}]} \right| \quad (13)$$

$$\lambda_2 = \left| \frac{\lambda\left[[(1-r_{11})(1-r_{22})-r_{12}r_{21}][r_{03}(1-r_{22})+r_{02}r_{23}]+[r_{01}(1-r_{22})+r_{02}r_{21}][r_{13}(1-r_{22})+r_{12}r_{13}]\right]}{[(1-r_{11})(1-r_{22})-r_{12}r_{21}][(1-r_{22})(1-r_{33})-r_{23}r_{32}]-[r_{13}(1-r_{22})+r_{12}r_{23}][r_{31}(1-r_{22})+r_{21}r_{32}]} \right| \quad (14)$$

$$\Pi_1 = \Pi_0 \lambda_j + \Pi_1[\lambda_j\beta_j + (1-\lambda_j)(1-\beta_j)] + \Pi_2[\beta_j(1-\lambda_j)] \quad (16)$$

$$\Pi_2 = \Pi_1[\lambda_j(1-\beta_j)] + \Pi_2[\lambda_j\beta_j + (1-\lambda_j)(1-\beta_j)] + \Pi_3[\beta_j(1-\lambda_j)] \quad (17)$$

We will generally obtain Equations (15-17) for *i* as shown below:

$$\Pi_i = \Pi_{i-1}[\lambda_j(1-\beta_j)] + \Pi_i[\lambda_j\beta_j + (1-\lambda_j)(1-\beta_j)] + \Pi_{i+1}[\beta_j(1-\lambda_j)] \quad (18)$$

Where i=2, 3, 4,..., H_j-2

$$\Pi_{Hj-1} = \Pi_{Hj-2}[\lambda_j(1-\beta_j)] + \Pi_{Hj-1}[\lambda_j\beta_j + (1-\lambda_j)(1-\beta_j)] + \Pi_{Hj}[\beta_j(1-\lambda_{j+3})] \quad (19)$$

$$\Pi_{Hj} = \Pi_{Hj-1}[\lambda_j(1-\beta_j)] + \Pi_{Hj}[\lambda_{j+3}\beta_j + (1-\lambda_{j+3})(1-\beta_j)] + \Pi_{Hj+1}[\beta_j(1-\lambda_{j+3})] \quad (20)$$

$$\Pi_{Hj+1} = \Pi_{Hj}[\lambda_j(1-\beta_j)] + \Pi_{Hj+1}[\lambda_{j+3}\beta_j + (1-\lambda_{j+3})(1-\beta_j)] + \Pi_{Hj+2}[\beta_j(1-\lambda_{j+3})] \quad (21)$$

Generally we obtain,

$$\Pi_i = \Pi_{i-1}[\lambda_{j+3}(1-\beta_j)] + \Pi_i[\lambda_{j+3}\beta_j + (1-\lambda_{j+3})(1-\beta_j)] + \Pi_{i+1}[\beta_j(1-\lambda_{j+3})] \quad (22)$$

Where i= H_j+1, H_j+2, H_j+3,..., K_j-2, K_j-1. Then finally we can obtain,

$$\Pi_{Kj} = \Pi_{Kj-1}\left[\lambda_{j+3}\left(1-\beta_j\right)\right] + \Pi_{Kj}\left[\lambda_{j+3}\beta_j + \left(1-\beta_j\right)\right] \quad (23)$$

where $K_j = H_j + I$.

After solving Equations (15-23) in terms of Π_0 we obtain generally Equation (24).

$$\Pi_i = \Pi_0\left[\lambda_j^i\left(1-\beta_j\right)^{i-1} \Big/ \beta_j^i\left(1-\lambda_j\right)^i\right] \quad (24)$$

Let

$$\gamma_i = \lambda_i\left[\left(1-\beta_i\right)\Big/\beta i\left(1-\lambda_i\right)\right],$$

where

Figure 5. The state transition diagram of queue node1,2 and 3 in the proposed analytical model

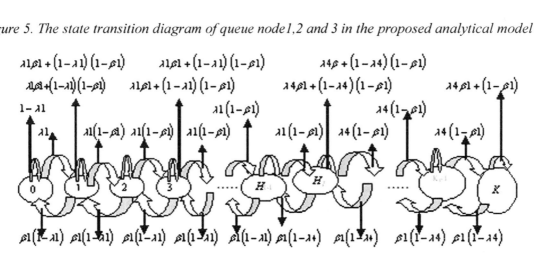

The state transition diagram of queue node 1 in the proposed analytical model.

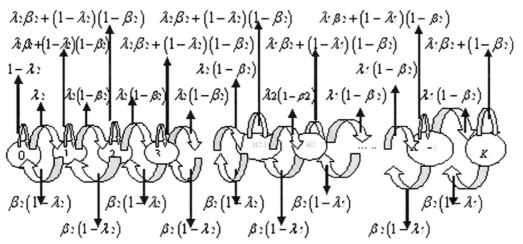

The state transition diagram of queue node 2 in the proposed analytical model

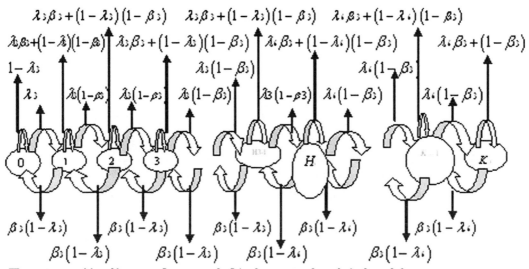

The state transition diagram of queue node 3 in the proposed analytical model

Figure 6. λ_{arr} *vs.* $Q_{avg,j}$

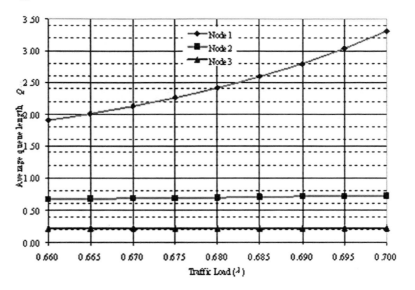

Figure 7. λ_{arr} *vs.* $P_{loss,j}$

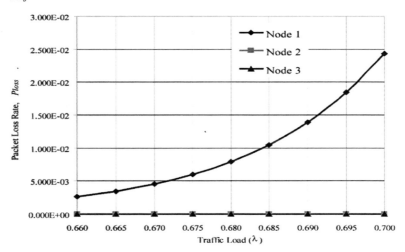

$$i = 1, 2, 3, 4, ..., H - 1 \qquad (25)$$

By substitution Equation (25) in Equations (15-23) to derive the equilibrium network queuing network probabilities in the network queuing, we achieve:

$$\Pi_i = \Pi_0 [\lambda_j^i (1 - \beta_j)^{i-1} / \beta_j^i (1 - \lambda_j)^i]$$
$$= \Pi_0 [\lambda_j (1 - \beta_j) / \beta_j (1 - \lambda_j)]^i (1 - \beta_j)^{-1}$$
$$= \Pi_0 [\gamma_j^i / (1 - \beta_j)]$$

$$(26)$$

Where j=1, 2, 3 and i=1, 2, 3, 4, ..., H_j-1

$$\Pi_{Hj+i} = \left[\frac{\lambda_j^{Hj} \lambda_{j+3}^{\ i} \left(1 - \beta_j\right)^{Hj+i-1}}{\beta_j^{Hj+i} \left(1 - \lambda_j\right)^{Hj-1} \left(1 - \lambda_{j+3}\right)^{i+1}} \right] \Pi_0$$

$$= \Pi_0 \left[\frac{\gamma_j^{Hj} \gamma_{j+3}^{\ i} \left(1 - \lambda_j\right)}{\left(1 - \lambda_{j+3}\right)\left(1 - \beta_j\right)} \right]$$

(27)

Where j=1, 2, 3 and I=1, 2, 3, 4,..., I

We also evaluate the probabilities of that queue nodes 1, 2 and 3 are idle (Π_0) using Equation (25), then by applying Equations (26, 27 in Equation (24), we obtain the final result for the Π_0 as in Equation (28).

$$\Pi_0 = (1 - \beta_j)\left[\left[1 - \gamma_j^{Hj} - \beta_j(1 - \gamma_j) / (1 - \gamma_j)\right] + \left[\frac{\gamma_j^{Hj}(1 - \lambda_j)(1 - \gamma_{j+3}^{\ I+1})}{(1 - \lambda_{j+3})(1 - \gamma_{j+3})}\right]\right]^{-1}$$

(28)

After computing Π_0, we estimate the performance measures ($Q_{avg,j}$, $P_{loss,j}$) for the proposed discrete-time queuing network analytical model. We start calculating the $Q_{avg,j}$ which is the first performance measure for the proposed model using the generating function P(z) as shown in Equation (29).

$$P(z) = \sum_{i=0}^{K} z^i \Pi_i$$

(29)

We utilize the $P(z)$ equation to obtain the $Q_{avg,j}$ by producing the first derivative for the generating function $P(z)$ at $z=1$. We can compute the $Q_{avg,j}$ for the queue nodes as shown in Equation (30).

$$Q_{avg,j} = P^{(1)}(1) = \left[\Pi_0 / (1 - \beta_j)\right]$$
$$\left[\frac{\gamma_j - \gamma_j^{Hj}\left[\gamma_j + Hj(1 - \gamma_j)\right]}{(1 - \gamma_j)^2}\right.$$
$$+ \left(\frac{\gamma_j^{Hj}(1 - \lambda_j)}{(1 - \lambda_{j+3})}\right)\left(\frac{Hj(1 - \gamma_{j+3})(1 - \gamma_{j+3}^{\ I+1})}{(1 - \gamma_{j+3})^2}\right.$$
$$+ \frac{\gamma_{j+3}}{(1 - \gamma_{j+3})^2} - \left.\left.\frac{\gamma_{j+3}^{\ I+1}\left[1 + I(1 - \gamma_{j+3})\right]}{(1 - \gamma_{j+3})^2}\right)\right]$$

(30)

Or using another equation to calculate average queue length $\sum_{i=0}^{Kj} i \Pi_i$ formula, where $j=1,2,$ and 3.

The second performance measure is the packets loss probability ($P_{loss,j}$), which can be defined as the proportion of packets that lost the service at the router buffer from all the packets that were arrived is evaluated. The router buffer begins dropping packets only after the $Q_{avg,j}$ reaches the H_j position. Hence, we compute the $P_{loss,j}$ according to the below Equation (31).

$$P_{loss,j} = \sum_{i=th}^{Kj} \Pi_i$$

(31)

The joint equilibrium probability of the queuing network system is expressed in Equation (32), where K_1 packets at node 1, K_2 packets at node 2 and K_3 packets at node 3.

$$\Pi(K_1, K_2, k_3) = \Pi_1(K_1)\Pi_2(K_2)\Pi_3(K_3)$$ (32)

RESULT AND DISCUSSION

The results have been produced according to diverse performance measures (Qavg,j, Ploss,j), beside that, an independent performance evaluation for each node in the three nodes system is conducted to figure out the node that has better performance. We also computed the dropping

Table 2. The $Q_{avg,j}$ results for the proposed three queue nodes analytical model in DRED

λ_{arr}	λ_j			$Q_{avg,j}$		
	Node1	Node2	Node3	Node1	Node2	Node3
0.660	0.825	0.528	0.297	1.902	0.670	0.224
0.665	0.831	0.532	0.299	2.005	0.677	0.224
0.670	0.838	0.536	0.302	2.122	0.683	0.225
0.675	0.844	0.540	0.304	2.257	0.690	0.225
0.680	0.850	0.544	0.306	2.412	0.697	0.225
0.685	0.856	0.548	0.308	2.591	0.704	0.226
0.690	0.863	0.552	0.311	2.798	0.711	0.226
0.695	0.869	0.556	0.313	3.038	0.718	0.226
0.700	0.875	0.560	0.315	3.313	0.725	0.226

Table 3. The $P_{loss,j}$ results for the proposed three queue nodes analytical model in DRED

λ_{arr}	λ_j			$P_{loss,jj}$		
	Node1	Node2	Node3	Node1	Node2	Node3
0.660	0.825	0.528	0.297	2.604E-03	8.288E-08	3.128E-11
0.665	0.831	0.532	0.299	3.423E-03	9.393E-08	3.422E-11
0.670	0.838	0.536	0.302	4.512E-03	1.065E-07	3.741E-11
0.675	0.844	0.540	0.304	5.964E-03	1.206E-07	4.090E-11
0.680	0.850	0.544	0.306	7.898E-03	1.367E-07	4.468E-11
0.685	0.856	0.548	0.308	1.048E-02	1.549E-07	4.880E-11
0.690	0.863	0.552	0.311	1.390E-02	1.756E-07	5.327E-11
0.695	0.869	0.556	0.313	1.842E-02	1.990E-07	5.813E-11
0.700	0.875	0.560	0.315	2.434E-02	2.256E-07	6.341E-11

probability (Pd(j)) for each node to determine the node that drops the highest number of packets. In the experiments, the following parameters' settings and assumptions are used: (λ_{arr}) values are shown in Table 1, i.e. [0.66 and 0.70], queue node 1 has higher priority than the queue node 2 and node 2 has higher priority than the node 3. Queue node 1 parameters r_{01} is set to 0.50, ($\lambda=0.75, \beta_1=0.90, K_1=20$), and the routing probabilities of queue node 1($r_{10}=0.4, r_{11}=0.3, r_{12}=0.2$ and $r_{13}=0.1$). Whereas, queue node 2 r_{02} is set to 0.30, ($\lambda=0.75, \beta_2=0.90, K_2=20$), and the routing probabilities of queue node 2 ($r_{20}=0.4, r_{21}=0.3, r_{22}=0.2$ and $r_{23}=0.1$). Finally queue node 3 r_{03} is set to 0.20, ($\lambda=0, 75$,

$\beta_3=0.90, K_3=20$), and the routing probabilities of queue node 3($r_{30}=0.4, r_{31}=0.3, r_{32}=0.2$ and $r_{33}=0.1$). λ_{arr} parameter has been set to [0.66-0.7] in order to see the performance effect for different probabilities of packet arrival (moderate and high traffic load). K_1, K_2 and K_3 were tuned to 20 to observe the queue node performances with small buffer sizes. The performance measure results are obtained based on the given setting values for the above parameters. Thus, the performance which node is provide better performance result is only given based on the above parameter setting values.

For our proposed analytical model, the performance measure results for the proposed three queue nodes (1, 2, and 3) analytical model, are shown in tables (2and 3). In particular, Tables 2 and 3 show the $Q_{avg,j}$ and $P_{loss,j}$ performance results for the three queue nodes, respectively.

The performance measures results for the proposed analytical model are depicted in Figures 6-7. Particularly, we illustrated the results of queue nodes 1, 2 and 3 with respect to λ_{arr} against $Q_{avg,j}$ and λ_{arr} against $P_{loss,j}$. This is since we want to determine the relationship between λ_{arr} and packet loss values.

Starting analyzing results of proposed analytical three nodes model through the first performance measure which is the relationship between arrival traffic load and average queue as shown in Figure6, we found out that queue node 1 achieved higher $Q_{avg,j}$ than node 2 and node 2 has higher $Q_{avg,j}$ than node 3, but queue node 2 achieved slightly higher $Q_{avg,j}$ results than queue node 3. In other words, queue node 3 stabilizes the average queue size better than queue node 2 and 2 is better than 3 in our analytical model.

Corresponding to the second performance measure which is the relationship between packet arrival probability and the packet loss rate $P_{loss,j}$ generally we discovered that the traffic load is linearly proportion with the loss rate. In other words, if the traffic loads for nodes1, 2 and 3 in-

crease the dropping rate will increase linearly. For the same traffic load, nodes 1 and 2 drop packets similarly, but in contrary node 1 has higher packet loss rate than nodes 2 and 3 as shown in Figure 7. This means node 3 is the best node since it has the least packet loss rate among the three nodes.

Finally, the $P_d(j)$ results of the three nodes are analyzed and shown in Figure 8. In $p_{d(j)}$ experiments, the parameters used for both models are set to the same values discussed at the beginning of this section with a single exception which is λ_{arr} which is set here to a fixed value 0.66. From Figure 8 we can note that node 3 has the lowest packet dropping probability among the nodes, also nodes 2 and 3 has nearly zero packets dropping probability.

This indicates that we can increase their traffic load to make a balance among the three nodes. Node 3 in the proposed model is better than nodes 1 and 2 with regard to p_d at all traffic load levels. In other words, node 1 increases its dropping probability due to the existence of congestion. Additionally, these numbers in Figure 9 indicate that node 3 drops fewer packets than the other nodes. Node 1 has the highest dropping probability results among nodes since it received the largest number of packets. Lastly, according to the tables and the figures, node 3 outperformed

Figure 8. K VS. $P_{d(j)}$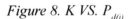

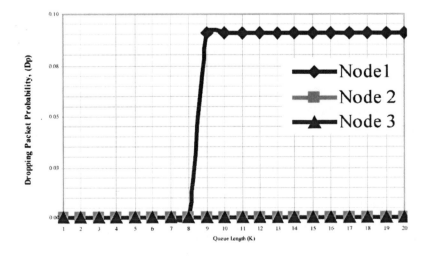

nodes 1 and 2 in our analytical model with reference to average queue length, and dropping packet rate.

CONCLUSIONS AND FUTURE RESEARCH DIRECTION

This chapter presents development of a discrete-time analytical model to evaluate and analyze the performance of multi-queue systems utilizing the DRED AQM algorithm to manage congestion at the network buffers. Calculating three performance measures in a new discrete-time queuing network analytical model that consists of three queue nodes a derivation of three analytical expressions are derived in this chapter. This chapter compare between the three nodes in the proposed analytical model using three performance measures; average queue length ($Q_{avg,j}$), packet dropping probability ($P_{loss,j}$), and packet dropping probability $P_d(j)$ in order to decide the one that provides better performance measure results. The performance of nodes is obtained based on the setting values for the parameters. Therefore the node which offers better performance results than others is solely based on the given values for the setting parameters. Specifically, we calculate every performance measure for each queue node. According to the performance of each queue node in the proposed model, the results indicated that queue node 3 has better performance than other nodes, and queue node 2 slightly outperformed queue node 1 with respect to packet loss rate. On the other hand, due to the fact that our model is not influenced by the traffic load level, its average queue length results in queue node 3 are better than those in queue nodes 1 and 2, and queue node 2 is better than queue node 1. It was achieved that queue node 3 dropped fewer packets than queue nodes 1 and 2. This is since queue node 1 has a higher priority than queue node 2 and queue node 2 has higher priority than queue node 3 in serving packets coming from outside the queuing network.

We planned in near future to propose comparative study between varieties discrete–time analytical models based on DRED algorithm that consist of one queue node, two queue nodes, and three queue nodes to know which one has better performance measures or better QoS than other, also extend our model by applying multiple arrivals and departures as rather than single arrival and departure. Also applied this ideas on other algorithms such as; RED, Gently RED, adaptive RED, BLUE. In addition to we intend to apply the three analytical models as congestion control methods in computer networks such as the internet and apply a priority policy between the nodes. In addition, analysis and modeling a queue node system is based on exponential and linear reduction of the value of packet arrival probability.

REFERENCES

Ababneh, D. (2009). *Development of a discrete-time DRED-based multi-queue nodes network analytical model*. Unpublished doctoral dissertation, The Arab Academy for Banking & Financial Sciences, Amman, Jordan.

Ababneh, J., Thabtah, F., Abdel-Jaber, H., Hadi, W., & Badarneh, E. (2010). Derivation of three queue nodes discrete-time analytical model based on DRED algorithm. *Proceedings of the 7th IEEE International Conference on Information Technology: New Generations* (ITNG 2010), Las-Vegas, USA, (pp. 885-890).

Abdel-Jaber, H., Woodward, M. E., Thabtah, F., & Abu-Ali, A. (2008). Performance evaluation for DRED discrete-time queuing network analytical model. *Journal of Network and Computer Applications*, *31*(4), 750–770. doi:10.1016/j.jnca.2007.09.003

Abdel-Jaber, H., Woodward, M. E., Thabtah, F., & Etbega, M. (2007). A discrete-time queue analytical model based on dynamic random early drop. *Proceedings of the 4th IEEE International Conference on Information Technology: New Generations* (ITNG 2007), Las-Vegas, USA, (pp. 71-76).

Athuraliya, S., Li, V. H., Low, S. H., & Yin, Q. (2001). REM: Active queue management. *IEEE Network, 15*(3), 48–53. doi:10.1109/65.923940

Aweya, J., Ouellette, M., & Montuno, D. Y. (2001). A control theoretic approach to active queue management. *Journal of Computer Networks, 36*(2–3), 203–235. doi:10.1016/S1389-1286(00)00206-1

Braden, R., Clark, D., Crowcroft, J., Davie, B., Deering, S., Estrin, D., et al. (1998). *Recommendations on queue management and congestion avoidance in the Internet*. RFC 2309.

Brakmo, L., & Peterson, L. (1995). TCP Vegas: End to end congestion avoidance on a global Internet. *IEEE JSAC, 13*(8), 1465–1480.

Brandauer, C., Iannaccone, G., Diot, C., Ziegler, T., Fdida, S., & May, M. (2001). Comparison of tail drop and active queue management performance for bulk-data and Web-like Internet traffic. In *Proceeding of ISCC, IEEE*, (pp. 122–9).

Chengyu, Z., Yang, O. W. W., Aweya, J., Ouellette, M., & Montuno, D. Y. (2002). A comparison of active queue management algorithms using the OPNET modeler. *Proceedings IEEE Communications Magazine, 40*(6), 158. doi:10.1109/MCOM.2002.1007422

Feng, W., Kandlur, D., Saha, D., & Shin, K. G. (1999). *Blue: A new class of active queue management algorithms. Technical report UM CSE-TR-387-99*. Ann Arbor, MI: University of Michigan.

Feng, W., Kandlur, D., Saha, D., & Shin, K. G. (2001). Stochastic fair blue: A queue management algorithm for enforcing fairness. *Proceedings - IEEE INFOCOM, 3*, 1520–1529.

Floyd, S., & Jacobson, V. (1993). Random early detection gateways for congestion avoidance. *IEEE/ACM Transactions on Networking, 1*(4), 397–413. doi:10.1109/90.251892

Floyd, S., Ramakrishna, G., & Shenker, S. (2001). *Adaptive RED: An algorithm for increasing the robustness of RED's active queue management. Technical report*. ICSI.

Kang, J., & Nath, B. (2004). *Resource-controlled MAC-layer congestion control scheme in cellular packet network*. IEEE 59th Vehicular Technology Conference, (pp. 1988–1992). VTC 2004-Spring.

Lapsley, D., & Low, S. (1999). Random early marking for Internet congestion control. In *Proceeding of GlobeCom '99*, (pp. 1747–52).

Leeuwaarden, J., Denteneer, D., & Resing, J. (2006). A discrete-time queuing model with periodically scheduled arrival and departure slots. *Journal of Performance Evaluation, 63*(4), 278–294. doi:10.1016/j.peva.2005.03.001

Pentikousis, K., & Badr, H. (2002). *On the resource efficiency of explicit congestion notification* (pp. 588–599). Berlin / Heidelberg, Germany: Springer.

Robertazzi, T. (2000). *Computer networks and systems: Queuing theory and performance evaluation* (3rd ed.). Springer Verlag.

Stevens, W. (2001). *TCP slow start, congestion avoidance, fast retransmit, and fast recovery algorithms*. IETF RFC, 1997.

Welzl, M. (2005). *Network congestion control: Managing Internet traffic*.

Woodward, M. E. (2003). *Communication and computer networks: Modeling with discrete-time queues*. London, UK: Pentech Press.

Wydrowski, B., & Zukerman, M. (2002). GREEN: An active queue management algorithm for a self managed Internet. *Proceedings of IEEE ICC, 4,* 2368–2372.

KEY TERMS AND DEFINITIONS

Active Queue Management (AQM): Is a technique that consists in dropping or marking packets before a router's queue is full, they operate by maintaining one or more drop/mark probabilities, and probabilistically dropping or marking packets even when the queue is short. There are a number of AQM algorithms that have been developed by many researchers; RED, DRED, GRED,GREEN,BLUE,SRED.

Average Queue Length: Is defined as the average number of packets that are waiting in the queue buffer

Congestion Control: It is the main problems in computer networks and data communications systems, which is occurred when the required resources by the sources at a router buffer is more than the available network resources

Discrete-Time Analytical Model: Non- continues time that can be utilized as a congestion control method in fixed and wireless networks, a basic time unit called slot is used, where in each slot, single or multiple events can occur.

Discrete-Time Queue Network: Is an approach that models and analyses the performance of the queuing system in communication and computer networks, where packets arrival takes place just after the beginning of the slot, while packets departure takes place just before the end of the slot

DRED: Dynamic Random Early Drop is an algorithm for controlling the problem of congestion in networks. It was proposed to reduce the main problem of RED, it stabilizes the queue length (Q) at a predetermined level called the target level of the queue length (Q_{tar})) independently from the number of TCP connections in the network.

Packets Loss Probability: The proportion of packets that lost the service at the router buffer from all the packets that were arrived is evaluated

Quality of Service: Is the ability to provide different priority to different applications, users, or data flows, or to guarantee a certain level of performance to a data flow. packet dropping probability, average queue length.

Chapter 19
Knowledge:
An Information Systems' Practical Perspective

Shadi Ettantawi
The Arab Academy for Banking & Financial Sciences, Jordan

Asim El-Sheikh
The Arab Academy for Banking & Financial Sciences, Jordan

ABSTRACT

A number of disciplines have approached the concept of knowledge. None of the existing definitions of knowledge can be generalized to other disciplines, and most importantly, none of such attempts fit the requirements of information systems (IS). This chapter suggests to perceive knowledge from the point of view of IS, as an attempt to answer IS requirements better. The proposed vision of knowledge is based on Information Systems' layers.

INTRODUCTION

There exist a number of attempts from different points of views to define the term knowledge. The concept of knowledge has been seen as a human related concept, therefore it has been approached philosophically, spiritually, psychologically, socially and organizational based, e.g. (Ayer, 1956), (Hrachovec & Pichler, 2007), (Bergson, 1999), (Bay & Backius, 1999), (Goswami, 2006), (Kotis & Vouros, 2006), (Singh & Jain, 2008), and (Choo, 1998). None of existing definitions of the term knowledge is yet the final one, i.e. none can be seen as true or complete enough.

Most importantly, none of the existing attempts to approach the concept of knowledge is practical enough for information systems' applications. For instance, the taxonomy of knowledge into both explicit knowledge and tacit knowledge has been widely spread, although this taxonomy is a false problem since it lacks to evidence (Bergson, 1999), (Popper, 1959), and (Styhre, 2004). Moreover, the concept of tacit knowledge is vague and produces problems to practical applications (Perraton & Tarrant, 2007), since tacit knowledge is seen as knowledge that the knower does not know he knows it, and cannot be systematically expressed, elicited, or codified (Polanyi, 1967).

Advising a framework for capturing and describing knowledge for conceptual design as

DOI: 10.4018/978-1-60960-887-3.ch019

sharable artifact among multiple domains and yet interpretable by the machine is still a research problem (Kitamura & Mizoguchi, 2004). Hence, from the point of view of computer-based information systems, there exists a need for perceiving the concept of knowledge with a more practical vision that relates the concept of knowledge more to information systems' applications.

In Literature Review, sample definitions of the concept of knowledge are surveyed as an attempt to show that existing definitions of knowledge, at least from an IS perspective, are:

a. *Ambiguous*, i.e. they do not come with a specific interpretation of the general statements used.
b. *Unpractical*, i.e. no direct link with IS applications can be well established.
c. *Incomplete*, i.e. not all IS layers are covered.
d. *Not true*, i.e. some knowledge theories lack evidence, e.g. the taxonomy of knowledge into explicit and tacit; hence, consequent theories of such false problems are redundant.

Then, The Relationship between Ontology, Knowledge, and Software Development provides an attempt to draw the relationship between the concept of knowledge and the concept of ontology. In Defining Knowledge: The WH Layers, a definition of knowledge is proposed based on the 'wh' approach as an attempt to provide a definition of the concept of knowledge that is:

a. *Unambiguous*, i.e. it has a clear and a specific interpretation.
b. *Practical,* i.e. it has a direct and pragmatic link to IS application.
c. *Complete,* i.e. it covers enough needed IS layers.
d. *Submissive to a validation procedure,* i.e. its truthiness can be shown with tangible experimentation.
e. *Domain-independent,* i.e. it can be applied on a wide enough set of application domains.

In Example on the Use of the 'WH' Approach, an example to demonstrate the application of the 'wh' approach is provided. Conclusion and Future Work concludes with final remarks highlighting main points in criticizing existing shape of the concept of knowledge, and summarizes main advantages of the proposed perception of knowledge from an IS perspective.

LITERATURE REVIEW

Knowledge is a) expertise, and skills acquired by a person through experience or education; the theoretical or practical understanding of a subject, b) what is known in a particular field or in total; facts and information or c) awareness or familiarity gained by experience of a fact or situation (The Oxford English Dictionary). This definition highlights the ambiguity surrounding the concept of knowledge; since it may refer to multiple distinct concepts such as facts or data, information or processed data, expertise, skills, a subject or a domain of theory, Etc.

Philosophy has its say about knowledge, e.g. Plato's, 427BC, sees knowledge as "*justified true belief*" (www.wikipedia.org). This definition does not provide any help for software developers, since it lacks to any pragmatic meaning concerning requirement elicitation. The much later Scientific Method assumes knowledge to be created by gathering observable, empirical and measurable evidence subject to specific principles of reasoning (www.wikipedia.org). The Scientific Method encompasses a set of principles and procedures for the systematic pursuit of knowledge involving the recognition and formulation of a problem, the collection of data through observation and experiment, and the formulation and testing of hypotheses (www.merriam-webster.com). Although this definition sets up the basic principles of thinking in gathering knowledge and hence requirements elicitation; however; requirement elicitation process needs more a more precise definition of the concept of knowledge.

(Davenport & Prusak, 1998) Relate knowledge to both individuals and organizations; hence define knowledge as a flow mix of framed experiences, values, contextual information, and expert insight that provides a framework for evaluating and incorporating new experiences and information. It originates and is applied in the mind of knower. In organizations, it often becomes embedded not only in documents or repositories but also in organizational routines, processes, practices, and norms.

The above definition is criticized by (Tsoukas, 2001) for being vague; i.e. it does not clearly distinguishes knowledge from information, also the definition packs into knowledge too many things, such as 'values', 'experiences' and 'contexts', without specifying their relationships. In addition, it does not clarify how individuals deal with organizational knowledge.

Other definitions of knowledge view it as digital information (Gates, 1999), (Lehner, 1990), and (Terrett, 1998). Attempts followed this path, have explored some practical issues as how knowledge is best stored, retrieved, transmitted, and shared (Brown & Duguid, 2000), and (Hendriks & Vriens, 1999) but not how it can be elicited.

Others, e.g. (Kay, 1993), saw knowledge as a pattern formed within, and drawn upon a firm over time. This vision has its own vagueness and impracticality, since it does not show the characteristic features of organizational knowledge, and does not specify the relationship between individual and organizational knowledge (Tsoukas, 2001).

With respect to the effect of both the individual role and the collective understandings, (Tsoukas, 2001) has defined knowledge as the individual ability to draw distinctions within a collective domain of action, based on an appreciation of context or theory, or both.

From an organizational perspective, knowledge has been seen as a primary resource for businesses.

In a knowledge-based view, an organization has been viewed as a site for the development, use of and dissemination of knowledge (Styhre, 2004).

Knowledge is still a vague concept (Tsoukas, 2001). The definition of knowledge affects its application's possible implementations. In the field of Information Systems (IS), we need a definition for knowledge that is

a. *Practical*: Provide direct link to implementation concepts.
b. *Unambiguous*: Clear and avoid philosophical and unspecific terms. And
c. *Comprehensive*: Covers all needed aspects of an IS applications.

THE RELATIONSHIP BETWEEN ONTOLOGY, KNOWLEDGE, AND SOFTWARE DEVELOPMENT

Bad ontology has led to badly perceive knowledge, and hence to badly develop software, e.g. Object-Oriented modeling approach has a number of limitations described in (Aguirre-Urreta & Marakas, 2008), (Potok, Vouk, & Rindos, 1999), (Cardelli, 1996), and (Armstrong, 2006). One source for those limitations originates from not providing a best match vision to the universe by neglecting global tasks performed on objects.

Ontology is the foundation of knowledge (Siricharoen, 2007). In philosophy, Ontology is the basic description of things in the world. In information systems, an ontology is an engineering artifact, constituted by a specific vocabulary used to describe a certain reality from a certain perspective (Fonseca, 2007).

From an information systems perspective, ontology can be seen as a part of knowledge. If knowledge is the answer of a set of questions, then ontology is the answer for the question: what exist in terms of a list of entities, their attributes and methods, a list of relationships joining those entities, and a set of applicable processes. That is, ontology is architecture of a systems, a Meta model of a knowledge base, or a global description of a possible body of knowledge.

Knowledge as a recursive and multidimensional concept can denote three types of knowledge; informative knowledge that describes a certain sphere or instance of the universe, a procedural knowledge that describes how to create and operate on such informative knowledge, and architectural or ontological knowledge that describes the hierarchy and arrangement of knowledge instances within a certain sphere of knowledge.

A philosophical question could be raised here; which is: can we have upper-level ontology (Smith, 2003)? That is, ontology that describes the hierarchy and arrangement of knowledge spheres. The answer is composed of two parts: the first part is 'no'. Since such ontology is beyond the perception of our limited minds. The second part, IS community should be pragmatic and to focus on practical issues that are applicable in IS.

Philosophy and social sciences as such, can be helpful to computer science field, but we should control the distance we drift in that direction keeping in mind that philosophy is a good source for inspiration, but not as a good source for practical applications.

DEFINING KNOWLEDGE: THE WH LAYERS

Introduction

Although there have been many attempts to distinguish between data, information, and knowledge, e.g. (Boisot, 1995), (Choo, 1998), (Davenport & Prusak, 1998), (Nonaka & Takeuchi, 1995), and (Bell, 1999); in essence, they are all knowledge with different levels of revelation. Hence, they can be all included within one definition of knowledge, at least from the point of view of IS.

There is a need to approach the concepts of ontology and knowledge in a more pragmatic

way than the one provided by philosophy or other social sciences, one such attempt can be found in (Fonseca, 2007).

Moreover, there is a need to extend the concept of knowledge to include more layers. One such attempt is found in (Geisler, 2006), in which knowledge has been classified into four types: *structure* ('*what*' are the entities and relations exist); *purpose* ('*why*' to use this knowledge); *function* ('*how*' to operate on this knowledge); and disciplinary *content* (Domain specifications).

In (Singh & Jain, 2008), the time dimension, i.e. '*when*' layer, is explored along with its relation to knowledge. One way used to describe the relation between time and knowledge is to divide time into three parts: past, present, and future (R. Mitch Casselman, 2005).

Modeling a software system is a nontrivial task due to the possible complexity in eliciting required knowledge, and possible large amount of details to care about. There is a need for a systematic way to structure analyzed knowledge and convert it into a logical model. Once a satisfactory logical model is constructed, it can be systematically converted into physical design, and hence construct an application. The key to be able to perform systematic analysis is to structure human knowledge in an unambiguous and practical way that covers IS applications layers.

Existing attempts to provide gaudiness in the knowledge extraction process in the analysis and design phases, have followed a domain-specific approach in order to specify a knowledge sphere, e.g. (Pinto, Marques, & Santos, 2009), (Abu-Hanna & Jansweiier, 1994), (Garcia, Tomas, & Breis, 2008), and (Eugenia & Isidro, 2008). Other attempts specified a domain of technology, e.g. Object-Oriented based programming, and constructed related knowledge ontology, e.g., (Siricharoen, 2007) and (Garzas & Piattini). In (Maillot, Thonnat, & Boucher, 2004), a domain-independent knowledge acquisition process was proposed, were the process of ontology development is divided into four phases:

Specification, Conceptualization, Formalization, and *Implementation.*

However, this article suggest that our vision to domains should be task-oriented rather than domain-specific in order to better perceive the universe and to provide a more reusable ontology and software components. We need to look after two levels of behavior; one, micro behavior within an object, i.e. a method, and two, macro behavior between objects, i.e. a task.

If functions are defined depending on operands or their realization method, few functions are reused and transferred to different domains. Innovative design can be facilitated by flexible application of knowledge or ideas across domains (Kitamura & Mizoguchi, 2004).

Software Systems are task-oriented in essence. One main reason why the development of software systems does not meet the required needs is the lack of understanding of the software's real objective, i.e. the task it should perform and how it should be performed (Zlot, Oliveira, & Rocha, 2002).

Approaching software development in a task-oriented fashion in the phases of analysis, design, and implementation of a software development process, can lead to a better attempt towards tackling the problem-implementation gap described in (France, 2007). Applying Task-Oriented approach in software development, e.g. (Cooperman, 1996), (Gaizauskas, 2006), (Kitamura Y. S., 2000), (Martinez, 2001), (El Sheikh, 1987), (Mizoguchi, 1996) (Mustafa), (Reichart, 2004), (Rosson, 1995), (Yen, 1993), and (Zlot F. O., 2002), can result into a software system that is adaptable, flexible, expandable, and open (Böttcher, 2004).

Layers of Knowledge: An Information Systems Perspective

Information systems are composed of layers. The most common layers are *'what'* and *'how'*. Those layers reflect the *'wh'* questions the system is build to answer:

a. A *'what is'* layer that describes what exists in such a system, i.e. what are the entities or data to care about.

b. A *'which'* layer that describes the relations or the hierarchy entities have among each other.

c. A *'how to'* procedural knowledge layer that describes functions of such a system, i.e. How to perform required operations on defined entities or data on a macro level, and how do objects behave on a micro level. A function on a macro level is a task that could be composed of smaller subtasks. A task is a unit of logical work, performed by a software system, applicable on a set of domains to fulfill a certain purpose. A task is *'how to'* knowledge applied on *'what is'* knowledge. For example, diagnosis is a logical task performed by a software system that shares the same ontology among different domains.

d. A *'why'* layer that describes the intended purpose of the system, hence, aids in defining its scope and constraints.

e. A *'when'* or a temporal layer that describes the time dimension, i.e. what was the value of this datum in a certain moment in the past, what is the current value of a certain datum, and what is the anticipated value of a certain datum in the future.

f. A *'where'* or a spatial layer that describes the place dimension, i.e. given a certain location, what was/is/will be the value of a certain datum there.

g. A *'who'* or audit layer that records which user did this or that, and which user is allowed to perform a certain operation, i.e. who did this, and who can do that.

h. An *aspects* layer or the *'physical how'* that describes desired performance characteristics of such a system, i.e. how to implement above layers giving some desired quality, such as speed, fragmentation, or/and security... etc. Aspect-Oriented programming is originally proposed by (Kiczales, 1997).

Hence, the concept of knowledge from an IS perspective can be extended to include not only the traditional *'what'* and logical *'how'* knowledge, but also to include *'when'*, *'where'*, *'why'*, *'who*, and *'aspect'* layers of knowledge. Knowledge as a general concept can be seen as *answers of a set of 'wh' questions*. The quality of knowledge depends on the number of questions it answers, and the degree of correctness and thoroughness each answer is.

EXAMPLE ON THE USE OF THE 'WH' APPROACH

Three spheres of knowledge are required when building an information system; knowledge about the platform, i.e. the machine, its operating system, and development technology, knowledge about the problem or domain in hand, and knowledge about how to build such a software system within required time, cost, and functionality.

Typically, for a software developer a gap exists with the domain-specific knowledge. The analysis phase aims to elicit such domain or problem related knowledge in such a way that is formal to be transformable to a formal logical model, and complete enough to cover hopefully all needed functionalities required by the targeted system. Constructing knowledge in this phase usually follows an ad-hoc approach and is subjective.

The difficulties faced by software developers in the analysis phase result from the vagueness of the goal they are seeking, i.e. the unclear concept of knowledge they are targeting to construct for this information system. Moreover, the existing definitions of knowledge recognize the *'what'* and *'how'* layers, i.e. they do not cover all information systems layers.

As an experiment, a group of undergraduate senior students were given an exercise to analysis

a certain information system. After spending four sessions on two weeks, the students were unable to come out with a satisfactory logical design. Difficulties reported were in the lack of practical guidance in exploring such a knowledge sphere. The similar difficulty was also reported by professional analysts, although with years of experience and practice each one of them has been able to develop his/her own ways, based on intuition and previous successful solutions, to conclude to a satisfactory design.

There is a lack of a proper approach for constructing knowledge in content-oriented domain; an ad-hoc approach is usually applied, and there is no methodology that enables knowledge to accumulate systematically (Kitamura & Mizoguchi, 2004).

The 'wh' approach is proposed to provide the needed guidance in exploring required domain or problem knowledge for an information system.

To demonstrate the approach, a case of a supermarket is given to be analyzed and modeled using an ERM or a Class Diagram. All IS layers of an IS are related to an ERM except the 'how to' layer which is depicted as a Data Flow Diagram for each module, as methods for each object in a class diagram, and as a system's architecture that describes how software components interacts. A supermarket IS keeps records of products, families of products, invoices, suppliers list, users, etc. The 'wh' knowledge constructing approach for IS suggests that knowledge construction should follow the layers of an IS.

a.   The *'what'* layer: This *declarative* knowledge layer describes what entities and attributes exist in an IS. Below a sample list of *'what'* questions is given that the targeted IS should answer, divided into two levels of abstraction; I. Entities, the macro level, and II. Attributes, the micro level:

 ◦   ***Entities***:

  1.   What products the supermarket has.

2. What is the list of suppliers the supermarket deals with.
3. What transaction the supermarket has, i.e. what is the list of invoices.

○ *Attributes* The '*product*' entity is picked as an example:

1. What is the primary key.
2. What is the amount of each product in store.
3. What is the product's name.
4. What is a certain product unit weight, size, or count.

b. The '*which*' layer: This *declarative* layer describes which relationship a certain entity set has with another entity set in the targeted IS, e.g.

1. Which Family this product belongs to.
2. Which products included in this invoice.
3. Which customer bought this certain product.
4. Which supplier provides this product.

c. The '*why*' layer: This layer describes the purpose of the targeted system in two levels:

○ The macro level, which describes the overall purpose of the system. As an answer for the question: why this system is being built for, or what is this system's task.

○ The micro level, which describes each components purpose. As an answer for what is the subtask or the function this component performs.

d. The '*when*' layer: This layer reflects the temporal nature of knowledge and hence of an IS database. To simplify this example, the database is assumed not be temporal. Typical question could be asked here are:

○ At a macro level, this describes the time dimension for the whole system. As an answer for the questions: What was/is/ will be the state of the system or one of its subsystems at a certain moment in time.

○ At micro level, to describe the time dimension for a variable as an answer for the question: What was/is/ will be the value of a given variable at a certain moment in time. E.g., what was the price for this product at this date.

e. The '*where*' layer: This layer describes space dimension for mobile objects and applications in which is reflected into spatial database. The selected case may not have space dimension since the supermarket is a single place. E.g. where was this entity when this event occurred or when this condition was satisfied.

f. The '*who*' layer: This layer is reflected by adding the audit aspect to a database, in which it mainly answers two questions: who did this, and who can do that.

g. The '*aspect*' layer. This layer turns the attention from logical design to physical design, concerning performance aspects. For example, which dataset is assumed to be large enough and should be defragmented horizontally or mirrored. Which algorithm is the best choice to apply for this function given a certain dataset.

CONCLUSIONS AND FUTURE WORK

Applying the 'wh' approach for exploring knowledge in analyzing phase provides a sort of guidance for the software developer to enable him/her to systematically elicit the application domain, and hence, produces a logical model. A logical model can be then transformed into a physical model that is applicable to the machine.

The concept of knowledge needs to be stabilized and extended to include more than the traditional 'what' and 'how' knowledge, to provide a better fit to IS needs as a practical, clear, and complete enough concept.

Knowledge is the answer of a question, it is composed of the seven layers: '*what*', '*which*', 'how', 'why', 'when', 'where', and 'who' layers. To describe an IS, we need to know the entities with their attributes and relations among them, the task and function perfumed by the system, the temporal and spatial dimension, and the audit dimension.

Unlike other disciplines' attempts to perceive knowledge, in Information Systems all layers of knowledge have a direct link to IS application. For instance, the 'how to' knowledge can be represented on the forms of an algorithm, a procedure, or a function.

Knowledge by nature is a recursive multidimensional concept; ontology can be seen as one part of knowledge that describes the structure of a certain sphere of knowledge.

As a future work, applying the 'wh' approach on domains knowledge can bring order to such expertise so that it can be explored, represented, and shared in a systematic manner. If so, IS can help in connecting and unifying such expertise into one domain-independent system of knowledge spheres.

REFERENCES

Abu-Hanna, A., & Jansweiier, W. (1994). Modeling domain knowledge using explicit conceptualization. *IEEE Expert, 9*(5). doi:10.1109/64.331490

Aguirre-Urreta, I., & Marakas, M. (2008). Comparing conceptual modeling techniques: A critical review of the EER vs. OO empirical literature. *ACM SIGMIS Database Newsletter, 39*(2).

Armstrong, J. (2006). *The quarks of object-oriented development.*

Ayer, A. (1956). *The problem of knowledge.* London, UK: Penguin.

Bay, T., & Backius, P. (1999). Reiterating experimentation: Inventing new possibilities of life. *Emergence, 1*(3). doi:10.1207/s15327000em0103_7

Bell, D. (1999). *Foreword to The coming of the post industrial society.* New York, NY: Basic Books, Special Anniversary Edition.

Bergson, H. (1999). *An introduction to metaphysics.* Indianapolis, IN: Hackett.

Boisot, M. (1995). *Information space: A framework for learning in organizations, institutions and culture.* London, UK: Routledge.

Böttcher, S. A. (2004). A task-oriented reconfigurable software architecture for e-commerce document exchange. *Proceedings of the 37th Hawaii International Conference on System Sciences,* IEEE Computer Society.

Brown, S., & Duguid, P. (2000). *The social life of information.* Boston, MA: Harvard Business School Press.

Cardelli, L. (1996). Bad engineering properties of object-oriented languages. *ACM Computing Surveys, 28*(4). doi:10.1145/242224.242415

Choo, C. (1998). *The knowing organization: How organizations use information to construct meaning, create knowledge, and make decisions.* New York, NY: Oxford University Press.

Cooperman, G. (1996). TOP-C: A task-oriented parallel C interface. *Proceedings of 5th IEEE International Symposium on High Performance Distributed Computing.* Syracuse, NY.

Davenport, T. H., & Prusak, L. (1998). *Working knowledge.*

El Sheikh, A. (1987). *Simulation modeling using a relational database package.* PhD Thesis. London, UK: The London School of Economics.

Eugenia, C., & Isidro, R. (2008). *The baseline: The milestone of software product lines for expert systems automatic development.* Baja, CA: Mexican International Conference on Computer Science, IEEE.

Fonseca, F. (2007). The double role of ontologies in information science research. *Journal of the American Society for Information Science and Technology, 58*(6). doi:10.1002/asi.20565

France, R. A. (2007). *Model-driven development of complex software: A research roadmap.*

Gaizauskas, R. A. (2006). Task-oriented extraction of temporal information: The case of clinical narratives. *Proceedings of the Thirteenth International Symposium on Temporal Representation and Reasoning,* IEEE Computer Society.

Garcia, R., Tomas, J., & Breis, F. (2008). A knowledge acquisition methodology to ontology construction for information retrieval from medical documents. *Expert Systems: International Journal of Knowledge Engineering and Neural Networks, 25*(3).

Garzas, J., & Piattini, M. (2006). An ontology for understanding and applying object-oriented design knowledge. *International Journal of Software Engineering and Knowledge Engineering, 17*(3).

Gates, B. (1999). *Business @ the speed of thought.* London, UK: Penguin Books.

Geisler, E. (2006). A taxonomy and proposed codification of knowledge and knowledge systems in organizations. *Knowledge and Process Management, 13*(4). doi:10.1002/kpm.265

Goswami, C. (2006). *What do we do with knowledge?* Hendriks, P., & Vriens, D. (1999). Knowledge-based systems and knowledge management: friends or foes? *Information & Management, 35*(2).

Hrachovec, H., & Pichler, A. (2007). *Philosophy of the information society. Proceedings of the 30th International Ludwig Wittgenstein Symposium Kirchberg am Wechsel.*

Kay, J. (1993). *Foundations of corporate success.* New York, NY: Oxford University Press.

Kiczales, G. (1997). *Aspect-oriented programming.* Finland, Springer-Verlag: European Conf. on Object-Oriented Programming.

Kitamura, Y., & Mizoguchi, R. (2004). Ontology-based systematization of functional knowledge. *Journal of Engineering Design, 15*(4). doi:10.1080/09544820410001697163

Kitamura, Y. S. (2000). Functional understanding based on an ontology of functional concepts. *Proceedings of Sixth Pacific Rim International Conference on AI,* Berlin.

Kotis, K., & Vouros, G. (2006). Human-centered ontology engineering: The HCOME methodology. *Knowledge and Information Systems Journal, 10* (1).

Lehner, F. (1990). Expert systems for organizational and managerial tasks. *Information & Management, 1*(23).

Maillot, N., Thonnat, M., & Boucher, A. (2004). Towards ontology-based cognitive vision. *Machine Vision and Applications,* 16.

Martinez, J. (2001). EZSTROBE- General-purpose simulation system based on activity cycle diagrams. *Proceedings of the 33rd Conference on Winter Simulation.* Arlington, VA: IEEE Computer Society.

Mizoguchi, R. S. (1996). *Task ontology design for intelligent educational/training systems.* Position Paper for ITS'96 Workshop on Architectures and Methods for Designing Cost-Effective and Reusable ITSs.

Mustafa, K. C. (2006). Task-oriented integrated use of biological Web data sources. *Proceedings of the 18th International Conference on Scientific and Statistical Database Management,* IEEE Computer Society.

Nonaka, I., & Takeuchi, H. (1995). *The knowledge-creating company: How Japanese companies create the dynamics of innovation.* New York, NY: Oxford University Press.

Perraton, J., & Tarrant, I. (2007). What does tacit knowledge actually explain? *Journal of Economic Methodology, 14*(3). doi:10.1080/13501780701562559

Pinto, F., Marques, A., & Santos, M. (2009). Ontology-supported database marketing. *Journal of Database Marketing and Customer Strategy Management, 16.*

Polanyi, M. (1967). *The tacit dimension.* London, UK: Routledge and Kegan Paul.

Popper, K. (1959). *The logic of scientific discovery.* London, UK: Hutchinson.

Potok, E., Vouk, M., & Rindos, A. (1999). Productivity analysis of object-oriented software developed in a commercial environment. *Software-Practice and Experience, 29*(10). Casselman, R. M., & Samson, D. A. (2005). Moving beyond tacit and explicit: Four dimensions of knowledge. *Proceedings of the 38th Annual Hawaii International Conference on System Sciences,* IEEE Computer Society.

Reichart, D. F. (2004). Task models as basis for requirements engineering and software execution. *Proceedings of the 3rd Annual Conference on Task Models and Diagrams.* Prague, Czech Republic: ACM.

Rosson, M. A. (1995). Integrating task and software development for object-oriented applications. *Proceedings of the SIGCHI Conference on Human Factors in Computing Systems.* Denver, CO: ACM Press.

Singh, R., & Jain, S. (2008). Knowledge mechanics and its temporal nature. *Ubiquity, 9*(15).

Siricharoen, W. (2007). Ontologies and object models in object oriented software engineering. *International Journal of Communication Systems, 33*(1).

Smith, B. (2003). *Blackwell guide to the philosophy of computing and information.* Oxford, UK: Blackwell.

Styhre, A. (2004). Rethinking knowledge: A Bergsonian critique of the notion of tacit knowledge. *British Journal of Management, 15*(2), 177–188. doi:10.1111/j.1467-8551.2004.00413.x

Terrett, A. (1998). Knowledge management and the law firm. Knowledge Management, *1*(2).

The Oxford English Dictionary. (n.d.). *OED.*

Tsoukas, H. (2001). What is organizational knowledge? *Journal of Management Systems, 7*(38).

Yen, J. a. (1993). *Task-based methodology for specifying expert systems.*

Zlot, F., Oliveira, K. D., & Rocha, A. (2002). Modeling task knowledge to support software development. *Proceedings of the 14th International Conference on Software Engineering and Knowledge Engineering.* Ischia, Italy: ACM.

Zlot, F. O. (2002). Modeling task knowledge to support software development. *Proceedings of the 14th International Conference on Software Engineering and Knowledge Engineering.* Ischia, Italy: ACM.

KEY TERMS AND DEFINITIONS

Domain: A sphere of knowledge with interest to humans, i.e. A set of vocabulary and a set of processes performed on those vocabularies in a certain human related functions; e.g. biology, medicine, pathology,.. Etc.

Domain-Specific Software Development: Developing software systems through the aid of software development tools with possibly modeling notations built and designed specifically for a certain domain, e.g. suggesting modeling notation and building tool for developing medical expert systems for diagnosis.

Knowledge Quality: The number of questions answered, and the degree of correctness and thoroughness each answer is.

Knowledge: As a general concept it can be seen as the answers of a set of questions. The concept of knowledge from an IS perspective can be extended to include not only the traditional 'what' and logical 'how' knowledge, but also to include 'when', 'where', 'why', 'who, and 'aspect' layers of knowledge.

Ontology: A part of knowledge. It is the answer for the question: what exist in terms of a list of entities, their attributes and methods, a list of relationships joining those entities, and a set of applicable processes.

Task: is a unit of logical work, performed by a software system, applicable on a set of domains to fulfill a certain purpose. A task is '*how to*' knowledge applied on '*what is*' knowledge. E.g. Diagnosing a phenomenon regardless the domain of application.

Task-Oriented Software Development: A programming paradigm where significant and recurring software tasks are defined and both tools and possibly modeling notations are suggested to enable rapid and quality software development; with emphasize on reusing those tools and notations among different domains.

Chapter 20
Virtual Reality in Architecture, in Engineering and Beyond

Nicoletta Sala
Università della Svizzera italiana, Switzerland

ABSTRACT

Virtual Reality (VR) is a technology which has various application fields (from video games to psychiatry). It is indispensable in critical simulation, for instance in military training, in surgical operation simulation, in creation of environments which could set off phobias (in psychiatry), or in realization of virtual prototypes, for instance in industrial design. The aim of this chapter is to present how the VR also finds excellent application fields in architecture and in engineering. For instance, in the teaching of the basic concepts, in techniques of graphic rebuilding for the building restoration, in realization of virtual visits inside buildings, and in urban generative processes simulated by computer. Another use of the virtual reality is in the introduction of a new kind of architecture: Virtual Architecture, strongly connected to the Information and Communication Technology (ITC), to the Internet, and in the virtual prototyping in engineering.

INTRODUCTION

Virtual reality (VR) is a technology which permits to create virtual objects and 3D virtual worlds which are hosted on the computer's memories. VR can be classified according to its methods of display; we have: a) immersive VR (which involves a high degree of interactivity and high cost peripheral devices, for example the head mounted displays), and b) non-immersive VR (often called desktop VR which is in the form of a windows into a virtual world displayed on a computer's monitor). A virtual reality system has the following three primary requirements (Rosenblum & Cross, 1998): a) immersion (which permits to the user the physical involvement, capturing exclusive visual attention and responding to three-dimensional input. For example, through a head-tracker, 3D mouse, data glove, or fully instrumented body suit); b) interaction

DOI: 10.4018/978-1-60960-887-3.ch020

Figure 1. VR in architecture: ARTHUR project (www.vr.ucl.ac.uk/projects/arthur/)

(through the three-dimensional control device to "navigate" in the virtual environment); c) visual realism (which is the representation of the virtual world using computer graphics techniques). The interface hardware components consist of a set of visual display apparatus (for example, head mounted display), input devices (for instance, data gloves), and position sensors (Sherman & Craig, 2002).

This chapter, that describes two application fields of VR, is organized as follows: first VR in architecture is discussed, then VR in engineering, which is followed by the conclusions and the future trends. The final section is dedicated to the references.

VIRTUAL REALITY IN ARCHITECTURE

Different disciplines involve the application of VR. Recent studies have recognized the potential of VR in education in different disciplines (Winn, 1993; Pantelidis, 1995; Byrne, 1996; Youngblut, 1998; Ainge, 2000; Kaufmann et al., 2000; Mantovani,

2001, Gerval et al., 2002; Shin, 2004; Sala & Sala, 2005; Popovici et al., 2009). Virtual Reality offers benefits that can support the education and the design project, in particular in the faculties of architecture.

Sala & Sala (2005) applied VR technology in a course of mathematics oriented for architectural studies and dedicated to the connections between this discipline and the world of arts and design. VR has been used firstly to help the students to visualise in three dimensions, since this is arguably the most difficult part of understanding architecture. In this case, the students have used the Virtual Reality: to study and to manipulate virtual polyhedra, to observe and to manipulate virtual molecules of C_{60} and C_{70} (fullerene molecules), to analyse the symmetry inside the architectural shapes, to observe and to manipulate 3D virtual models of geodesic domes, and to study their analogy with the fullerene molecules (Sala & Sala, 2005).

In the same faculty, VR is also integrated in multimedia presentation to describe "virtual 3D walkthroughs" in the buildings, in virtual museums and in archaeological sites. This educational approach, based on the use of VR, is important,

Figure 2. NYSE 3DTF (http://www.archphoto.it/images/asymptote/nyse2.htm)

because in the next future the architects will design themselves virtual environments, virtual buildings and reorganize new spaces (Prestinenza Puglisi, 1999). Figure 1 shows an application of VR in architecture: the ARTHUR project which developed an augmented reality interface for round table design meetings, organised by University College London (www.vr.ucl.ac.uk/projects/arthur/).

The powerful of computer systems connected to the evolution of virtual reality technologies have introduced new opportunities in the architectural field, for example to realize braver designs. Figure 2 shows the virtual environment conceived by Hani Rashid and Lise Anne Couture (Asymptote group): The New York Stock Exchange 3D Trading Floor (NYSE 3DTF). It was conceived as a reinterpretation and transformation of the existing physical trading environment, and as provocation for a new, physically augmented architecture. The architectural idealization had to provide absolute flexibility; NYSE 3DTF took full advantage of opportunities in virtual space to manipulate spatial and temporal dimensions. About the project Rashid and Couture stated: «*The idea was to create a visual environment through*

which traders can navigate, analyze, and act upon at-a-glance. Trade actions are very dynamic» (Andia & Busch, 1999).

NYSE was the first business application of an interactive virtual architecture. Rashid (2003) affirmed:

The virtual architecture is an evolving discipline which results from the convergence of mapped data and of simulation, production of the digital form, architecture of the information, buildings and theory of the virtual reality [...]. The conventional architectures base themselves on the stay and geometrical certainty, while virtual architecture uses digital technologies to widen the real events, the time and the space. (Rashid, 2003, p. 169)

All of the information that was relevant to the NYSE and its daily activity of trades and transactions was mapped into this fully navigable multi-dimensional world, and the complex relationship between financial "events" and media news-reporting of cause-and-effect was made transparent (as shown in figure 2).

Figure 3. An example of automatic generative method to generate 3D-Urban Model based on L-systems (Saleri, 2002)

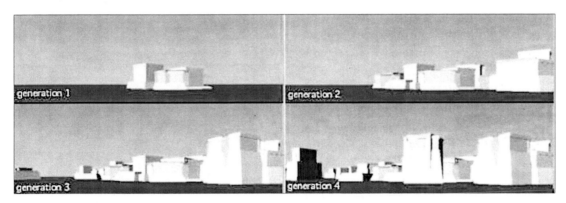

Other applications of VR is in the urban design which involve the use of "virtual models" in the following ways: a) to model proposed urban "guidelines" for newly developed areas. For example, different housing proposals could be compared for a vacant city block; b) to model existing urban zones which require constant reappraisal. For example, "virtual urban designers" could walk through urban spaces and observe how they might be better used. City commercial spaces may be replaced by urban dwellings.

Saleri (2002, 2006, 2008) studied different automatic generative methods able to produce architectural and urban 3D-models. He used fractal algorithms based on L-systems replacing the brackets by specific 3D operations, typically affine transformations, such as rotations or translations, and the "0" and "1" occurrences by 3D pre-defined objects. Iterative instructions written in VRML (Virtual Reality Modelling Language) generated urban virtual models, an example is shown in figure 3.

The use of VRML is not casual. In fact, it is a language that specifies the parameters to create virtual worlds networked together via the Internet and accessed via the World Wide Web hyperlinks. A VRML file is an ASCII file (with the suffix. wrl), which is interpreted by the browsers, for example Cosmoplayer® or Cortona®, and converted into a 3D display of the described world.

VIRTUAL REALITY IN ENGINEERING

The applications of VR technologies in engineering can involve the education, the training and the design. Virtual reality can help to train the students to avoid the laboratory accidents. In fact, every year hundreds of students are injured in laboratory accidents, in spite of all safety precautions. One reason for this is that people tend to become forgetful and complacent, and stop observing safe laboratory procedures. In agreement to this consideration, VR's technologies become tools to realize a set of based accident simulations, to allow users to experience first-hand the consequences of not following safe laboratory procedures, for example in the chemical laboratory (Bell & Fogler, 2001, 2004).

VR technologies can also help the applications designed to provide for the development and practice of work-related skills. Antonietti et al. (1999; 2001) proposed a prototypal system for machine tools teaching in a Virtual Reality environment integrated with hypermedia. The goal was to lead students not only to understand the structure and functioning of the lathe, but also to use such a machine (Antonietti et al., 1999; Antonietti et al., 2001).

Virtual reality also supports the creation of 3D environment to facilitate engineering learning activities (Popovici et al., 2009).

VR's technologies is applied in various domains like mechanical engineering, for example in the vehicle design, in particular to visualise the vehicle architecture during the upstream phase. This is possible because the development of Computer Aided Design (CAD) and the geometric based design have reached a high level of maturity and affordability. CAD systems permit to the designer to evaluate the geometry of his virtual design. At this step of the design process, modifications are still quite cheap, compared with changes to a physical prototype or, even worse, the final product. Many companies use this "virtual" approach to improve the effectiveness and efficiency of the design process. This can be a very much time-consuming and expensive process.

Thus, the designer should be able to define and to test the desired behaviour of a forthcoming product in such a way that the corresponding geometry is created automatically by the CAD system. The designers can interact with a virtual prototype as physical one. VR technologies are applied for producing a Virtual Prototyping (VP), instead of or as a supplement to development of physical prototypes. This is a process which uses a virtual prototype, instead of a physical prototype, to test and to evaluate of specific characteristics of an object of design. A virtual prototype can be defined as a computer-based simulation of a system or subsystem with a degree of functional realism comparable to a physical prototype. An example of virtual prototype is shown in figure 4. In VR prototyping, the 3D image is often viewable using a sophisticated set of glasses that imperceptibly shift the wearer's vision from one eye to the other using a system synchronized to a computer. The technology can be used on platforms ranging from PCs to four, five, or six-wall CAVE (Cave Automatic Virtual Environment) systems for fully immersive visualization environments (Elliott, 2005). Virtual reality is connected with the "true" reality through Projective VR which permits to the users to project actions carried out in the virtual world into the real world by means of robots or

Figure 4. Virtual Prototype: an example

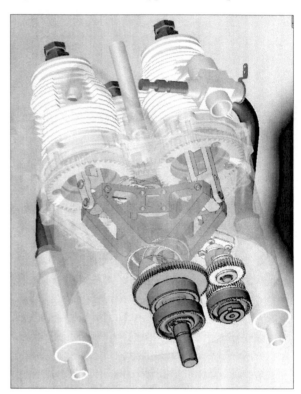

other means of automation (Freund & Rossmann, 2005). Figure 5 shows an example of Projective VR (http://www.ciros-engineering.com/en/products/virtual-reality/projective-vr.html).

CONCLUSIONS AND FUTURE TRENDS

Virtual reality will have potential future applications in architecture and engineering which will include: a) marketing tool (for example, interactive adaptive displays, to demonstrate the use of different solutions of a building); b) communication tool (cross distance and language barriers; for example, between the architect and client, and to educate architects and engineers); c) evaluation modelling tool (for example, to study effects of lighting, natural and artificial, to evaluate acoustics phenomena, to simulate the properties' of

Figure 5. Projective VR

Figure 6. Virtual building in Second Life realised using fractal algorithms

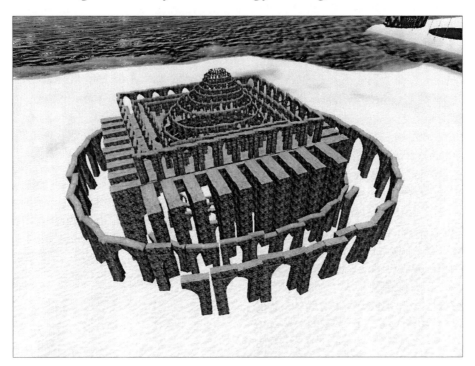

the material), and d) modelling/design tool (for example, to analyse spaces by actually "getting inside them", to incorporate rational data during schematic design stages, then look at different design solutions, to design "virtual architecture", and "virtual prototype" in collaboration).

Some of the possible benefits of VR on the design process and practice of architecture and engineering could be: a) the ability to test ideas in "real time" in a "three-dimensional" space during the design process; b) communication of ideas; c) the elimination of much of the guesswork in design, d) braver and better designs, and e) the integration of the design processes.

For the architectural education, virtual reality will become the place to go to do things that you could not normally do in architect-designed buildings. Spaces created using Fractal and non-Euclidean geometry will exist and they could be modified using algorithms and computer programs (Garofalo, 1999; Ervin & Hasbrouck, 2001; Sala, 2009a, 2009b).

Recent application of desktop VR is to create virtual worlds and virtual environments connected to the World Wide Web (WWW). Second Life (SL) is an example in this field. SL could begin a "territory" where is possible to try interesting virtual building, an example is shown Figure 6. It has been realized using fractal algorithms in the course of New Media at the Academy of Architecture of Mendrisio (USI, Switzerland).

A crucial point for future applications of VR could be: how to create the virtual worlds. Some virtual worlds will be oriented certainly to the educational field and other for training, works or fun. Architects will potentially help to make the virtual world a pleasant and stimulating place to work and live in, with a good quality of life. This will require people who understand the psychological effects of the spaces, generated by the computer, on people inside them, and the architects have to prepare themselves to this new work op-

portunity. Architects as designers of Virtual Worlds will be required to make these environments interesting, rich, and engaging places. Therefore, it is important to prepare a correct training on the use of VR in the faculties of architecture and engineering.

Virtual Reality will help us to change our concept of space. Bachelard wrote (1964):

By changing space, by leaving the space of one's usual sensibilities, one enters into communication with a space that is psychically innovating. ... For we do not change place, we change our nature.

REFERENCES

Ainge, D. (2000). Introducing primary students to VR with Do 3D. *VR in the Schools, 4*(3). Retrieved June 6, 2001, from http://www.coe.ecu. edu/vr/ vrits/4-3Ainge.htm

Andia, A., & Busch, C. (1999). *Architects design in cyberspace and real space for the NYSE*. Retrieved December 10, 2009, from http://www.members. tripod.com /wwwandia/art/100.html

Antonietti, A., Imperio, E., Rasi, C., & Sacco, M. (2001). Virtual reality and hypermedia in learning to use a turning lathe. *Journal of Computer Assisted Learning, 17*, 142–155. doi:10.1046/j.0266-4909.2001.00167.x

Antonietti, A., Rasi, C., Imperio, E., & Sacco, M. (1999). Virtual reality in engineering instruction: In search of the best learning procedures. *Proceedings World Conference on Educational Multimedia, Hypermedia and Telecommunications 1999*, (pp. 663-668).

Bachelard, G. (1964). *The poetics of space*. New York, NY: Orion Press.

Bell, J. T., & Fogler, H. S. (2001). Virtual reality laboratory accidents. *Proceedings of American Society for Engineering Education Annual Conference*, Albuquerque, NM, American Society for Engineering Education, 2001. Retrieved January 2, 2008, from http://www.cs.uic.edu/~jbell /Professional/Papers/ASEE2001.pdf

Bell, J. T., & Fogler, H. S. (2004). The VRUPL lab - Serving education on two fronts. *Proceedings of the Special Interest Group on Computer Science Education Annual Conference*, Norfolk, VA, March 2004. Retrieved January 3, 2010, from http://www.cs.uic.edu/~jbell/ Professional/ Papers/ Bell_sigcse2004.pdf

Byrne, C. M. (1996). *Water on tap: The use of virtual reality as an educational tool.* Ph.D. Dissertation. University of Washington, Seattle, WA.

Elliott, L. (2005). *See it, feel, it, hear* it. Retrieved December 20, 2009, from http://www.deskeng. com/ articles/aaaapn.htm

Ervin, S., & Hasbrouck, H. (2001). *Landscape modeling.* New York, NY: McGraw-Hill.

Freund, E., & Rossmann, J. (2005). Projective virtual reality as a basis for on-line control of complex systems-not only-over the internet. *Journal of Robotic Systems, 22*(3), 147–155. doi:10.1002/rob.20055

Galofaro, L. (1999). *Digital Eisenman-An office of the electronic era.* Basel, Switzerland: Birkhäuser.

Gerval, J. P., Popovici, M., Ramdani, M., El Kalai, O., Boskoff, V., & Tisseau, J. (2002). Virtual environments for children. *Proceedings International Conference on Computers and Advanced Technology Education (CATE),* (pp. 416-420). Cancun, Mexico.

Kaufmann, H., Schmalstieg, D., & Wagner, M. (2000). Construct3D: A virtual reality application for mathematics and geometry. [Springer Netherlands.]. *Education and Information Technologies, 5*(4), 263–276. doi:10.1023/A:1012049406877

Mantovani, F. (2001). VR learning: Potential and challenges for the use of 3D environments in education and training. In Riva, G. (Ed.), *Towards CyberPsycology: Mind, cognition and society in the Internet age* (Vol. 2, pp. 207–225). Amsterdam, The Netherlands: IOS Press.

Pantelidis, V. S. (1995). Reasons to use virtual reality in education. *VR in the Schools, 1*(1). Retrieved February 15, 2004, from http://eastnet. educ.ecu.edu /vr/vr1n1a.txt

Popovici, D. M., Gerval, J. P., Hamza-Lup, F., Querrec, R., Polceanu, M., Popovici, N., & Zăgan, R. (2009). 3D virtual spaces supporting engineering learning activities. *International Journal of Computers, Communications & Control, 4*(4), 401–414.

Prestinenza Puglisi, L. (1999). *Hyper-architecture spaces in the electronic age.* Basel, Switzerland: Birkhäuser.

Rashid, H. (2003). Architettura virtuale, spazio reale. In L. Sacchi & M. Unali (Eds.), *Architettura e cultura digitale,* (pp. 169-178). Milano, Italia: Skirà.

Rosemblum, L. J., & Cross, R. A. (1997). The challenge of virtual reality. In Earnshaw, W. R., Vince, J., & Jones, H. (Eds.), *Visualization & modeling* (pp. 325–399). San Diego, CA: Academic Press.

Sala, N. (2009a). Fractal geometry in virtual world and in Second Life. [New York, NY: Nova Science.]. *Chaos and Complexity Letters, 4*(1), 23–34.

Sala, N. (2009b). Fractals in architecture, hyper-architecture and…beyond. [New York, NY: Nova Science.]. *Chaos and Complexity Letters, 4*(3), 147–180.

Sala, N., & Sala, M. (2005). Virtual reality in education. In Carbonara, D. (Ed.), *Technology literacy applications in learning environments* (pp. 358–367). Hershey, PA: Idea Group Inc. doi:10.4018/978-1-59140-479-8.ch025

Saleri, R. (2002). Pseudo-urban automatic pattern generation, *Proceedings 5ᵗʰ Generative Art Conference,* Milano, Italy. Retrieved March 10, 2003, from http://hal.archives-ouvertes.fr/docs/00/26/74/00/PDF/GA2002.pdf

Saleri, R. (2006). Urban and architectural 3D fast processing. *Proceedings 9ᵗʰ Generative Art Conference,* Milano, Italy. Retrieved January 15, 2008, from http://hal.archives-ouvertes.fr/docs/00/26/73/63/PDF/Renato_Saleri.pdf

Saleri, R. (2008). 3D automatic processing of architectural and urban artifacts. In Orsucci, F., & Sala, N. (Eds.), *Reflexing interfaces: The complex coevolution of Information Technology ecosystems* (pp. 278–289). Hershey, PA: IGI Global.

Sherman, W., & Craig, A. (2002). *Working with virtual reality.* New York, NY: Morgan Kaufmann Publishers.

Shin, Y. S. (2004). Virtual experiment environment's design for science education. *International Journal of Distance Education Technologies, 2*(2), 62–76. doi:10.4018/jdet.2004100104

Winn, W. (1993). *A conceptual basis for educational applications of virtual reality* (HITL Technical Report No. TR-93-9). Seattle, WA: Human Interface Technology Laboratory. Retrieved March 10, 2006, from http://www.hitl.washington.edu/publications/r-93-9

Youngblut, C. (1998). *Educational uses of virtual reality technology.* Institute for Defence Analyses, IDA Document D-2128. Retrieved June 5, 2001, from http://www.hitl.washington.edu/scivw/youngblut-edvr/D2128.pdf

KEY TERM AND DEFINITIONS

Cave Automatic Virtual Environment (CAVE): It is an immersive virtual reality environment where projectors are directed to three, four, five or six of the walls of a room-sized cube. The name is an allegory of the Cave in Plato's Republic where a philosopher contemplates perception, reality and illusion.

Fractal Geometry: It is the geometry used to describe the irregular pattern and the irregular shapes present in the nature. Fractals display the characteristic of self-similarity, an unending series of motifs within motifs repeated at all length scales.

Projective Virtual Reality: It is to allow users to "project" actions carried out in the virtual world into the real world by means of robots or other means of automation.

Second Life: It is a virtual world accessible on the Internet. It has been developed by Linden Lab that launched on June 23, 2003.

Virtual Architecture: It is an evolving discipline which results from the convergence of mapped data and of simulation, production of the digital form, architecture of the information, buildings and theory of the virtual reality

Virtual Prototyping: It is a process which uses a virtual prototype, instead of a physical prototype, to test and to evaluate of specific characteristics of an object of design.

Virtual Reality: A form of human-computer interaction in which a real or imaginary environment is simulated and users interact with and manipulate that world. It is a modern technology which gives to its users the illusion to be immersed in a computer generated virtual world with the ability to interact with it.

Virtual World: A kind of online community that takes the form of a computer-based simulated environment, through which users can interact with one another and use and create objects

VRML (Virtual Reality Modeling Language, originally known as the Virtual Reality Markup Language): A language that specifies the parameters to create virtual worlds networked together via the Internet and accessed via the World Wide Web hyperlinks. The aim of VRML is to bring to the Internet the advantages of 3D spaces, known in VRML as worlds whether they compromise environments or single objects. It was conceived in the spring of 1994 and it has been presented to the first annual WWW conference in Geneva, Switzerland. Mark Pesce (b. 1962) was one of the inventors of this language, and he has recognized as the man who brought virtual reality into the World Wide Web.

World Wide Web (WWW or "Web"): A system of Internet servers that support particular formatted documents. The documents are written using a markup language called HTML (HyperText Markup Language) that supports links to other documents, as well as graphics, audio, and video files. The World Wide Web uses the HTTP (HyperText Transfer Protocol) protocol to transmit data over the Internet. The Web was invented around 1990 by the Englishman Timothy (Tim) Berners-Lee (b. 1955) and the Belgian Robert Cailliau (b. 1947) working at CERN (the European Center for Nuclear Research) in Geneva, Switzerland.

Chapter 21

Effects of Packet–Loss and Long Delay Cycles on the Performance of the TCP Protocol in Wireless Networks

Hussein Al-Bahadili
The Arab Academy for Banking and Financial Sciences, Jordan

Haitham Y. Adarbah
Gulf College, Sultanate of Oman

ABSTRACT

Many analytical models have been developed to evaluate the performance of the transport control protocol (TCP) in wireless networks. This chapter presents a description, derivation, implementation, and comparison of two well-known analytical models, namely, the PFTK and PLLDC models. The first one is a relatively simple model for predicting the performance of the TCP protocol, while the second model is a comprehensive and realistic analytical model. The two models are based on the TCP Reno flavor, as it is one of the more popular implementations on the Internet. These two models were implemented in a user-friendly TCP performance evaluation package (TCP-PEP). The TCP-PEP was used to investigate the effect of packet-loss and long delay cycles on the TCP performance measured in terms of sending rate, throughput, and utilization factor. The results obtained from the PFTK and PLLDC models were compared with those obtained from equivalent simulations carried-out on the widely used NS-2 network simulator. The PLLDC model provides more accurate results (closer to the NS-2 results) than the PFTK model.

DOI: 10.4018/978-1-60960-887-3.ch021

INTRODUCTION

The TCP is the dominant transport layer protocol in the Internet Protocol (IP) suite. It carries a significant amount of the Internet traffics, such as Web browsing, files transfer, e-mail, and remote access. It is a reliable connection-oriented protocol that allows a byte stream originating on one machine to be delivered without error to any other machine on the Internet. (Forouzan 2007). An Internet work differs from a single network because different parts may have different topologies, delays, bandwidths, packet sizes, and other parameters. The TCP was designed to be dynamically adaptable to outfit Internet work, robust in the face of many kinds of failures, handle flow control to make sure a fast sender cannot swamp a slow receiver with more messages than it can handle, and support full duplex and point-to-point connections (Tekala & Szabo 2008).

The TCP does not support multicasting or broadcasting, and it only support point-to-point connection in which the sending and receiving TCP entities exchange data in the form of TCP segments. The size of the segments are decided by the TCP software, which decides how big the segments should be, and can accumulate data from several writes into one segment or can split data from one write over multiple segments. Two parameters restrict the segment size, these are: the IP payload (each segment must fit in the 65515 Byte IP payload), and the network maximum transfer unit (MTU) (each segment must fit in one MTU).

In practice, the MTU is generally 1500 bytes (the Ethernet payload size) and thus defines the upper bound on segment size. Segments can arrive out of order, so that some segment arrives but cannot be acknowledged because earlier segment has not turned up yet. Segments can also be delayed so long in transit that the sender times-out and retransmits them. The retransmissions may include different segment size than the original transmission. TCP must be prepared to

deal with these problems and solve them in an efficient way. The definition of the components of the TCP segment header and more details on TCP protocol can be found in many computer networks textbooks and literatures (Tanenbaum 2003).

The TCP was initially designed for wired networks for which a number of notable mechanisms have been proposed in the literature to improve the performance of TCP in such networks. For most of these mechanisms, analytical models have developed to predict and investigate their performance in wired networks in terms of the sending rate (S) and throughput (T), and utilization factor (U). Unfortunately, these mechanisms demonstrated a poor performance in wireless networks due to presence of packet-loss (PL) and long delay cycles (LDC) in such environment (Adarbah 2008).

Presence of LDC leads to spurious retransmissions (STs) or spurious fast retransmissions (SFRs), which produce serious end-to-end TCP performance degradation. However, since the emergent of wireless networks, new mechanisms have developed to enhance the performance of TCP in presence of STs and SFRs (Abouzeid & Roy 2003, Chen et. al. 2008). Consequently, new and adequate analytical models need to be developed to accommodate these new TCP mechanisms. This is because none of the existed models for wired networks considers the effect of STs and SFRs on the steady- state performance of TCP. This due to the fact that STs and SFRs do not occur frequently in wired networks, and also STs and SFRs are considered to be a transient state in a wired network, and thus cannot produce much impact on the steady-state performance of TCP (Yang 2003). Practically, in wireless networks, STs and SFRs are more frequent and must be explicitly modeled to accurately estimate S, T, and U of the TCP.

There are a number of mathematical models that have been developed throughout the years for evaluating the performance of a TCP connection in wireless networks (Abouzeid & Roy 2003, Dunaytsev 2006, Fu & Atiquzzaman 2003). In this chapter, we present a description, derivation, implementation, and comparison of two well-known analytical models, these are;

1. The PFTK analytical model. This model is a relatively simple model developed for predicting the performance of the TCP protocol. It is call the PFTK model after the initials of the last names of its authors, namely, Padhye, Firoiu, Towsley, and Kurose (Padhye et. al. 2000). It considers parameters such as: PL, acknowledgment behavior, receiver's maximum congestion window size, and average round trip time (*RTT*).

2. The PLLDC analytical model. This model is a comprehensive and realistic analytical model for evaluating the performance of the TCP protocol in wireless networks as it handles both PL and LDC; therefore, it was referred to as the PLLDC model (Adarbah 2008). It considers parameters, such as: duration of LCD, interval between long delays, timeout, and slow-start threshold.

The two models are based on the TCP Reno flavor as it is one of the most popular implementation on the Internet. These two models are implemented in a user-friendly TCP performance evaluation package (TCP-PEP). The TCP-PEP was used to investigate the effect of PL and LDC on the TCP performance measured in terms of *S*, *T*, and *U*. The results obtained from the PFTK and PLLDC models were validated against results obtained from equivalent simulations carried on the NS-2 simulator (Fall & Varadhan 2008). The PLLDC model provides results closer to those obtained from the NS-2 simulator than the PFTK model.

BACKGROUND

Congestion of TCP

TCP is the dominant transport protocol in the internet, and the current stability of the Internet depends on its end-to-end congestion control. Therefore, applications sharing a best-effort network need to positively respond to congestion to ensure network stability and high performance. Traditionally, congestion control algorithms have been implemented at the transport layer; therefore, it is referred to as congestion of TCP. One of the key elements for any TCP congestion control algorithm is the congestion signal that informs senders that congestion has or is about to occur. There is no explicit way that can be adopted by a TCP sender for congestion signal detection (Kung et. al. 2007). However, there are two implicit approaches that have been identified for congestion signal detection, these are: PL-based and delay-based approaches (Widmer 2001).

It is often not possible to draw sound conclusions on congestion from network delay measurements. Because it is difficult to find characteristic measures, such as the path's minimum *RTT*, due to persistent congestion at the bottleneck link or because of route changes. Consequently, PL is the only signal that senders can confidently use as an indication of congestion. A perceived PL is implemented either as a direct or an indirect trigger to throttle *S*; such flows are referred to as PL-responsive. In this sense, a TCP-based flow is a reliable PL-responsive flow.

One disadvantage of PL is that it is not unmistakable. Packets can get lost because of packet drops due to a buffer overflow at the bottleneck link or because of packet corruption due to a transmission error. The former indicates congestion, the latter does not. A sender is not able to discriminate among these events, because packet corruption usually leads to a frame checksum error and subsequent discard of the packet at the link layer.

Hence, transmission errors inevitably lead to an underestimation of available bandwidth for loss-responsive flows. As a consequence, applications can only fully utilize their share of bandwidth along the path if transmission errors are rare events. Due to the high error rate in wireless links, wireless links are often problematic, and the PL process and its consequences cannot be safely neglected as in wire line links.

There are two types of windows can be identified in a TCP connection, the congestion and advertised windows. The congestion window determines the number of bytes that can be outstanding at any time, or the maximum number of bytes can be transmitted without ACK that being received. This is a means of stopping the link between two places from getting overloaded with too much traffic. The size of this window is calculated by estimating how much congestion there is between the two places. Basically the size of the window, to a large degree, controls the speed of transmission as transmission pauses until there is ACK. The advertised window determines the number of bytes than can be sent over the TCP connection, which is imposed by the receiver. It is related to the amount of available buffer space at the receiver for this connection.

TCP congestion control consists of four mechanisms; these are (Voicu et. al. 2007):

Additive Increase/Multiplicative Decrease (AIMD)

This algorithm is a feedback control algorithm used in TCP congestion avoidance. Basically, AIMD represents a linear growth of the congestion window, combined to an exponential reduction when congestion takes place. The approach taken is to increase the transmission rate (window size), probing for usable bandwidth, until loss occurs. The policy of additive increase basically says to increase the congestion window by 1 maximum segment size (MSS) every RTT until a loss is detected. When loss is detected, the policy is changed

to be one of multiplicative decrease which is to cut the congestion window in half after loss. The result is a saw tooth behavior that represents the probe for bandwidth (Altman et. al. 2005).

Slow-Start (SS)

SS is part of the congestion control strategy used by TCP in many Internet applications, such as hyper text transfer protocol (HTTP), it is also known as the exponential growth phase. SS is used in conjunction with other algorithms to avoid sending more data than the network is capable of transmitting, that is, network congestion. The basic SS algorithm begins in the exponential growth phase initially with a congestion window size ($cwnd$) of 1 or 2 segments and increases it by 1 segment size ($SSize$) for each ACK received. This behavior effectively doubles the window size each RTT of the network.

This behavior continues until the $cwnd$ reaches the size of the receivers advertised window or until a loss occurs. When a loss occurs half of the current $cwnd$ is saved as a SS threshold (SST) and SS begins again from its initial $cwnd$. Once the $cwnd$ reaches the SST TCP goes into congestion avoidance mode where each ACK increases the $cwnd$ by $SSize^2/cwnd$. This results in a linear increase of the $cwnd$.

Fast Retransmit (FR)

Modifications to the congestion avoidance algorithm were proposed in 1990. Before describing the change, realize that TCP may generate an immediate ACK or a DUPACK, when an out of order segment is received. This DUPACK should not be delayed. The purpose of this DUPACK is to let the other end knows that a segment was received out of order, and to tell it what sequence number is expected.

Since TCP does not know whether a DUPACK is caused by a lost segment or just a reordering of segments, it waits for a small number of DUPACKs to be received. It is assumed that if there is just a reordering of the segments, there will be only one or two DUPACKs before the reordered segment is processed, which will then generate a new ACK. If three or more DUPACKs are received in a row, it is a strong indication that a segment has been lost. TCP then performs a retransmission of what appears to be the missing segment, without waiting for a retransmission timer to expire.

Fast Recovery

An improvement that allows a high throughput under moderate congestion, especially for large windows, is implemented as follows: After FR sends what appear to be the missing segment, then congestion avoidance but not SS is performed. This is called the fast recovery algorithm. The reason for not performing SS in this case is that the receipt of the DUPACKs tells TCP more than just a packet has been lost. Since the receiver can only generate the DUPACK when another segment is received, that segment has left the network and is in the receiver's buffer. In other words, there is still data flowing between the two ends, and TCP does not want to reduce the flow abruptly by going into SS.

The FR and fast recovery algorithms are usually implemented together as follows:

1. When the third DUPACK in a row is received, set *SST* to one-half the current *cwnd* but not less than two segments. Retransmit the missing segment. Set *cwnd* to *SST* plus 3 times the *SSize*. This inflates the congestion window by the number of segments that have left the network and which the other end has cached.

2. Each time another DUPACK arrives, increment *cwnd* by the *SSize*. This inflates the congestion window for the additional segment that has left the network. Transmit a packet, if allowed by the new value of *cwnd*.

3. When the next ACK arrives that acknowledges new data, set *cwnd* to *SST* (the value set in step 1). This ACK should be the ACK of the retransmission from step 1, one *RTT* after the retransmission. Additionally, this ACK should acknowledge all the intermediate segments sent between the lost packet and the receipt of the first DUPACK. This step is congestion avoidance, since TCP is down to one-half the rate it was at when the packet was lost.

The FR algorithm first appeared in the Tahoe release, and it was followed by SS. The fast recovery algorithm appeared in the Reno release. Since, in this work we concern with TCP Reno, a description of this TCP flavor is given the next section.

TCP Reno

The TCP Reno implementation retained the enhancements incorporated into Tahoe, but modified the FR operation to include fast recovery. The new algorithm prevents the communication path from going empty after FR, thereby avoiding the need to SS to refill it after a single PL (Lulling 2004, Lai & Yao 2002, Vendicits et. al. 2003).

Fast recovery operates by assuming each DUPACK received represents a single packet having left the communication path. Thus, during fast recovery the TCP sender is able to make intelligent estimates of the amount of outstanding data. Fast recovery is entered by a TCP sender after receiving an initial threshold of DUPACKs. Once the threshold of DUPACKs is received, the sender retransmits one packet and reduces its congestion window by one half. Instead of SS, as is performed by a Tahoe TCP sender, the Reno sender uses additional incoming DUPACKs to clock subsequent outgoing packets.

During fast recovery the sender "inflates" its window by the number DUPACKs it has received, according to the observation that each DUPACK indicates some ACK has been removed from the network and is now cached at the receiver. After entering fast recovery and retransmitting a single packet, the sender effectively waits until half a window of DUPACKs have been received, and then sends a new packet for each additional DUPACK that is received.

Upon receipt of an ACK for new data (called a "recovery ACK"), the sender exits fast recovery. Reno's fast recovery algorithm is optimized for the case when a single packet is dropped from a window of data. The Reno sender retransmits at most one dropped packet per *RTT*. Reno significantly improves upon the behavior of Tahoe TCP when a single packet is dropped from a window of data, but can suffer from performance problems when multiple packets are dropped from a window of data (Xin & Jamalipour 2006, Lai & Yao 2002)

Performance of TCP

A wireless network may suffer from extensive data loss due to: transmission errors in noisy environment, non-reliable wireless communication links, variable capacity links, frequent disconnections, limited communication bandwidth, broadcast nature of the communications, etc. Therefore, a wireless data communication session may involve a lot of data retransmission that degrades the performance of the networks. Data retransmissions reduce bandwidth utilization, and on the other hand increase delay and power consumption. These retransmissions are unavoidable and they are referred to them as factual retransmissions, because the data is lost and it will not reach the destination and they have to be retransmitted (Fu & Atiquzzaman 2003).

There is another form of data retransmission that is initiated by the TCP sender, which is referred to it as SFR, which is occurred when segments get re-ordered beyond the DUPACK-threshold in the network before reaching the receiver, i.e. the reordering length is greater than the DUPACK threshold (three for TCP). There are two main

reasons for SFR; these are (Ho 2008, Weigle et. al. 2005):

1. Timeout-based retransmission
2. DUPACK-based retransmission

DUPACK-based retransmission is triggered when three successive (triple) DUPACKs (TD) for the same sequence number have been received, i.e., without waiting for the retransmission timer to expire.

Timeout-based retransmission can be explained as follows. Since, TCP was initially designed for wired networks, and hence performs poorly in the presence of delay spikes which are especially more frequent in wireless networks than in traditional wired network (Gurtov & Reiner 2002).

These delay spikes may exhibit a sudden increase in the instantaneous *RTT* beyond the sender's retransmission timeout (RTO) (or simply abbreviated as TO) value causes retransmission ambiguity, resulting in spurious timeout (ST), which is defined as a TO which would not have happened if the sender waited long enough, and it results in retransmission due to a segment being delayed (but NOT lost) beyond TO. This produces serious end-to-end TCP performance degradation (Ludwig & Katz 2000, Kesselman & Mansour 2005).

One of the main reasons for the delay spikes to occur in a wireless environment is congestion and the lack of mechanisms through which the sender can detect or be informed about these congestions, and consequently prohibits SRs. However, there are other reasons for the delay spikes to occur in a wireless environment (Fu & Atiquzzaman 2003).

MODELING OF TCP

There are three basic techniques for performance evaluation of TCP connection in wireless networks; these are (Lulling & Vaughan 2004, Al-Bahadili 2010, Fall & Floyd 1996):

1. Experimental measurement
2. Computer simulation
3. Mathematical modeling

Experimental measurement is not widely used to evaluate the performance of TCP connection, due to its difficultly, cost, limited flexibility, etc. On the other hand, due to the enormous development in computing resources and tools, computer simulation and mathematical modeling are extensively used to evaluate the performance of TCP connection algorithms. They have an equivalent importance as they are usually validated each other. In this work, we concern with analytical model of TCP connection, in particular TCP Reno.

Essential of TCP Modeling

A number of notable models for TCP have been developed, which either shed light on a particular aspect of the protocol or add a new level of generality to the process of modeling transport control within the Internet. It is very useful, however, to consider the similarities of all these models before focusing attention on any particular model, as this allows keeping the key features of the model in mind and not getting lost in the details of a specific model.

Earlier versions of the protocol do exist, which may not necessarily contain the features that are classed here as essential for proceeding with TCP modeling. Future TCP flavors may also depart from these essentials. In addition, there are some important dynamics of TCP, such as its SS procedure, and other phenomena, such as loss of ACK packets in queues, which are not generally included in mathematical models.

All TCP connections commence in SS and many spend their entire lives in SS, because only a few kilobytes of data are being transferred. Thus, it is important to understand that models do have their limitations in reflecting reality. There are two key processes that a model of TCP needs to include (Hassan & Jain 2003):

1. The dynamics of the window that defines the number of packets a TCP source can convey into the network.
2. The PL process that indicates current traffic loads or congestion within the network.

One thing to notice about these processes is that they are both observed from the reference point of the TCP sender. This is obvious for window size, which is controlled by an algorithm within the source itself. The PL is also observed by the source. The loss process does not arise from any one particular node in the network but can be triggered by any node along the path of the TCP connection, with the source node observing the loss process as aggregation of information being generated along the connection path.

Window dynamics: The essential dynamics of this window size are its AIMD (Kesselman & Mansour 2005). During the interval in which TCP receives information (i.e., packets are not being lost in the network), TCP increases its window linearly. When the source deduces that a packet has been lost, it reduces its window by a factor of the current W (i.e., multiplicatively). Implementation of TCP normally increases W by one packet each round trip (in the linear increase phase) and reduce W size by half in the event of a PL. Although, W can be generalized in mathematical models of TCP, some models have been developed using S, as S can ease the analysis that follows the development of the model. The standard assumption in this case is that W is related to S by RTT:

$$S = \frac{W}{RTT} \qquad (1)$$

where S is the sending rate in packets/sec, W is the window size, and RTT is the RTT. This does assume that increasing S has negligible effect on queuing delays at nodes within the network, so that RTT is effectively constant. Regardless of whether a model uses W or S, all models incorporate the AIMD dynamics of TCP.

Packet-loss (PL) process: The other main component of a TCP model is the PL process, which triggers the TCP source to reduce its W. As previously mentioned, this process aggregates information regarding network conditions at all nodes along the path of the TCP connection. The particular TCP connection being considered is competing for network resources, along its path, with other TCP connections that have routes intersecting with this path. It is also competing for network bandwidth with other network traffic in general. These variations in traffic load introduce uncertainty into the arrival of PL information at the TCP source.

This typically can be modeled as a stochastic process, either with regard to the probability p of losing a particular packet in the network or the intervals between instances when lost packets are detected (Abouzeid & Roy 2003). The key point is that models usually incorporate the arrival of PL information, with the TCP source responding by decreasing its window. In fact, they do not necessarily need to consider the information being returned from the network as confirmation that packets have not been lost. Network information can take the form of explicit notification regarding congestion within the network, although individual congestion messages are most likely to still be coded as binary information. Regardless of whether the information is PL or explicit congestion information, TCP models must respond to the stream of network load information that is aggregated along the connection path.

THE PFTK MODEL

This section presents a description and derivation of the TCP congestion control and avoidance model that was developed by J. Padhye, V. Firoiu, D. Towsley, and J. Kurose; therefore, it was referred to it as the PFTK model after the initials of the last names of the authors (Padhye et. al. 2000).

Derivation of the TCP Sending Rate (S)

Many TCP receiver implementations send one cumulative ACK for consecutive packets (delayed ACK). Let b be the number of packets that are delayed and acknowledged by a single ACK. So that $b=2$ if one ACK is sent for two consecutive received packets. If W packets are sent in the first round and are all received and acknowledged correctly, then W/b ACKs will be received. Since each ACK increases W by $1/W$, W at the beginning of the next round is then $W_r = W_{r-1} + 1/b$, where W_r is W at round r. That is, during congestion avoidance and in the absence of loss, W increases linearly in time, with a slope of $1/b$ packets each RTT.

A PL can be detected at the TCP sender in one of two ways (Hassan & Jain 2003):

1. Reception of triple-DUPACK (TD), which is denoted as a TD loss indication.
2. Timeout (TO), which is denoted as a TO loss indication.

The PFTK model is developed in three steps in accordance with its operating regimes, these are:

1. Loss indications are exclusively TD
2. Loss indications are both TD and TO
3. The congestion W is limited by the receiver's advertised window

For each of the above operating regimes, an expression was derived for estimating S. A brief description and summary of the expressions are given below, while a detail description and derivation of equations can be found in (Padhye et. al. 2000).

Loss indications are exclusively TD. At this stage, the loss indications are assumed to be exclusively of type TD, and that W is not limited by the receiver's advertised flow control window.

Figure 1. Evolution of W over time when loss indications are TD

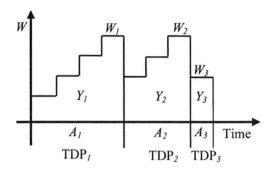

Figure 2. Packets sent during a TDP

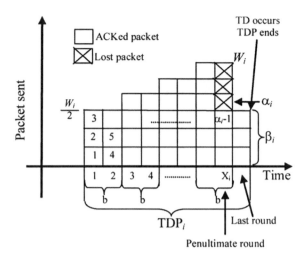

A sample path of the evolution of congestion W is given in Figure 1. Between two TD loss indications, the sender is in congestion avoidance, and W increases by $1/b$ packets per round, and immediately after the loss indication occurs, W is reduced by a factor of two.

The period between two TD loss indications is referred to it as a TD period (TDP) and denoted by (D_{TDP}) as shown in Figure 1. A TDP starts immediately after a TD loss indication as shown in Figure 2. For the ith TDP, define S_{TDP} to be the number of packets sent in the period D_{TDP}, which is the duration of the period, W_i the window size at the end of the period, and p the probability that a packet is lost, given that either it is the first packet in its round or the preceding packet in its round is not lost. Due to the stochastic nature of the loss process, the long-term steady-state TCP S as a function of p presented in average form, $E[S]$, can be expressed as:

$$E[S] = \frac{E[S_{TDP}]}{E[D_{TDP}]} \qquad (2)$$

where $E[S_{TDP}]$ is the average number of packets sent during a TDP, and $E[D_{TDP}]$ is the average duration of a TDP.

It is clear from the above equation that in order to calculate S, it is important, first, to derive an expression for calculating the average values of S_{TDP} and D_{TDP}. Mathematical expressions were derived in (Padhye et. al. 2000) for calculating S_{TDP} and D_{TDP}, these are:

$$E[S_{TDP}] = \frac{1-p}{p} + \frac{2+b}{3b} + \sqrt{\frac{8(1-p)}{3bp} + \left(\frac{2+b}{3b}\right)^2} \qquad (3)$$

$$E[D_{TDP}] = RTT\left(\frac{2+b}{6} + \sqrt{\frac{2b(1-p)}{3p} + \left(\frac{2+b}{6}\right)^2} + 1\right) \qquad (4)$$

Thus, S of a TCP source when loss indications are exclusively TD is expressed as:

$$E[S] = \frac{\dfrac{1-p}{p} + \dfrac{2+b}{3b} + \sqrt{\dfrac{8(1-p)}{3bp} + \left(\dfrac{2+b}{3b}\right)^2}}{RTT\left(\dfrac{2+b}{6} + \sqrt{\dfrac{2b(1-p)}{3p} + \left(\dfrac{2+b}{6}\right)^2} + 1\right)} \qquad (5)$$

For small PL rate, the above equation can be simplified to:

$$E[S] = \frac{1}{RTT} \sqrt{\frac{3}{2bp}} + o\left(1/\sqrt{p}\right) \qquad (6)$$

Furthermore, for $b=1$, Equation (6) can be reduced to the equation of the inverse square-root p law or periodic model (Hassan & Jain 2003).

Loss indications are TD and TO. In practice, it has been realized that in many cases the majority of window decreases are due to TO, rather than FR. Therefore, a good mathematical model should capture TO loss indications. This occurs when packets (or ACKs) are lost, and less than TD are received. In normal operation, the sender waits for a TO period denoted as T_o, and then retransmits non acknowledged packets. Following a TO, the congestion window is reduced to one, and one packet is thus resent in the first round after a TO. In the case that another TO occurs before successfully retransmitting the packets lost during the first TO, the period of TO doubles to $2T_o$; this doubling is repeated for each unsuccessful retransmission until a TO period of $64T_o$ is reached, after which the TO period remains constant at $64T_o$.

Figure 3 illustrates an example on the evolution of congestion W. In this case, due to the stochastic nature of the PL process, S can be calculated as:

$$E[S] = \frac{E[S_T]}{E[D_T]} = \frac{E[S_{TD} + S_{TO}]}{E[D_{TD} + D_{TO}]} \qquad (7)$$

where D_{TO} is the duration of a sequence of TOs, D_{TD} is the time interval between two consecutive TO sequences, D_T is the sum of D_{TO} and D_{TD}, S_{TO} is the number of packets sent during D_{TO}, S_{TD} is the number of packets sent during D_{TD}, and S_T is the number of packets sent during D_T.

Another form of Equation (7) was derived in (Padhye et. al. 2000), which is expressed as:

Figure 3. Evolution of window size when loss indications are TD and TO.

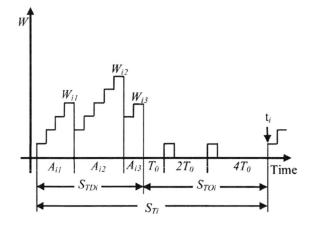

$$E[S] = \frac{E[S_{TDP}] + Q * E[R_{NP}]}{E[D_{TDP}] + Q * E[D_{TO}]} \qquad (8)$$

In Equation (8), $E[S_{TDP}]$ and $E[D_{TDP}]$ are as given in Equations (3) and (4), while Q and $E[R_{NP}]$ can be calculated as:

$$Q \approx \hat{Q}(E[W]) = \min\left(1, \frac{3}{E[W]}\right) \qquad (9)$$

and

$$E[W] = \frac{2+b}{3b} + \sqrt{\frac{8(1-p)}{3bp} + \left(\frac{2+b}{3b}\right)^2} \qquad (10)$$

$E[R_{NP}]$ denotes the number of retransmitted packets during the TO period in one normal period (*NP*), and it is computed by:

$$E[R_{NP}] = \frac{1}{1-p} \qquad (11)$$

$E[D_{TO}]$ denotes the average duration of a TOs sequence excluding retransmissions, which is computed by:

$$E[D_{TO}] = T_o \frac{f(p)}{1-p} \quad (12)$$

$$f(p) = 1 + p + 2p^2 + 4p^3 + 8p^4 + 16p^5 + 32p^6 = 1 + \sum_{i=1}^{6} 2^{i-1} p^i \quad (13)$$

Substituting Eqns. (3), (4), (11), and (12) into Equation (8) yields the expression for S shown in Equation (14) (see Box 1).

Impact of window limitation. So far, no limitation is considered on W. At the beginning of TCP flow establishment, the receiver advertises a maximum buffer size which determines a maximum congestion window size, W_m. As a consequence, during a period without loss indications, W can grow up to W_m, but will not grow further beyond this value. An example of the evolution of W is depicted in Figure 4. To simplify the analysis of the PFTK model, the following assumption is assumed. Let W_u denote the unconstrained W, the mean of which is derived by (Padhye et. al. 2000) and it is given by Equation (10) with W_u replaces W.

Figure 4. Evolution of W limited by W_m

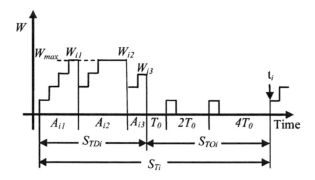

It is assumed that if $E[W_u] < W_m$, then the approximation $E[W] \approx E[W_u]$ is satisfactory. In other words, if $E[W_u] < W_m$, the receiver window limitation has negligible effect on S, and thus S is given by Equation (14).

On the other hand, if $W_m \leq E[W_u]$, the approximation $E[W] \approx W_m$ can be considered as a satisfactory approximation. In this case, consider an interval D_{TD} between two TO sequences consisting of a series of TDPs as in Figure 5. During the first TDP, the window grows linearly up to W_m for H rounds, afterwards remains constant for L rounds, and then a TD indication occurs. The window then drops to $W_m/2$, and the process repeats. According to the above discussion the following expression for S can be derived:

Box 1.

$$E[S] = \frac{\dfrac{1-p}{p} + \dfrac{2+b}{3b} + \sqrt{\dfrac{8(1-p)}{3bp} + \left(\dfrac{2+b}{3b}\right)^2} + Q \cdot \dfrac{1}{1-p}}{RTT \left(\dfrac{2+b}{6} + \sqrt{\dfrac{2b(1-p)}{3p} + \left(\dfrac{2+b}{6}\right)^2} + 1\right) + Q \cdot T_o \dfrac{f(p)}{1-p}} \quad (14)$$

$$E[S] =$$

$$\frac{\dfrac{1-p}{p} + W_m + Q_m \cdot \dfrac{1}{1-p}}{RTT\left(\dfrac{b}{8}W_m + \dfrac{1-p}{pW_m} + 2\right) + Q_m \cdot T_O \dfrac{f(p)}{1-p}}$$

(15)

Where all variables are as defined above, except Q_m, which is different from Q and it is given as a function of W_m and it is expressed as:

$$Q_m \approx \hat{Q}(w) =$$

$$\min\left(1, \frac{\left(1-(1-p)^3\left[1+(1-p)^3\left(1-(1-p)^{w-3}\right)\right]\right)}{1-(1-p)^w}\right)$$

(16)

In conclusion, the complete characterization of S can be expressed as Equation (17) (see Box 2).

Equation (17) is referred to as the "PFTK full model", which can be approximated to Equation (18) (see Box 3).

Equation (18) is referred to as the "PFTK approximate model".

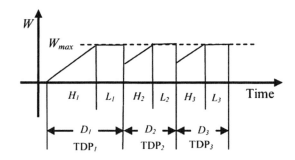

Figure 5. Fast retransmit with window limitation

Derivation of the TCP Throughput (*T*)

The steady-state performance of TCP Reno is also be characterized by T, which is the amount of data received per unit time. The same analysis that has been used to derive Equation (17) for S can be easily modified to calculate T. All it needs is to modify the numerator of Equation (8), so that, the number of packets that make it to the receiver in a TDP ($E[S'_{TDP}]$) and the number of packets sent in the TO sequence ($E[R'_{NP}]$) are needed to be calculated. T can be calculated as:

Box 2.

$$E[S] =$$

$$\begin{cases}
\dfrac{\dfrac{1-p}{p} + \dfrac{2+b}{3b} + \sqrt{\dfrac{8(1-p)}{3bp} + \left(\dfrac{2+b}{3b}\right)^2} + Q \cdot \dfrac{1}{1-p}}{RTT\left(\dfrac{2+b}{6} + \sqrt{\dfrac{2b(1-p)}{3p} + \left(\dfrac{2+b}{6}\right)^2} + 1\right) + Q \cdot T_o \dfrac{f(p)}{1-p}} & \text{for } E[W_u] < W_m \\[4em]
\dfrac{\dfrac{1-p}{p} + W_m + Q_m \cdot \dfrac{1}{1-p}}{RTT\left(\dfrac{b}{8}W_m + \dfrac{1-p}{pW_m} + 2\right) + Q_m \cdot T_O \dfrac{f(p)}{1-p}} & \text{Otherwise}
\end{cases}$$

(17)

Box 3.

$$E[S] =$$

$$\min\left(\frac{W_m}{RTT}, \frac{1}{RTT\sqrt{\frac{2bp}{3} + T_o \min\left(1, 3\sqrt{\frac{3bp}{8}}\right) p\left(1 + 32p^2\right)}}\right) \quad (18)$$

$$E[T] = \frac{E[S'_{TDP}] + Q \cdot E[R'_{NP}]}{E[D_{TDP}] + Q \cdot E[D_{TO}]} \quad (19)$$

$$E[S'_{TDP}] = \frac{1-p}{p} + \frac{E[W]}{2} \quad (21)$$

Since only one packet makes it to the receiver in a TO sequence (i.e., the packet that ends the TO sequence), it is evident that

$$E[R'_{NP}] = 1 \quad (20)$$

To calculate the number of packets that reach the receiver in a TDP, consider Figure 2. The TD event is induced by the loss of packet. Assume the window size is W, when the loss occurs. Then, the number of packets received by the receiver is:

From Equations (20) and (21) along with the analysis for $E[W]$ and Q, T can be expressed as Equation (22) (see Box 4).

Where $E[W]$ and $f(p)$ are defined in Equations (10) and (13), respectively, and $Q(p,w)$ was derived in (Padhye et. al. 2000) as:

$$Q(p, w) =$$

$$\min\left(1, \frac{\left(1 - (1-p)^3\left[1 + (1-p)^3\left(1 - (1-p)^{w-3}\right)\right]\right)}{1 - (1-p)^3}\right) \quad (23)$$

Box 4.

$$E[T] =$$

$$\begin{cases} \dfrac{\dfrac{1-p}{p} + \dfrac{E[W]}{2} + Q(p, E[W])}{RTT\left(E[W] + 1\right) + Q(p, E[W]) \cdot T_o \dfrac{f(p)}{1-p}} & \text{for } E[W_u] < W_m \\[6ex] \dfrac{\dfrac{1-p}{p} + \dfrac{W_m}{2} + Q(p, W_m)}{RTT\left(\dfrac{W_m}{4} + \dfrac{1-p}{pW_m} + 2\right) + Q(p, W_m) \cdot T_o \dfrac{f(p)}{1-p}} & \text{Otherwise} \end{cases} \quad (22)$$

Features of the PFTK Model

The main features of the PFTK model include:

1. It is a relatively simple analytical model for predicting S of a saturated TCP sender, i.e., a steady-state flow with an unlimited amount of data to send.

2. It illustrates the congestion avoidance behavior of TCP and its impact on S, taking into account the dependence of congestion avoidance on:
   a. ACK behavior.
   b. The manner in which PL is inferred, whether by DUPACK detection and FR or by TO.
   c. Limited W.
   d. Average RTT.

3. The model is based on the TCP Reno flavor as it is one of the more popular implementations in the Internet today.

4. In this model, W is increased by $1/W$ each time an ACK is received. Conversely, the window is decreased whenever a lost packet is detected, with the amount of the decrease depending on whether PL is detected by DUPACK or by TO.

5. The model represents the congestion avoidance behavior of TCP in terms of rounds. A round starts with transmission of W packets. Once all packets falling within the congestion window have been sent, no other packets are sent until the first ACK is received for one of these W packets. This ACK reception marks the end of the current round and the beginning of the next round.

6. The duration of a round is equal to RTT and is assumed to be independent of W.

7. The time needed to send all the packets in a window is smaller than the RTT.

8. A PL in a round is entirely independent of PL in other rounds.

9. PLs are correlated among the back-to-back transmissions within a round: if a packet is lost, all remaining packets transmitted until the end of that round are also lost.

THE PLLDC MODEL

This section presents the mathematical derivation of the PLLDC model (Adarbah 2008, Fu & Atiquzzaman 2003). Before we proceed with the derivation of the PLLDC model for estimating TCP Reno S and T in presence of PL and LDC, we introduce, first, the dynamics of sender window around a long delay (LD), and then the statistical modeling of the LD pattern.

Dynamics of sender window around a LD. In order to analyze the dynamics of the sender window around a LD, it is important to be familiar with the evolution of sender's W as represented by the number of packets that can be sent. This is shown in Figure 6. At each round W is increased by $1/b$. After X_i rounds, the LD begin, when some of the packets in the X_i-th round are delayed (packets marked "d") (Tekala & Szabo 2008).

Since the LD is of a much larger timescale than a round, any extra packets that were sent in round X_{i+1}, corresponding to the ACKs of successfully delivered packets of round X_i, are also delayed. After T_o, which is the converged value of the TO when the RTT is stable for a relatively long period of time, the sender will TO and reduce the window to one and retransmit the first delayed packet. If it is not acknowledged within $2T_o$, the sender will retransmit it again, and so on.

The number of retransmissions during the LD is denoted by R_D; all these retransmitted packets are also delayed. Eventually, when the ACK for the first delayed packets comes back after the LD has cleared, the sender will enter SS and spuriously retransmit all the delayed packets. The sender will exit SS when the window hits the *SST*.

The LD period (LDP) is defined as consisting of two consecutive TDPs, one LD, and one SS as shown in Figure 6. Note that even though the first period, labeled with TDP$_i$ in the figure, does not end with a TD loss indication, the number of

Figure 6. Packet sent during one LDP

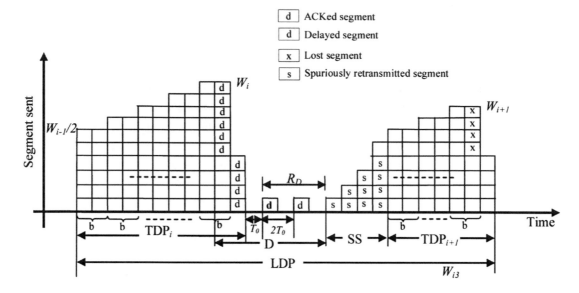

packets sent and the duration of TDP$_i$ is the same as other TDPs, so just TDP will be used for convenience. The sender's window was $W_{i-1}/2$ at the end of TDP$_{i-1}$; after the FR, it has been reduced to $W_{i-1}/2$, which is the sender's window at the start of TDP$_i$. TCP Reno starts FR after receiving TD.

Statistical modeling of the LD pattern. It has been shown in Equation (1) that S is obtained by dividing W by the RTT, where RTT is the expected value of RTT when there is no LD. However, in the presence of LD spikes, the actual time is much more than RTT, and the time measured by the sender is referred to as the LDs as illustrated in Figure 7a.

A two-state Markov chain is used to model the start and end of a LD as shown in Figure 7b (Altman et. al. 2005). The two states are: interval between LDs (S_I) and duration of LD (S_D). Here, it is assumed that the length of the S_I and S_D states are both exponentially distributed, with u and v being the transition probabilities from state S_I to state S_D and state S_D to state S_I, respectively. By solving the Markov chain in Figure 7b, the relationship between I and D can be expressed as:

$$E[D] = \frac{u\,E[I]}{u + v} \qquad (24)$$

Given a model for the lower layer events (such as link layer retransmission, mobile handoff, etc.)

Figure 7. (a) Variation of RTT showing four LDs, and (b) model of LDs

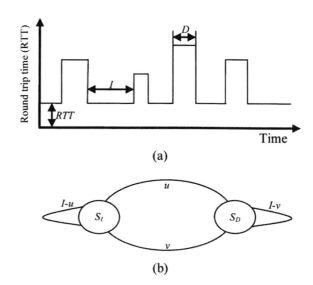

(a)

(b)

that cause LDs, the values of D, I, u, and v can be obtained to be used in Equation (24).

Derivation of the TCP Sending Rate (S)

In order to develop a satisfactory mathematical model for estimating S of wireless TCP source over an unreliable link, with certain PL probability (p), and in the presence of LD spikes, then S of the TCP source can be expressed as:

$$E[S] = \frac{E[S_{LDC}]}{E[D_{LDC}]} = \frac{mE[S_{NP}] + E[S_{LDP}]}{mE[D_{NP}] + E[D_{LDP}]}$$

(25)

Where S_{LDC} is the number of packets sent during one LDC, D_{LDC} is the duration of the LDC, m is the number of NPs in one LDC, S_{NP} is the number of packets sent during one NP, S_{LDP} is the number of packets sent during one LDP, D_{NP} is the duration of one NP, and D_{LDP} is the duration of one LDP.

In the above equation, the numerator denotes the number of packets sent during one LDC and the denominator is the duration of an LDC. We first look at the macroscopic behavior of one LDP, which will then be used to determine the number of packets sent and the duration of an LDC.

Analysis of a long delay period (LDP). The total number of packets sent during one LDP is the sum of packets sent during two TDPs, the TO period, and the SS stage as shown in Figure 6 can be expressed as:

$$E[S_{LDP}] = 2E[S_{TDP}] + E[R_D] + E[S_{SST}]$$

(26)

The duration of LDP can be written as the sum of the time duration of the two TDP, LD, and one SS stage, minus the overlapping area ($2RTT$) between D and TDP_i:

$$E[D_{LDP}] = 2E[D_{TDP}] + E[D] + k\, RTT - 2\, RTT$$

(27)

$$E[D_{LDP}] = 2E[D_{TDP}] + E[D] + (k-2)\, RTT$$

(28)

$E[S_{TDP}]$ and $E[D_{TDP}]$ in Equations (26) and (28) can be determined from Equations (3) and (4), respectively. Next, the three unknown variables $E[R_D]$ and $E[S_{SST}]$ in Equation (26), and k in Equation (27) are derived below.

Deriving $E[R_D]$. Since D is assumed to be exponentially distributed with mean $E[D]$, and the sender experiences a LD of D, then the probability that there is one TO can be expressed as:

$$Pr(T_o < D \leq 2T_o) = Pr(D \leq 2T_o) - Pr(D \leq T_o)$$

(29)

$$Pr(T_o < D \leq 2T_o) = e^{-\frac{T_o}{E[D]}} - e^{-\frac{2T_o}{E[D]}}$$

(30)

The probability that there are two or more TOs is:

$$Pr(D \leq 2T_0) = e^{-\frac{2T_0}{E[D]}}$$

(31)

Because the sender sends out a packet when a TO occurs, the number of packets sent during D is the same as the number of TOs. Since the sender can back-off a maximum of 6 times to get a TO of $64T_o$, the number of packets sent can be expressed as Equation (32) (see Box 5) or

$$E[R_D] = \sum_{j=0}^{5} \left(e^{-\frac{2^j T_o}{E[D]}} - e^{-\frac{64T_o}{E[D]}} \right)$$

(33)

Deriving k. After the LD, the SST value will be $max(W/2, 2)$ if there is only one TO during D, otherwise, it will be two for two or more TOs. Therefore, the expected value of *SST* after the long delay is:

Box 5.

$$E[R_D] =$$
$$Pr(T_o < D \le 2T_o) + 2Pr(2T_o < D \le 3T_o) + \ldots + 6Pr(32T_o < D \le 64T_o) \tag{32}$$

$$E[SST] =$$

$$\max\left(\frac{W_i}{2}, 2\right)\left(e^{-\frac{T_o}{E[D]}} - e^{-\frac{2T_o}{E[D]}}\right) + 2e^{-\frac{2T_o}{E[D]}} \tag{34}$$

During the SS, if the receiver adopts delayed acknowledgment, the sender's congestion window will grow by half of W in the previous round according to the following rule:

$$cwnd_{j+1} = cwnd_j + \left\lfloor\left\lfloor\frac{cwnd_j}{2}\right\rfloor\right\rfloor \tag{35}$$

with $cwnd_j = 1$ and $j = 1, 2, 3, \ldots$

This can be approximated as:

$$cwnd_j = \left(\frac{3}{2}\right)^j \text{ and } j = 1, 2, 3 \ldots \tag{36}$$

End of the SS stage at $E[SST]$ after k rounds implies that $cwnd_k = E[SST]$; the number of rounds needed to complete this stage is approximately expressed as:

$$k = \left\lceil\frac{\ln(E[SST])}{\ln(1.5)}\right\rceil \tag{37}$$

Deriving $E[S_{SST}]$. The number of packets sent in each round of the SS stage in Figure 6 is given in Equation (36). So the number of packets sent

during SS can be approximated by the sum of the packets sent during these k rounds:

$$E[S_{SST}] = \sum_{j=1}^{k}\left(\frac{3}{2}\right)^j \approx 3\left(\frac{3}{2}\right)^k - 3 \tag{38}$$

By substituting $E[R_D]$, k, and $E[S_{SST}]$ from Equations (33), (36), and (38) into Equations (26) and (28), we can obtain the number of packets sent and the duration of one LDP.

Analysis of one long delay cycle (LDC). It can be seen from Figure 8 that the total number of packets sent during one LDC (S_{LDC}) is the sum of packets sent during m instances of NP periods and an LDP period. Thus, $E[S_{LDC}]$ can be expressed as:

$$E[S_{LDC}] = m E[S_{NP}] + E[S_{LDP}] \tag{39}$$

Where $E[S_{LDP}]$ represents the total number of packets sent during one LDP and it can be calculated using Equation (26), and $E[S_{NP}]$ represents the number of packets sent during the ith NP ($i=1, 2, 3, \ldots, m$), which can be obtained from (Padhye et. al. 2000) as given below:

$$E[S_{NP}] = \frac{1}{Q} E[S_{TDP}] + E[R_{NP}] \tag{40}$$

Where $E[S_{TDP}]$, Q, and $E[R_{NP}]$ can be determined using Equations (3), (9), and (11), respectively. By substituting these equations into Equation (40) yields the expression for $E[S_{NP}]$ shown in Equation (41) (see Box 6).

Where $E[W]$ can be obtained from Equation (10). In Equation (25), $E(D_{LDP})$ has already been

Figure 8. Sender window evaluation in one LDC

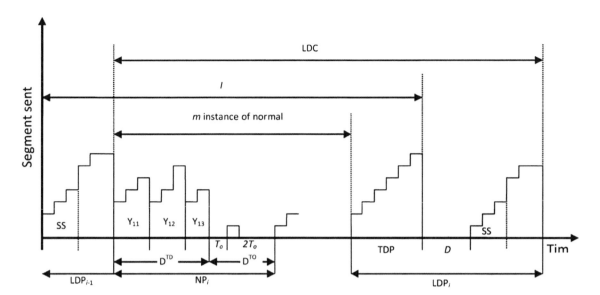

Box 6.

$$E[S_{NP}] =$$

$$\frac{1}{\min(1, \dfrac{3}{E[W]})} \left(\frac{1-p}{p} + \frac{2+b}{3b} + \sqrt{\frac{8(1-p)}{3bp} + \left(\frac{2+b}{3b}\right)^2} \right) + \frac{1}{1-p} \qquad (41)$$

developed in Equation (28). Another parameter in Equation (25) that needs to be determined is $E[D_{NP}]$, which denotes the duration of the ith NP (i=1, 2,, 3, ..., m). $E[D_{NP}]$ is given as:

$$E[D_{NP}] = \frac{1}{Q} E[D_{TDP}] + E[D_{TO}] \qquad (42)$$

Where $E[D_{TDP}]$, Q, and $E[D_{TO}]$ can be determined using Equations (4), (9), and (12), respectively. By substituting these equations into Equation (42) yields the expression for $E[D_{NP}]$ shown in Equation (43) (see Box 7).

At this stage, to estimate S using Equation (25), only one parameter remains and needs to be determined which is m. In order to derive an expression for determining m, let us go back to the definition of LDC in Figure 8. It starts with the end of the previous LDP. An LDC consists of several instances of NPs at the beginning and an LDP at the end. In the PLLDC model, the normal period, $E[D_{NP}]$, denotes the time interval with no LD, which is given by Equation (3.43). It is equal to the sum of D_{TD} and D_{TO}.

Referring to Figure 8 the interval between LDs (I) consists of a SS phase following the previous long delay, m instances of NP and a TDP. We can calculate m as:

Box 7.

$$E[D_{NP}] =$$

$$\frac{RTT}{\min(1, \dfrac{3}{E[W]})} \left(\frac{2+b}{6} + \sqrt{\frac{2b(1-p)}{3p} + \left(\frac{2+b}{6}\right)^2} + 1 \right) + T_o \frac{f(p)}{1-p} \qquad (43)$$

$$m = \frac{E[I] - 2E[D_{TDP}] - k \cdot RTT}{E[D_{NP}]} \qquad (44)$$

$$E[S] = \frac{E[S_{LDC}]}{E[D_{LDC}]} = \frac{mE[S_{NP}] + E[S_{LDP}]}{mE[D_{NP}] + E[D_{LDP}]} \qquad (49)$$

As it has been stated earlier that the total number of packets sent during one LDC is the sum of packets sent during m instances of NP period and an LDP period:

$$E[S_{LDC}] = \sum_{i=1}^{m} (S_{NP})_i + E[S_{LDP}] \qquad (45)$$

Similarly, since one LDC consists of m instances of NP and ends with one LDP, the duration of one LDC can be obtained as:

$$E[D_{LDC}] = \sum_{i=1}^{m} (D_{NP})_i + E[D_{LDP}] \qquad (46)$$

Due to the randomness of the process, the above two equations can be written in expected forms as follows:

$$E[S_{LDC}] = m\, E[S_{NP}] + E[S_{LDP}] \qquad (47)$$

$$E[D_{LDC}] = m\, E[D_{NP}] + E[D_{LDP}] \qquad (48)$$

By substituting $E[S_{LDC}]$ from Equation (47), and $E[D_{LDC}]$ from Equation (48) into Equation (25), the long-term steady-state S of the TCP sender can be expressed as:

Derivation of the TCP Throughput (*T*)

T can be determined by subtracting the spuriously retransmitted and lost packets from S. Referring to Figure 6, the delayed packets in the X_i and X_{i+1}-th rounds of the first TDP are subsequently spuriously retransmitted, therefore, one window of packets $E[W]$ must be subtracted from $E[S_{TDP}]$. So that the net number of packets send during the first TDP, $E[S'_{TDP,1}]$, can be expressed as:

$$E[S'_{TDP,1}] = E[S_{TDP}] - E[W] \qquad (50)$$

Substituting $E[S_{TDP}]$ and $E[W]$ from Equations (3) and (10) into Equation (50) yields the following value for $E[S'_{TDP,1}]$:

$$E[S'_{TDP,1}] = \frac{1-p}{p} \qquad (51)$$

In the second TDP of the LDP period, the lost the delayed packets need to be subtracted from S, i.e., on the average, $E[W]/2$ must be subtracted. So that the net number of packets send during the second TDP, $E[S'_{TDP,2}]$, can be expressed as:

$$E[S'_{TDP,2}] = E[S_{TDP}] - \frac{E(W)}{2} \qquad (52)$$

Substituting $E[S_{TDP}]$ and $E[W]$ from Equations (3) and (10) into Equation (52) yields the following value for $E[S'_{TDP,2}]$:

$$E[S'_{TDP,2}] =$$
$$\frac{1-p}{p} + \frac{1}{2}\left[\frac{2+b}{3b} + \sqrt{\frac{8(1-p)}{3bp} + \left(\frac{2+b}{3b}\right)^2}\right] \qquad (53)$$

Because the packets retransmitted during the TO period are discarded by the receiver, $E[R_{NP}]$ in Equation (40) can be replaced with $E[R'_{NP}] = 1$. Similarly, $E[R_D]$ in Equation (26) can be replaced with $E[R'_D] = 1$. . Replacing $E[S_{TDP}]$, $E[R_{NP}]$ and $E[R_D]$ in Equations (26) and (40) with $E[S'_{TDP,1}]$, $E[S'_{TDP,2}]$, $E[R'_{NP}]$ and $E[R'_D]$, the following expressions can be found for $E[S'_{LDP}]$ and $E[S'_{NP}]$:

$$E[S'_{LDP}] = E[S'_{TDP,1}] + E[R'_D] + E[S_{SST}] + E[S'_{TDP,2}] \qquad (54)$$

$$E[S'_{NP}] = \frac{1}{Q} E[S'_{TDP,2}] + E[R'_{NP}] \qquad (55)$$

Therefore, the average T during one LDC can be calculated as the total number of packets delivered to the receiver $E[S'_{LDC}]$ divided by the duration of one LDC $E[D_{LDC}]$. $E[S'_{LDC}]$ can be obtained by replacing $E[S_{LDP}]$ and $E[S_{NP}]$ in Equation (47) with $E[S'_{LDP}]$ and $E[S'_{NP}]$, so that

$$E[S'_{LDC}] = m \ E[S'_{NP}] + E[S'_{LDP}] \qquad (56)$$

Although, the spuriously retransmitted and lost packets are subtracted from the total number of packets received, the duration of an LDC remains unchanged. Therefore, T can be determined as:

$$E[T] = \frac{E[S'_{LDC}]}{E[D_{LDC}]} = \frac{m \ E[S'_{NP}] + E[S'_{LDP}]}{m \ E[D_{NP}] + E[D_{LDP}]} \qquad (57)$$

Where $E[S'_{LDC}]$ is given in Equation (56) and it represents the actual number packets delivered to the receiver, and $E[D_{LDC}]$ is given in Equation (48) and it represents the duration of one LDC.

Derivation of the TCP Utilization Factor (U)

U is determined as T over S and it is expressed mathematically as:

$$U = \frac{T}{S} \times 100 \qquad (58)$$

Substituting T and S from Equations (57) and (49) into Equation (58), yields the following equation for U:

$$E[U] = \frac{E[T]}{E[S]} \times 100 = \frac{m \ E[S'_{NP}] + E[S'_{LDP}]}{m \ E[S_{NP}] + E[S_{LDP}]} \times 100 \qquad (59)$$

Where m, $E[S'_{NP}]$, $E[S'_{LDP}]$, $E[S_{NP}]$, and $E[S_{LDP}]$ are given by Equations (44), (55), (54), (41), and (28), respectively.

IMPLEMENTATION

The PFTK and PLLDC models described earlier are implemented in an interactive and user-friendly package using VB environment. The package can be used for TCP performance evaluation; therefore, we refer to this package as TCP-PEP. It can be used easily by professionals, researchers, and

Table 1. Definition of the code input parameters

Sym.	Description
p	Packet-loss rate.
D	Duration of the long delay.
I	Interval between long delays.
T_0	Timeout (TO)
SST	Value of slow-start threshold at the end of a long delay D.
b	Number of packets acknowledged by one ACK packet.
RTT	Expected value of round trip time (RTT) when there is no long delay.
W_m	The receiver's maximum advertised congestion window size.

students to analyze the performance of the TCP Reno protocol in a realistic wireless environment and to investigate the effect of a number of parameters on the performance. The input parameters for TCP-PEP are listed in Table 1, while the code main computed parameters are given in Table 2. Each parameter is computed by the equation indicated in column #3 in Table 2. The computed parameters are ordered in the sequence by which they are computed. They are categorized into four main groups as shown in Table 2.

RESULTS AND DISCUSSIONS

In order to assess the effectiveness of the TCP-PEP package in evaluating the performance of the TCP Reno protocol in a wireless networks suffering from PL and LDCs, it was used to investigate the variation of S, T, and U against p for various values of D. Also, to provide an insight into the behavior of TCP Reno in various wireless environments, the number of packets sent during different stages of the communication process was calculated, such as: S_{LDC}, S_{LDP}, S_{NP}, S_{TDP}, R_{NP}, R_D, and S_{SST}. All input parameters are listed in Table 3.

Calculation of S. The TCP-PEP package was used to compute the variation of S with p for four various values of D (6, 8, 10, and 12 sec). The results obtained are plotted in Figure 9, which shows that S is inversely proportional to p, i.e., it

is decreasing as p increases for all values of D. If all other network parameters are remained unchanged, S is slightly decreasing as D increases, in other word; D has only a slight effect on S.

The obtained results demonstrate that D has a recognizable effect on S at low p ($p \leq 0.1$) and has no effect at high p ($p > 0.1$). In other words the dominant parameter at low p is D and vice-versa. To numerically illustrate the variation of S as D changed, we define the following formula:

$$V = \frac{S_{D=x} - S_{D=6}}{S_{D=6}} \times 100 \qquad (60)$$

Where V represents the variation with respect to a reference value ($S_{D=6}$). $S_{D=6}$ is the reference value (D=6 sec). $S_{D=x}$ is the sending rate at the same p and D takes any other valid value. For example, it can be deduced from Figure 9 that when p=0.01, and D is doubled (increased from 6 to 12 sec), S is reduced by 14.5% (from 32.80 to 28.09 packets/sec). A negative value indicates that S is decreasing as D increases.

In order to validate the accuracy of the models, their results are compared against results obtained by the NS-2 network simulator as shown in Figure 10. It shows that the PLLDC model can predict S more accurately than the PFTK model. It also shows that when D increases, the gap between the PFTK model and the simulation results increases,

Table 2. Definition of the code computed parameters

Sym.	Description	Equation
	Group 1: Length of each duration	
D_{TDP}	Duration of triple duplicate period (TDP), i.e. the time between two successive triple duplicate loss indications.	4
$f(p)$	Polynomial function.	13
D_{TO}	Duration of the TO period in one normal point (NP).	12
W	TCP sender window size.	10
Q	The probability that a loss indication in W is TO.	9
Q_m	The probability that a loss indication in W_m is TO.	16
D_{NP}	Duration of one NP, where NP consists of n instances of TDP and one instance of TO period.	42
k	The number of rounds needed to complete the SS stage after a long delay.	37
m	Number of NPs in one long delay cycle (LDC).	44
D_{LDP}	Duration of long delay period (LDP), which consists of one TDP, one long delay, one SS, and a second TDP.	28
D_{LDC}	Duration of LDC, which consists of m NPs and one LDP.	46
	Group 2: Number of packets sent during each duration	
S_{TDP}	Number of packets sent from the sender during one TDP.	3
R_{NP}	Number of packets sent during the TO period in one NP.	11
S_{NP}	Number of packets sent during i^{th} NP, $i = 1, 2, …, m$.	41
R_D	Number of packets sent during long delay (D).	33
S_{SST}	Number of packets sent during the SS stage in an LDP.	38
S_{LDP}	Number of packets sent during one LDP.	26
S_{LDC}	Number of packets sent during one LDC.	47
	Group 3: Net number of packets sent during each duration	
$S'_{TDP,1}$	Net number of packets sent during the first TDP.	51
$S'_{TDP,2}$	Net number of packets sent during the second TDP.	52
R'_D	Number of packets sent during long delay (R'_D =1).	-
R'_{NP}	Number of packets sent during TO period in one NP (R'_{NP} =1).	-
S'_{LDP}	Net number of packets sent during one LDP.	54
S'_{NP}	Net number of packets sent during i^{th} NP, $i = 1, 2, …, m$.	55
S'_{LDC}	Net number of packets delivered to the receiver during one LDC.	56
	Group 4: Main computed parameters	
S	Long-term steady-state sending rate of TCP connection.	49
T	Long-term steady-state throughput of TCP connection.	57
U	Long-term steady-state utilization factor.	59

Table 3. Input parameter

Parameter	Value
Packet-loss rate *(p)*	0.001 to 0.5
Duration of the long delay (*D*)	6, 8, 10, 12 sec
Interval between long delays (*I*)	30 sec
Round Trip Time *(RTT)*	0.200 sec
Retransmission timeout or timeout (*T$_o$*)	1.00 sec
Slow start threshold at the end of a long delay *D* (*SST*)	2
Number of packets acknowledged by one ACK packet (*b*).	2
Receiver's maximum congestion window size (*W$_m$*)	800 packet

Figure 9. Variation of S with p for various values of D

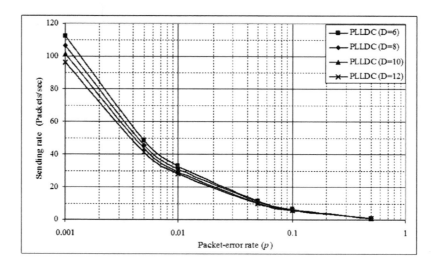

but the PLLDC model accommodates the increase of *D* well.

In order to provide an insight into the behavior of TCP Reno in a PL environment and during LCD, the values of the number of packets sent during each time period within the LDC, namely, S_{LDC}, S_{LDP}, S_{NP}, S_{TDP}, R_{NP}, R_D, and S_{SST} were computed and plotted in Figure 11. It can be seen from the results in Figure 11 that S_{NP}, S_{TDP}, R_{NP}, and S_{SST} are not affected by the variation of *D* and only vary with *p*. This can be clarified by looking back to Equations (41), (3), (11), and (38), respectively. The value of R_D is increasing with *D*;

consequently, the values of S_{LDC} (Equation (47)) and S_{LDP} (Equation (26)) are also increased.

Calculation of *T*. The TCP-PEP also used to calculate *T* that is associated with each *S*. The results obtained are shown in Figure 12. Once again *T* is inversely proportional to *p*, and if all other network parameters are remained unchanged, *T* is slightly decreasing as *D* increases. Using Equation (60), our calculations show that *T* is reduced by nearly 14% if *D* is doubled, for all values of *p*. For example, when *p*=0.01 and *D* is doubled (increased from 6 to 12 sec), *T* is reduced by 14.44% (from 30.81 to 26.36 packets/ sec). It can be deduced from Figure 9 and 12 that

Figure 10. Comparison of S

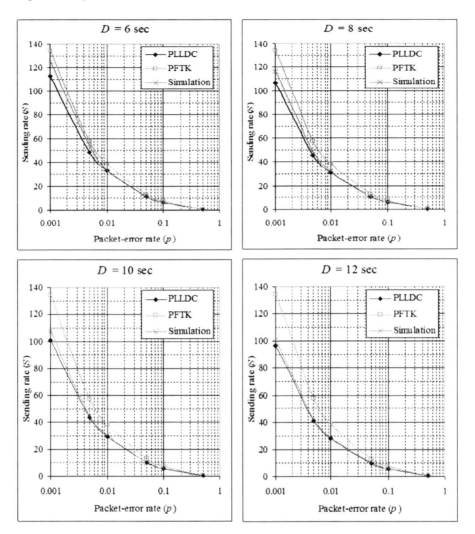

the difference between S and T is increasing as p increases for all values of D. This is mainly because, as p is increasing, S_{NP} is drastically decreased (for example, for D=6, S_{NP} decreases from 12838.90 to 5.00 packets as p increases from 0.001 to 0.5. While the number of packets set during other stages of the LDC is either remained unchanged or slightly changed, as shown in Figure 11.

Next, we compare the predicted throughput from the PLLDC and PFTK models against the values obtained from NS-2 simulation as shown in Figure 13. The results obtained for T show that

the PLLDC model can predict T more accurately than the PFTK model. It is also shown that the difference between the PFTK model and the simulation result is always higher than the difference between the PLLDC model and the simulation results.

Calculation of U. The variation of U with p for various values of D is shown in Figure 14, which shows that U decreases as p increases due the fact that the difference between S and T is increasing as p increases, which has discussed above. For example, U decreases from around 97% to 60% as p increases from 0.001 to 0.5. In

Figure 11. Packets send during the different stages of LDC for various values of D

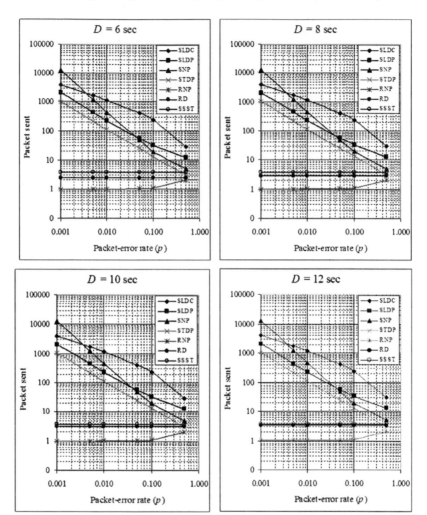

Figure 12. Variation of T with p for various values of D

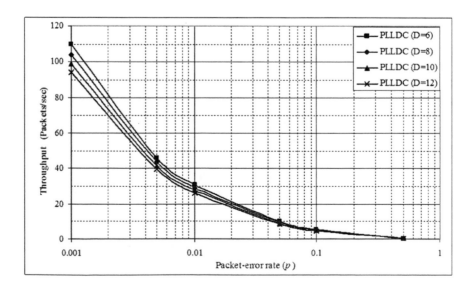

Figure 13. Comparison of T

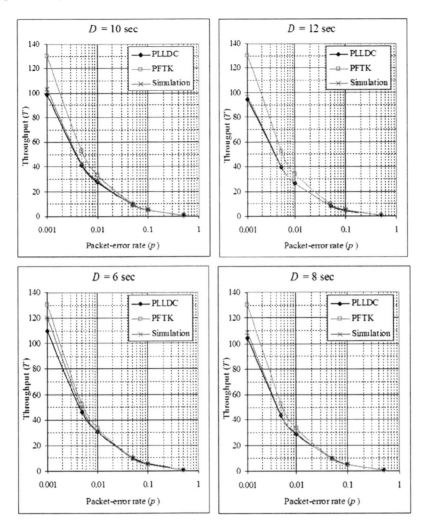

addition, presence of LCD almost equally effects *S* and *T* for all values of *D*, then the values of *U*, and consequently the performance of TCP Reno, are unaffected by the variation in *D*. The results are presented in semi-logarithmic scale for clarity.

CONCLUSION

This chapter presented a description, derivation, implementation, and comparison of two well-known analytical models, which were developed for evaluating the performance of the TCP

Reno protocol, namely, the PFTK and PLLDC models. These two models were implemented in a user-friendly TCP performance evaluation package, which was referred to as TCP-PEP. Then, the TCP-PEP was used to investigate the effect of PL and LDC on the performance of the TCP protocol in wireless network. The obtained results demonstrated that analytical modeling can be used effectively and efficiently to accurately and reliably evaluate the performance of the TCP protocol in realistic wireless environment, or, in other words, a wireless environment suffers from PL and LDC.

Figure 14. Variation of U with p for various values of D

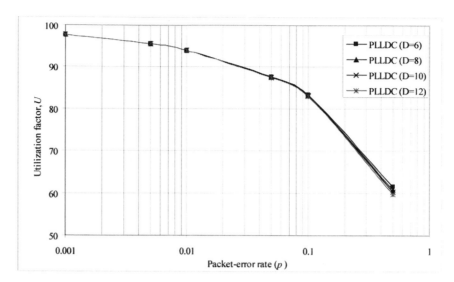

Implementation of these two analytical models or any other analytical models in an interactive and user-friendly package can be done easily, and can be of great help to professional, researchers, and students to analyze the performance of TCP in realistic wireless environment, and to investigate the effect of all parameters that may affect the TCP performance. Furthermore, implementing these models in a well structured package using object-oriented programming methodology can easily help with accepting other analytical models, and also modify the codes to accommodate other TCP flavors, such TCP New Reno, TCP Tahoe, etc.

The TCP-PEP package is expected to significantly contribute to a number of future studies, such as: assisting in determining an appropriate value of minimum T_o, evaluating the impact of any modifications on the current models, comparing the performance of the modified TCP with previous versions of TCP, and estimating the TCP performance with SFRs for finite receiver case.

REFERENCES

Abouzeid, A., & Roy, S. (2003). Stochastic modeling of TCP in networks with abrupt delay variations. *Journal of Wireless Networks, 9,* 509–524. doi:10.1023/A:1024644301397

Adarbah, H. (2008). *Modeling and analysis of TCP Reno with packet-loss and long delay cycles in wireless networks.* M.Sc Thesis, Amman Arab University for Graduate Studies, Amman, Jordan.

Al-Bahadili, H. (2010). On the use of discrete-event simulation in computer networks analysis and design. In Abu-Taieh, E. M. O., & El-Sheikh, A. A. (Eds.), *Handbook of research on discrete-event simulation environments: Technologies and applications* (pp. 418–442). Hershey, PA: Information Science Reference.

Altman, E., Barakat, C., & Ramos, V. (2005). Analysis of AIMD protocols over paths with variable delay. *Journal of Computer Networks, 48*(6), 960–971. doi:10.1016/j.comnet.2004.11.013

Chen, J., Gerla, M., Zhong Lee, Y., & Sanadidi, M. Y. (2008). TCP with delayed ACK for wireless networks. *Journal of Ad Hoc Networks, 6*(7), 1098–1116. doi:10.1016/j.adhoc.2007.10.004

Dunaytsev, R., Koucheryavy, Y., & Harju, J. (2006). The PFTK-model revised. *Journal of Computer Communications, 29*(13-14), 2671–2679. doi:10.1016/j.comcom.2006.01.035

Fall, K., & Floyd, S. (1996). Simulation-based comparison of Tahoe, Reno, and SACK TCP. *ACM Computer Communication Review, 26*(3), 5–21. doi:10.1145/235160.235162

Fall, K., & Varadhan, K. (2008). *The NS manual.* The VINT Project.

Forouzan, B. A. (2007). *Data communications and networking* (4th ed.). McGraw-Hill.

Fu, S., & Atiquzzaman, M. (2003). *Modeling TCP Reno with spurious timeout in wireless mobile environment.* International Conference on Computer Communication and Network. Dallas.

Gurtov, A., & Reiner, L. (2002). *Making TCP robust against delay spikes.* Internet Draft, draft-gurtov-tsvwg-tcp-delay-spikes-00.txt.

Hassan, M., & Jain, R. (2003). *High performance TCP/IP networking: Concepts, issues, and solutions.* Prentice-Hall.

Ho, C. Y., Chen, Y. C., Chan, Y. C., & Ho, C. Y. (2008). Fast retransmit and fast recovery schemes of transport protocols: A survey and taxonomy. *Journal of Computer Networks, 52*(6), 1308–1327. doi:10.1016/j.comnet.2007.12.012

Kesselman, A., & Mansour, Y. (2005). Adaptive AIMD congestion control. *Special Issue on Network Design, 43*(1-2), 97–111.

Kesselman, A., & Mansour, Y. (2005). Optimizing TCP retransmission timeout. *Proceedings of the 4th International Conference on Networking (ICN'05), 2,* (pp. 133-140).

Kung, H. T., Tan, K. S., & Hsiao, P. H. (2003). TCP with sender-based delay control. *Journal of Computer Communications, 26*(14), 1614–1621. doi:10.1016/S0140-3664(03)00110-5

Lai, Y. C., & Yao, C. L. (2002). Performance comparison between TCP Reno and TCP Vegas. *Journal of Computer Communications, 25*(18), 1765–1773. doi:10.1016/S0140-3664(02)00092-0

Lestas, M., Pitsillinds, A., Ioannou, P., & Hadjipollas, G. (2007). Adaptive congestion protocol: A congestion control protocol with learning capability. *Journal of Computer Networks, 51*(13), 3773–3798. doi:10.1016/j.comnet.2007.04.002

Ludwig, R., & Katz, R. H. (2000). The Eifel algorithm: Making TCP robust against spurious retransmissions. *ACM Computer Communications Review, 1*(30).

Lulling, M., & Vaughan, J. (2004). A simulation-based performance evaluation of Tahoe, Reno and SACK TCP as appropriate transport protocols for SIP. *Journal of Computer Communications, 27*(16), 1585–1593. doi:10.1016/j.comcom.2004.05.013

Padhye, J., Firoiu, V., Towsley, D., & Kurose, J. (2000). Modeling TCP Reno performance: A simple model and its empirical validation. *IEEE/ACM Transactions on Networking, 8*(2), 133–145. doi:10.1109/90.842137

Tanenbaum, A. (2003). *Computer networks* (4th ed.). Prentice Hall.

Tekala, M., & Szabo, R. (2008). Dynamic adjustment of scalable TCP congestion control parameters. *Journal of Computer Communications, 31*(10), 1890–1900. doi:10.1016/j.comcom.2007.12.035

Vendicits, A., Baiocchi, A., & Bonacci, M. (2003). Analysis and enhancement of TCP Vegas congestion control in a mixed TCP Vegas and TCP Reno network scenario. *Journal of Performance Evaluation*, *53*(3-4), 225–253. doi:10.1016/S0166-5316(03)00064-6

Voicu, L., Bassi, S., & Labrador, M. A. (2007). Analytical and experimental evaluation of TCP with an additive increase smooth decrease (AISD) strategy. *Journal of Computer Communications*, *30*(2), 479–495. doi:10.1016/j.comcom.2006.09.010

Weigle, M. C., Jaffay, K., & Simith, D. (2005). Delay-based early congestion detection and adaptation in TCP: Impact on Web performance. *Journal of Computer Communications*, *28*(8), 837–850. doi:10.1016/j.comcom.2004.11.011

Widmer, J., Denda, R., & Mauve, M. (2001). A survey on TCP-friendly congestion control. *Special Issue of the IEEE Network Magazine Control of Best Effort Traffic*, *15*(3), 28–37.

Xin, F., & Jamalipour, A. (2006). TCP performance in wireless networks with delay spike and different initial congestion window sizes. *Journal of Computer Communications*, *29*(8), 926–933. doi:10.1016/j.comcom.2005.06.012

Yang, Y. R., Kim, M. S., & Lam, S. S. (2003). Transient behaviors of TCP-friendly congestion control protocols. *Journal of Computer Networks*, *41*(2), 193–210. doi:10.1016/S1389-1286(02)00374-2

Compilation of References

A|D|S. (n.d.). *Home page*. Retrieved from http://www.adsgroup.org.uk/

AAPA. (2006). *Environment*. Association of Asia Pacific Airlines (AAPA). Retrieved from http://www.aapairlines.org/ Environment.aspx

Ababneh, D. (2009). *Development of a discrete-time DRED-based multi-queue nodes network analytical model*. Unpublished doctoral dissertation, The Arab Academy for Banking & Financial Sciences, Amman, Jordan.

Ababneh, J., Thabtah, F., Abdel-Jaber, H., Hadi, W., & Badarneh, E. (2010). Derivation of three queue nodes discrete-time analytical model based on DRED algorithm. *Proceedings of the 7th IEEE International Conference on Information Technology: New Generations* (ITNG 2010), Las-Vegas, USA, (pp. 885-890).

Abbott, D. W., & Wise, M. A. (1999). Underpinnings of system evaluation. In Kantowitz, B. H., Daniel, J. A., & Garland, D. J. (Eds.), *Handbook of aviation human factors* (pp. 51–52). Mahwah, NJ & London, UK: Lawrence Erlbaum Associates, Publishers.

Abdel-Jaber, H., Woodward, M. E., Thabtah, F., & Abu-Ali, A. (2008). Performance evaluation for DRED discrete-time queuing network analytical model. *Journal of Network and Computer Applications*, *31*(4), 750–770. doi:10.1016/j.jnca.2007.09.003

Abdel-Jaber, H., Woodward, M. E., Thabtah, F., & Etbega, M. (2007). A discrete-time queue analytical model based on dynamic random early drop. *Proceedings of the 4th IEEE International Conference on Information Technology: New Generations* (ITNG 2007), Las-Vegas, USA, (pp. 71-76).

Abouzeid, A., & Roy, S. (2003). Stochastic modeling of TCP in networks with abrupt delay variations. *Journal of Wireless Networks*, *9*, 509–524. doi:10.1023/A:1024644301397

Abu-Hanna, A., & Jansweiier, W. (1994). Modeling domain knowledge using explicit conceptualization. *IEEE Expert*, *9*(5). doi:10.1109/64.331490

Abu-Taieh, E. M. O. (2009). Information Technology and aviation industry: Marriage of convenience. In Abu-Taieh, E., El-Sheikh, A., & Abu-Tayeh, J. (Eds.), *Utilizing Information Technology systems across disciplines: Advancements in the application of computer Science* (pp. 153–164). Hershey, PA: Information Science Reference. doi:10.4018/978-1-60566-616-7.ch011

ACARE. (n.d.). *Home page*. Retrieved from http://www.acare4europe.org

ACI. (2009). *ACI World Director General, Angela Gittens gives a speech at the Aviation & Environment Summit*. 4th Aviation & Environment Summit. Geneva, 31 March, 2009, Opening Session. Retrieved from http://www.aci.aero/ cda/ aci_common/ display/ main/ aci_content07_c.jsp ?zn=aci& cp=1-7-3475^ 28415_666_2

Adarbah, H. (2008). *Modeling and analysis of TCP Reno with packet-loss and long delay cycles in wireless networks*. M.Sc Thesis, Amman Arab University for Graduate Studies, Amman, Jordan.

AENA SNA development. (2006). *Integrated tool PITOT for the analysis of air navigation systems*. Retrieved from http://www.aena.es/

Aguirre-Urreta, I., & Marakas, M. (2008). Comparing conceptual modeling techniques: A critical review of the EER vs. OO empirical literature. *ACM SIGMIS Database Newsletter, 39*(2).

AIA. (2008). *Aerospace Industries Association, the environment and civil aviation-Ensure environmental standards and policies that are global in development and application. Election 2008 Issues.* Keeping America Strong.

Aida, N., Nishijima, T., Hayashi, S., Yamada, H., & Kawakami, T. (2005). Combustion of lean prevaporized fuel–air mixtures mixed with hot burned gas for low-NO_x emissions over an extended range of fuel–air ratios. *Proceedings of the Combustion Institute, 30*, 2885–2892. doi:10.1016/j.proci.2004.08.040

Ainge, D. (2000). Introducing primary students to VR with Do 3D. *VR in the Schools, 4*(3). Retrieved June 6, 2001, from http://www.coe.ecu.edu/vr/ vrits/4-3Ainge.htm

Airbus Deutschland GmbH. (2003). *Cryoplane: Liquid hydrogen fuelled aircraft – System analysis.* Final Technical Report.

Airports Council International (ACI). (2010). *Year to date aircraft movements, April 2010.* Retrieved June 20, 2010, from http://www.aci.aero/ cda/ aci_common/ display/main/

Ajovalastit, A., Barone, S., & Petrucci, G. (1995, September). Towards RGB photoelasticity: Full-field automated photoelasticity in white light. *Experimental Mechanics, 35*, 193–200. doi:10.1007/BF02319657

Akour, S., Nayfeh, J., & Nicholson, D. (2003). Design of defense hole system for shear loaded plate. *The Journal of Strain Analysis for Engineering Design, 38*(6), 507–518. doi:10.1243/030932403770735872

Akour, S., Nayfeh, J., & Nicholson, D. (2010). Defense hole design for shear dominant loaded plate. *International Journal of Applied Mechanics, 2*(2). doi:10.1142/S1758825110000548

Al-Bahadili, H. (2010). On the use of discrete-event simulation in computer networks analysis and design. In Abu-Taieh, E. M. O., & El-Sheikh, A. A. (Eds.), *Handbook of research on discrete-event simulation environments: Technologies and applications* (pp. 418–442). Hershey, PA: Information Science Reference.

Albinsson, P.-A., & Alfredson, J. (2002). *Reflections on practical representation design (FOI-R-0716-SE).* Linköping, Sweden: Swedish Defence Research Agency.

Alfredson, J., Oskarsson, P.-A., Castor, M., & Svensson, J. (2003). Development of a meta instrument for evaluation of man-system interaction in systems engineering. In G. L. Rafnsdóttir, H. Gunnarsdóttir, & Þ. Sveinsdóttir (Eds.), *Proceedings of the 35th Annual Congress of the Nordic Ergonomics Society* (pp.77-79). Reykjavík, Island: Svansprent.

Allen, H. G. (1969). *Analysis and design of structural Sandwich Panels.* Oxford, UK: Pergamon Press.

Alm, T., Alfredson, J., & Ohlsson, K. (2008). Business process reengineering in the automotive area by simulator-based design. In El Sheikh, A., Al Ajeeli, A. T., & Abu-Taieh, E. M. (Eds.), *Simulation and modeling: Current technologies and applications* (pp. 337–358). Hershey, PA: IGI Global.

Alm, T. (2007). *Simulator-based design: Methodology and vehicle display applications.* Doctoral dissertation (No. 1078). Linköping, Sweden: Linköping University.

Alm, T., Ohlsson, K., & Kovordanyi, R. (2005). Glass cockpit simulators: Tools for IT-based car systems design and evaluation. In *Proceedings of Driving Simulator Conference – North America 2005*, Orlando, FL.

Aloi, D., Sleity, M., & Kiran, S. (2004). Analysis of LAAS integrated multipath limiting antenna using high fidelity electromagnetic models. *Proceedings of the Institute of Navigation Annual Meeting*, (pp. 441-470).

Altman, E., Barakat, C., & Ramos, V. (2005). Analysis of AIMD protocols over paths with variable delay. *Journal of Computer Networks, 48*(6), 960–971. doi:10.1016/j.comnet.2004.11.013

Anda, B. (2007). *Assessment of software system evolvability*. Ninth International Workshop on Principles of software evolution: In conjunction with the 6th ESEC/FSE (pp. 71–74).

Anderson, B. E., Chen, G., & Blake, D. R. (2006). Hydrocarbon emissions from a modern commercial airliner. *Atmospheric Environment, 40*, 3601–3612. doi:10.1016/j.atmosenv.2005.09.072

Andia, A., & Busch, C. (1999). *Architects design in cyberspace and real space for the NYSE*. Retrieved December 10, 2009, from http://www.members.tripod.com /wwwandia/art/100.html

Anonymous,. (2009). Directive 2008/101/EC of the European Parliament and of the Council. *Official Journal of the European Union. L&C, 52*(L8), 3–21.

Anonymous. (2010a). *de Havilland ghost*. Wikipedia. Retrieved June 23, 2010, from http://en.wikipedia.org/wiki/ De_Havilland_Ghost

Anonymous. (2010b). *General Electric J79*. Wikipedia. Retrieved June 23, 2010, from http://en.wikipedia.org/wiki/ J79#Specifications_.28J79- GE-17.29

Antonietti, A., Imperio, E., Rasi, C., & Sacco, M. (2001). Virtual reality and hypermedia in learning to use a turning lathe. *Journal of Computer Assisted Learning, 17*, 142–155. doi:10.1046/j.0266-4909.2001.00167.x

Antonietti, A., Rasi, C., Imperio, E., & Sacco, M. (1999). Virtual reality in engineering instruction: In search of the best learning procedures. *Proceedings World Conference on Educational Multimedia, Hypermedia and Telecommunications 1999*, (pp. 663-668).

AODB. (2009). *Airport Operation Data Base (AODB)*. Retrieved October 1, 2009 from http://www.aims.com.mo/upload/ brochure-AODB- final-eng-061204.pdf

API/EI. (2007). *Considerations for electronic sensors to monitor free water and/or particulate matter in aviation fuel*. 1598 Draft Standard, 2007.

Armstrong, J. (2006). *The quarks of object-oriented development*.

Association of Flight Attendants. CWA, AFL - CIO. (2010). *Counterterrorism initiative, legislative update*. Retrieved August 10, 2010, from http://www.afausairways.org /index.cfm?zone=/unionactive/ view_article.cfm& homeID=168976

Association of Flight Attendants (AFA). (2010). *Discreet, secure, hands-free, wireless communications for flight attendants*. Retrieved August 10, 2010, from http://www.afanet.org/pdf/ AFA_CabinCrewWirelessDevice.pdf

ASTM. (1995). Standard test method for flexural properties of sandwich constructions. *Annual Book of ASTM Standards, 15*(3), C393–C394.

ASTM. (1999). Standard test method for two-dimensional flexural properties of simply supported composite Sandwich Plates subjected to a distributed load. *Annual Book of ASTM Standards, 15*(3), D6419–D6499.

ASTM. (2005). *Standard practice for automatic sampling of petroleum and petroleum products*. D4177-95 (2005), 2007.

ASTM. (2007). *Standard specification for aviation turbine fuels*. D1655-05, 2007.

ASTM. (2007). *Standard test method for particulate contamination in aviation fuel by line sampling*. D2276-05, 2007.

ASTM. (2007). *Standard test method for undissolved water in aviation turbine fuels*. D3240-05, 2007.

ATA. (2010). *Emission limits*. Air Transport Association of America, Washington, DC. Retrieved July 25, 2010, from http://www.airlines.org/ Environment/ LocalAirQuality/ Pages/ EmissionsLimits.aspx

Athuraliya, S., Li, V. H., Low, S. H., & Yin, Q. (2001). REM: Active queue management. *IEEE Network, 15*(3), 48–53. doi:10.1109/65.923940

Aweya, J., Ouellette, M., & Montuno, D. Y. (2001). A control theoretic approach to active queue management. *Journal of Computer Networks, 36*(2–3), 203–235. doi:10.1016/S1389-1286(00)00206-1

Ayer, A. (1956). *The problem of knowledge*. London, UK: Penguin.

Bachelard, G. (1964). *The poetics of space*. New York, NY: Orion Press.

Bartone, C., & Van Graas, F. (1997). *Airport pseudolite for precision approach applications* (pp. 1841–1850). Institute of Navigation GPS Proceedings.

Baughcum, S. L., Tritz, T. G., Henderson, S. C., & Pickett, D. C. (1996). *Scheduled aircraft emission inventories for 1992: Database development and analysis. NASA contract report no 4700*. Hanover, MD, U.S.: NASA Center for Aerospace Information.

Bay, T., & Backius, P. (1999). Reiterating experimentation: Inventing new possibilities of life. *Emergence*, *1*(3). doi:10.1207/s15327000em0103_7

Bell, D. (1999). *Foreword to The coming of the post industrial society*. New York, NY: Basic Books, Special Anniversary Edition.

Bell, J. T., & Fogler, H. S. (2001). Virtual reality laboratory accidents. *Proceedings of American Society for Engineering Education Annual Conference*, Albuquerque, NM, American Society for Engineering Education, 2001. Retrieved January 2, 2008, from http://www.cs.uic.edu/~jbell /Professional/Papers/ASEE2001.pdf

Bell, J. T., & Fogler, H. S. (2004). The VRUPL lab - Serving education on two fronts. *Proceedings of the Special Interest Group on Computer Science Education Annual Conference*, Norfolk, VA, March 2004. Retrieved January 3, 2010, from http://www.cs.uic.edu/~jbell/ Professional/ Papers/ Bell_sigcse2004.pdf

Benini, E., Pandolfo, S., & Zoppellari, S. (2009). Reduction of NO emissions in a turbojet combustor by direct water/steam injection: Numerical and experimental assessment. *Applied Thermal Engineering*, *29*, 3506–3510. doi:10.1016/j.applthermaleng.2009.06.004

Bergson, H. (1999). *An introduction to metaphysics*. Indianapolis, IN: Hackett.

Bessee, G. (2008). *Research into electronic sensors for "guaranteed clean fuel."* ExxonMobil Aviation Technical Meeting, October, 2008.

Blakey, M. C. (2007, February 15). *Testimony – Statement of Marion C. Blakey*. Retrieved July 19, 2010, from http://www.faa.gov/ news/ testimony/ news_story.cfm? newsId=8184

Boeing. (2010a). *Current market outlook 2009-2028*. Market analysis, Boeing Commercial Airplanes, Seattle, WA. Retrieved July 25, 2010, from http://www.boeing.com/ commercial/ cmo/ pdf/ Boeing_Current_Market_Outlook_2009_to _2028.pdf

Boeing. (2010b). *2010 environment report*. Retrieved July 25, 2010, from http://www.boeing.com/ aboutus/ environment/ environment_report_10/ boeing-2010-environment- report.pdf

Boisot, M. (1995). *Information space: A framework for learning in organizations, institutions and culture*. London, UK: Routledge.

Böttcher, S. A. (2004). A task-oriented reconfigurable software architecture for e-commerce document exchange. *Proceedings of the 37th Hawaii International Conference on System Sciences*, IEEE Computer Society.

Boyce, M. P. (2002). *Gas turbine engineering handbook* (2nd ed.). Houston, TX: Gulf Professional Publishing.

Braden, R., Clark, D., Crowcroft, J., Davie, B., Deering, S., Estrin, D., et al. (1998). *Recommendations on queue management and congestion avoidance in the Internet*. RFC 2309.

Brady, C. (2010). Auxiliary power unit. In *The Boeing 737 technical guide* (57th ed., pp. 62–68). Tech Pilot Services Ltd.

Brakmo, L., & Peterson, L. (1995). TCP Vegas: End to end congestion avoidance on a global Internet. *IEEE JSAC*, *13*(8), 1465–1480.

Brandauer, C., Iannaccone, G., Diot, C., Ziegler, T., Fdida, S., & May, M. (2001). Comparison of tail drop and active queue management performance for bulk-data and Web-like Internet traffic. In *Proceeding of ISCC, IEEE*, (pp. 122–9).

Brasseur, G. P., Cox, R. A., Hauglustaine, D., Isaksen, I., Lelieveld, J., & Lister, D. H. (1998). European scientific assessment of the atmospheric effects of aircraft emissions. *Atmospheric Environment*, *32*(13), 2329–2418. doi:10.1016/S1352-2310(97)00486-X

Brown, L., & Rantz, W. (2010). *The efficacy of flight attendant/pilot communication, in a post 9/11 environment: Viewed from both sides of the fortress door*. International Journal of Applied Aviation Studies. IJAAS.

Brown, S., & Duguid, P. (2000). *The social life of information*. Boston, MA: Harvard Business School Press.

Brown, L. (2010). *In-flight security onboard commercial aircraft: Critical improvements needed.* 27th International Congress of Aeronautical Sciences. Nice, France.

Brown, L., & Niehaus, J. (2009). *Improving pilot/flight attendant communications to enhance aviation safety.* Flight Safety Foundation, Corporate Aviation Safety Seminar. Orlando, Florida, April 2009.

Bureau of Transportation Statistics (BTS). (2010). *Airline fuel cost and consumption (U.S. carriers-Scheduled): January 2000 - May 2010.* Research and Innovative Technology Administration, U.S. Department of Transportation, Washington, DC. Retrieved July 4, 2010, from http://www.transtats.bts.gov/ fuel.asp? pn=0& display=data4

Burki-Cohen, J., Go, T. H., et al. (2003). Simulator fidelity requirements for airline pilots training and evaluation continued: An update on motion requirements research. In *Proceedings of the Twelfth Annual International Symposium on Aviation Psychology*, Dayton, OH: ISAP

Byrne, C. M. (1996). *Water on tap: The use of virtual reality as an educational tool.* Ph.D. Dissertation. University of Washington, Seattle, WA.

Caprino, G., & Langelan, A. (2000). Study of 3 pt. bending specimen for shear characterization of sandwich cores. *Journal of Composite Materials, 34*(9), 791–814.

Cardelli, L. (1996). Bad engineering properties of object-oriented languages. *ACM Computing Surveys, 28*(4). doi:10.1145/242224.242415

Carslaw, D. C., Ropkins, K., Laxen, D., Moorcroft, S., Marner, B., & Williams, M. L. (2008). Near-field commercial aircraft contribution to nitrogen oxides by engine, aircraft type, and airline by individual plume sampling. *Environmental Science & Technology, 42*, 1871–1876. doi:10.1021/es071926a

Castor, M. (2009). *The use of structural equation modeling to describe the effect of operator functional state on air-to-air engagement outcomes.* Doctoral dissertation (No. 1251). Linköping, Sweden: Linköping University.

Castor, M., Hanson, E., Svensson, E., Nählinder, S., Le Blaye, P., Macleod, I., et al. Ohlsson, K. (2003). *GARTEUR handbook of mental workload measurement* (GARTEUR TP 145). The Group for Aeronautical Research and Technology in Europe.

Chapanis, A. (1996). *Human factors in systems engineering.* New York, NY: John Wiley & Sons.

Chappell, D. (2004). Introduction to the enterprise service bus. In M. Hendrickson (Ed.), *Enterprise service bus* (pp. 1-22). California: O'Reilly Media, Inc.

Chappell, D. (2005). *ESB myth busters: 10 enterprise service bus myths debunked. Clarity of definition for a growing phenomenon.* Retrieved September 1, 2009, from http://soa.sys-con.com/node/48035

Chapra, S., & Canal, R. (2006). *Numerical methods for engineers* (pp. 355–360). McGraw Hill.

Chen, J., Gerla, M., Zhong Lee, Y., & Sanadidi, M. Y. (2008). TCP with delayed ACK for wireless networks. *Journal of Ad Hoc Networks, 6*(7), 1098–1116. doi:10.1016/j.adhoc.2007.10.004

Cheng, N. J. (2001). *An integration framework for airport automation systems.* Retrieved October 10, 2009, from http://www.mitre.org/work/ tech_papers/tech_papers_01/ cheng_integration/ cheng_integration.pdf

Chengyu, Z., Yang, O. W. W., Aweya, J., Ouellette, M., & Montuno, D. Y. (2002). A comparison of active queue management algorithms using the OPNET modeler. *Proceedings IEEE Communications Magazine, 40*(6), 158. doi:10.1109/MCOM.2002.1007422

Choo, C. (1998). *The knowing organization: How organizations use information to construct meaning, create knowledge, and make decisions.* New York, NY: Oxford University Press.

Civil Aviation Authority. (2010). *ICAO aircraft engine emissions databank definitions.* Retrieved July 25, 2010, from http://www.caa.co.uk/ docs/ 02/ 070716% 20Introduction.pdf

Climate Committee. (2009). *The Committee on Climate Change* (CCC). Retrieved from http://www.theccc.org.uk/ topics/ international-action-on- climate-change/ international-aviation

Cooperman, G. (1996). TOP-C: A task-oriented parallel C interface. *Proceedings of 5th IEEE International Symposium on High Performance Distributed Computing*. Syracuse, NY.

CORDIS News. (2009). *A more realistic World Wide Web.* Retrieved from http://cordis.europa.eu/

Coulouris, G., Dollimore, J., & Kindberg, T. (2005). *Distributed systems concepts and design* (4th ed., p. 19). Harlow, UK: Addison Wesley.

Cross, N. (1995). Discovering design ability. In Buchanan, R., & Margolin, V. (Eds.), *Discovering design* (pp. 105–120). Chicago, IL: The University of Chicago Press.

Cusson, J. (2000). *The Xonon combustion system.* Presented at Industrial Center - TMAC Meeting, San Francisco. Retrieved July 25, 2010, from http://www.energysolutionscenter.org/distgen/AppGuide/DataFiles/Xonon.pdf

Daggett, D. L. (2004). *Water misting and injection of commercial aircraft engines to reduce airport NO$_x$. NASA/CR—2004-212957.* Hanover, MD: NASA Center for Aerospace Information.

Darses, F., & Wolff, M. (2006). How do designers represent to themselves the users' needs? *Applied Ergonomics*, *37*(6), 757–764. doi:10.1016/j.apergo.2005.11.004

Dartmouth College. (n.d.). *The clinician's black bag of quality improvement tools.* Retrieved March 11, 2008, from www.darthmouth.edu

Davenport, T. H., & Prusak, L. (1998). *Working knowledge.*

Defense Acquisition University. (2010, May 17). *Community browser.* Retrieved July 19, 2010, from www.acc.dau.mil/communitybrowser.aspx?id=314774&lang=en-US

Dekker, S. W. A. (2003). Illusions of explanation: A critical essay on error classification. *The International Journal of Aviation Psychology*, *13*(2), 95–106. doi:10.1207/S15327108IJAP1302_01

Dekker, S. W. A. (2005). *Ten questions about human error: A new view of human factors and systems safety.* Mahwah, NJ: Erlbaum.

Dekker, S. W. A. (2007). Doctors are more dangerous than gun owners: A rejoinder to error counting. *Human Factors*, *49*(2), 177–184. doi:10.1518/001872007X312423

Dekker, S. W. A., & Nyce, J. M. (2004). How can ergonomics influence design? Moving from research findings to future systems. *Ergonomics*, *47*(15), 1624–1639. doi:10.1080/00140130412331290853

Dekker, S. W. A., & Woods, D. D. (1999). Extracting data from the future: Assessment and certification of envisioned systems. In Dekker, S., & Hollnagel, E. (Eds.), *Coping with computers in the cockpit* (pp. 131–143). Aldershot, England: Ashgate.

Dekker, S. W. A. (2005). Why we need new accident models. In Harris, D., & Muir, H. (Eds.), *Contemporary issues in human factors and aviation safety* (pp. 181–198). Burlington, VT: Ashgate Publishing Co.

Dekker, S. W. A. (1996). Cognitive complexity in management by exception: Deriving early human factors requirements for an envisioned air traffic management world. In D. Harris (Ed.), *Engineering psychology and cognitive ergonomics, volume I: Transportation systems* (pp. 201-210). Aldershot, England: Ashgate.

Department of Defense and Department of Transportation. (2001). *Federal radionavigation plan.* Technical Specification (FRP) DOT-VNTSC-RSPA-01-3/DOD-4650.5

Desart, B. (2007). *A window to airport system information sharing.* PACS Workshop. Retrieved from http://www.eurocontrol.int/aim/public/standard_page/pacs.html

Dhir, S. K. (1981). Optimization in a class of hole shapes in plate structures. *Journal of Applied Mechanics*, *48*(4), 905–909. doi:10.1115/1.3157754

Dings, J. M. W., Wit, R. C. N., Leurs, B. A., Davidson, M. D., & Fransen, W. (2003). *External costs of aviation. (Research Report 299 96 106: UBA-FB 000411), Federal Environmental Agency.* Berlin: Umweltbundesamt.

Dobson, G. (2006). Using WS-BPEL to implement software fault tolerance for Web services. *Proceedings of the 32nd EUROMICRO Conference Software Engineering and Advanced Applications (SEAA 2007)*, (pp. 126-133).

Dong-Shan, F., & Xing, Z. (1995). Analytical-variational method of analysis about a finite plate with a cracked hole stiffened by a ring with rivet joints. *Engineering Fracture Mechanics, 50*(3), 325–344. doi:10.1016/0013-7944(94)00205-V

Dunaytsev, R., Koucheryavy, Y., & Harju, J. (2006). The PFTK-model revised. *Journal of Computer Communications, 29*(13-14), 2671–2679. doi:10.1016/j.comcom.2006.01.035

Durelli, A. J., Brown, K., & Yee, P. (1978). Optimization of geometric discontinuities in stresses fields. *Experimental Mechanics, 18*(8), 303–308. doi:10.1007/BF02324161

EEA. (n.d.). *European Environment Agency website.* Retrieved from http://www.eea.europa.eu/

Ehhalt, D., Dentener, F., Derwent, R., Dlugokencky, E., Holland, E., Isaksen, I., et al. Wang, M. (2001). *Atmospheric chemistry and greenhouse gases.* Climate Change 2001: The Scientific Basis. Cambridge, United Kingdom: Cambridge University Press.

El Sheikh, A. (1987). *Simulation modeling using a relational database package.* PhD Thesis. London, UK: The London School of Economics.

Elliott, L. (2005). *See it, feel, it, hear* it. Retrieved December 20, 2009, from http://www.deskeng.com/articles/aaaapn.htm

Endsley, M. R. (1995b). Toward a theory of situation awareness in dynamic systems. *Human Factors, 37*(1), 32–64. doi:10.1518/001872095779049543

Endsley, M. R., Bolté, B., & Jones, D. G. (2003). *Designing for situation awareness: An approach to user-centered design.* London, UK: Taylor & Francis.

Endsley, M. (1999). Situation awareness in aviation systems. In Kantowitz, B. H., Daniel, J. A., & Garland, D. J. (Eds.), *Handbook of aviation human factors* (pp. 257–259). Mahwah, NJ: Lawrence Erlbaum Associates, Publishers.

Endsley, M. R. (1988). Situation awareness global assessment technique (SAGAT). In *Proceedings of the IEEE National Aerospace and Electronics Conference* (pp. 789-795). New York, NY: IEEE.

Endsley, M. R., Bolstad, C. A., Jones, D. G., & Riley, J. M. (2003). Situation awareness oriented design: From user's cognitive requirements to creating effective supporting technologies. In *Proceedings of the Human Factors and Ergonomics Society 47th Annual Meeting* (pp. 268-272). Santa Monica, CA: Human Factors and Ergonomics Society.

Enge, P. (1999). Local area augmentation of GPS for the precision approach of aircraft. *Proceedings of the IEEE, 87*(1), 111–132. doi:10.1109/5.736345

Enge, P. (1996). Wide area augmentation of the global positioning system. *Proceedings of the IEEE, 84*(8), 1063–1088. doi:10.1109/5.533954

Erickson, P. E., & Riley, W. F. (1978). Minimizing stress concentration around circular holes in uniaxially loaded plates. *Experimental Mechanics, 18*, 97–100. doi:10.1007/BF02325003

Ervin, S., & Hasbrouck, H. (2001). *Landscape modeling.* New York, NY: McGraw-Hill.

Eugenia, C., & Isidro, R. (2008). *The baseline: The milestone of software product lines for expert systems automatic development.* Baja, CA: Mexican International Conference on Computer Science, IEEE.

Eurocontrol. (2009). *Medium-term forecast update: IFR flight movements (2009-2015).* Eurocontrol forecast reports, September 2009. Retrieved from http://www.eurocontrol.int/

Eurocontrol. (2010). Communications, navigation and surveillance at the heart of the future ATM system. *Eurocontrol Skyway Magazine, (Winter, 2010).*

EUROCONTROL. (n.d.). *Homepage.* Retrieved from http://www.eurocontrol.int/

European Commission. (August 2005). *Environment fact sheet: Climate change.*

European Commission. (13 July 2006). *Science for environment policy, DG environment news alert service: Minimizing the climate impact of aviation.*

European Commission. (March 2006). *Environment fact sheet: Energy for sustainable development.*

European Commission. (16 May 2007). *Science for environment policy, DG environment news alert service: Regulating international aviation and shipping emissions.*

European Commission. (2009). *EU action against climate change, leading global action to 2020 and beyond.*

Ewell, C. D., & Chidester, T. (1996). *American Airlines converts CRM in favor of human factors and safety training. The Flightdeck, July/August, 1996. Flight Department, American Airlines.* DFW Airport.

Eyers, C. J., Addleton, D., Atkinson, K., & Broomhead, M. J. Christou, R., Elliff, T., ... Stanciou, N. (2004). *AERO2k global aviation emissions inventories for 2002 and 2025.* European Commission, Contract No.G4RD-CT-2000-00382, QinetiQ ltd.

Fall, K., & Floyd, S. (1996). Simulation-based comparison of Tahoe, Reno, and SACK TCP. *ACM Computer Communication Review, 26*(3), 5–21. doi:10.1145/235160.235162

Fall, K., & Varadhan, K. (2008). *The NS manual.* The VINT Project.

Faudi Aviation. (2009, November). *AFGuard.* Retrieved November 10, 2009, from http://www.faudi-aviation.com/images/ stories/ ds_afguard_en.pdf

Federal Aviation Administration. (2007, February 28). *AC 150/5200-37.* Retrieved March 11, 2008, from www.faa.gov

Federal Aviation Administration. (2004, September 27). *AC 120-35C: Line operational simulations: Line oriented flight training, special purpose operational training, line operational evaluation.*

Federal Aviation Administration. (2004, January 22). *AC 120-51E: Crew resource management.*

Federal Aviation Administration. (2004, June 23). *AC 120-54A: Advanced qualification program.*

Federal Aviation Administration (FAA). (1999). *Federal Aviation Administration specification for the wide area augmentation system* (WAAS). (DTFA01-96-C-00025, FAA-E- 2892b).

Federal Aviation Administration (FAA). (2002). *Navigation and landing transition strategy.* FAA Office of Architecture and Investment Analysis, ASD-1.

Federal Regulations (CFR). (2003). *Title 14, aeronautics and space- SFAR 58: Advanced qualification program.*

Federal Regulations (CFR). (2005). *Title 14, aeronautics and space- Part 121- Operating requirement: Domestic flight and supplemental operations, note sub-part Y.*

Feng, W., Kandlur, D., Saha, D., & Shin, K. G. (1999). *Blue: A new class of active queue management algorithms. Technical report UM CSE-TR-387-99.* Ann Arbor, MI: University of Michigan.

Feng, W., Kandlur, D., Saha, D., & Shin, K. G. (2001). Stochastic fair blue: A queue management algorithm for enforcing fairness. *Proceedings - IEEE INFOCOM, 3,* 1520–1529.

Fenn, M., Poth, M., & Meixner, T. (2005). *Atmospheric nitrogen deposition and habitat alteration in terrestrial and aquatic ecosystems in Southern California: Implications for threatened and endangered species,* (pp. 269-271). (USDA Forest Service Gen. Tech. Rep. PSW-GTR-195).

Fitch, E. C. (1988). *Fluid contamination control. Oklahoma State University.* Stillwater, OK: FES.

Fitts, P. M., Jones, R. E., & Milton, J. L. (1950). Eye movements of aircraft pilots during instrument-landing approaches. *Aeronautical Engineering Review, 9*(2), 24–29.

Fitts, P. M. (1951). *Human engineering for an effective air navigation and traffic control system.* Ohio State University Research Foundation Report, Columbus, OH.

Flamme, M. (2004). New combustion systems for gas turbines (NGT). *Applied Thermal Engineering, 24,* 1551–1559. doi:10.1016/j.applthermaleng.2003.10.024

Fleisher, L. (2005, June 5th). Terror response is tested at Boston's Logan Airport in Operation Atlas. *Boston Globe.* Retrieved April 5, 2009, from http://www.cra-usa.net/1inthenews-Boston.htm

Flexman, R. H., & Stark, E. A. (1987). Training simulators. In Salvendy, G. (Ed.), *Handbook of human factors* (pp. 1012–1038). New York, NY: John Wiley and Sons.

Flin, R., Martin, L., et al. (2003, 2005). Development of the NOTECHS (non-technical skills) system for assessing pilots' CRM skills. In D. Harris & H. C. Muir (Eds.), *Contemporary issues in human factors and aviation safety* (pp. 133-154). Burlington, VT: Ashgate Publishing Co.

Flippone, A. (2010). Cruise altitude flexibility of jet transport aircraft. *Aerospace Science and Technology, 14,* 283–294. doi:10.1016/j.ast.2010.01.003

Florides, G. A., & Christodoulides, P. (2009). Global warming and carbon dioxide through sciences. *Environment International, 35,* 390–401. doi:10.1016/j.envint.2008.07.007

Floyd, S., & Jacobson, V. (1993). Random early detection gateways for congestion avoidance. *IEEE/ACM Transactions on Networking, 1*(4), 397–413. doi:10.1109/90.251892

Floyd, S., Ramakrishna, G., & Shenker, S. (2001). *Adaptive RED: An algorithm for increasing the robustness of RED's active queue management. Technical report.* ICSI.

Folland, C. K., Karl, T. R., Christy, J. R., Clarke, R. A., Gruza, G. V., & Jouzel, J. ... Wang, S. W. (2001). *Observed climate variability and change.* Climate Change 2001: The Scientific Basis. Cambridge, United Kingdom: Cambridge University Press.

Fonseca, F. (2007). The double role of ontologies in information science research. *Journal of the American Society for Information Science and Technology, 58*(6). doi:10.1002/asi.20565

Forouzan, B. A. (2007). *Data communications and networking* (4th ed.). McGraw-Hill.

Forster, P., Ramaswamy, V., Artaxo, P., Berntsen, T., & Betts, R. Fahey, ... Dorland, R. (2007). *Changes in atmospheric constituents and in radiative forcing.* Climate Change 2007: The Physical Science Basis. Cambridge, United Kingdom: Cambridge University Press.

Forzatti, P. (2003). Status and perspectives of catalytic combustion for gas turbines. *Catalysis Today, 83,* 3–18. doi:10.1016/S0920-5861(03)00211-6

France, R. A. (2007). *Model-driven development of complex software: A research roadmap.*

Freund, E., & Rossmann, J. (2005). Projective virtual reality as a basis for on-line control of complex systems-not only-over the internet. *Journal of Robotic Systems, 22*(3), 147–155. doi:10.1002/rob.20055

Friend, P. (2007). *Aviation security part II: A frontline perspective on the need for enhanced human resources and equipment.* Testimony of Patricia A. Friend, before the subcommittee on transportation and infrastructure. November 1, 2007, U.S. House of Representatives, Washington, D.C. Retrieved August 1, 2010, from http://homeland.house.gov/ hearings/index.asp?ID=101

Friends of the Earth. (n.d.). *Aviation and global climate change.* Retrieved from www.foe.co.uk

Frostig, Y., Baruch, M., Vilnai, O., & Sheinman, I. (1992). Higher-order theory for sandwich beam behavior with transversely flexible core. *Journal of Engineering Mechanics, 118*(5), 1026–1043. doi:10.1061/(ASCE)0733-9399(1992)118:5(1026)

Fu, S., & Atiquzzaman, M. (2003). *Modeling TCP Reno with spurious timeout in wireless mobile environment.* International Conference on Computer Communication and Network. Dallas.

Fuchs, H. O., & Stephens, R. I. (1980). *Metal fatigue in engineering* (pp. 115–118). John Wiley & Sons.

Gaffney, J. S., & Marley, N. A. (2009). The impacts of combustion emissions on air quality and climate – From coal to biofuels and beyond. *Atmospheric Environment, 43,* 23–36. doi:10.1016/j.atmosenv.2008.09.016

Gaizauskas, R. A. (2006). Task-oriented extraction of temporal information: The case of clinical narratives. *Proceedings of the Thirteenth International Symposium on Temporal Representation and Reasoning,* IEEE Computer Society.

Galofaro, L. (1999). *Digital Eisenman-An office of the electronic era.* Basel, Switzerland: Birkhäuser.

GAO. (n.d.). *United States Government Accountability Office website.* Retrieved from http://www.gao.gov/

GAO-09-37. (November 2008). *Aviation and the environment.*

Garcia, R., Tomas, J., & Breis, F. (2008). A knowledge acquisition methodology to ontology construction for information retrieval from medical documents. *Expert Systems: International Journal of Knowledge Engineering and Neural Networks, 25*(3).

Gardner, R. M., Adams, K., Cook, T., Deidewig, F., Ernedal, S., & Falk, R. (1997). The ANCAT/EC global inventory of NO_x emissions from aircraft. *Atmospheric Environment*, *31*(12), 1751–1766. doi:10.1016/S1352-2310(96)00328-7

Garzas, J., & Piattini, M. (2006). An ontology for understanding and applying object-oriented design knowledge. *International Journal of Software Engineering and Knowledge Engineering*, *17*(3).

Gates, B. (1999). *Business @ the speed of thought*. London, UK: Penguin Books.

Gdoutos, E. E., Daniel, I. M., Wang, K.-A., & Abot, J. L. (2001). Non-linear behavior of composite Sandwich Beams in three-point bending. *Experimental Mechanics*, *41*(2), 182–189. doi:10.1007/BF02323195

Geisler, E. (2006). A taxonomy and proposed codification of knowledge and knowledge systems in organizations. *Knowledge and Process Management*, *13*(4). doi:10.1002/kpm.265

Gersh, J. R. (2005). *Cognitive engineering: Understanding human interaction with complex systems*. Baltimore, MD: John Hopkins APL.

Gerval, J. P., Popovici, M., Ramdani, M., El Kalai, O., Boskoff, V., & Tisseau, J. (2002). Virtual environments for children. *Proceedings International Conference on Computers and Advanced Technology Education (CATE)*, (pp. 416-420). Cancun, Mexico.

Giampaolo, T. (2003). *The gas turbine handbook: Principles and practices* (2nd ed.). Lilburn, GA: Fairmont Press.

Givoni, M., & Rietveld, P. (2010). The environmental implications of airlines' choice of aircraft size. *Journal of Air Transport Management*, *16*, 159–167. doi:10.1016/j.jairtraman.2009.07.010

Glover, B. (2005). *Fuel cell opportunity*. Presented at AIAA/AAAF Aircraft Noise and Emissions Reduction Symposium, Monterey, CA. Retrieved July 25, 2010, from http://www.aiaa.org/ events/ aners/ Presentations/ ANERS-Glover.pdf

Go, T. H., Burki-Cohen, J., et al. (2003). The effects of enhanced hexapod motion on airline pilot recurrent training and evaluation. In *Proceedings of American Institute of Aeronautics and Astronautics AIAA Modeling and Simulation Technical Conference*. Dallas, TX: AIAA

GOA. (2009). *Aviation and climate change*. Report to congressional committees, Number GAO-09-554, United States Government Accountability Office, June.

Gohlke, O., Weber, T., Seguin, P., & Laborel, Y. (2010). A new process for NO_x reduction in combustion systems for the generation of energy from waste. *Waste Management (New York, N.Y.)*, *30*, 1348–1354. doi:10.1016/j.wasman.2010.02.024

Goold, R. (2009). *Airport management systems for the 21st century*. Retrieved September 25, 2009, from http://www.airport-int.com/ categories/airport -management-systems /airport-management-systems -for-the-21st-century.asp

Goswami, C. (2006). *What do we do with knowledge?* Hendriks, P., & Vriens, D. (1999). Knowledge-based systems and knowledge management: friends or foes? *Information & Management*, *35*(2).

Graham, A., & Raper, D. W. (2006). Transport to ground of emissions in aircraft wakes. Part II: Effect on NO_x concentrations in airport approaches. *Atmospheric Environment*, *40*, 5824–5836. doi:10.1016/j.atmosenv.2006.05.014

Green, J. E. (2006). Civil aviation and the environment - The next frontier for the aerodynamicist. *Aeronautical Journal*, *110*(1110), 469–486.

Groob, J., Bruhl, C., & Peter, T. (1998). Impact of aircraft emissions on tropospheric and stratospheric ozone. Part 1: Chemistry and 2D model results. *Atmospheric Environment*, *32*(18), 3173–3184. doi:10.1016/S1352-2310(98)00016-8

Gungor, S., Nurse, A. D., & Patterson, E. A. (1995). Experimental determination of stress intensity factors of cracks in sheet structures with bolted stiffeners. *Engineering Fracture Mechanics*, *32*(17/18), 2423–2445.

Gupta, A. K. (1997). Gas turbine combustion: Prospects and challenges. *Energy Conversion and Management*, *38*(10-13), 1311–1318. doi:10.1016/S0196-8904(96)00160-4

Gurtov, A., & Reiner, L. (2002). *Making TCP robust against delay spikes.* Internet Draft, draft-gurtov-tsvwg-tcp-delay-spikes-00.txt.

Haglind, F. (2008). Potential of lowering the contrail formation of aircraft exhausts by engine re-design. *Aerospace Science and Technology*, *12*(6), 490–497. doi:10.1016/j.ast.2007.12.001

Hamman, W. R., Seamster, T. L., Lofaro, R. J., & Smith, K. M. (1992). The future of LOFT scenario design and validation. *Proceedings of the Seventh International Symposium on Aviation Psychology.* Columbus, OH: ISAP

Hassan, M., & Jain, R. (2003). *High performance TCP/IP networking: Concepts, issues, and solutions.* Prentice-Hall.

Hayashi, S., Yamada, H., & Makida, M. (2000). Short-flame/quick quench: A unique ultralow emissions combustion concept for gas turbine combustors. *Proceedings of the Combustion Institute*, *28*, 1273–1280. doi:10.1016/S0082-0784(00)80340-9

Herndon, S. C., Rogers, T., Dunlea, E. J., Jayne, J. T., Miake-Lye, R., & Knighton, B. (2006). Hydrocarbon emissions from in-use commercial aircraft during airport operations. *Environmental Science & Technology*, *40*, 4406–4413. doi:10.1021/es0512091

Herndon, S. C., Shorter, J. H., Zahniser, M. S., Nelson, D. D., Jayne, J., & Brown, R. C. (2004). NO and NO_2 emission ratios measured from in-use commercial aircraft during taxi and takeoff. *Environmental Science & Technology*, *38*, 6078–6084. doi:10.1021/es049701c

Hinds, W. C. (1982). *Aerosol technology.* New York, NY: Wiley-Interscience.

Ho, C. Y., Chen, Y. C., Chan, Y. C., & Ho, C. Y. (2008). Fast retransmit and fast recovery schemes of transport protocols: A survey and taxonomy. *Journal of Computer Networks*, *52*(6), 1308–1327. doi:10.1016/j.comnet.2007.12.012

Hollnagel, E. (2007). Flight decks and free flight: Where are the system boundaries? *Applied Ergonomics*, *38*(4), 409–416. doi:10.1016/j.apergo.2007.01.010

Hollnagel, E., & Woods, D. D. (1983). Cognitive systems engineering: New wine in new bottles. *International Journal of Man-Machine Studies*, *18*(6), 583–600. doi:10.1016/S0020-7373(83)80034-0

Hollnagel, E., & Woods, D. D. (2005). *Joint cognitive systems: Foundations of cognitive systems engineering.* Boca Raton, FL: Taylor & Francis. doi:10.1201/9781420038194

Hollnagel, E. (1999). From function allocation to function congruence. In Dekker, S., & Hollnagel, E. (Eds.), *Coping with computers in the cockpit* (pp. 29–53). Aldershot, England: Ashgate.

Homeland Security News. (2008, July 28). Is U.S. bioterror attack just a matter of time? *Homeland Security News*, 2008.

Howard, M. (2002). *Usefulness in representation design.* Doctoral dissertation (No. 753). Linköping, Sweden: Linköping University.

Hrachovec, H., & Pichler, A. (2007). *Philosophy of the information society. Proceedings of the 30th International Ludwig Wittgenstein Symposium Kirchberg am Wechsel.*

Hult, L., Irestig, M., & Lundberg, J. (2006). Design perspectives. *Human-Computer Interaction*, *21*(1), 5–48. doi:10.1207/s15327051hci2101_2

Hunecke, K. (2003). *Jet engines, fundamentals of therory, design and operation* (6th ed.). Osceola, FL: Motorbooks International Publishers & Wholesalers.

Husban, M. (2009). *Design and optimization of stress relief system for laminate composite plate.* Unpublished doctoral dissertation, University of Jordan, Amman, Jordan.

IATA. (2009). *A global approach to reducing aviation emissions. First stop: carbon-neutral growth from 2020.* Report, International Air Transport Association.

IATA. (2004). *Guidance material for aviation turbine fuels specifications,* part III: *Cleanliness and handling,* 5th edition, 2004.

ICAO. (2007). *Discreet wireless communication for civil aviation cabin crewmembers.* 36th Session of the Assembly [ICAO.]. *Agenda (Durban, South Africa)*, 15.

ICAO. (2007). *ICAO airport air quality guidance manual.* Doc. 9889, Preliminary Edition, International Civil Aviation Organization, Montreal. Retrieved July 25, 2010, from http://www.icao.int/ icaonet/ dcs/ 9889/ 9889_en.pdf

ICAO. (2008). *ICAO annual report of the council.* Doc. 9916, International Civil Aviation Organization, Montreal. Retrieved June 20, 2010, from http://www.icao.int/ icaonet/ dcs/ 9916/ 9916_en.pdf

ICAO. (2010a). *CAEP.* International Civil Aviation Organization, Air Transport Bureau, Environment Branch. Retrieved July 5, 2010, from http://www.icao.int/ icao/ en/ nv/ caep.htm

ICAO. (2010b). *Aircraft engine emissions.* International Civil Aviation Organization, Air Transport Bureau, Environment Branch. Retrieved June 25, 2010, from http:// www.icao.int/ icao/ en/ env/ aee.htm

ICAO. (2010c). *Aircraft noise.* International Civil Aviation Organization, Air Transport Bureau, Environment Branch. Retrieved June 25, 2010, from http://www.icao. int/ icao/ en/ env/ noise.htm

IEEE. (1990). *IEEE std 610.12-1990, IEEE standard glossary of software engineering terminology.* The IEEE Standards Association. Retrieved October 2, 2009, from http://amutiara.staff.gunadarma.ac.id /Downloads/ files/7081/ IEEE+Standard+Glossary+ of+Softwar.pdf

IPCC. (1999). *Intergovernmental Panel on Climate Change- IPCC special report, aviation and the global atmosphere, summary for policymakers* (Eds. J. E. Penner, D. H. Lister, D. J. Griggs, D. J. Dokken, & M. McFarland, p. 13). Retrieved from http://www.ipcc.ch/ ipccreports/ sres/ aviation/ index.php?idp=0

Isaksen, I. S. A., Granier, C., Myhre, G., Berntsen, T. K., Dalsoren, S. B., & Gauss, M. (2009). Atmospheric composition change: Climate-chemistry interactions. *Atmospheric Environment, 43*(33), 5138–5192. doi:10.1016/j. atmosenv.2009.08.003

ISO. (2008). *Hydraulic fluid power-Fluids-Method for coding the level of contamination by solid particles.* 4406:1999, 2008.

Janic, M. (2008). The potential of liquid hydrogen for the future "carbon-neutral" air transport system. *Transportation Research Part D, Transport and Environment, 13,* 428–435. doi:10.1016/j.trd.2008.07.005

Jensen, R. S. (1982). Pilot judgment: Training and evaluation. *Human Factors, 2.*

Jindal, U. C. (1983). Reduction of stress concentration around a hole in uniaxially loaded plate. *Journal of Strain Analysis, 18,* 135–141. doi:10.1243/03093247V182135

Jindal, U. C. (1983). Stress distribution around an oblong hole in a uniaxially loaded plate. *Journal of the Institute of Engineers India. Part ME, 64,* 66–70.

Jones, D. G., Endsley, M. R., Bolstad, M., & Estes, G. (2004). The designer's situation awareness toolkit: Support for user-centered design. In *Proceedings of the Human Factors Society 48th Annual Meeting* (pp. 653-657). Santa Monica, CA: Human Factors Society.

Kamel, B., & Liaw, M. (1991). Boundary element analysis of cracks at a fastener hole in anisotropic sheet. *International Journal of Fracture, 50,* 263–280.

Kang, J., & Nath, B. (2004). *Resource-controlled MAC-layer congestion control scheme in cellular packet network.* IEEE 59th Vehicular Technology Conference, (pp. 1988–1992). VTC 2004-Spring.

Kanninen, M. F., & O'Donognue, P. E. (1995). Research challenges arising from current and potential applications of dynamic fracture mechanics to the integrity of engineering structures. *International Journal of Solids and Structures, 32*(17/18), 2423–2445. doi:10.1016/0020-7683(94)00275-2

Kaufmann, H., Schmalstieg, D., & Wagner, M. (2000). Construct3D: A virtual reality application for mathematics and geometry. [Springer Netherlands.]. *Education and Information Technologies, 5*(4), 263–276. doi:10.1023/A:1012049406877

Kay, J. (1993). *Foundations of corporate success.* New York, NY: Oxford University Press.

Keen, M., Acharya, A., Bishop, S., Hopkins, A., Milinski, S., Nott, C., et al. (2004). *Patterns: Implementing an SOA using an enterprise service bus*. IBM Redbook. Retrieved September 25, 2009, from http://ck20.com/MQ/WBIMB/sg246346%20Implementing%20SOA%20using%20ESB.pdf

Kesselman, A., & Mansour, Y. (2005). Adaptive AIMD congestion control. *Special Issue on Network Design, 43*(1-2), 97–111.

Kesselman, A., & Mansour, Y. (2005). Optimizing TCP retransmission timeout. *Proceedings of the 4ᵗʰ International Conference on Networking (ICN'05), 2*, (pp. 133-140).

Kiczales, G. (1997). *Aspect-oriented programming*. Finland, Springer-Verlag: European Conf. on Object-Oriented Programming.

Kim, B., Fleming, G., Balasubramanian, S., Malwitz, A., Fleming, G., & Lee, J. … Gillette, W. (2005). *System for assessing aviation's global emissions (SAGE), version 1.5, global aviation emissions inventories for 2000 through 2004*. Report FAA-EE-2005-02, Federal Aviation Administration, Office of Environment and Energy, Washington, DC.

Kitamura, Y., & Mizoguchi, R. (2004). Ontology-based systematization of functional knowledge. *Journal of Engineering Design, 15*(4). doi:10.1080/09544820410001697163

Kitamura, Y. S. (2000). Functional understanding based on an ontology of functional concepts. *Proceedings of Sixth Pacific Rim International Conference on AI*, Berlin.

Kitson-Smith, A., & Hughes, V. (2007, October). *The use of electronic sensors in field measurements of aviation jet fuel cleanliness*. Paper presented at the 10ᵗʰ Meeting of the International Conference on Stability, Handling and Use of Liquid Fuels, Tucson, AZ.

Klein, G., Orasanu, J., Calderwood, R., & Zsambok, C. (Eds.). (1993). *Decision making in action: Models and methods*. Norwood, NJ: Ablex.

Knudsen, F. B. (2004). *Defining sustainability in the aviation sector. EEC/SEE/2004/003*. EUROCONTROL Experimental Centre.

Kohler, I., Sausen, R., & Reinberger, R. (1997). Contributions of aircraft emissions to the atmospheric NO_x content. *Atmospheric Environment, 31*(12), 1801–1818. doi:10.1016/S1352-2310(96)00331-7

Koppenjan, J., & Groenewegen, J. (2005). Institutional design for complex technological systems. *International Journal of Technology. Policy and Management, 5*(3), 40–257.

Koroneos, C., Dompros, A., Roumbas, G., & Moussiopoulos, N. (2005). Advantages of the use of hydrogen fuel as compared to kerosene. *Resources, Conservation and Recycling, 44*, 99–113. doi:10.1016/j.resconrec.2004.09.004

Kotis, K., & Vouros, G. (2006). Human-centered ontology engineering: The HCOME methodology. *Knowledge and Information Systems Journal, 10* (1).

Kung, H. T., Tan, K. S., & Hsiao, P. H. (2003). TCP with sender-based delay control. *Journal of Computer Communications, 26*(14), 1614–1621. doi:10.1016/S0140-3664(03)00110-5

Kuper, W. J., Blaauw, M., Berg, F., & Graaf, G. H. (1999). Catalytic combustion concept for gas turbines. *Catalysis Today, 47*, 377–389. doi:10.1016/S0920-5861(98)00320-4

Lai, Y. C., & Yao, C. L. (2002). Performance comparison between TCP Reno and TCP Vegas. *Journal of Computer Communications, 25*(18), 1765–1773. doi:10.1016/S0140-3664(02)00092-0

Langhans, W., Scheiflinger, C., et al. (2007). *WAM Austria-Innsbruck Eurocontrol Conference 2007*. Retrieved from http://www.eurocontrol.int/ surveillance/gallery/content/ public/documents/WAM/ 13.3.%20WAM%20Innsbruck%20Full%20Set.pdf

Lapsley, D., & Low, S. (1999). Random early marking for Internet congestion control. In *Proceeding of GlobeCom'99*, (pp. 1747–52).

Lebedev, A. B., Secundov, A. N., Starik, A. M., Titova, N. S., & Schepin, A. M. (2009). Modeling study of gas-turbine combustor emission. *Proceedings of the Combustion Institute, 32*, 2941–2947. doi:10.1016/j.proci.2008.05.015

Lee, D. S. (2004). The impact of aviation on climate. *Issues in Environmental Science and Technology* [Royal Society of Chemistry.]. *Transport and the Environment, 20*, 1–23. doi:10.1039/9781847552211-00001

Lee, D. S., Fahey, D. W., Forster, P. M., Newton, P. J., Wite, R. C. N., & Lim, L. L. (2009). Aviation and global climate change in the 21st century. *Atmospheric Environment, 43*, 3520–3537. doi:10.1016/j.atmosenv.2009.04.024

Lee, D. S., Pitari, G., Grewe, V., Gierens, K., Penner, J. E., & Petzold, A.... Sausen, R. (in press). Transport impacts on atmosphere and climate: Aviation. *Atmospheric Environment*. doi:.doi:10.1016/j.atmosenv.2009.06.005

Leeuwaarden, J., Denteneer, D., & Resing, J. (2006). A discrete-time queuing model with periodically scheduled arrival and departure slots. *Journal of Performance Evaluation, 63*(4), 278–294. doi:10.1016/j.peva.2005.03.001

Lehner, F. (1990). Expert systems for organizational and managerial tasks. *Information & Management, 1*(23).

Lestas, M., Pitsillinds, A., Ioannou, P., & Hadjipollas, G. (2007). Adaptive congestion protocol: A congestion control protocol with learning capability. *Journal of Computer Networks, 51*(13), 3773–3798. doi:10.1016/j.comnet.2007.04.002

Levy, Y., Sherbaum, V., & Arfi, P. (2004). Basic thermodynamics of FLOXCOM, the low-NO$_x$ gas turbines adiabatic combustor. *Applied Thermal Engineering, 24*, 1593–1605. doi:10.1016/j.applthermaleng.2003.11.022

Lofaro, R. J., Gibb, G. D., & Garland, D. (1994). *A protocol for selecting airline passenger baggage screeners. DOT/FAA/CT-94/110*. Springfield, VA: National Technical Information Service.

Lofaro, R. J., & Smith, K. M. (1998). *Rising risk? Rising safety? The millennium and air travel. Transportation Law Journal, 25(2)*. Denver, CO: University of Denver Press.

Lofaro, R. J., & Smith, K. M. (2008). Flight simulators and training. In Vincenzi, D. A., Mouloua, M., Hancock, P. A., & Wise, J. A. (Eds.), *Human factors in training and simulation* (pp. 257–286). Philadelphia, PA: Taylor and Francis.

Lofaro, R. J. (1992). A small group Delphi paradigm. *The Human Factors Society Bulletin, 35*(2).

Lofaro, R. J. (1992). *Workshop on Integrated Crew Resource Management (CRM)*). DOT/FAA/RD-95/5. Springfield, VA: National Technical Information Service.

Lofaro, R. J., & Smith, K. M. (1993). The role of LOFT in CRM integration. In the *Proceedings of the Seventh International Symposium of Aviation Psychologists*. Columbus, OH: ISAP.

Lofaro, R. J., & Smith, K. M. (1999). Operational decision-making (ODM) and risk management (RM): Rising risk, the critical mission factors and training. In *Proceedings of Tenth International Symposium on Aviation Psychology*. Columbus, OH: ISAP.

Lofaro, R. J., & Smith, K. M. (2000). Operational decision-making: Integrating new concepts into the paradigm. In *Proceedings of Eleventh International Symposium on Aviation Psychology*. Columbus, OH: ISAP.

Lofaro, R. J., & Smith, K. M. (2010). Collaborative risk identification and management between cockpit and dispatch: Operational decision-making. In *HCI-Aero 2010 Conference Proceedings*.

Ludwig, H., Gimpel, H., Dan, A., & Kearney, R. D. (2005). Template-based automated service provisioning - Supporting the agreement-driven service life-cycle. In *Proceedings International Conference on Service Oriented Computing (ICSOC'05)*, (pp. 283-295).

Ludwig, R., & Katz, R. H. (2000). The Eifel algorithm: Making TCP robust against spurious retransmissions. *ACM Computer Communications Review, 1*(30).

Lulling, M., & Vaughan, J. (2004). A simulation-based performance evaluation of Tahoe, Reno and SACK TCP as appropriate transport protocols for SIP. *Journal of Computer Communications, 27*(16), 1585–1593. doi:10.1016/j.comcom.2004.05.013

Lyon, A., Westbrook, J., & Guida, U. (2005). *Operating EGNOS. Proceedings of the 18ᵗʰ International Technical Meeting of the Satellite Division of the Institute of Navigation* (ION GNSS 2005), California, USA, (pp. 419-423).

Maillot, N., Thonnat, M., & Boucher, A. (2004). Towards ontology-based cognitive vision. *Machine Vision and Applications, 16*.

Maliko-Abraham, H., & Lofaro, R. J. (2003). Usability testing: Lessons learned and methodology. In *Proceedings of the 2003 International Military Testing Association (IMTA) Conference* (pp. 1-5). Retrieved from www. International IMTA.org

Mannstein, H., Spichtinger, P., & Gierens, K. (2005). A note on how to avoid contrail cirrus. *Transportation Research Part D, Transport and Environment, 10*(5), 421–426. doi:10.1016/j.trd.2005.04.012

Mantovani, F. (2001). VR learning: Potential and challenges for the use of 3D environments in education and training. In Riva, G. (Ed.), *Towards CyberPsycology: Mind, cognition and society in the Internet age* (*Vol. 2*, pp. 207–225). Amsterdam, The Netherlands: IOS Press.

Marquart, S., Sausen, R., Ponater, M., & Grewe, V. (2001). Estimate of the climate impact of cryoplanes. *Aerospace Science and Technology, 5*, 73–84. doi:10.1016/S1270-9638(00)01084-1

Martinez, J. (2001). EZSTROBE- General-purpose simulation system based on activity cycle diagrams. *Proceedings of the 33rd Conference on Winter Simulation*. Arlington, VA: IEEE Computer Society.

Matthews, R. (2004, November). Ramp accidents and incidents constitute a significant safety issue. *ICAO Journal*, 4–25.

Matweb. (n.d.). *Homepage*. Retrieved on February 12, 2008, from http://www.matweb.com

Maurice, L. Q., Lander, H., Edwards, T., & Harrison, W. E. (2001). Advanced aviation fuels: A look ahead via a historical perspective. *Fuel, 80*, 747–756. doi:10.1016/S0016-2361(00)00142-3

Maurino, D. (1999). Crew resource management: A time for reflection. In Garland, D., Wise, J., & Hopkin, V. D. (Eds.), *Handbook of aviation human factors* (pp. 215–234). Mahwah, NJ: Lawrence Erlbaum Publishers.

Mazaheri, M., Johnson, G. R., & Morawska, L. (2009). Particle and gaseous emissions from commercial aircraft at each stage of the landing and takeoff cycle. *Environmental Science & Technology, 43*, 441–446. doi:10.1021/es8013985

McDonell, V. (2008). Lean combustion technology and control. In Rankin, D. D. (Ed.), *Lean combustion in gas turbines* (pp. 121–160). Elsevier.

McSweeney, J., & Pray, B. C. (2009). *Integration of human factors engineering in to design–An applied approach*. Houston, TX: ABS.

Meguid, S. A. (1986). Finite element analysis of defense hole systems for the reduction of stress concentration in a uniaxially loaded plate with two coaxial holes. *Engineering Fracture Mechanics, 25*(4), 403–413. doi:10.1016/0013-7944(86)90254-7

Meguid, S. A. (1989). *Engineering fracture mechanics*. Elsevier Science Publishers Ltd.

Mehta, N., & Medvidovic, N. (2002). Architectural style requirements for self-healing systems. *Proceedings of the first Workshop on Self-Healing systems*, (pp. 49-54).

Melton, H. K. (2003). *The U.S. government guide to surviving terrorism*. New York, NY: Barnes and Noble Books.

Mercado, L. L., & Sikarskie, D. L. (1999). On response of a sandwich panel with a bilinear core. *Mechanics of Composite Materials and Structures, 6*, 57–67.

Mercado, L. L., Sikarskie, D. L., & Miskioglu, I. (2000). Higher order theory for sandwich beams with yielded core. *Proceedings of ICSS-5 Conference*, (pp. 141–53). Zurich.

Miers, S. A. (2001). *Analysis and design of edge inserts in Sandwich Beams*. Master of Science Thesis, Michigan Technological University.

Ministry of Defence (UK). (2008). *Turbine fuel, aviation kerosene type, jet A-1*. Defence Standard 91-91, Issue 6, 2008.

Minoli, D. (2008). *Enterprise architecture A to Z frameworks, business process modeling, SOA, and infrastructure technology*. New York, NY: Taylor & Francis Group. doi:10.1201/9781420013702

Misra, P., & Enge, P. (2004). *Global positioning system: Signals, measurements and performance*. Lincoln, MA: Ganga-Jamuna Press.

Mittal, N. D., & Jain, N. K. (2008). Finite element analysis for effect of fiber orientation on stress concentration factor in a laminated composite plate with central hole under in-plane static loading. *Material Science and Engineering, 498*(1-2: A), 115-124.

Miyoshi, C., & Mason, K. J. (2009). The carbon emissions of selected airlines and aircraft types in three geographic markets. *Journal of Air Transport Management, 15*, 138–147. doi:10.1016/j.jairtraman.2008.11.009

Mizoguchi, R. S. (1996). *Task ontology design for intelligent educational/training systems.* Position Paper for ITS'96 Workshop on Architectures and Methods for Designing Cost-Effective and Reusable ITSs.

Mohammadi, B., Najafi, A., & Ghannadpour, S. A. M. (2006). Effective widths of compression-loads of perforated cross ply laminated composite. *Science and Direct/Composite Structure Journal, 75*, 7-13.

Morrell, P. (2009). The potential for European aviation CO_2 emissions reduction through the use of larger jet aircraft. *Journal of Air Transport Management, 15*, 151–157. doi:10.1016/j.jairtraman.2008.09.021

Morrison, B., & Levin, A. (2004, November 22). Flights into USA get new security. *USA Today*. Retrieved December 1, 2004, from http://www.usatoday.com/ travel/ news/2004-11-22 -flights-security_x.htm

Mustafa, K. C. (2006). Task-oriented integrated use of biological Web data sources. *Proceedings of the 18th International Conference on Scientific and Statistical Database Management*, IEEE Computer Society.

Nagel, D. C. (1988). Human error in aviation operations. In Weiner, E. L., & Nagel, D. C. (Eds.), *Human factors in aviation* (pp. 263–303). San Diego, CA: Academic Press.

National Highway Traffic Safety Administration. (2008). *Home.* Retrieved July 19, 2010, from http://www-fars.nhtsa.dot.gov/ Main/ index.aspx

National Security Institute. (2004). *Bomb threats and physical security planning.* Retrieved December 1, 2004, from http://nsi.org/Library/Terrorism /bombthreat.html

National Transportation Safety Board. (1997). *Bus collision with pedestrians.* Washington, DC: National Transportation Safety Board.

National Transportation Safety Board. (2003). *Derailment of Northeast Illinois regional commuter railroad train.* Washington, DC: National Transportation Safety Board.

National Transportation Safety Board. (2007). *Attempted takeoff from wrong runway: Comair Flight 5191.* Washington, DC: National Transportation Safety Board.

National Transportation Safety Board. (2010). *Loss of control on approach, Colgan Air, Inc., operating as Continental connection flight 3407.* Washington, DC: National Transportation Safety Board.

National Transportation Safety Board. (2010, April 8th). *Press release.* Retrieved 19 July, 2010, from http://www.ntsb.gov/ pressrel/ 2010/ 100408.html

Neven, W. H. L., Quilter, T. J., Weedon, R., & Hogendoorn, R. A. (2005). *Wide area multilateration. Report on EATMP TRS 131/04, Version 1.1.* Eurocontrol.

Nojoumi, H., Dincer, I., & Naterer, G. F. (2009). Greenhouse gas emissions assessment of hydrogen and kerosene-fueled aircraft propulsion. *International Journal of Hydrogen Energy, 34*, 1363–1369. doi:10.1016/j.ijhydene.2008.11.017

Nonaka, I., & Takeuchi, H. (1995). *The knowledge-creating company: How Japanese companies create the dynamics of innovation.* New York, NY: Oxford University Press.

North, D. M. (2000). Finding common ground in envelope protection systems. *Aviation Week & Space Technology, 153*(6), 23–24.

NTSB. (2007, February 10). *AAR-07/06.*

NTSB. (2009, February 12). *AAR-10/11.*

Office of Homeland Security. (2004). *Recognizing terrorist activity.* Retrieved December 1, 2004, from http://www.national terroristalert.com/ readyguide/activity.htm

Padhye, J., Firoiu, V., Towsley, D., & Kurose, J. (2000). Modeling TCP Reno performance: A simple model and its empirical validation. *IEEE/ACM Transactions on Networking, 8*(2), 133–145. doi:10.1109/90.842137

Pantelidis, V. S. (1995). Reasons to use virtual reality in education. *VR in the Schools, 1*(1). Retrieved February 15, 2004, from http://eastnet.educ.ecu.edu /vr/vr1n1a.txt

Papantonopoulos, S. (2004). How system designers think: A study of design thinking in human factors engineering. *Ergonomics*, *47*(14), 1528–1548. doi:10.1080/0014013 0412331290916

Papazoglou, M., & Heuvel, W. V. D. (2007). Service oriented architectures: approaches, technologies and research issues. *The VLDB Journal*, *16*(3), 389–415. doi:10.1007/ s00778-007-0044-3

Pejovic, T., Noland, R. B., Williams, V., & Toumi, R. (2009). A tentative analysis of the impacts of an airport closure. *Journal of Air Transport Management*, *15*, 241–248. doi:10.1016/j.jairtraman.2009.02.004

Penner, J. E., Lister, D. H., Griggs, D. J., Dokken, D. J., & McFarland, M. (2001). *Aviation and the global atmosphere*. Special Report of Working Groups I and III of the Intergovernmental Panel on Climate Change.

Pentikousis, K., & Badr, H. (2002). *On the resource efficiency of explicit congestion notification* (pp. 588–599). Berlin / Heidelberg, Germany: Springer.

Perraton, J., & Tarrant, I. (2007). What does tacit knowledge actually explain? *Journal of Economic Methodology*, *14*(3). doi:10.1080/13501780701562559

Pham, V. V., Tang, J., Alam, S., Lokan, C., & Abbass, H. A. (in press). Aviation emission inventory development and analysis. *Environmental Modelling & Software*. doi:. doi:10.1016/j.envsoft.2010.04.004

Pinto, F., Marques, A., & Santos, M. (2009). Ontology-supported database marketing. *Journal of Database Marketing and Customer Strategy Management*, *16*.

Plantema, F. J. (1966). *Sandwich construction*. New York, NY: John Wiley and Sons.

Polanyi, M. (1967). *The tacit dimension*. London, UK: Routledge and Kegan Paul.

Ponater, M., Pechtl, S., Sausen, R., Schumann, U., & Huttig, G. (2006). Potential of the cryoplane technology to reduce aircraft climate impact: A state-of-the-art assessment. *Atmospheric Environment*, *40*, 6928–6944. doi:10.1016/j.atmosenv.2006.06.036

Popovici, D. M., Gerval, J. P., Hamza-Lup, F., Querrec, R., Polceanu, M., Popovici, N., & Zăgan, R. (2009). 3D virtual spaces supporting engineering learning activities. *International Journal of Computers, Communications & Control*, *4*(4), 401–414.

Popper, K. (1959). *The logic of scientific discovery*. London, UK: Hutchinson.

Potok, E., Vouk, M., & Rindos, A. (1999). Productivity analysis of object-oriented software developed in a commercial environment. *Software-Practice and Experience*, *29*(10). Casselman, R. M., & Samson, D. A. (2005). Moving beyond tacit and explicit: Four dimensions of knowledge. *Proceedings of the 38th Annual Hawaii International Conference on System Sciences*, IEEE Computer Society.

Pratt & Whitney. (2010a). *PurePower PW1000G engine*. Retrieved July 5, 2010, from http://www.purepowerengine.com/ noise.html

Pratt & Whitney. (2010b). *PurePower PW1000G engine: Photos. Engine cross section and fan drive gear system*. Retrieved July 5, 2010, from http://www.purepowerengine.com/ photos.html

Prestinenza Puglisi, L. (1999). *Hyper-architecture spaces in the electronic age*. Basel, Switzerland: Birkhäuser.

Quilty, S. M. (2005). *Airport security and response to emergencies*. Alexandria, VA: American Association of Airport Executives.

Rajaiah, K., & Naik, N. K. (1984). Hole shape optimization in a finite plate in the presence of auxiliary holes. *Experimental Mechanics*, *24*(2), 157–161. doi:10.1007/ BF02324999

Rashid, H. (2003). Architettura virtuale, spazio reale. In L. Sacchi & M. Unali (Eds.), *Architettura e cultura digitale*, (pp. 169-178). Milano, Italia: Skirà.

Reichart, D. F. (2004). Task models as basis for requirements engineering and software execution. *Proceedings of the 3rd Annual Conference on Task Models and Diagrams*. Prague, Czech Republic: ACM.

Reinach, S. (2007). *An introduction to human systems integration (HSI) in the railroad industry*. Washington, DC: Federal Railroad Administration.

Report for Congress. (2006). *Issues and options for combating terrorism and counterinsurgency.* Order Code, RL32737.

Ribeiro, K. S., Kobayashi, S., Beuthe, M., Gasca, J., Greene, D., & Lee, D. S. … Zhou, P. J. (2007). *Transport and its infrastructure.* Climate Change 2007: Mitigation. Cambridge, United Kingdom: Cambridge University Press.

Robertazzi, T. (2000). *Computer networks and systems: Queuing theory and performance evaluation* (3rd ed.). Springer Verlag.

Robertson, P., & Williams, B. (2006). Automatic recovery from software failure. *Communications of the ACM, 49*(3), 41–47. doi:10.1145/1118178.1118200

Rogers, H. L., Lee, D. S., Raper, D. W., Forster, P. M. F., Wilson, C. W., & Newton, P. (2002). *The impact of aviation on the atmosphere.* Report QINETIQ/FST/CAT/TR021654, Centre for Aerospace Technology, Cody Technology Park, Farnborough, United Kingdom.

Rosemblum, L. J., & Cross, R. A. (1997). The challenge of virtual reality. In Earnshaw, W. R., Vince, J., & Jones, H. (Eds.), *Visualization & modeling* (pp. 325–399). San Diego, CA: Academic Press.

Rosson, M. A. (1995). Integrating task and software development for object-oriented applications. *Proceedings of the SIGCHI Conference on Human Factors in Computing Systems.* Denver, CO: ACM Press.

Rowe, D., & Leaney, J. (1997). Evaluating evolvability of computer based systems architectures-an ontological approach. *Proceedings International Conference and Workshop on Engineering of Computer-Based Systems* (pp. 360-367).

Rowe, M. R., & Ladd, G. T. (1945, May 23). Water injection for aircraft engines. *Flight,* 517-518. Retrieved July 25, 2010, from http://www.flightglobal.com/ pdfarchive/view/ 1946/ 1946% 20-% 201007.html

Ruffell-Smith, H. S. (1979). *A simulator study of the interactions of pilot workload with errors, vigilance and decision*s. NASA Technical Memorandum 78482. Washington, DC: NASA Scientific and Technical Information Office

Ruffles, P. C. (2003). Aero engines of the future. *Aeronautical Journal, 107*(1072), 307–321.

Sala, N. (2009a). Fractal geometry in virtual world and in Second Life. [New York, NY: Nova Science.]. *Chaos and Complexity Letters, 4*(1), 23–34.

Sala, N. (2009b). Fractals in architecture, hyper-architecture and…beyond. [New York, NY: Nova Science.]. *Chaos and Complexity Letters, 4*(3), 147–180.

Sala, N., & Sala, M. (2005). Virtual reality in education. In Carbonara, D. (Ed.), *Technology literacy applications in learning environments* (pp. 358–367). Hershey, PA: Idea Group Inc.doi:10.4018/978-1-59140-479-8.ch025

Saleri, R. (2008). 3D automatic processing of architectural and urban artifacts. In Orsucci, F., & Sala, N. (Eds.), *Reflexing interfaces: The complex coevolution of Information Technology ecosystems* (pp. 278–289). Hershey, PA: IGI Global.

Saleri, R. (2002). Pseudo-urban automatic pattern generation, *Proceedings 5th Generative Art Conference,* Milano, Italy. Retrieved March 10, 2003, from http://hal.archives-ouvertes.fr/docs/ 00/26/74/00/PDF/GA2002.pdf

Saleri, R. (2006). Urban and architectural 3D fast processing. *Proceedings 9th Generative Art Conference,* Milano, Italy. Retrieved January 15, 2008, from http://hal.archives-ouvertes.fr/docs/00 /26/73/63/PDF/Renato_Saleri.pdf

SBIR. (2009). Success. *Navy Transitions, 6,* 6–7.

Schafer, K., Jahna, C., Sturmb, P., Lechnerb, B., & Bacher, M. (2003). Aircraft emission measurements by remote sensing methodologies at airports. *Atmospheric Environment, 37,* 5261–5271. doi:10.1016/j.atmosenv.2003.09.002

Scheelhaase, J. D., & Grimme, W. G. (2007). Emissions trading for international aviation—An estimation of the economic impact on selected European airlines. *Journal of Air Transport Management, 13,* 253–263. doi:10.1016/j.jairtraman.2007.04.010

Schipper, Y. (2004). Environmental costs in European aviation. *Transport Policy, 11,* 141–154. doi:10.1016/j.tranpol.2003.10.001

Schumann, U. (1997). The impact of nitrogen oxides emissions from aircraft upon the atmosphere at flight altitudes: Result from the AERONOX project. *Atmospheric Environment, 31*(12), 1723–1733. doi:10.1016/S1352-2310(96)00326-3

Schumann, U. (2000). Influence of propulsion efficiency on contrail formation. *Aerospace Science and Technology, 4*, 391–401. doi:10.1016/S1270-9638(00)01062-2

Schumann, U. (2005). Formation, properties and climatic effects of contrails. *C.R. Physique, 6*, 549–565. doi:10.1016/j.crhy.2005.05.002

Schumann, U., Busen, R., & Plohr, M. (2000). Experimental test of the influence of propulsion efficiency on contrail formation. *Journal of Aircraft, 37*(6), 1083–1087. doi:10.2514/2.2715

Schurmann, G., Schafer, K., Jahn, C., Hoffmann, H., Bauerfeind, M., Fleuti, E., & Rappengluck, B. (2007). The impact of NO$_x$, CO and VOC emissions on the air quality of Zurich airport. *Atmospheric Environment, 41*, 103–118. doi:10.1016/j.atmosenv.2006.07.030

Sharawi, M. S. (2009). *Interference mitigation for local area augmentation navigation systems, methods and analysis.* Saarbrucken, Germany: VDM Verlag.

Shell Oil Company. (2009, November). *Shell water detector.* Retrieved November 4, 2009, from http://www.shell.com/ home/ content/ aviation/ products_and_services/ products/ shell_water_detector/

Shellard, H. C. (1949). Humidity of the lower stratosphere. *Meteorological Magazine, 78*(390), 341–349.

Sheridan, T. B. (1993). My anxieties about virtual environments. *Presence (Cambridge, Mass.), 2*(2), 141–142.

Sherman, W., & Craig, A. (2002). *Working with virtual reality.* New York, NY: Morgan Kaufmann Publishers.

Shin, Y. S. (2004). Virtual experiment environment's design for science education. *International Journal of Distance Education Technologies, 2*(2), 62–76. doi:10.4018/jdet.2004100104

Shneiderman, B. (1992). *Designing the user interface* (2nd ed.). Reading, MA: Addison-Wesley.

Simons, R. (2005). *Levers of organization design: How managers use accountability systems for greater performance and commitment.* Boston, MA: Harvard Business School Press.

Singer, G. (2002). *Methods for validating cockpit design: The best tool for the task.* Doctoral dissertation. Stockholm, Sweden: Royal Institute of Technology.

Singh, R., & Jain, S. (2008). Knowledge mechanics and its temporal nature. *Ubiquity, 9*(15).

Siricharoen, W. (2007). Ontologies and object models in object oriented software engineering. *International Journal of Communication Systems, 33*(1).

SJU. (2008). *European ATM master plan.* Retrieved from http://www.eurocontrol.int /sesar/

Slanina, S. (2008). Impact of ozone on health and vegetation. In C. J. Cleveland (Ed.), *Encyclopedia of Earth.* Environmental Information Coalition, National Council for Science and the Environment, Washington, DC. Retrieved June 21, 2010, from http://www.eoearth.org/article/ Impact_of_ozone_ on_health_and _vegetation

Sloan, A. W. (1975). Adaptation and failure of adaptation to extreme natural environments. *Forensic Science, 5*(1), 81–89. doi:10.1016/0300-9432(75)90091-6

Smallman, H. S., & St.John, M. (2005). Naïve realism: Misplaced faith in realistic displays. *Ergonomics in Design, 13*(3), 6–13. doi:10.1177/106480460501300303

Smith, B. (2003). *Blackwell guide to the philosophy of computing and information.* Oxford, UK: Blackwell.

Smith, K. M., & Hastie, R. (1992). Airworthiness as a design strategy. In *Proceedings of the FSI Air Safety Symposium,* San Diego, CA.

Smith, K. M., & Lofaro, R. J. (2001). A paradigm for developing operational decision-making (ODM). In *Proceedings of 2001 SAE World Aviation Congress (WAC) Conference.* Seattle, WA: SAE WAC.

Smith, K. M., & Lofaro, R. J. (2003). The finalized paradigm for operational decision-making (ODM) paradigm: Components and placement. In *Proceedings of the 12th International Symposium on Aviation Psychology.* Dayton, OH: ISAP.

Smith, K. M., & Lofaro, R. J. (2009) Professionalism in airline operations…and accident investigations? In *Proceedings of the 15ᵗʰ International Symposium on Aviation Psychology,* Dayton, OH: ISAP.

Soares, C. (2008). *A handbook of air, land, and sea applications*. Elsevier.

Sommerville, I. (2007). *Software engineering* (8th ed.). London, UK: Pearson Education Limited.

SPADE consortium. (2006). *Project objectives.* Retrieved from http://spade.nlr.nl/pobj.htm/

Sprenger, G., Stevens, J., White, M., Hillis, R., Lavenberg, J., & Templeman, C. (2009). *U.S. patent no 7,518,719 B2*. Washington, DC: U.S. Patent and Trademark Office.

Sprogis, H. (1997). Is the aviation industry expressing CRM failure? In *Proceedings of Ninth International Symposium on Aviation Psychology.* Columbus, OH: ISAP.

Staehelin, J., Harris, N. R. P., Appenzeller, C., & Eberhard, J. (2001). Ozone trends: A review. *Reviews of Geophysics, 39*, 231–290. doi:10.1029/1999RG000059

Stanney, K. (1995). *Realizing the full potential of virtual reality: Human factors issues that could stand in the way* (pp. 28–33). Orlando, FL: IEEE.

Stevens, W. (2001). *TCP slow start, congestion avoidance, fast retransmit, and fast recovery algorithms.* IETF RFC, 1997.

STG Aerospace Ltd. (2010). *Website.* Retrieved from http://www.stgaerospace.com/

Strauch, B. (2002). *Investigating human error: Incidents, accidents, and complex systems.* Brookfield, VT: Ashgate Publishing.

Styhre, A. (2004). Rethinking knowledge: A Bergsonian critique of the notion of tacit knowledge. *British Journal of Management, 15*(2), 177–188. doi:10.1111/j.1467-8551.2004.00413.x

Sutkus, D. J., Baughcum, S. L., & DuBois, D. P. (2001). *Scheduled civil aircraft emission inventories for 1999: Database development and analysis.* Report NASA/CR-2001-211216, NASA Center for Aerospace Information, Hanover, MD.

Svensson, F., Hasselrot, A., & Moldanova, J. (2004). Reduced environmental impact by lowered cruise altitude for liquid hydrogen-fuelled aircraft. *Aerospace Science and Technology, 8*, 307–320. doi:10.1016/j.ast.2004.02.004

Svensson, E., Angelborg-Thanderz, M., & Sjöberg, L. (1993). Mission challenge, mental workload and performance in military aviation. *Aviation, Space, and Environmental Medicine, 64*(11), 985–991.

Svensson, E., Angelborg-Thanderz, M., Sjöberg, L., & Olsson, S. (1997). Information complexity: Mental workload and performance in combat aircraft. *Ergonomics, 40*(3), 362–380. doi:10.1080/001401397188206

Svensson, E., & Wilson, G. F. (2002). Psychological and psychophysiological models of pilot performance for systems development and mission evaluation. *The International Journal of Aviation Psychology, 12*(1), 95–110. doi:10.1207/S15327108IJAP1201_8

Tanenbaum, A. (2003). *Computer networks* (4th ed.). Prentice Hall.

Tbone Aviation. (2010). *Website.* http://www.tboneaviation.com/

Tekala, M., & Szabo, R. (2008). Dynamic adjustment of scalable TCP congestion control parameters. *Journal of Computer Communications, 31*(10), 1890–1900. doi:10.1016/j.comcom.2007.12.035

Terrett, A. (1998). Knowledge management and the law firm. Knowledge Management, *1*(2).

The 9/11 Commission Staff Monograph. (2004). National Commission on Terrorist Attacks upon the United States. *9/11 Commission Final Report,* p. 54.

The Oxford English Dictionary. (n.d.). *OED.*

Thomsen, O. T. (1993). Analysis of local bending effects in sandwich plates with orthotropic face layers subjected to localized loads. *Composite Structures, 25*(1-4), 511–520. doi:10.1016/0263-8223(93)90199-Z

Thomsen, O. T. (1995). Theoretical and experimental investigation of local bending effects in Sandwich Plates. *Composite Structures, 30*(1), 85–101. doi:10.1016/0263-8223(94)00029-8

Thornberg, D. B. (2003). LAAS integrated multipath limiting antenna. *Navigation: Journal of the Institute of Navigation, 50*(2), 117–130.

Transportation Security Administration. (2008, March). *Electronic code of regulations.* Retrieved March 10, 2008, from www.tsa.gov

Tsoukas, H. (2001). What is organizational knowledge? *Journal of Management Systems, 7*(38).

U.S. Congress. (2010). *H.R. 5900: Airline Safety and Federal Aviation Administration Extension Act: Counterterrorism Enhancement and Department of Homeland Security Authorization Act.* 111th Congress. Washington, DC: *GPO.*

Ulrich, T. W., & Moslehy, F. A. (1995). Boundary element method for stress reduction by optimal auxiliary holes. *Engineering Analysis with Boundary Elements, 15*(3), 219–223. doi:10.1016/0955-7997(95)00025-J

Unal, A., Hu, Y., Chang, M. E., Odman, M. T., & Russell, A. G. (2005). Airport related emissions and impacts on air quality: Application to the Atlanta International Airport. *Atmospheric Environment, 39,* 5787–5798. doi:10.1016/j.atmosenv.2005.05.051

US Congress. (2009). *HR. 2200: Transportation Security Administration Authorization Act. 111th Congress. § 234.* Washington, DC: GPO.

US Department of Defense. (2003, June). *News no. 411-03.* Retrieved October 28, 2009, from http://www.defenselink.mil/ contracts/ contract.aspx? contractid=2533

Vatcha, S. R. (1997). Low-emission gas turbines using catalytic combustion. *Energy Conversion and Management, 38*(10-13), 1327–1334. doi:10.1016/S0196-8904(96)00162-8

Velcon Filters, L. L. C. (2009, November) *HYDROKIT,* 1752-R9 11/08. Retrieved November 4, 2009, from http://www.velcon.com/ aviation/ hydrokit.html

Vendicits, A., Baiocchi, A., & Bonacci, M. (2003). Analysis and enhancement of TCP Vegas congestion control in a mixed TCP Vegas and TCP Reno network scenario. *Journal of Performance Evaluation, 53*(3-4), 225–253. doi:10.1016/S0166-5316(03)00064-6

Vespermann, J., & Wald, A. (in press). Much ado about nothing? – An analysis of economic impacts and ecologic effects of the EU-emission trading scheme in the aviation industry. *Transportation Research Part A.* doi:. doi:10.1016/j.tra.2010.03.005

Voicu, L., Bassi, S., & Labrador, M. A. (2007). Analytical and experimental evaluation of TCP with an additive increase smooth decrease (AISD) strategy. *Journal of Computer Communications, 30*(2), 479–495. doi:10.1016/j.comcom.2006.09.010

Wang, Y. D., Huang, Y., Wright, D. M., McMullan, J., Hewitt, N., Eames, P., & Rezvani, S. (2006). A techno-economic analysis of the application of continuous staged-combustion and flameless oxidation to the combustor design in gas turbines. *Fuel Processing Technology, 87*(8), 727–736. doi:10.1016/j.fuproc.2006.02.003

Wanlin, G. (1993). Stress intensity factors for corner cracks at holes subjected to biaxial and pin loads. *Engineering Fracture Mechanics, 46*(3), 473–479. doi:10.1016/0013-7944(93)90239-O

Watt, A. (n.d.). *EUROCONTROL Environment Domain Manager.*

Web Service MQ. (n.d.). *Providing a messaging backbone for SOA connectivity.* Retrieved October 10, 2009, from ftp://ftp.software.ibm.com/software /integration/wmq/ WS_MQ_ Messaging_Backbone _for_SOA.pdf

Weigle, M. C., Jaffay, K., & Simith, D. (2005). Delay-based early congestion detection and adaptation in TCP: Impact on Web performance. *Journal of Computer Communications, 28*(8), 837–850. doi:10.1016/j.comcom.2004.11.011

Wells, A. T., & Young, S. B. (2004). *Commercial aviation safety* (3rd ed.). New York, NY: McGraw-Hill.

Welzl, M. (2005). *Network congestion control: Managing Internet traffic.*

Westerdahl, D., Fruin, S. A., Fine, P. L., & Sioutas, C. (2008). The Los Angeles International Airport as a source of ultrafine particles and other pollutants to nearby communities. *Atmospheric Environment, 42,* 3143–3155. doi:10.1016/j.atmosenv.2007.09.006

White, M., Templeman, C., Frish, M., & Nebolsine, P. (2008). *U.S. Patent No. 7,450,234*. Washington, DC: U.S. Patent and Trademark Office.

Wickens, C. D., Lu, L., & Gordon, S. (1998). *An introduction to human factors engineering*. Prentice Hall.

Wickens, C. D. (2002). Situation awareness and workload in aviation. *Current Directions in Psychological Science*, *11*(4), 128–133. doi:10.1111/1467-8721.00184

Widmer, J., Denda, R., & Mauve, M. (2001). A survey on TCP-friendly congestion control. *Special Issue of the IEEE Network Magazine Control of Best Effort Traffic*, *15*(3), 28–37.

Wiegmann, D. S. (2004). *HFACS analysis of military and civilian aviation accidents: A North American comparison*. Oklahoma City, OK: ISASI.

Wikforss, M. (2008). *Usability design principles in JAS39 Gripen*. Master Thesis. Stockholm: Royal Institute of Technology. In Swedish.

Wilkes, R. (2008). *U.S. Air Force VEMSO product evaluation program final report – Fuel containment analyzer project no. E07-02. Air Force Petroleum Office*. AFPET.

William, L. K. (1985). *NASA- Stress concentration around a small circular hole in the HiMAT composite plate*. NASA Technical Memorandum 88038, Dec. 1985.

Williams, V., & Noland, B. R. (2006). Comparing the CO_2 emission and contrail formation from short and long haul air traffic routes from London Heathrow. *Environmental Science & Policy*, *9*, 487–495. doi:10.1016/j.envsci.2005.10.004

Williams, V., Noland, R. B., & Toumi, R. (2002). Reducing the climate change impacts of aviation by restricting cruise altitudes. *Transportation Research Part D, Transport and Environment*, *7*, 451–464. doi:10.1016/S1361-9209(02)00013-5

Williams, K. W. (2006). *Human factors implications of unmanned aircraft accidents: Flight control problems*. Oklahoma City, OK: FAA Civil Aerospace Medical Institute.

Williams, J. (2007). *Pragmatist philosophy and the global environmental crisis*.

Williams, V. (2007). The engineering options for mitigating the climate impacts of aviation. *Philosophical Transactions of the Royal Society A- Mathematical, Physical, and Engineering Sciences, 365*(1861), 3047-3059.

Wilson, E. K. (2009). Ozone's health impact. *Chemical and Engineering News, 87*(11), 9. doi:10.1021/cen-v087n011.p009a

Winn, W. (1993). *A conceptual basis for educational applications of virtual reality* (HITL Technical Report No. TR-93-9). Seattle, WA: Human Interface Technology Laboratory. Retrieved March 10, 2006, from http://www.hitl.washington.edu /publications/r-93-9

Winther, M., Kousgaard, U., & Oxbel, A. (2006). Calculation of odour emissions from aircraft engines at Copenhagen Airport. *The Science of the Total Environment, 366*(1), 218–232. doi:10.1016/j.scitotenv.2005.08.015

Woltjer, R. (2009). *Functional modeling of constraint management in aviation safety and command and control*. Doctoral dissertation (No. 1249). Linköping, Sweden: Linköping University.

Wood, E. C., Herndon, S. C., Timko, M. T., Yelvington, P. E., & Miake-Lye, R. C. (2008). Speciation and chemical evolution of nitrogen oxides in aircraft exhaust near airports. *Environmental Science & Technology, 42*, 1884–1891. doi:10.1021/es072050a

Woods, D. D., & Hollnagel, E. (2006). *Joint cognitive systems: Patterns in cognitive systems engineering*. Boca Raton, FL: Taylor & Francis. doi:10.1201/9781420005684

Woods, D. D. (1995). Toward a theoretical base for representation design in the computer medium: Ecological perception and aiding human cognition. In Flach, J. M., Hancock, P. A., Caird, J., & Vicente, K. (Eds.), *Global perspectives on the ecology of human-machine systems* (pp. 157–188). Hillsdale, NJ: Erlbaum.

Woodward, M. E. (2003). *Communication and computer networks: Modeling with discrete-time queues*. London, UK: Pentech Press.

WSROC Ltd. (2009). *Response to the national aviation policy green paper*. February 2009, Western Sydney Regional Organisation of Councils Ltd. ISBN 186271 0193

Wulff, A., & Hourmouziadis, J. (1997). Technology review of aeroengine pollutant emissions. *Aerospace Science and Technology, 8*, 557–572. doi:10.1016/S1270-9638(97)90004-3

Wunning, J. A., & Wunning, J. G. (1997). Flameless oxidation to reduce thermal NO-formation. *Progress in Energy and Combustion Science, 23*, 81–94. doi:10.1016/S0360-1285(97)00006-3

Wydrowski, B., & Zukerman, M. (2002). GREEN: An active queue management algorithm for a self managed Internet. *Proceedings of IEEE ICC, 4*, 2368–2372.

Xin, F., & Jamalipour, A. (2006). TCP performance in wireless networks with delay spike and different initial congestion window sizes. *Journal of Computer Communications, 29*(8), 926–933. doi:10.1016/j.comcom.2005.06.012

Xu, N., Cassell, R., Evers, C., Hauswald, S., & Langhans, W. (2010). Performance assessment of multilateration systems - A solution to Nextgen surveillance. *Proceedings of Integrated Communications Navigation and Surveillance Conference, 2010* (pp. D2-1-D2-8). Herdon, VA, USA. ISSN: 2155-4943

Yang, Y. R., Kim, M. S., & Lam, S. S. (2003). Transient behaviors of TCP-friendly congestion control protocols. *Journal of Computer Networks, 41*(2), 193–210. doi:10.1016/S1389-1286(02)00374-2

Yen, J. a. (1993). *Task-based methodology for specifying expert systems.*

Yoshiaki, Y., & Kiyoshi, T. (1987). Stress concentration for FRP plates with a circular hole under bi-axial tension. *Proceeding of the Faculty of Engineering of Tokyo University, 27*(2), 91-97.

Youngblut, C. (1998). *Educational uses of virtual reality technology.* Institute for Defence Analyses, IDA Document D-2128. Retrieved June 5, 2001, from http://www.hitl.washington.edu/ scivw/youngblut-edvr/D2128.pdf

YVR. (2004). *Vancouver International Airport Authority website.* Retrieved from http://www.yvr.ca/

Zenkert, D. (1995). *An introduction to sandwich construction, (EMAS).* London, UK: The Chameleon Press Ltd.

Zenkert, D. (1997). *The handbook of sandwich construction.* United Kingdom: Engineering Materials Advisory Services.

Zlot, F., Oliveira, K. D., & Rocha, A. (2002). Modeling task knowledge to support software development. *Proceedings of the 14th International Conference on Software Engineering and Knowledge Engineering.* Ischia, Italy: ACM.

About the Contributors

Evon Abu-Taieh, PhD, Editor-in-Chief of the *International Journal of Aviation Technology, Engineering and Management* and has been a guest editor for the *Journal of Information Technology Research*. She has more than 26 published works Aviation, IT, PM, KM, GIS, AI, simulation, security and ciphering; in addition, Dr. Abu-Taieh holds positions on the editorial board of the *International Journal of E-Services and Mobile Applications*, *International Journal of Information Technology Project Management*, and *International Journal of Information Systems and Social Change*. Dr. Abu-Taieh served as Chair, Track chair, & reviewer in many renowned conferences.

Mostafa Jafari was born in year 1956 in Tehran, Iran. He started in school in year 1962 and obtained Natural Diploma in 1974 (Tehran, after 12 years). He finished his first degree (B.Sc.) in Forest and Range, graduated in 1978 (Iran), and finished his Ph.D. in Plant Science (Ecology) in 1990 (UK). His Post doctorate research was in Plant Ecophysiology Methodology in 1997 (Japan). He is member of scientific board since 1990. His interest is in plant ecology, forestry, and climate change, and giving lectures in Universities on Ecophysiology, Ecology, Range Rehabilitation, Plant Geography. He is an advisor of several post graduate students in different universities. He enjoy from Membership to the different Professional Organizations. He was Director (President) of Research Institute of Forest and Rangelands, I.R. Iran, from 1992 to Nov. 1997 (With 600 staff including 250 scientific researcher and 900 research projects, publishing about 200 books in this period). He was Director of First Vice President Office of WMO, from August 2004 to 2009. He is Head of TP Secretariat of Low Forest Cover Countries (LFCCs, International Intergovernmental Organization). Since, March 2003, he is also International Affair Advisor to the Deputy Minister and Head of IRIMO since August 2004. He published 69 articles, 8 books, and 2 university textbooks. He is Managing Director of National Research Project on Climate Change in RIFR entitled: "Investigation on Climate Change Effects on Forest Ecosystems in Hyrcanian forests with Emphasize of Wood Dendrology Studies."

Asim El Shiekh, PhD(LSE): Dean of Faculty of Information Systems & Technology, Arab Academy For Banking & Financial Sciences, Jordan. Author of more than 5 books in the Information technology, 60 published research papers.

* * *

Jafar Ababneh is an Assistant Professor. He received his PhD degree from Arab Academy for Banking & Financial Sciences (Jordan) in 2009. He received his M.Sc degree in computer engineering from

University of the Yarmouk (Jordan) in 2005. He earned his B.Sc in Telecommunication engineering from University of Mu'ta (Jordan) in 1991. He has published many research papers in different fields of science in refereed journal and international conference proceedings. His field research lies in development and performance evaluation of multi-queue nodes queuing systems for congestion avoidance at network routers using discrete and continuous time; also his research interests include computer networks design and architecture, wired and wireless communication, artificial intelligence and expert system, knowledge based systems, security systems, data mining, and information.

Hussein Abdel-Jaber graduated and was awarded a BSc in Computer Sciences and Computer Information Systems from the Philadelphia University, Jordan in 2003. Moreover, Dr. Hussein Abdel-Jaber received his MSc in mobile computing from University of Bradford, U.K. in 2004 and from the same university he received his PhD, with research on congestion control of networks and network performance engineering. In 2009, he joined The World Islamic Sciences and Education (WISE) University as a head of the departments of computer Information Systems and network systems in the school of Information Technology. His research interests are in congestion control of networks (i.e. internet), queueing networks analysis using discrete-time queues or continuous-time queues, networks performance engineering, fuzzy logic control, and data mining. He has several research papers in the previous research interest's retrieval, and software engineering.

Haitham Y. Adarbah is a Lecturer at the Department of Computer Science, Faculty of Computing Studies, Gulf College, Muscat, Sultanate of Oman since 2008. He received his M.Sc degree in Computer Science from Amman Arab University for Graduate Studies, Amman, Jordan in 2008. His thesis title is: Modeling and Analysis of TCP Reno with Packet-Loss and Long Delay Cycles in Wireless Networks. His current research interests are studying the performance of TCP protocol in wired and wireless networks, developing efficient dynamic routing protocols for mobile ad hoc networks, wireless networks management and security, and ad hoc networks modeling and simulation. In addition, he is interested in data compression, signal processing, and distributed system architecture.

Salih N. Akour is an Associate Professor of Mechanical Engineering at Sultan Qaboos University (SQU) since September 2008. He is on leave from the Mechanical Engineering Department at the Faculty of Engineering and Technology in the University of Jordan. He received his Ph.D. in Mechanical Systems from University of Central Florida (UCF) in 2000. Dr. Akour is a member of Phi Kappa Phi Honor Society and Jordan Engineering Association. Dr. Akour's current fields of research include design optimization, computer aided design/computer aided manufacturing/finite element analysis, rapid prototyping, rapid manufacturing and reverse engineering, 3-D biomedical modeling, structural dynamics, and composite materials and structures technology. He has published 16 journal papers and many conference proceedings papers. He has graduated 3 Ph.D. Students. Dr. Akour serves on a number of national professional committees. Dr. Akour is a reviewer of a number of international technical journals. He is a Consultant to some national industries: Seabird Aviation Jordan, Jordan Aerospace Industries, PREFAB (Prefabricated Buildings, Jordan), and also to some international industries in US through UCF- Lockheed Martin Missiles and Fire Control, US Filters, Siemens Westinghouse Power Corporation /Florida.

Jens Alfredson is, since 2006 and in an earlier period from 1996-2001, employed by Saab, developing and evaluating novel presentations for fighter aircraft displays. He is Head of Technical Discipline for Human-Machine Interaction and appointed Technical Fellow, Human Capability, since 2008. He received an MSc in Industrial Ergonomics from Luleå University of Technology in 1995. He received a Ph D in Human-Machine Interaction from Linköping University of Technology in 2007. Since 1999, he is certificated as an Authorized European Ergonomist, Eur.Erg. (CREE). He has previously (2001-2006) worked as a researcher at the department of man-system-interaction, Swedish Defense Research Agency.

Rikard Andersson is, since 1997, employed by Saab. He is manager of area Human-Machine Interaction Design and Technology and is, and has been, involved in several projects regarding the design and development of fighter aircraft cockpit and support systems. He received an MSc in Industrial Ergonomics from Luleå University of Technology in 1998. He has previously (2000-2003) worked as a technical leader of the HMI domain including insertion and development of design- and prototyping methods and tools.

Miles Austin joined Velcon as a Project Manager in September 2009. Miles was a Cofounder of Sensornet, a company that in 2002 pioneered telemetry based contamination monitoring of diesel fuel using early particle counting technologies. Miles has a wealth of R&D, operational, and technical experience in the field of contamination sciences. Miles is also an Information Technology integration specialist, a holder of several patents, and an entrepreneur. Combining focused tribological approaches, a grasp of multiple disciplines, and pioneering business acumen, Miles is currently developing new contamination detection systems and services for industrial applications.

Emran Badarneh received his PhD and Master's degrees in Computer Information Systems from Arab Academy for banking and financial sciences–Amman in 2009 and 2004 respectively and his B.S. in Computer Science Mutt'a University Jordan –Karak in 1989. He is a highly motivated Consulting Manager with over 20 years of experience in IT industry, possessing exceptional communication and leadership skills. He has a strong background in enterprise solutions development and enterprise architecture that enhances business efficiency and stream lines operations, and is an expert in system architecture analysis and modeling (ERM & UML). He has comprehensive knowledge of Rational Unified Process and Oracle CDM development methodology, and a strong understanding of mechanisms required for satisfying architectural attributes of a system. He is an expert in a broad set of IT development technologies and tools: Oracle database administration (9i DBA OCP) and tools, data conversion using ETL methods, database replication, Oracle distributed databases, data warehouse building, data access auditing for critical systems, and object relational mapping.

Hussein Al-Bahadili received his B.Sc degree in Engineering from University of Baghdad, Iraq, in 1986. He received the M.Sc and PhD degree from University of London (Queen Mary College), UK, in 1988 and 1991, respectively. He is currently an Associate Professor at the Arab Academy for Banking & Financial Sciences (AABFS), Jordan. He is also a visiting researcher at the Centre of Wireless Networks and Communications (WNCC), University of Brunel (UK). He has published many papers in leading journals and world-level scholarly conferences. He recently published two chapters in books on IT. His research interests include computer networks design and architecture, routing protocols optimizations, parallel and distributed computing, cryptography and network security, and data compression.

Lori Brown is an Airline Transport Pilot and faculty member at Western Michigan University, College of Aviation. Ms. Brown's work in applied Aerospace Science and Education provides the framework required for the conception and execution of interdisciplinary research and education. Her educational interests include: critical infrastructure vulnerability analysis and protection, technology for homeland security inspection and detection technologies, and research process and methods in homeland security. Her work is published in a number of academic journals, conference publications, and books. Her research interests include high lux lights to mitigate fatigue among crewmembers, wireless technologies, and pilot/flight attendant communication. She is an appointed member of the Battle Creek Homeland Security Committee, the Civil Aviation Consulting Team, for the Spectrum Group in the Nation's Capital, as well as, a civil and military evaluator of the American Council on Education (ACE), to provide a collaborative link between the U. S. Department of Defense and higher education.

Anthony Cerullo is an undergraduate student at Western Michigan University. He is majoring in English with a minor in History, at the college of education. He is currently employed at Western Michigan University as a Research Assistant for Lori Brown and as a Teaching Assistant. Mr. Cerullo has worked extensively alongside Ms. Brown researching the current climate of communication between pilots and flight attendants in China. This groundbreaking research will add to the limited body of scientific knowledge in this area, as aviation in China continues to grow. His areas of interest include postmodern fiction and interactive fiction, as well as 20th century history. Following his baccalaureate, he plans to continue his education in graduate school to pursue a Master's degree in Literature studies. He resides in Kalamazoo, Michigan.

Liang Dong is an Associate Professor of Electrical and Computer Engineering at Western Michigan University. Dr. Dong received his Ph.D. degree in electrical and computer engineering from University of Texas at Austin in 2002. Prior to joining the faculty of Western Michigan University, he was a Research Associate with the Department of Electrical Engineering at University of Notre Dame, and worked at various industry sectors such as IBM, CWill Telecommunications, and Navini Networks. His work is in the fields of information theory, wireless communications, digital signal processing, and wireless ad hoc and sensor networks. His current research thrusts include energy-efficient wireless network applications, localization and tracking in ad hoc networks, and energy harvesting for sustainable wireless sensor networks. Dr. Dong is a senior member of Institute of Electrical and Electronics Engineers (IEEE), and a member of American Society for Engineering Education (ASEE).

Thabtan Fadi is a Senior Lecturer at Philadelphia University. His research interests lie in the investigation and development of new data mining, scheduling, and machine learning techniques, which bridge the gap between the theory and practice of decision making using artificial intelligence techniques. In the last few years Dr. Fadi participated in several technical committees board for international journals and conferences, i.e. Chair for the IEEE ITNG '08 conference, AI-05, AI-06, AI-07, DBA-06, Journal of Applied Soft Computing, Journal of Software Engineering and Knowledge Engineering, et cetera. Fadi was working as a Senior Lecturer at the University of Huddersfield. He successfully published multiple journal and conference papers in data mining and computer networks, many of which were published by IEEE and Springer. Dr. Fadi was active in research as a member of the Modeling Optimization Scheduling and Intelligent Computing (MOSAIC) research group at Bradford.

Shadi Ettantawi holds a PhD in MIS, a Master degree in CIS, and a BSc degree in Computer Science. He worked as a programmer and as an IT manager in the government of Jordan before pursuing an academic career. He has been teaching at the Arab Academy for Banking and Financial Sciences since 2005. His main interests are in the fields of Software Engineering and Information Retrieval.

Jaime García Sáez is SW Architecture Specialization Area Manager in INECO since 2008. He conducts active research in European funded projects (SPADE2, AAS) and innovation-related activities (ePlanAirport, Flyability Simulator). His areas of interest are service engineering, software development in aeronautical field, development and validation of SESAR concepts, specification, and quality-assurance methodologies. He got MSc in Telecommunications Engineering (1996) and started working in Telefonica RTD in Simulations Department for 2 years. After, he joined Fujitsu as Software Analyst working in different projects for Spanish mobile operators and national public administration for 8 years. Also, he has international experience in RTD projects for FP6 and FP7 in Semantics and Service Engineering fields as he worked for Atos RTD for 2 years. Lately, he joined INECO as System Engineer heading the specialized area of SW Architecture in the ATM Management Directorate.

Wa'el Hadi currently has a PhD in Computer Information Systems from the Arab Academy for Banking and Financial Sciences, Amman-Jordan. He is an Assistant Professor in the Department of Management Information System at the Philadelphia University, Jordan. His research interest includes knowledge management, customer relationship management, customer knowledge management, risk management, strategic management, Information Systems, data mining and information retrieval, and congestion control. Dr. Hadi has published 17 articles in refereed journal as well as international conference proceedings.

Issam Hamad Al Hadid, is an Assistant Professor. He received his Ph.D. degree from University of Banking and Financial Sciences (Jordan) in 2010. He received his MSc degree in Computer Science from Amman Arab University (Jordan) in 2005. He earned his BSc in Computer Science from Al Zaytoonah University (Jordan) in 2002. He has published many research papers in different fields of science in refereed journal and international conference proceedings. His field research lies in design and architecture of self-healing; also his research interests include artificial intelligence and expert system, knowledge base systems, security systems, simulation, information retrieval, software engineering, and data base management systems.

Mohammad Mahmoud Al-Husban is the Director of airworthiness standards in Civil Aviation Regulatory Commission (CARC) of Jordan since May 2009. Previously, he was the Director of Flight Safety for three years. He has joined CARC since August 1992. He has worked as Service Engineer at Royal Jordanian Airline on CF6 and CFM engines (1990-1991). Dr. Husban received his Ph.D. in Mechanical Engineering in the field of Mechanical Engineering / Design and Optimization from Jordan University in 2009. Dr. Husban is a member of Jordan Engineers Association. Dr. Husban current field of profession is aircraft airworthiness which includes aircraft design organizations approvals, aircraft production organizations approvals, aircraft modification and repairs approval, aircraft certification, aircraft incident /accident investigation, repair stations approvals, and aircraft maintenance program approval. He has attended more than 60 technical courses, seminars, and workshops in aviation safety. Dr. Husban's research interests are design optimization, computer aided design, and composite materials and structures technology.

Werner Langhans joined ERA as Director of ATM Solutions for the Europe, Middle East, and Africa (EMEA) region in February 2008 and is now responsible for Global Business Development and International Relations for ERA. Previously, Dr. Langhans was with Austro Control, where he held positions such as Head of Technology Development and Head of Strategic Business Development for all Austro Control Engineering Services. In 1994, Dr. Werner Langhans received a Master's of Science in Telecommunications and High Frequency Engineering from Vienna Technical University. He worked at the "European Laboratory for Particle Physics, CERN, Geneva" on Multi-Quantum-Well Electro Optical Modulators" and received his Ph.D. from Vienna Technical University in 1997. In 2002, he received an MBA from Danube-University, Krems, Austria. In 2007, Dr. Langhans received the ATC Maastricht Award for "Enabling Technologies."

Ronald John Lofaro, PhD, aviation psychologist, is currently an Associate Professor at ERAU NAS Pensacola campus. He spent 16 years with the FAA to include a post as the FAA's sole liaison to the USAF Research Labs/USAF Aeronautical Systems Center at WPAFB, OH. Prior to that, he spent 6 years at the FAA's Technical Center as manager in the FAA's Airworthiness Assurance R&D Division and almost 4 years at FAA HQ, as project manager on a variety of R&D efforts with NASA, USN, USAF, and universities. Dr. Lofaro came to the FAA from the Army Aviation Command, where his 5-plus year stint included R&D in helicopter pilot air-to-air combat training and pilot selection. Ron has published 72 articles and 4 book chapters on decision-making in the operational environment (ODM), crew resource management (CRM), line oriented flight training (LOFT) to include design and development of LOFTs, aviator selection/classification, aviation safety, aircrew/pilot training, and performance evaluation.

Hussein Maaitah is a Lecturer of Mechanical Engineering at Prince Faisal Technical College (PFTC) (in conjunction with Al-Balqa Applied University) since September 2008 as part time lecturer. He received his Ph.D. in Mechanical Engineering from University of Jordan in 2008. Dr. Maaitah is a member of Jordan Engineering Association. Dr. Maaitah's current fields of research include design optimization and computer aided design finite element analysis. Dr. Maaitah has served in many positions on the national level: a Maintenance Engineer at the Royal Jordanian Air force (RJAF) during the period from 1991 to 1998, Chief of Maintenance section as Air Base Commander - Assistance for all electromechanical affairs, 2006 to present. Also Dr. Maaitah has severed as full time Lecturer at Prince Faisal Technical College (PFTC) that grants Diploma Certificates (in conjunction with Al-Balqa Applied University) in automobile mechanics, air conditioning, and aerospace ground equipments, during the period from 1998 to 2006.

Jamal F. Nayfeh is the Dean of the College of Engineering and a Professor of Mechanical Engineering at Prince Mohammad Bin Fahd University (PMU) since September 2009. Previously, he was the

Associate Dean for Academics, Marketing, and Outreach in the College of Engineering and Computer Science and a Professor of Mechanical Engineering in the Department of Mechanical, Materials, and Aerospace Engineering (MMAE) at the University of Central Florida (UCF). He received his Ph.D. in Engineering Mechanics from Virginia Tech in 1990. Dr. Nayfeh is a member of Tau Beta Pi Engineering Honor Society, American Society of Mechanical Engineers, American Institute of Aeronautics and Astronautics, Society of Automotive Engineers, and American Society for Engineering Education. Dr. Nayfeh's current fields of research include design optimization, computer aided design/computer aided manufacturing/finite element analysis, rapid prototyping and reverse engineering, 3-D biomedical modeling, structural dynamics, composite/smart materials and structures technology, and undergraduate engineering education. He has published 24 journal papers and 52 conference proceedings papers, presented 51 papers at national and international conferences and meetings, and conducted 15 short courses and workshops. He graduated 16 M.S. and 6 Ph.D. students. State, federal, and industrial grants have funded his research at UCF (40 grants totaling $5,178,526). Dr. Nayfeh organized national and international technical sessions and serves on a number of national professional committees. Dr. Nayfeh is a reviewer of a number of international technical journals. Dr. Nayfeh is the President and CEO of Nayfeh Engineering Consulting Serves (NECS). He is a Consultant to US Navy SPAWAR, Lockheed Martin Missiles and Fire Control, Walt Disney World Ride and Show Engineering, Siemens Power Corporation, and US Corps of Engineers Water Ways Experiment Station.

Tim Quilter joined ERA as Director of Product Management in January 2008 and is responsible for strategic product development across ERA's range of air traffic management products. He has been working in multilateration research and development since 1996 and has been closely involved in the development of multilateration and ADS-B standards within Europe. Previously, Tim Quilter was a Product Manager at Roke Manor Research where he was responsible for the development of multilateration systems. In 1993, Tim Quilter received a Bachelor of Engineering Degree in Electronic Engineering from Brunel University in London. He is a Chartered Engineer and Member of the Institute of Engineering and Technology.

William Rankin, PhD, currently is the Chair and Professor of the Department of Aviation at the University of Central Missouri. Prior to this position he served as an Assistant Professor of Aviation at Florida Memorial University in Miami, Florida and Deputy Director of Aviation for the Fort Lauderdale Hollywood International Airport. Dr. Rankin joined the Broward County Aviation Department in 1998, overseeing the operations, maintenance, planning, and development sections of BCAD. He has 24 years of airport management experience including the Director of Aviation for the El Paso International Airport, Texas; the Director of Operations, Washington National Airport, Washington, DC; Director of the Eastern Iowa Regional Airport, Cedar Rapids, Iowa; and Manager of Smith-Reynolds Airport, Winston-Salem, NC. Dr. Rankin has a PhD in Business Administration from Northcentral University, Prescott, AZ. In addition, he is an Accredited Airport Executive in the American Association of Airport Executives and has a Commercial Pilot and Advanced Ground Instructor License.

Marc A. Rosen is a Professor in the Faculty of Engineering and Applied Science at UOIT, where he was founding Dean. His research interests focus on energy sustainability, environmental impact of industrial systems, and exergy. He previously served as Chair of the Department of Mechanical, Aerospace and Industrial Engineering, and as Director of Aerospace Engineering, at Ryerson University in Toronto. Dr. Rosen has also worked for such organizations as Imatra Power Company in Finland, Argonne National Laboratory near Chicago, and the Institute for Hydrogen Systems near Toronto. He has served as President of the Engineering Institute of Canada and of the Canadian Society for Mechanical Engineering, and is a fellow of these and other societies. He received an Award of Excellence in Research and Technology Development from the Ontario Ministry of Environment and Energy.

Nicoletta Sala received the degree in Physics ("Laurea") and applied cybernetics at the University of Milan (Italy), and his Ph.D. in Communication Science at University of Lugano (USI, Lugano, Switzerland). Postgraduate degrees were earned (two years for each) in: "Didactics of the communication and multimedia technologies" and "Journalism and mass media." She is Professor in Information Technology and in Electronics, and she teaches "New Media for Architecture" at University of Lugano (Mendrisio, Switzerland). She is founder and co-editor of Chaos and Complexity Letters - International Journal of Dynamical System Research (Nova Science, New York). Her research interests concern various scientific topics from an interdisciplinary point of view and comprise the following areas: fractal geometry and complexity; mathematics in arts, architecture, industrial design; new media and IT in the learning environments; virtual reality in architecture. She has authored 16 mathematics and Information Technology books, and edited 10 others. She has written 290 scientific papers. Her last book as author and editor is: Orsucci F. and Sala N. (2008) Reflexing Interfaces: The Complex Coevolution of Information Technology Ecosystems, IGI Global, Hershey.

Dujuan B. Sevillian currently works for the Boeing Company as a Human Factors and Systems Safety Engineer within the Defense, Space, and Security division; he is also a member of the Air Safety Investigation Organization developing air safety investigation processes. Mr. Sevillian has worked for the Boeing Commercial Airplanes division in the Airplane Maintenance Engineering group supporting airworthiness practices and processes in maintenance. Prior to joining Boeing, Mr. Sevillian worked for the airline industry as an aviation safety specialist and regulatory compliance evaluator, supporting the Aviation Safety Action program, Dispatch Safety Action Program, Maintenance Safety Action Program, Air Transportation Oversight System, and the Department of Defense model for air carrier safety. Sevillian worked as an air safety investigator trainee at the National Transportation Safety Board. Dujuan is a PhD candidate at Cranfield University in Bedfordshire, UK-School of Engineering studying to receive his degree in Flight Deck Design-Aviation Safety Human Factors. Sevillian received his Bachelors and Masters degree within the field of human factors and systems safety. Dujuan is an adjunct professor at Embry-Riddle Aeronautical University, a private pilot, Rudolf Kapustin award recipient, and has multiple aviation safety certifications.

Mohammad S. Sharawi is an Assistant Professor of Electrical Engineering at King Fahd University of Petroleum and Minerals, Dhahran, Saudi Arabia. He was a Research Scientist with the Applied Electromagnetics and Wireless Laboratory in the Electrical and Computer Engineering Department, Oakland University, Michigan, USA, during 2008-2009. Dr. Sharawi was a faculty member in the Computer Engineering Department at Philadelphia University, Amman, Jordan, from 2007-2008. He served as the Organization Chair of the IEEE Conference on Systems, Signals and Devices that was held in Jordan in July 2008. He obtained his Ph.D in Systems Engineering from Oakland University, Michigan, USA, in 2006. During 2002-2003 he was a hardware design engineer with Silicon Graphics Inc., California, USA. Dr. Sharawi has more than 40 refereed international journal and conference paper publications. His research interests include RF electronics and antenna arrays, applied electromagnetics, wireless communications and navigation systems, and hardware/embedded systems design.

Kevin M. Smith, type-rated in the B-737, 747, 75/767 and the 777, had multiple tours of duty in the United Airlines Training Center, and was a line-check airman (United's 767 fleet). He was internationally qualified in both Atlantic and Pacific operations. Captain Smith was Task Force leader/manager of the industry's first Advanced Qualification Program (AQP) for pilots, and Chairman of the crew resource management/line oriented flight training (CRM/LOFT) integration sub-committee of the ATA (Air Transport Association) AQP working group. Flying continuously for forty years, including both military (retired as a USN Captain) and civil flight experience, Captain Smith has accumulated over 25,000 flight hours (and over 200 carrier landings). He was a United Air Lines Senior Captain, reaching mandatory retirement age in 2003. He was a USN consultant on the F-22 development and was the driving force, with Ron Lofaro, in developing the Operational Decision-Making model, of which numerous articles have been written and published.

Greg Sprenger has a BS in Physics. He has been working in the fuel filtration field for over 30 years, in a variety of positions. Most recently he is General Manager of Technical Services for Velcon Filters, a world leader in the filtration of aviation fuel. He holds over 10 US and international patents for various filtration devices, including electronic instruments to provide fuel quality information. Greg is a contributing member of many industry fuel and filtration committees, including ASTM, API, and EI (Energy Institute). He has provided technical support to many fuel filtration specifications over many years.

Jed Stevens has a BS in Mechanical Engineering and has been a Product Development Engineer with Velcon since 2005. He is an expert on particle sensing in fuels and is responsible for R&D, testing, assembly, installation, and industry coordination of the VCA (Velcon Contaminant Analyzer.) He has collaborated with major oil company R&D personnel on development and testing of contaminant analyzers. Jed was a finalist for the 2008 Energy Institute Innovation Award and has a US Patent for "Contaminant Analyzer for Fuel," along with six other patents pending for sensing and fuel related methods and products. Jed has recently spoken at technical forums including, IATA Technical Group, ExxonMobil Aviation Technical Group, and the API/EI 1598 Workshop; relating to contaminant analyzer development and requirements.

Enis T. Turgut is an Assistant Professor in the Anadolu University, School of Civil Aviation. He received his BSc and MSc in the School of Civil Aviation and Graduate School of Sciences at Anadolu University in 1999 and 2003, respectively. He received his PhD in 2007.

Index

CPSIA information can be obtained at www.ICGtesting.com
Printed in the USA
BVOW03*0243020714

357817BV00014B/235/P

9 781609 608873